Evolving Geographical Structures

NATO ASI Series

Advanced Science Institutes Series

A Series presenting the results of activities sponsored by the NATO Science Committee, which aims at the dissemination of advanced scientific and technological knowledge, with a view to strengthening links between scientific communities

The Series is published by an international board of publishers in conjunction with the NATO Scientific Affairs Division

A	Life Sciences	Plenum Publishing Corporation
B	Physics	London and New York
C	Mathematical and Physical Sciences	D. Reidel Publishing Company Dordrecht and Boston
D	Behavioural and Social Sciences	Martinus Nijhoff Publishers The Hague/Boston/Lancaster
E	Applied Sciences	
F	Computer and Systems Sciences	Springer-Verlag Berlin/Heidelberg/New York
G	Ecological Sciences	

Series D: Behavioural and Social Sciences – No. 15

Evolving Geographical Structures
Mathematical Models and Theories for Space-Time Processes

edited by

Daniel A. Griffith, B.S., M.A., Ph.D.
Associate Professor of Geography
State University of New York at Buffalo
Buffalo, N.Y. 14260, U.S.A.

and

Anthony C. Lea, B.A., M.A., Ph.D.
Assistant Professor of Geography
University of Toronto
Ontario M5S 1A1, Canada

1983 **Martinus Nijhoff Publishers**
The Hague / Boston / Lancaster
Published in cooperation with NATO Scientific Affairs Division

Proceedings of the NATO Advanced Study Institute on Evolving Geographical
Structures, i Cappuccini, San Miniato, Italy, July 18-30, 1982

Library of Congress Cataloging in Publication Data

NATO Advanced Study Institute on Evolving Geographical
     Structures (1982 : San Miniato, Italy)
     Evolving geographical structures.

     (NATO advanced science institutes series. Series D,
Behavioural and social sciences ; no. 15)
     "Proceedings of the NATO Advanced Study Institute on
Evolving Geographical Structures, i Cappuccini, San
Miniato, Italy, July 18-30, 1982."
     "Published in cooperation with NATO Scientific Affairs
Division."
     Includes index.
     1. Regional economics--Mathematical models--Con-
gresses. 2. Regional planning--Mathematical models--
Congresses. 3. Migration, Internal--Mathematical
models--Congresses. 4. Diffusion of innovations--
Mathematical models--Congresses. 5. Finance, Public--
Mathematical models--Congresses. I. Griffith, Daniel A.
II. Lea, Anthony C. III. North Atlantic Treaty
Organization. Scientific Affairs Division. IV. Title.
V. Series.
HT391.N354 1982   361.6'0724   83-11373
ISBN-13:978-94-009-6895-0   e-ISBN-13:978-94-009-6893-6
DOI: 10.1007/978-94-009-6893-6

ISBN-13:978-94-009-6895-0

Distributors for the United States and Canada: Kluwer Boston, Inc., 190 Old Derby
Street, Hingham, MA 02043, USA

Distributors for all other countries: Kluwer Academic Publishers Group, Distribution
Center, P.O. Box 322, 3300 AH Dordrecht, The Netherlands

PREFACE

The development of evolutionary models for space-time processes seems to be a natural extension to dynamic analysis, and a timely next step in the evolution of formal spatial analysis. Geographical systems, such as school systems, and geographic networks, such as grain elevator and gas station networks, experiencing rationalization, growth or contraction and decline have demonstrated empirically the asymmetry of life-cycle trajectories. Linear models have proven inappropriate in the wake of dominant non-linear trends and relationships, and independent activities and actions are all but non-existent. The capacity to predict and to select, or at least control, in a rational way, future time paths is an important endeavor. Successful selection of and movement onto a particular trajectory requires an understanding of the asymmetric nature of its life-cycle path, its latent non-linearities, and its accompanying spatial interactions. In other words, formulation of evolutionary models will help lead to an understanding of geographical organization with its many complexities. To date, the theories and tools necessary for such formulations have not fully crystalized. Because coming to grips with this problem requires a firm foundation in dynamic spatial modelling, it seemed appropriate to provide a forum for intensive interaction among scholars of spatial dynamics. A NATO Advanced Study Institute was held in July of 1980 at the Chateau de Bonas, France, to address critical issues associated with dynamic spatial modelling research. The successful and fruitful cooperation of European and North American scientists at this first Institute provided an impetus for organizing a second one that would focus on evolving geographical structures. This second Institute, the basis for this volume, held at ⟨i Cappuccini⟩, Italy, drew upon the scholarship and findings of the preceding Institute in order to take another small step in unravelling the mysteries of space-time processes and patterns or structures.

This second Institute had as its general objective a better understanding of socioeconomic space-time processes and patterns, and how these patterns may be better described, predicted and controlled. Four more specific goals were:

(1)   to continue the identification and description of significant real world space-time processes and planning problems,

(2)   to exchange ideas and viewpoints held by geographers, economists, regional scientists, planners and applied mathematicians from different NATO countries concerning geographical systems analysis and mathematical models of spatial systems dynamics and evolution,

(3) to assess the predictability of outcomes of dynamic spatial processes, and

(4) to discuss the possibility of planning and control of dynamic and evolving spatial strcutures.

Accordingly, the Institute focused on theories and mathematical models for space-time processes. International exchanges of ideas are an efficient way of reviewing both the state-of-the-art and stimulating new ideas that lead to more rapid progress. In pursuit of this goal the lectures were organized around the following five principal themes: (a) equilibrium analysis in dynamic spatial modelling, (b) the application of catastrophe and bifurcation theory to spatial dynamical problems, (c) spatial dynamics involving flows and interactions over space, (d) statistical analysis for spatial dynamics, and (e) economic development, planning and policy.

These five Institute themes fell into two generic classes, namely dynamic models--the logical prior step for evolutionary models--and evolutionary models themselves. The dynamic models tended to be policy oriented and focussed on a variety of problems including dynamic models of simplified general space economies, the growth and decline of particular systems, problems associated with spatial pricing, housing search, job search, regional inequalities, capital accumulation, unemployment and public sector systems. The evolutionary models focussed upon space-time trajectories, catastrophes and bifurcations, instabilities and varieties of steady states. They were often closely analogous with models of the physical sciences. These models generally dealt with problems involving flows through space and competitive spatial interaction processes.

We are indebted to the Scientific Affairs Division of the North Atlantic Treaty Organization for funding the Institute, to Drs. Peter Nijkamp, Giovanni Rabino and Ross MacKinnon for serving as advisors to the Institute, and to Drs. Leslie Curry, Ross MacKinnon and Grant Thrall for helping in the editing of this volume. Special appreciation goes to Cassa di Risparmio di San Miniato for its hospitality during the Institute, and for providing an excellent physical environment, to the Geography Department of the State University of New York at Buffalo for providing clerical help and word processing facilities for the typesetting of this volume, and to our wives for being understanding about our two-week sojourn at ⟨i Cappuccini⟩, and the efforts devoted to putting together this volume.

Finally, we apologize for the changes in notation we had to make in the manuscripts. These changes were necessitated by the limited number of symbols available to us on the word processor system that was used to prepare this volume. Most notably are the substitution of Δ for the lower case Greek letter nu in Haag and Weidlich's paper, upper case sigma for a bold upper case sigma (denoting a matrix) in Ancot and Paelinck's paper, and Ω for upper case Greek letter lambda in Sheppard's paper. Hopefully these alterations will not hinder a reader's understanding of the text. One should keep in mind that all matrices appear in bold type.

Daniel A. Griffith

and

Anthony C. Lea

Buffalo, New York
January 13, 1983

TABLE OF CONTENTS

INTRODUCTION

Most spatial models that are concerned with process tend to be dynamic as opposed to truly evolutionary in nature. A distinction has been made between static and dynamic spatial models by Griffith and MacKinnon (1981) in the introduction to the companion volume to this one. The purpose of this introductory section is three-fold. First, a distinction will be made between dynamic and evolutionary spatial models. Second, key concepts associated with the topic of this book--evolving geographical structures--will be explored briefly. Third, each of the papers of this volume will be related to these concepts.

Dynamic refers to motion. Often, then, dynamic spatial models are written in terms of a set of differential or difference equations in order to describe an entire geographic system, with clock time as the independent variable (cf., Georgescu-Roegen, 1971). The time variable permits a description of change in the spatial distribution of one or more variables to be represented by a transformation between times t and t+1. Feedback effects are introduced into this type of model with lag structures, moving averages, and the use of systems of simultaneous equations. One interesting but problematical feature of such models is that the processes described are necessarily time reversible; this is illustrated simply by placing a minus sign on the subscript t, and then moving backwards through time (Carlstein, Parkes and Thrift, 1978). This reversibility implies that the movement taking place is symmetric in time, and that while limiting cases exist at $\pm\infty$, neither the initial condition nor the final state is of special interest. If convergence occurs, it is to a unique equilibrium or steady state, which is context (model and parameter) specific. Divergence implies that the trajectory of the system does not achieve a limiting steady state. Consequently, most of the literature on dynamics is strongly methodological, even technical (see, for instance, Baumol, 1970; Neher, 1971; Gandolfo, 1971; Chow, 1975; Aoki, 1976; Isard and Liossatos, 1979; Bennett, 1979; Miller, 1979). The greatest development of dynamic models has occurred in the physical and biological sciences. In the social sciences most of the literature is by economists and relates to aspatial economic problems; the authors just cited here tend to reflect this bias. Until very recently the spatial dimension of problems had not been treated dynamically.

Evolution refers to the gradual development, through progressive and rather systematic changes, of more complex entities. The focus is on movement along a trajectory, which may be described by a dynamic model, towards equilibrium, and the disappearance of anomolies from and the increasing disorder within a system (Allen, 1980a). In other words, it addresses a set of

questions related to the way in which new kinds of structures
emerge. In the context of evolving spatial systems, evolution
generally refers to increasing complexity, which may be measured by
the entropy of a system (see Curry, this volume), and increasing
geographic specialization and organization. As Glansdorff and
Prigogine (1971, p. xx) comment upon

> ... instabilities which lead to space organization ...
> Such symmetry breaking instabilities are of special
> interest as they lead to a spontaneous 'self-
> organization' of the system both from the point of view
> of its space order and its function...

In contradistinction to pure dynamic models, then, evolutionary
models attempt to take into account the indelible memory of the
system, which implies that time must be considered to be
irreversible. Although movement may be towards an equilibrium,
fluctuations and perturbations can cause short-term divergence,
which in the long-run can lead to convergence upon a new stable
steady state, which may be far from the original one. Thus when
significant shocks drive the system

> ... far from ... equilibrium, the competition between
> homogenization of ... components by free diffusion and
> spatial localization due to local disturbances ... give
> rise to instabilities and lead ultimately to the
> appearance of stable non-uniform distributions.
> (Glansdorff and Prigogine, 1971, p. 247)

Hence, in the context of self-organizing systems and the theory of
dissipative structures, multiple equilibria can exist. The
critical points at which the system attaches itself to a new
trajectory leading to new equilibria are called bifurcation points
(Allen, 1980a; Wilson, 1981). In models of these processes the
focus is upon the history of the system, the initial conditions,
and the variety of possible space-time trajectories and final
steady states, and no longer is upon the transformation that takes
place between times t and t+1. This theory of self-organizing
evolutionary models has been most fully developed by the Prigogine
or Brussels school, and is exemplified by Prigogine (1980),
Glansdorff and Prigogine (1971), Nichols and Prigogine (1977),
Allen (1980b), and Jantsch (1980a, 1980b).

Given the preceding discussion, the idea of an evolving
geographical structure now will be clarified. Consider a two-
dimensional surface over which some phenomenon is distributed.
Items can be located on this surface by noting their Cartesian
coordinates (x,y). Changes in this surface can be captured by
adding the third coordinate of time, giving (x,y,t). Such a
distribution is called geographical or spatial because each of its

constituent items has a specific location. Structure refers to the way in which values of the phenomenon taken on by one location relate to those taken on at other locations. The relationships between values at different points in space transcend the simple property of juxtaposition, and are attributable to the relative rather than the absolute nature of space. Consequently, models of evolving geographical structures emphasize increasing complexities in the spatial organization of phenomena, the spatial development and progressive change depicted by the space–time trajectories of the variables involved, descriptions of these different trajectories, the sensitivity of any equilibrium or equilibria to initial conditions or parameter values as well as perturbations and fluctuations that materialize, and the asymmetries of change through time and over space. Turner (1980, p. 47) points out

> ... how the homogeneity and isotropy of space and time are destroyed by the appearance of ... dissipative structures ... Subject to constraints imposed by the environment (boundary conditions), the system is found in one of possibly several space–time (dissipative) structures made possible by the specific non-linearities of the mechanism ... If fluctuations ... are large enough ... the structure may become unstable. The system can then bifurcate to a new space–time structure.

Illustrative of these types of models are the works of Allen and Sanglier (e.g., 1981).

This book ultimately is concerned with better understanding and modelling evolving geographical structures. As static analysis frequently is a prerequisite for dynamic analysis, in turn, dynamic analysis tends to be a prerequisite for evolutionary analysis. Unfortunately, although a considerable body of theory and a variety of methods are available to construct a sound foundation for dynamic modelling, relatively little theory and few models exist for evolutionary modelling (Wilson, 1981). Thus the papers of this volume seek to improve the understanding of the evolution of phenomena through time and over space, either in terms of explicit evolutionary modelling attempts, or in terms of dynamic models that lay a foundation for evolutionary ones. The ultimate goal of this line of research is the development of fully evolutionary models. These models will go beyond standard and sometimes mechanistic differential and difference equation formulations in which systems with no memory tend to reach a unique stable equilibrium, with the emergence of one of a set of possible final, complex states being of little interest. While first–order and second–order optimality conditions are important, the structural properties of the system and the way in which the system follows one particular trajectory out of a set of possible ones are equally important. Judging from the papers in this volume, modelling evolving geographical structures involves the following:

(1)   the development of a set of differential or difference equations that describes the dynamic behavior of the spatial system,

(2)   identification of the sources and roles of perturbations and fluctuations in movement of the spatial system along trajectories,

(3)   investigation of slow and smooth parameter changes in a dynamic spatial model that lead to sudden and discrete (and possibly catastrophic) changes in the trajectory being followed or the spatial organization that is manifested in a final state,

(4)   analysis of the non-linear nature of changes and the interdependence of locations within a geographical landscape that lead to bifurcation possibilities, and

(5)   analysis of the geographic differentiation that emerges in a spatial system through time.

One recurrent theme in the papers of this volume is concerned with evolution in a strict Prigoginian sense. The major contributions to this topic, for a spatial setting, are made by Leslie Curry, Dimitrios Dendrinos and Henry Mullally, Guenter Haag and Wolfgang Weidlich, Peter Nijkamp, and Michael Sonis. Curry's paper addresses the evolution of a spatial pricing system, with regard to both its general properties and specific conditions, and seeks to analyze the nature and degree of fluctuations that shock the system from an equilibrium of efficient prices to one of inefficient prices. Dendrinos and Mullally use a Volterra-Lotka description of individual urban area trajectories, in an attempt to identify critical points at which bifurcations as well as catastrophes occur. In contrast, Haag combines non-linear with migration interactions in order to capture the conditions leading to these critical points. He combines his migration model with a birth-death predator-prey model, in a generalized Volterra-Lotka dynamic model, and among other things discusses the possibility of limit cycles characterizing geographical population structures. Similarly, Sonis models non-linear diffusion and competitive interaction processes and studies the stability of spatio-temporal equilibria. After reviewing the various theories of long wave cycles in Western economies, Nijkamp developes a simple catastrophe theoretic optimal control model of regional growth. Particular emphasis is placed upon the asymmetric behavior, or the irreversibility, of the catastrophes. Each of these writers casts his analysis within the context of continuous time.

As was mentioned earlier, few purely evolutionary models presently exist. Many dynamic modelling attempts aspire to be evolutionary, however. Papers by Professor Dejon and Daniel Griffith fall into this category. Dejon formulates a dynamic spatial interaction equilibrium model in continuous time, but is not centrally concerned with self-organization and the possibility of multiple equilibria or new types of geographic structures that could emerge. Griffith focuses on the non-linearities and interactions of a particular spatial system, but again does not deal with possible bifurcations or catastrophes associated with the evolution of the system. This latter study is couched in a discrete time framework.

Another theme running through the papers of this volume is the dynamics that lead to end states (often equilibria) having certain properties (typically efficiency). Writers who address this theme include Robert Bennett, Anthony Lea, Giorgio Leonardi, Bruce Ralston, Eric Sheppard and Claus Shoenebeck. Bennett, for example, reviews a set of dynamic econometric models of expenditure determination in public finance. Typically the dynamics of the process are modelled using lagged (endogenous) variables. Lea addresses the problem of optimal jurisdiction size and public good capacity. Much of his paper developes a static equilibrium model in preparation for a discrete time dynamic optimal control model. In Leonardi's paper, the dynamics of choice in a complex setting are treated. He constructs an optimal control model for a situation in which actors compete under uncertainty for a scarce resource, such as a house. The essential problems in the papers by Lea and Leonardi can be traced to externalities. Ralston is interested in processes of the supply of information to the public, and its dynamic diffusion over space. After describing a range of processes, he makes use of optimal control theory to find strategies for optimizing long-run objectives relating to the spatial distribution of information. Sheppard's concern is with the problem of urban accumulation and the efficiency of classical neo-Marxian and neo-classical resource allocation mechanisms of multicommodity production and trade in a spatial context. Sheppard uses traditional dynamic analysis to examine the existence of equilibria under a variety of assumptions, and the existence, stability and efficiency of equilibria under a variety of interesting conditions. Shoenebeck's paper is concerned primarily with comparative statics of economic, technical and social change in a region. His dynamic simulation model is directed towards producing a most likely final state. None of these papers relate directly to the evolution of self-organizing systems in the same sense as those noted in the preceding paragraph.

Keeping in mind the paucity of tools and theories for constructing evolutionary models, the last salient theme of this volume is concerned with specification of the dynamics that eventually will lead to evolutionary models. The models here tend to be based upon comparative statics analysis to a large degree, and can be considered to be in the formative stage of dynamic modelling. Papers in this class include those by William Clark, Robert Haining, Jean Paul Ancot and Jean Paelinck, and Peter Rogerson. Clark's paper reviews the literature on housing search in a spatial context, and stresses the importance of various kinds of thresholds. Although there has been some recent progress, to date theories of housing search in space are insufficiently sophisticated to underpin powerful dynamic models. Haining, on the other hand, does formulate a type of dynamic model, but is concerned with comparing its equilibria at different points in time. His model shows promise of developing into an evolutionary model because of its stress on the important relationship between spatial structure and spatial interaction. Ancot and Paelinck discuss a 'spatial econometric model of the European Community,' and pay special attention to how it should be specified. The problem they emphasize has to do with classifying regions into the two categories of growth and non-growth. Finally, Rogerson develops a dynamic model of job search, system vacancies and unemployment, and analyzes a variety of initial conditions, primarily in a comparative static manner. It seems that little additional research would be required to extend his analysis to a fully evolutionary model.

The process of creating sections of this book and placing papers within them has been a difficult one. On the one hand, all papers deal with spatial or geographical problems so that common problems could be used as the organizing principle. On the other hand, the papers tend to stress method and models and, as has been noted above, several communalities here could be used as a basis for classification. Since all attempted classification schemes based upon both model type (e.g., goal, evolutionary, dynamic, comparative static, continuous versus discrete time) and primary focus produced a fuzzy taxonomy in which many papers could be placed in more than one class, it was decided to base the sections on a classification of substantive problems. Although the fuzziness is much reduced, some of the papers still address more than one of the substantive problem areas used as section headings.

Any collection of papers, such as this one, almost necessarily will put forth a number of divergent as well as convergent themes. It is believed that fruitful further research will grow out of these similarities and differences. Because the topics addressed here are very much on a research frontier, the success of this volume should be evaluated on the quality of the research that it stimulates.

1. REFERENCES

Allen, P., 1980a, The Evolutionary Paradigm of Dissipative Structures, in The Evolutionary Vision: Toward A Unifying Paradigm of Physical, Biological and Sociocultural Evolution, edited by E. Jantsch, Boulder Colorado: Westview Press, pp. 25-72.

_____, 1980b, Self Organization in Human Systems, Revue Belge de Statistique, d'Informatique et de Recherche Operationnelle, 20 (No. 4): 21-76.

_____ and M. Sanglier, 1981, Urban Evolution, Self Organization and Decision Making, Environment and Planning A, 13: 167-183.

Aoki, M., 1976, Optimal Control and System Theory in Dynamic Economic Analysis, Amsterdam: North-Holland.

Baumol, W., 1970, Economic Dynamics, 3rd ed., New York: Macmillan.

Bennett, R., 1979, Spatial Time Series: Analysis-Forecasting-Control, London: Pion.

Carlstein, T., D. Parkes and N. Thrift (eds.), 1978, Time and Regional Dynamics, Vol. 3 of Timing Space and Spacing Time, New York: Wiley.

Chow, G., 1975, Analysis and Control of Dynamic Economic Systems, New York: Wiley.

Gandolfo, G., 1971, Mathematical Methods and Models in Economic Dynamics, Amsterdam: North-Holland.

Georgescu-Roegen, N., 1971, The Entropy Law and the Economic Process, Cambridge, Mass.: Harvard University Press.

Glansdorff, P. and I. Prigogine, 1971, Structure, Stability and Fluctuations, New York: Wiley Interscience.

Griffith, D. and R. MacKinnon (eds.), 1981, Dynamic Spatial Models, Alphen aan den Rijn: Sijhoff and Noordhoff, NATO Advanced Study Institutes Series D, Behavioural and Social Sciences No. 7.

Isard, W. and P. Liossatos, 1979, Spatial Dynamics and Optimal Space-Time Development, Amsterdam: North-Holland, Studies in Regional Science and Urban Economics No. 4.

8

Jantsch, E. (ed.), 1980a, <u>The Evolutionary Vision: Toward A Unifying Paradigm of Physical, Biological and Sociocultural Evolution</u>, Boulder Colorado: Westview Press, American Association for the Advancement of Science.

_____, 1980b, <u>The Self-organizing Universe</u>, Oxford: Pergamon.

Miller, R., 1979, <u>Dynamic Optimization and Economic Applications</u>, New York: McGraw-Hill.

Neher, P., 1971, <u>Economic Growth and Development: A Mathematical Introdution</u>, New York: Wiley.

Nicolis, G. and I. Prigogine, 1977, <u>Self Organization in Non Equilibrium Systems</u>, New York: Wiley.

Prigogine, I., 1982, <u>From Being To Becoming</u>, San Francisco: W. H. Freeman.

Turner, J., 1980, Non-equilibrium Thermodynamics, Dissipative Structures, and Self-organization: Some Implications for Biomedical Research, in <u>Dissipative Structures and Spatio-temporal Organization Studies in Biomedical Research</u>, edited by G. Scott and J. McMillan, Ames, Iowa: Iowa State University Press, pp. 13-52.

Wilson, A., 1981, <u>Catastrophe Theory and Bifurcation: Applications to Urban and Regional Systems</u>, London: Croom Helm.

SECTION 1

TOWARDS EVOLUTIONARY MODELS OF MIGRATION

It is fair to say that amongst the very first models formulated by geographers and regional economists were models of migration. The literature treating these models is vast indeed. Yet only a very small portion of this literature pictures migration as the explicitly dynamic process that it really is. Given the tremendous variety in static treatments of the problem, it is likely that an equally diverse family of dynamic models is now in the process of evolution. Certainly, since human migration processes bear great similarity (at least superficially) to many processes in the physical sciences in which dynamic and evolutionary models are well-developed, presumably an even greater variety of models will develop. Part of this development already has taken place. The three papers in this section posit differential equation formulations of migration processes, and each represents an extension of existing models. However, both the specific problems examined and their accompanying models show some of the variety noted above. Whereas the models developed by Haag and Weidlich are squarely within the evolutionary paradigm of dissipative structures, the dynamic models of Dejon and Leonardi are not particularly evolutionary.

The first paper in this section, by Dejon, presents a dynamic equilibrium analysis of a class of fairly well-known multiplicative, attraction-constrained gravity-type interaction models. Dejon shows how the mover-pool or potential migrants may be captured in a dynamic model as a function of both an intrinsic mover-pool, and the differential attractiveness of the origin. He further shows how at any time the migrant flow between an origin-destination pair can be obtained as the equilibrium flows of a properly chosen network. A dynamic Lowry problem with migratory households and service employment is presented as an illustration. This problem then is generalized to one with n different migration processes. The general structure of the problem is discussed and various modelling approaches are noted. The particular model treated most fully in this paper is one based upon general network equilibrium theory. This formulation allows the vast body of knowledge of this theory to be exploited, and yields some important insights into the dynamics of migration.

The second paper, by Haag and Weidlich, summarizes some of their work on evolutionary models of birth, death, interaction and migration processes. Although these models derive from, as well as extend, those for biological species, they appear to furnish at least some implications for simpler human processes. The models

are non-linear differential equation models that generalize
Volterra-Lotka dynamics by adding migration, depicted as a Markov
process, to the standard birth, death and predation processes. A
fundamental master equation is used to represent the evolution of
the probability distribution of states over time. The model first
is illustrated on a relatively simple problem involving only
migration between two parts of a city. A stationary solution to
the master equation is found, and a 'principle of detailed balance'
is invoked. The solution is in the form of approximate closed
equations of motion, and the authors analyze the stability of these
time paths and discover that the solution is in the form of a
'limit cycle.' That a limit cycle could exist for a problem of
this type is interesting and may have important implications for
certain real world migration processes. The interesting features
of this migration model have been nicely illustrated using
numerical simulations with graphical summaries. In the final part
of the paper the authors develop a model that combines non-linear
migration with the classical predator-prey interaction for two
species. They show that this new model leads to a large number of
different solution cases, including a limit cycle that resembles
the classical predator-prey cycle, but is independent of the
initial conditions. The modelling efforts of Haag and Weidlich
indicate that considerable progress has been made in building truly
evolutionary models of migration, and this work is likely to
stimulate further research. Some parallels to their models are
found in the papers by Curry, Sonis and Dendrinos, which appear
later in this volume.

In the third paper in this section, Leonardi presents an
optimal control model of a stochastic multi-actor choice process.
The exemplary problem considered is the migration associated with
changing residence. Each actor in the model is assumed to choose a
sequence of actions over time that maximizes utility, subject to
random noise on the utility evaluation. There is competition for a
limited stock, which here is vacant houses, so that the shortage
generates an externality. The multistage problem first is
structured directly as a nested random utility model, the solution
of which is a set of differential equations describing the optimal
transition rates between states in the system. Leonardi then shows
how the dual of this problem may be cast as a continuous time
optimal control problem whose Hamiltonian can be interpreted as
total social benefit. This problem then is solved using
Pontryagin's maximum principle, and its first-order conditions are
interpreted in terms of the direct random utility formulation. A
consistent micro-economic interpretation of the conditions and
costate variables is discussed in terms of a landlord who is trying
to adjust his rent optimally, while the migration process proceeds.
There is still some theoretical work to be done on the existence,
uniqueness and stability of the solutions Leonardi derives. It
might be noted that in the last paper in this volume Lea also

treats a control theoretic model of a process that involves an externality.

12

ATTRACTION-REGULATED DYNAMIC EQUILIBRIUM MODELS OF MIGRATION OF THE MULTIPLICATIVE TYPE

Bruno Dejon

University of Erlangen-Nuernberg
Federal Republic of Germany

1. INTRODUCTION

This paper focuses on multiplicative push-pull models of migration. If M_{ij} denotes migrants going from region j to region i (per unit of time), multiplicative push-pull models are of the following type:

$$M_{ij} = V_i W_j b_{ij} \qquad , \qquad (1.1)$$

where W_j and V_i are locational factors associated with the origin and destination locations, respectively, and b_{ij} is an interlocational factor known as the inverse friction of space. As an aside, 'additive' push-pull models of migration having the form

$$M_{ij} = (V_i + W_j)/b_{ij}$$

are being studied elsewhere [see Dorigo and Tobler(1982)].

W_i represents a push factor and V_i represents a pull factor. V_i will be called the attractiveness of alternative i. It will often be expedient to use a more explicit representation of W_j by introducing the mover-pool M_j, where

---

Acknowledgement: This manuscript was prepared with the support of the Conselho Nacional de Desenvolvimento Cientifico, e Tecnologico (CNP$_q$) of the Republic of Brazil during the author's visiting professorship at the Laboratorio de Optimizacao de Sistemas of PUC, Rio de Janeiro, in August and September of 1982.

$$M_j = \sum_i M_{ij} \qquad , \qquad (1.2)$$

or, by use of equation (1.1),

$$M_j = W_j \sum_i V_i b_{ij} \qquad . \qquad (1.3)$$

Solving equation (1.3) for W_j and inserting this result into equation (1.1) gives

$$M_{ij} = M_j [V_i b_{ij} / \sum_i V_i b_{ij}] \qquad . \qquad (1.4)$$

This type of migration model has been proposed in various forms in the literature [e.g., Alonso, 1973,1977,1978; Ginsberg, 1972; Wilson, 1974; Dejon and Graef, 1982].

The structure of equation (1.4) is such that the matrix **B** of b_{ij}s may be written as a stochastic matrix, with

$$0 \leq b_{ij} \leq 1 \qquad \text{and} \qquad \sum_i b_{ij} = 1 \quad , \text{ for all } j. \qquad (1.5)$$

The j-th column of matrix **B** provides the transition probabilities of the mover-pool M_j, in case there exist no attraction differentials between alternative regions (i.e., all the V_i are equal). Therefore, the entries of matrix **B** may be called intrinsic transition probabilities, while the term after M_j in equation (1.4) represents the actual transition probability. The attraction variables V_i modify the intrinsic transition probabilities b_{ij} to produce the actual probabilities.

The aim of this paper is twofold. First, a model will be formulated, and discussed, of the mover-pool M_j as the product of an intrinsic mover-pool ^OM_j and a function $\tilde{u}$ representing the differential attractiveness of alternative regions j. Second, it will be shown that, at any time t, the migrants M_{ij}, the mover-pools M_j, and the time rates of change $\dot{P}_j$ of stocks P_j — still to be introduced — may be obtained as the equilibrium flows of a properly chosen network (hence the wording 'dynamic equilibrium models' in the title of this paper). Because of this first purpose, and equation (1.4), the models discussed in the sequel may be called attraction-regulated. The work by Gueldner (1982), to be referred to in more detail in the third section of this paper, appears to provide strong empirical evidence of the usefulness of the concept of attraction-regulation. A natural way of extending the class of models under study here will be sketched. The reader should consult Wenzel (1982) for ways of doing stability analyses on these types of models.

2. ILLUSTRATIVE EXAMPLE: A DYNAMIC LOWRY MODEL

In a highly aggregated Lowry model [see Lowry, 1964; Putman, 1979; Oppenheim, 1980; Wilson, 1974; Batty, 1976], there are two spatial activities, namely residences and service employment, that are being spatially allocated, with respect to some localized basic employment, by acts of discrete choice carried out by two populations. The first of these populations consists of households seeking residences (H), whereas the second of these populations consists of entrepreneurs seeking locations for their service activities (S). Basic employment is assumed to be allocated by some exogenous process. These populations possess discrete locational choice sets, I^H and I^S, respectively, and both of these choice sets, for the sake of simplicity, will be assumed to be comprised of a set of zones into which the urban area under study is partitioned. At any given time, t, let $P_i^H(t)$ and $P_i^S(t)$ denote the number of households and service employment units that are allocated to $i \epsilon I^H$ and $i \epsilon I^S$, respectively. The related static equilibrium Lowry model requires functions

$$A_i^H(P^H, P^S) \qquad \text{and} \qquad A_i^H(P^H, P^S) \qquad ,$$

or, more briefly,

$$A_i^H(P) \qquad \text{and} \qquad A_i^S(P) \qquad ,$$

which can be called, respectively, residential and service employment share factors. Accordingly, the equilibrium stock vectors $^OP^H$ and $^OP^S$ are characterized, respectively, by

$$A_1^H(^OP)/^OP_1^H = \ldots = A_i^H(^OP)/^OP_i^H = \ldots \qquad , \text{ and} \quad (2.1)$$

$$A_1^S(^OP)/^OP_1^S = \ldots = A_i^S(^OP)/^OP_i^S = \ldots \qquad . \quad (2.2)$$

In a typical Lowry model, the term $A_i^H(P)$ could be an expression like

$$A_i^H(P) \quad \propto \quad \sum_j (P_j^S + P_j^B) \, d_{ij}^\lambda \qquad , \quad \lambda < 0, \qquad (2.3)$$

where P_j^B denotes basic employment in region j (hence, $P_j^S + P_j^B$ equals total employment in region j), and d_{ij} is the generalized travel cost between regions i and j [cf., Batty, 1976].

A generalization of equation (2.3) is introduced in Dejon and Graef (1982). It is of the following form:

$$A_i^H(P) = \{\omega^E [\sum_j \overline{(P_j^S + P_j^P)} d_{ij}^\lambda]^{1/\lambda} + \omega^H [\sum_j \overline{P_j^H} d_{ij}^\sigma)^{1/\sigma}\}^{-\gamma} \quad , \quad (2.4)$$

$$\text{where} \quad \overline{P_j^H} = P_j^H / \sum_i P_i^H \qquad ,$$

$$\overline{P_j^S + P_j^B} = (P_j^S + P_j^B)/\underset{i}{\Sigma}\,(P_i^S + P_i^B) \qquad ,$$

ω^E and ω^H are non-negative constants, γ is a positive constant, and λ and σ, in principle, may be any elements of the real number system, even $+\infty$ or $-\infty$. Equation (2.3) may be obtained from equation (2.4) by setting $\omega^H=0$ and $\lambda=-\gamma$.

The time rates of change $\dot{P}^H(t)$ and $\dot{P}^S(t)$ of the stock vectors $P^H(t)$ and $P^S(t)$, respectively, are given by the following accounting equation:

$$\dot{P}_i(t) = \underset{j}{\Sigma}\,M_{ij}(t) - M_i(t) \qquad . \qquad (2.5)$$

Equation (2.5) must hold for each zone i, and for households and service employment separately. Using equation (1.4), equation (2.5) may be rewritten as

$$\dot{P}_i = \underset{j}{\Sigma}\,[V_ib_{ij}/(\underset{i}{\Sigma}\,V_ib_{ij})]M_j - M_i \qquad . \qquad (2.6)$$

The class of models to be studied in this paper links $V_i(t)$ with what might be called the reduced share factor $A_iP(t)/P_i(t)$, by some isotonic function. An example of such a function would be

$$V_i(t) = [A_iP(t)/P_i(t)]^\pi \qquad , \quad \pi \geq 0. \qquad (2.7)$$

For $\pi=0$, the intrinsic transition probabilities b_{ij} are not modified at all by the attractions of the zones. The modifying effect will be stronger as π increases.

3. CLOSED MIGRATION SYSTEMS WITH ATTRACTION-REGULATION

In the highly aggregated Lowry model of the preceding section there are only two migration processes going on, namely that of households and that of service employment. Generally speaking, there will be a large number of simultaneous migration processes. Therefore, consider N populations of a system, at an urban, regional or national level, each population having a finite choice set I_n, $n\varepsilon N$, with migration M_{ij}^n from $j\varepsilon I^n$ to $i\varepsilon I^n$ simultaneously going on in all N populations. In general, time rates of changes, $\dot{P}_i^n$, of stocks, P_i^n, are not related to migration by the simple accounting equation (2.5), for that would presuppose the populations P^n, $n\varepsilon N$, to be closed in the sense that no individuals are entering or leaving them. However, for the sake of simplicity, this assumption will be made in this paper. Thus, equation (2.5), or more specifically, equation (2.6) is assumed to hold for each population in the system.

In order to achieve attraction-regulation in the sense

intended in this paper, let the mover-pools M_j be defined as the following products:

$$M_j = \Omega(V_j/\bar{V}_j)^O M_j \quad , \tag{3.1}$$

where $\bar{V}_j = \sum_i a_{ij} V_i \quad , \tag{3.2}$

$$\sum_i a_{ij} = 1 \ , \quad a_{ij} \geq 0 \quad , \tag{3.3}$$

and $\Omega: R^+ \to R^+$ is a function that possesses an inverse function Ω^{-1} as well as the normalizing property

$$\Omega(1) = 1 \quad . \tag{3.4}$$

The coefficients of the matrix $\mathbf{A}$, as well as $^O M_j$, in general will depend upon the stock vector $P(t)$. In the case of static equilibrium, when all V_j are equal per population of the system (i.e., $V_j/\bar{V}_j = 1$ for all alternative regions j), then the factor Ω in equation (3.1) takes on the value of 1 [because of equation (3.4)], and the actual mover-pool M_j of alternative region j turns out to be equal to the intrinsic mover-pool $^O M_j$. Therefore, the normalizing property (3.4) is only seemingly restrictive. If $\Omega(1)$ were different from unity, equation (3.1) would be replaced by

$$M_j = [\Omega(V_j/\bar{V}_j)/\Omega(1)][\Omega(1)^O M_j] \quad , \tag{3.5}$$

in which the second bracketted expression is the intrinsic mover-pool while the first bracketted expression is the normalized modification factor.

Although it is not the purpose of this paper to propose particular functions or classes of functions for the functional dependence of the a_{ij}, the b_{ij}, the $^O M_j$ or the V_j terms on the stock vector $P(t)$, a few comments still may be illuminating. The intrinsic mover-pool $^O M_j$ might be chosen according to the equation

$$^O M_j = \xi P_j \quad , \tag{3.6}$$

and ξ then could be considered as the intrinsic mobility of the respective populations. Some of the most recent results on transition probabilities and, in a way, also on the intrinsic transition probabilities b_{ij}, are given in Smith and Slater (1981) and Smith (1981). These authors have examined procedures for incorporating information flow processes and choice set constraints into the models. It also is important to be able to deal directly with the intransitivity of many empirically observed migration flows. It can be shown that with multiplicative push-pull models of the type shown by equation (1.1), the matrix $\mathbf{B}$ must not be chosen to be biproportional to a symmetric matrix if intransitivity is to be achieved. A particular functional dependence of the matrix **A** on the stock vector P(t) must be specified. In a first approach one presumably would choose the matrix **A** equal to the corresponding matrix **B**, on the grounds that, in a way, the j-th columns of matrices **B** and **A** both serve to express the attraction of the various alternatives as perceived by an individual in region j.

Finally, it should be noted that there is strong empirical evidence that some form of attraction-regulation is needed in migration models. Gueldner (1982) has shown that effects reported by Lansing and Mueller (1967), by Gleave and Cordey-Hayes (1974), and in a non-quantitative form as early as Ravenstein (1885), may be well-captured by attraction-regulated multiplicative migration models, provided the mover-pool modifying factor Ω is chosen within some class of not too steeply rising isotonic functions of $V_j/\bar{V}_j$. Isotonicity of Ω implies that the out-migration of the individuals associated with alternative region j increases with increasing relative attractiveness $V_j/\bar{V}_j$. This feature may appear to be counter-intuitive, but it is what has been reported by the above listed authors. In this context it also should be remarked that such comparatively attractive alternative regions reveal even higher in-migration.

Gueldner (1982) has analyzed the relation of out-migration to in-migration as a function of the steepness of Ω, which may be expressed by a single parameter τ if Ω chosen is of the type

$$\Omega(V_j/\bar{V}_j) = (V_j/\bar{V}_j)^\tau \quad . \tag{3.7}$$

Strict isotonicity prevails if $\tau > 0$. Wenzel (1982) has shown that for $\tau \geq 1$ stability fails, while for $\tau < 1$ the system is asymptotically stable, provided some additional assumptions are satisfied. Gueldner (1982) shows numerically and, under certain conditions, also analytically that the solutions of equations (2.6), (3.1) and (3.7) are located in the mobility plane (where out-migration per capita is plotted against in-migration per capita) within a narrow strip with slope τ. This finding corresponds to the empirical observations alluded to above, which may be roughly summarized by the statement that the observation points of samples drawn in several countries all may be fitted well by a positively sloping straight line.

4. A DYNAMIC EQUILIBRIUM FORMULATION OF THE MIGRATION MODELS

Let the function f: $R^+ \to R^+$ be defined by

$$\bar{V}_j/\{V_j \exp [\{f(M_j/{}^O M_j)\}]\} = \Omega(V_j/\bar{V}_j) \quad , \tag{4.1}$$

or, equivalently,

$$\exp\{-f[\Omega(V_j/\bar{V}_j)]\} = [\Omega(V_j/\bar{V}_j)](V_j/\bar{V}_j) \quad . \quad (4.3)$$

With

$$\tilde{V}_j = \sum_i V_i b_{ij} \quad , \quad (4.3)$$

equation (1.4) may be rewritten in the following way:

$$M_{ij} = V_i b_{ij} \, \Omega(V_j/\bar{V}_j) \, {}^OM_j/\tilde{V}_j$$

$$= V_i b_{ij} \, (\bar{V}_j/\tilde{V}_j) \, {}^OM_j/\{V_j \exp[f(M_j/{}^OM_j)]\} \quad . \quad (4.4)$$

Introducing

$$U_j = V_j \exp[f(M_j/{}^OM_j)] \quad , \quad (4.5)$$

and rearranging terms, one may rewrite equation (4.4) in the following form:

$$M_{ij}/(b_{ij} \, {}^OM_j \bar{V}_j/\tilde{V}_j) = V_i/U_j \quad . \quad (4.6)$$

Taking the logarithms of equations (4.5) and (4.6) yields

$$f(M_j/{}^OM_j) = u_j - v_j \quad , \text{ and} \quad (4.7)$$

$$\ln(M_{ij}) - \ln(b_{ij} \, {}^OM_j \bar{V}_j/\tilde{V}_j) = v_i - u_j \quad , \quad (4.8)$$

with

$$u_j = \ln(U_j) \quad \text{and} \quad v_j = \ln(V_j) \quad . \quad (4.9)$$

Equations (1.2), (2.5), (4.7) and (4.8) constitute a set of network equilibrium conditions. The underlying digraph (M,A) of the network consists of as many components as there are populations in the system, namely

$$(M,A) = \bigcup_{n \varepsilon N} (M^n, A^n) \quad . \quad (4.10)$$

Each component (M^n, A^n) is of the type shown in Figure 1. M^n constitutes the node set and A^n the arc set. The nodes of $I^n \subset M^n$ correspond to the various alternatives of the choice set I^n of the n-th population. $\underline{I}^n$ is just a copy of I^n. Arcs $(0^n, i)$, $i \varepsilon I^n$, carry P_i^n as flow quantities. The mover-pool M_j^n is carried by arcs $(j,\underline{j})$, $j \varepsilon I^n$ and $j \varepsilon \underline{I}^n$, while the migrants M_{ij}^n are carried by arcs (j,i), $i \varepsilon I^n$ and $j \varepsilon \underline{I}^n$. The P_i^n, M_j^n and M_{ij}^n collected over all arcs of all components of the digraph (M,A), constitute what may be called the rate vector **R** of the system. Equations (1.2) and (2.5) are Kirchhoff's nodal conditions for the rate vector **R**, with equation (1.2) being associated, for each population $n \varepsilon N$, with the nodes of I^n and equation (2.5) with the nodes of $\underline{I}^n$.

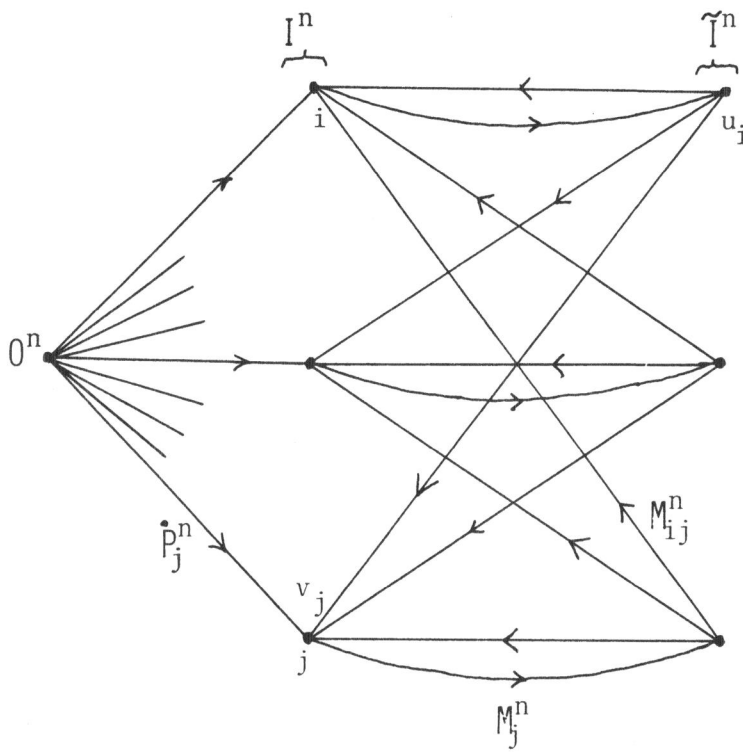

Figure 1.   Typical component of the digraph of the dynamic
equilibrium network.

The real variables u_i and v_j constitute nodal potentials,
with the u_i associated, for each population $n \varepsilon N$, with the nodes of
I^n, while the v_j are associated with the nodes of $\tilde{I}^n$.   The
potentials at the nodes 0^n, $n \varepsilon N$, are all zero.

For each arc there is an equilibrium condition.   Equation
(4.7) constitutes the equilibrium conditions for arcs of type
$(j,\tilde{j})$, and equation (4.8) constitutes those for arcs of type (j,i),
$j \varepsilon \tilde{I}^n$ and $i \varepsilon I^n$.   Therefore, the inductance functions of arcs $(j,\tilde{j})$,
$\tilde{j}$ being the copy of j, are

$$y_{j\tilde{j}} = f(M_j / {}^O M_j) \quad , \qquad (4.11)$$

and of arcs (j,i), $j \varepsilon \tilde{I}^n$ and $i \varepsilon I^n$, are

$$y_{ji} = \ln(M_{ij}) - \ln(b_{ij} \, {}^O M_j \bar{V}_j / \tilde{V}_j) \quad . \qquad (4.12)$$

As an aside, while in static equilibrium models one speaks of impedance functions of arcs, for dynamic equilibrium models one similarly may make use of electrical engineering terminology, and speak of inductance functions. Dejon (1980) provides a very brief introduction to the basic notions of network theory for these types of purposes.

For exactness, the superscript n will be used here, whereas it was omitted earlier. The equilibrium conditions for arcs $(0^n, i)$, $i \varepsilon I^n$, have to be formulated yet. These conditions simply and trivially read as

$$0 - v_i^n = -v_i^n \quad , \qquad (4.13)$$

which means that the inductance functions of the arcs $(0^n, i)$ are of the type

$$y_{0^n i} \equiv -v_i^n \qquad (4.14)$$

(i.e., of a constant type; independently of the value of the flow variable $\overset{\bullet}{P}_i^n$, $y_{0^n i}$ always takes the same value $-v_i^n$).

5. CONCLUSIONS AND EXTENSIONS

What is to be gained from the dynamic equilibrium formulation of attraction-regulated multiplicative migration models presented here? Generally speaking, there exists a vast body of knowledge in network equilibrium theory that may be exploited. Wenzel (1982), for instance, has used network theoretic methods for stability analyses of the type of migration models presented here (as well as of, in a sense, more general systems of differential inclusions in the place of equations). Another advantage of the dynamic network equilibrium formulation is that it points to some natural ways of extending the class of available models. For example, one may introduce parameters λ and ρ into the inductance functions (4.11) and (4.12), so that

$$y_{j\tilde{j}} = \lambda \; f(M_j / {}^O M_j), \text{ and} \qquad (5.1)$$

$$y_{ij} = \rho \; \ln(M_{ij}) - \rho \; \ln(b_{ij} \; {}^O M_j \bar{V}_j / \tilde{V}_j) \qquad . \qquad (5.2)$$

This alteration will change the speed with which the migration processes evolve.

A further extension of the class of models is achieved when the simple inductance function (4.14) is replaced by

$$y_{0^n i} = \omega_i \overset{\bullet}{P}_i - v_i \qquad , \quad \omega_i > 0, \qquad (5.3)$$

or, more generally, by

$$y_{0n_i} = \tilde{t}_{0n_i}(\dot{P}_i) - v_i \quad , \qquad (5.4)$$

where the $\tilde{t}_{0n_i}(\cdot)$ are isotonic functions, possibly set-valued, with $0 \varepsilon \tilde{t}_{0n_i}(0)$. Dejon (1980) deals with the set of inductance functions (5.1), (5.2) and (5.4) in which the assumption is made that $\lambda=\rho=0$. Under this assumption the migration streams M_{ij}, and the mover-pools M_j, in a dynamic equilibrium rate vector $\dot{R}$, may be chosen arbitrarily, except that the nodal conditions (1.2) and (2.5) should be satisfied. It is just the $\dot{P}_i$ that are uniquely defined [at least if the $\tilde{t}_{0n_i}(\cdot)$ are strictly isotonic]. This means that these models are not complete migration models, since only the rates of change of stocks are being modelled.

To conclude, the central question that still should be asked is: Which ones of all these models are the good ones (i.e., models that are good fits to empirical data)? Gueldner's (1982) observations appear to help in answering part of this question. Certainly the introduction of attraction-regulation appears to be highly desirable.

6. REFERENCES

Alonso, W., 1973, National Interregional Demographic Accounts, Monograph No. 17, Institute of Urban and Regional Development, University of California, Berkeley.

_____, 1977, Policy-Oriented Interregional Demographic Accounting and a Generalization of Population Flow Models, in Internal Migration: A Comparative Perspective, edited by A. Brown and E. Neuberger, New York: Academic Press, pp. 75-89.

_____, 1978, A Theory of Movements, in Human Settlement Systems: International Perspective on Structure, Change, and Public Policy, edited by N. Hansen, Cambridge, Mass.: Ballinger, pp. 197-211.

Batty, M., 1976, Urban Modelling: Algorithms, Calibrations, Predictions, Cambridge: Cambridge University Press.

Dejon, B., 1980, Dynamics of Equilibrating Processes in Disequilibrated Spatial Interaction Networks, Papers of the Regional Science Association, 44: 161-173.

_____ and F. Graef, 1982, Eine Klasse von Modellen zur Beschreibung von Staedtischen Agglomerations-und Deglomerationsprozessen, forthcoming in Stadt, Region, Land, Institut fuer Stadt-bauwesen, Rheinisch-Westfaelische TH Aachen.

Dorigo, G. and W. Tobler, 1982, Push-Pull Migration Laws, unpublished paper, Geographical Institute of the University of Zuerich, Department of Geography of the University of California, Santa Barbara.

Ginsberg, R., 1972, Incorporating Causal Structure and Exogenous Information With Probabilistic Models: With Special Reference to Choice, Gravity, and Markov Models, Journal of Mathematical Sociology, 2: 83-103.

Gleave, D. and M. Cordey-Hayes, 1974, Migration Dynamics and Labour Market Turnover, Progress in Planning, 8: 1-95.

Gueldner, B., 1982, Attraction-regulated Migration Models With Isotone Mobility Modulators, unpublished paper, Institute of Applied Mathematics of the University of Erlangen-Nuernberg.

Lansing, J. and E. Mueller, 1967, The Geographic Mobility of Labor, Ann Arbor: Survey Research Center, Institute for Social Research, University of Michigan.

Lowry, I., 1964, A model of metropolis, Research Memorandium RM-4035-RC, Santa Monica, California: RAND Corporation.

Oppenheim, N., 1980, Applied Models in Urban and Regional Analysis, Englewood Cliffs, N. J.: Prentice-Hall.

Putman, S., 1979, Urban Residential Location Models, Leiden: Martinus Nijhoff.

Ravenstein, E., 1885, The Laws of Migration, Journal of the Royal Statistical Society, 48: 167-235.

Smith, T., 1981, Transition Probabilities and Behavior of Master Equation Descriptions of Population Movement, in Dynamic Spatial Models, edited by D. Griffith and R. MacKinnon, Alphen van den Rijn, The Netherlands: Sijthoff and Noordhoff, pp. 49-66.

_____ and P. Slater, 1981, A Family of Spatial Interaction Models Incorporating Information Flows and Choice Set Constraints Applied to U. S. Interstate Labor Flows, International Regional Science Review, 6: 15-31.

Wenzel, G., 1982, Zur Existenz und Stabilitaet von Loesungen Gewisser Impliziter Differentialinklusionen zur Beschreibung des Dynamischen Gleichgewichts in Abstrakten Netzwerken, unpublished doctoral dissertation, Institute of Applied Mathematics of the University of Erlangen-Nuernberg.

Wilson, A., 1974, <u>Urban and Regional Models in Geography and Planning</u>, New York: Wiley.

A NON-LINEAR DYNAMIC MODEL
FOR THE MIGRATION OF HUMAN POPULATIONS

Gunter Haag
Wolfgang Weidlich
University of Stuttgart
Federal Republic of Germany

1.   INTRODUCTION

The purpose of this paper is to present a model of migration and birth-death processes for interacting populations. This model is a typical application of the general concepts set out in the book Concepts and Models of a Quantitative Sociology (Weidlich and Haag, 1982), while a similar type of model has been reported on elsewhere (Weidlich and Haag, 1980a). There exists an extensive literature on birth-death and migration processes of biological species, of which the books of May (1973), Rosen (1970), Pielou (1969), and Haken (1977) are only a small selection. This literature includes stochastic models as well as mean value models.

Although there is, of course, some overlap with these models, the approach introduced here contains several important new features. First, the migration of animals between habitats is usually described as a linear diffusion process with constant transition rates between neighboring sites. This description seems to be appropriate for modelling 'random migration' of populations of animals (May, 1973; Jorne and Carmi, 1977). For human populations, however, such a description is insufficient. Human decisions to move to another place, for instance to another part of a city, tend to take into account the existing distribution of different populations in the city (i.e., the existing socio-configuration). Furthermore, these decisions usually depend upon valuations of the quality of neighboring or distant quarters. Thus, human migration processes can be expected to be non-linear, with transition rates depending upon the existing socio-configuration, as well as non-diffusionary, with transitions to non-neighboring quarters. In the third section of this paper it

will be shown that even the simplest non-linear migraion process,
considering only two interacting populations between two parts of a
city, can lead to a variety of effects, such as ghetto formation or
cyclic migration. These effects depend upon the values of the
parameters describing the psychology of interactions between the
populations.

Another mathematically interesting and biologically realistic
model arises if the well-known Volterra-Lotka model is combined
with non-linear migration of the predatory and prey species between
habitats. In contrast to the pure Volterra-Lotka model, which is
sensitive to small perturbations, the modified Volterra-Lotka model
presented here may give rise to a limit cycle that is a stable
solution independent of initial conditions.

2. THE GENERAL MODEL

In this section the master equation and the mean value
equations for the general migration and birth-death process of
interacting populations will be set up. It is assumed that there
are P distinct populations P_α ($\alpha=1,2,...,P$) and L possible regions,
quarters or habitats ($i=1,2,...,L$). The socio-configuration can
be represented as a point in the C-dimensional socio-configuration
space L, where C=PL and

$$\mathbf{n} \equiv \{n_{\alpha i}\} \; \varepsilon \; L \qquad , \qquad (2.1)$$

where $n_{\alpha i}$ is the number of members of P_α living in quarter i. The
contributions to the transition probability $w[\mathbf{k},\mathbf{n}] \geq 0$ of change from
a socio-configuration $\mathbf{n}\varepsilon L$ to another $(\mathbf{n} + \mathbf{k})\varepsilon L$ are introduced by
assuming, for the sake of simplicity, individual migration and
birth-death processes. In other words, the number of group members
simultaneously undergoing transition is 1. In the general case the
total transition probability is the sum over all contributions made
by migration, birth-death and Volterra-Lotka processes.

The transition probability $w^\alpha_{ij}[\mathbf{k},\mathbf{n}]$ relating to migration of
members of population P_α from quarter i to quarter j ($i \neq j$) may be
defined as

$$n_{\alpha i}\mu^\alpha_{ji}(\mathbf{n},\mathbf{k}) \text{ for}$$

$$w^\alpha_{ji}[\mathbf{k};\mathbf{n}] = \{ \quad \mathbf{k} = \{...0...1_{\alpha j}...0...(-1)_{\alpha i}...\} \quad (2.2)$$

$$0 \text{ for all other } \mathbf{k} \quad ,$$

where $\mu^\alpha_{ji}(\mathbf{n},\mathbf{k})$ is the corresponding individual transition
probability depending in general on $\mathbf{n}$ and some trend parameters Ω.
The transition probabilities $w^\alpha_{i+}[\mathbf{k},\mathbf{n}]$ and $w^\alpha_{i-}[\mathbf{k},\mathbf{n}]$ with respect to
birth and death, respectively, of members of population P_α in

quarter i may be defined as

$$
w_{i+}^{\alpha} [k;n] = \begin{cases} n_{\alpha i} \beta_i^{\alpha} & \text{for } k = \{\ldots 0..1_{\alpha i}\ldots 0\ldots\} \\ 0 & \text{for all other } k \end{cases} \tag{2.3}
$$

, and

$$
w_{i-}^{\alpha} [k;n] = \begin{cases} n_{\alpha i}\delta_{i+}^{\alpha} n_{\alpha i}^2 \gamma_i^{\alpha} & \text{for } k=\{\ldots 0\ldots(-1)_{\alpha i}\ldots 0\ldots\} \\ 0 & \text{for all other } k \end{cases} \tag{2.4}
$$

where β_i^{α}, δ_i^{α} and γ_i^{α} are the corresponding birth, death and saturation rates. Definitions (2.3) and (2.4) can, if necessary, be generalized to take into account, in a much more complicated way, birth or death rates that are dependent upon the existing socio-configuration. The Volterra-Lotka transition probabilities for the interaction of prey (δ) and predator (β) populations of animals in habitat i may be defined as

$$
w_i^{\delta\beta}[k;n] = \begin{cases} n_{\delta i} \, n_{\beta i} \, v_{\delta\beta} & \text{for } k = \{0\ldots(-1)_{\delta i}\ldots 0\ldots 1_{\beta i}\ldots 0\} \\ 0 & \text{for all other } k \end{cases} \tag{2.5}
$$

By definition all the transition probabilities introduced in these four preceding equations are positive semi-definite.

The total transition probabilities are given by the sum over equations (2.2) thru (2.5), and therefore are given by

$$
w[k,n] = \sum_{j=1}^{j=L} \sum_{i=1}^{i=L} \sum_{\alpha=1}^{\alpha=P} w_{ji}^{\alpha}[k,n] +
$$

$$
\sum_{i=1}^{i=L} \sum_{\alpha=1}^{\alpha=P} (w_{i+}^{\alpha}[k,n] + w_{i-}^{\alpha}[k,n]) +
$$

$$
\sum_{i=1}^{i=L} w_i^{\delta\beta}[k,n] \tag{2.6}
$$

In the last term on the right-hand side of equation (2.6) the sum over (δ,β) extends over pairs of species for which δ is the prey of predator β. The fundamental master equation

$$
dp(n,t)/dt = \sum_{\{k\}} \{w[-k,n+k]p(n+k,t) - w[k,n]p(n,t)\} , \tag{2.7}
$$

determines the evolution through time of the probability distribution $p(n,t)$ of a given socio-configuration n. Inserting equation (2.6) into equation (2.7) yields the master equation of the model in a general form, namely

$$dp(\mathbf{n},t)/dt = (\partial p/\partial t)_M(\mathbf{n},t) + (\partial p/\partial t)_{BD}(\mathbf{n},t) + (\partial p/\partial t)_{VL}(\mathbf{n},t), (2.8)$$

where the three terms on the right-hand side of equation (2.8) refer to migration, birth-death and Volterra-Lotka processes, respectively, and arise from the contributions of equations (2.2) thru (2.5) to $w[\mathbf{k},\mathbf{n}]$ in equation (2.6). For the explicit form of these terms it is convenient to introduce a transition operator, denoted by $E_{\alpha i}$, which acts on a function of the socio-configuration as follows:

$$E_{\alpha i}^{\pm 1} f(n_{11}\cdots n_{\alpha i}\cdots n_{PL}) = f[n_{11}\cdots(n_{\alpha i}\pm 1)\cdots n_{PL}] \quad . \quad (2.9)$$

Inserting equations (2.2) thru (2.6) into equation (2.7), and applying the transition operator renders explicit forms of the terms of the right-hand side of the master equation (2.8) as:

$$(\partial p/\partial t)_M(\mathbf{n},t) = \sum_{i=1}^{i=L} \sum_{j=1}^{j=L} \sum_{\alpha=1}^{\alpha=P} (E_{\alpha i}E_{\alpha j}^{-1}-1)[n_{\alpha i}\mu_{ji}^\alpha(\mathbf{n},k)p(\mathbf{n},t)] \quad (2.10)$$

$$(\partial p/\partial t)_{BD}(\mathbf{n},t) = \sum_{i=1}^{i=L} \sum_{\alpha=1}^{\alpha=P} \{(E_{\alpha i}^{-1} - 1)[n_\alpha \beta_i^\alpha p(\mathbf{n},t)] +$$

$$(E_{\alpha i}-1)[(n_{\alpha i}\delta_i^\alpha + n_{\alpha i}^2\gamma_i^\alpha)p(\mathbf{n},t)]\}, \text{ and } (2.11)$$

$$(\partial p/\partial t)_{VL}(\mathbf{n},t) = \sum_{i=1}^{i=L} \sum_{\delta=1}^{\delta=P} \sum_{\beta=1}^{\beta=P} (E_{\delta i}E_{\beta i}^{-1}-1)[n_{\delta i}n_{\beta i}v_{\delta\beta}p(\mathbf{n},t)] \quad . \quad (2.12)$$

It is possible to derive the following equations of motion for the mean value directly from the master equation (2.8):

$$\langle n_{\delta k}\rangle_t = \sum_{\{\mathbf{n}\}} n_{\delta k}\, p(\mathbf{n},t) \quad . \quad (2.13)$$

Here equation (2.8) has been multiplied by $n_{\delta k}$, and then summed over all configurations $\mathbf{n}$. Using transition formulae such as

$$n_{\gamma k}E_{\alpha i}E_{\alpha j}^{-1}f(\mathbf{n}) \stackrel{=}{=} E_{\alpha i}E_{\alpha j}^{-1}(n_{Gkk} - \delta_{\gamma\alpha}\delta_{ki} + \delta_{\gamma\alpha}\delta_{kj})f(\mathbf{n}) \quad (2.14)$$

derived from equation (2.8) yields the exact result

$$d\langle n_{\gamma k}\rangle/dt \;=\; \sum_{i=1}^{i=L} \langle n_{\gamma i}\mu_{ki}^{\gamma}(\mathbf{n},\mathbf{\Omega})\rangle_t \;-\; \sum_{j=1}^{j=L} \langle n_{\gamma k}\mu_{jk}^{\gamma}(\mathbf{n},\mathbf{\Omega})\rangle_t \;+$$

$$[(\beta_k^{\gamma} - \delta_k^{\gamma})\langle n_{\gamma k}\rangle_t \;-\; \gamma_k^{\gamma}\langle n_{\gamma k}^2\rangle_t] \;+$$

$$\sum_{\delta} v_{\delta\gamma}\langle n_{\delta k}n_{\gamma k}\rangle_t \;-\; \sum_{\beta} v_{\gamma\beta}\langle n_{\gamma k}n_{\beta k}\rangle_t \quad . \quad (2.15)$$

The three terms on the right-hand side of equation (2.15) refer to migration, birth-death and prey-predatory processes, respectively. In the last of these terms the first sum extends over species δ, which are the prey of the predator γ, while the second sum extends over the predator species β with prey γ.

For distributions $p(\mathbf{n},t)$ with only one sharp peak the exact equations (2.15) can be replaced by the approximate closed set of equations for the mean values $\langle n_{\gamma k}\rangle_t \overset{\sim}{=} n_{\gamma k}$, namely

$$(d\tilde{n}_{\gamma k}/dt) \;=\; \sum_{i=1}^{i=L} \tilde{n}_{\gamma i}\mu_{ki}^{\gamma}(\mathbf{n},\mathbf{\Omega}) \;-\; \sum_{j=1}^{j=L} \tilde{n}_{\gamma k}\mu_{jk}^{\gamma}(\mathbf{n},\mathbf{\Omega}) \;+$$

$$[(\beta_k^{\gamma} - \delta_k^{\gamma})\tilde{n}_{\gamma k} \;-\; \gamma_k^{\gamma}\tilde{n}_{\gamma k}^2] \;+$$

$$\sum_{\delta} v_{\delta\gamma}\tilde{n}_{\delta k}\tilde{n}_{\gamma k} \;-\; \sum_{\beta} v_{\gamma\beta}\tilde{n}_{\gamma k}\tilde{n}_{\beta k} \quad . \quad (2.16)$$

Equations for the variances also can be derived in the same manner. This will be done in the following sections, where special cases of the general model will be explicitly investigated.

3. MIGRATION OF TWO INTERACTING POPULATIONS BETWEEN TWO PARTS OF A CITY

In this section birth-death processes and Volterra-Lotka processes will be neglected, and the simple but nevertheless non-trivial case of two distinct, non-linearly interacting human popualtions P_μ and P_Δ settling in and migrating between two parts of a city will be modelled (Weidlich and Haag, 1980b). In this model the two populations considered are assumed to be only a small fraction of a main 'background population.'

The number of members of population P_μ and P_Δ in the city area i ($i=1,2$) are denoted by m_i and n_i, respectively. The total number of the members of P_μ and P_Δ is to be conserved, thus,

$$(m_1 + m_2) = 2\bar{m} \qquad \text{and} \qquad (n_1 + n_2) = 2\bar{n} \quad . \quad (3.1)$$

Therefore there are two rather than four relevant integer variables for each of m and n characterizing the socio-configuration, where

$$m_1 = \bar{m} + m \quad \text{and} \quad m_2 = \bar{m} - m \quad \text{with} \quad -\bar{m} \le m \le \bar{m} \text{ , and}$$
$$n_1 = \bar{n} + n \quad \text{and} \quad n_2 = \bar{n} - n \quad \text{with} \quad -\bar{n} \le n \le \bar{n} \text{ .}$$
$$\} \ (3.2)$$

Using the individual transition probabilities $\mu_{ji}^{\mu}(m,n)$ and $\mu_{ji}^{\Delta}(m,n)$ for members of P_μ and P_Δ, respectively, moving from area i to area j of the city, the probabilities $w(k,l;m,n)$ for a transition from the socio-configuration, denoted by (m,n), to the configuration $(m+k,n+l)$ are obtained from equation (2.2) as:

$$w(-1,0;m,n) = (\bar{m} + m)\mu_{21}^{\mu}(m,n) \quad ,$$

$$w(1,0;m,n) = (\bar{m} - m)\mu_{12}^{\mu}(m,n) \quad ,$$

$$w(0,-1;m,n) = (\bar{n} + n)\mu_{21}^{\Delta}(m,n) \quad , \text{ and}$$

$$w(0,1;m,n) = (\bar{n} - n)\mu_{12}^{\Delta}(m,n) \quad .$$
$$\} \ (3.3)$$

In order to make the model developed so far fully explicit as well as applicable, the individual transition probabilities $\mu_{ji}^{\mu}(m,n)$ and $\mu_{ji}^{\Delta}(m,n)$ of the members of the two populations P_μ and P_Δ have to be adequately represented. The model adopted here, which is sufficiently flexible in that several possible effects of the psychology of interacting populations on the migration process are included, is given by the set of equations

$$\mu_{12}^{\mu}(m,n) = \gamma \ \exp[u(m,n)] \quad ,$$

$$\mu_{21}^{\mu}(m,n) = \gamma \ \exp[-u(m,n)] \quad ,$$

$$\mu_{12}^{\Delta}(m,n) = \gamma \ \exp[v(m,n)] \quad , \text{ and}$$

$$\mu_{21}^{\Delta}(m,n) = \gamma \ \exp[-v(m,n)] \quad ,$$
$$\} \ (3.4)$$

where

$$u(m,n) = \pi_\mu + \hat{\Omega}_\mu m + \hat{\sigma}_\mu n \quad , \text{ and}$$

$$v(m,n) = \pi_\Delta + \hat{\Omega}_\Delta n + \hat{\sigma}_\Delta m \quad .$$
$$\} \ (3.5)$$

An interpretation of the trend or propensity parameters γ, π_μ, π_Δ, $\hat{\Omega}_\mu$, $\hat{\Omega}_\Delta$, $\hat{\sigma}_\mu$, and $\hat{\sigma}_\Delta$ can be easily deduced by examining equations (3.4) and (3.5) along with equation (3.3). In general, the individual transition probabilities characterize the propensity of members of the populations P_μ and P_Δ to move to areas of the city where the existing number of members P_μ and P_Δ as well as the qualitative 'value' of these areas are taken into account. As the interpretations of γ, π_μ, $\hat{\Omega}_\Delta$, and $\hat{\sigma}_\Delta$ are synonymous with those of γ, π_μ, $\hat{\Omega}_\mu$, and $\hat{\sigma}_\mu$, only these latter four parameters need to be discussed.

The parameter γ determines the general time scale in which migration processes take place, and thus it may be called the flexibility parameter. The parameter π_μ describes the level of preference given to one of the two city areas by the members of P_μ: for $\pi_\mu > 0$ ($\pi_\mu < 0$) area 1 (area 2) is preferred by members of P_μ. Hence π_μ may be called the preference parameter of P_μ. The parameter $\hat{\Omega}_\mu$ ($\hat{\Omega}_\mu > 0$ or $\hat{\Omega}_\mu < 0$) describes the strength of the (positive or negative) propensity of members of P_μ to cluster or group in the same area of the city. In other words, it is this population group's propensity to 'live together.' Thus $\hat{\Omega}_\mu$ may be called the internal sympathy parameter for P_μ. Finally, the parameter $\hat{\sigma}_\mu$ ($\hat{\sigma}_\mu > 0$ or $\hat{\sigma}_\mu < 0$) describes the strength of the (positive or negative) propensity of members of P_μ to live with members of the population P_Ω in the same area of the city. Consequently, σ_μ may be called the external sympathy parameter for P_μ.

In equation (3.4) the trends described by π_μ, $\hat{\Omega}_\mu$ and $\hat{\sigma}_\mu$ are automatically aggregated. All trend parameters are assumed to remain constant in the time interval in which the migration process is to be investigated. In principle, however, it is possible to take into account a time dependency for these trend parameters. Despite their suggestive connotations, it is neither intended nor necessary to give an explicit psychological explanation of the motivations giving rise to these parameters. For the validity of the model, it is sufficient to have a bundle of individual motivations generating migration that can be described by the transition probabilities defined in equations (3.4) and (3.5), or further generalizations of these two equations.

3.1. The Master, Mean Value and Variance Equations

The master equation, corresponding to equation (2.7), for the probability distribution $p(m,n,t)$ over the two relevant variables m and n, using equation (3.3), is:

$$[dp(m,n,t)/dt] = (E_m-1)[w(-1,0;m,n)p(m,n,t)] +$$

$$(E_m^{-1}-1)[w(1,0;m,n)p(m,n,t)] +$$

$$(E_n-1)[w(0,-1;m,n)p(m,n,t)] +$$

$$(E_n^{-1}-1)[w(0,1;m,n)p(m,n,t)] \quad . \quad (3.6)$$

Applying the same technique as in the derivation of equation (2.15) from (2.8), the equations of motion for the mean values of m, n, m^2, mn and n^2 can be derived directly, where mean values $\langle f(m,n)\rangle_t$ of $f(m,n)$ are defined as

$$\langle f(m,n) \rangle_t \ = \ \sum_{m=-\bar{m}}^{m=\bar{m}} \ \sum_{n=-\bar{n}}^{n=\bar{n}} \ f(m,n) \ p(m,n,t) \qquad . \qquad (3.7)$$

The exact equations

$$(d\langle m \rangle/dt) \ = \ \langle k_m(m,n) \rangle_t \qquad , \ \text{and}$$

$$(d\langle n \rangle/dt) \ = \ \langle k_n(m,n) \rangle_t \qquad , \qquad \left. \right\} \ (3.8)$$

and

$$(dm\langle^2\rangle/dt) \ = \ 2\langle mk_m(m,n) \rangle_t \ + \ \langle q_m(m,n) \rangle_t \ ,$$

$$(d\langle mn \rangle/dt) \ = \ \langle mk_n(m,n) \rangle_t \ + \ \langle nk_m(m,n) \rangle_t \qquad , \ \text{and} \ \left. \right\} \ (3.9)$$

$$(d\langle n^2 \rangle/dt) \ = \ 2\langle nk_n(m,n) \rangle_t \ + \ \langle q_n(m,n) \rangle_t \ ,$$

are obtained after introducing the 'drift coefficients'

$$k_m(m,n) \ = \ w(1,0;m,n) \ - \ w(-1,0,m,n) \qquad , \ \text{and}$$

$$k_n(m,n) \ = \ w(0,1;m,n) \ - \ w(0,-1;m,n) \qquad , \qquad \left. \right\} \ (3.10)$$

and the 'fluctuation coefficients'

$$q_m(m,n) \ = \ w(1,0;m,n) \ + \ w(-1,0;m,n) \qquad , \ \text{and}$$

$$q_n(m,n) \ = \ w(0,1;m,n) \ + \ w(0,-1;m,n) \qquad . \qquad \left. \right\} \ (3.11)$$

Using equation (3.9) and the following definitions of the variances

$$\sigma_{mm} \ = \ \langle m^2 \rangle_t \ - \ \langle m \rangle_t^2 \qquad ,$$

$$\sigma_{mn} \ = \ \langle mn \rangle_t \ - \ \langle m \rangle_t \langle n \rangle_t \qquad , \ \text{and} \qquad \left. \right\} \qquad (3.12)$$

$$\sigma_{nn} \ = \ \langle n^2 \rangle_t \ - \ \langle n \rangle_t^2 \qquad ,$$

explicit forms for the equations of motion for the variances of the probability distribution may be derived.

3.2. A Stationary Solution to the Master Equation

The stationary solution $p_{st}(m,n)$ to the master equation must be obtained first. This solution takes into account the most probable configurations (m,n), and the fluctuations around these configurations in a stationary situation established by the migration process, after the relaxation of a possible initial imbalance. Haken (1977) has derived the explicit form for the solution $p_{st}(m,n)$ to the general stationary master equation

$$dp_{st}(i)/dt = \sum_{j} [w(j;i)p_{st}(i) - w(i;j)p_{st}(j)] \overset{!}{=} 0 \; , \quad (3.13)$$

where i and j can be multidimensional indices, for the case in which the transition probabilities $w(j,i)$ from i to j fulfill the condition

$$w(j;i)p_{st}(i) \; = \; w(i;j)p_{st}(i) \qquad\qquad (3.14)$$

for every index pair (j,i). this case is referred to as one of 'detailed balance.'

The principle of detailed balance requires that there are as many transitions per unit time from state i to state j as there are from state j to state i by the inverse process. For a solution, starting from an arbitrarily chosen site, denoted by $i_o = 0$, there has to exist at least one chain L, but in general many chains, of sites

$$i_1 = 1, \; i_2 = 2, \; \ldots, \; i_{n-1} = (n - 1), \; i_n = j$$

with non-vanishing probabilities of making transitions to every other site

$$w(2,1), \; w(1,2), \; \ldots, \; w(j,n-1), \; w(n-1,j) \qquad , \quad j \neq 0 \qquad .$$

Considering one of these chains, the stationary solution $p_{st}(j)$ is obtained by repeated application of equation (3.14), and takes the form

$$p_{st}(j) \; = \; p_{st}(0) \; \prod_{\Lambda=0}^{\Lambda=n-1} w(\Lambda+1,\Lambda)/w(\Lambda,\Lambda+1) \qquad . \qquad (3.15)$$

By using equation (3.14) this result can be proved to be unique and therefore independent of the choice of the chain L. The normalization of equation (3.15), such that the sum of the $p_{st}(i)$ equals unity, finally determines the magnitude of $p_{st}(0)$. Conditions for $w(j,i)$ that are equivalent to equation (3.14) but that do not contain the stationary solution can be derived simultaneously from equation (3.15). In other words, for any closed loop L [i=0, 1, 2, ..., (1-1), 1=0] of sites with $w(i,i+1) \neq 0$ and $w(i+1,i) \neq 0$, applying equation (3.15) for j=0 and with L substituted for L gives

$$(L) \; \prod_{\Lambda=0}^{\Lambda=1-1} w(\Lambda+1,\Lambda)/w(\Lambda,\Lambda+1) \; = \; 1 \qquad . \qquad (3.16)$$

The model cases for which this condition of detailed balance is fulfilled will now be checked.

As there exist transition probabilities defined by equation (3.3) only between neighboring sites in (m,n) space, the necessary and sufficient condition for detailed balance is that equation (3.16) holds for all elementary closed loops, namely

$$\mathtt{L}[(m,n),(m+1,n),(m+1,n+1),\ (m,n+1),\ (m,n)] \qquad .$$

This result implies that

$$\frac{w(1,0;m,n)\ w(0,1;m,n)\ w(-1,0;m+1,n+1)\ w(0,-1;m,n+1)}{w(-1,0;m+1,n)\ w(0,-1;m+1,n+1)\ w(1,0;m,n+1)\ w(0,1;m,n)} = 1 \quad .(3.17)$$

Substituting equations (3.4) and (3.5) into (3.3), and combining the result with equation (3.17) yields the condition

$$\exp[2(\hat{\sigma}_\mu - \hat{\sigma}_\Delta)] = 1 \qquad , \quad \text{or}$$

$$\hat{\sigma}_\mu = \hat{\sigma}_\Delta = \hat{\sigma} \qquad \text{and} \qquad \bar{m}\sigma_\mu = \bar{n}\sigma_\Delta \qquad . \qquad (3.18)$$

This last equation implies that detailed balance only holds when the external sympathy parameters are equal. In this case the exact stationary solution to equation (3.6) can be set up by using equation (3.15) and by choosing, for example, the chain

$$\mathtt{L}[(0,0),(1,0),\dots,(m,0),(m,1),\dots,(m,n)] \qquad .$$

Hence,

$$p_{st}(m,n) = p_{st}(0,0)[\prod_{\mu=0}^{\mu=m-1} w(1,0;\mu,0)/w(-1,0;\mu+1,0)] \times$$

$$[\prod_{\Delta=0}^{\Delta=n-1} w(0,1,m;\Delta)/w(0,-1,m;\Delta+1) \qquad . \qquad (3.19)$$

Substituting equation (3.3) into equations (3.4) and (3.5) and taking the product of these two results yields

$$p_{st}(m,n) = C(\tfrac{2\bar{m}}{m+m})(\tfrac{2\bar{n}}{n+n})\ \exp[\zeta(m,n)] \qquad , \qquad (3.20)$$

where

$$\zeta(m,n) = 2\pi_\mu m + 2\pi_\Delta n + \hat{\Omega}_\mu m^2 + 2\hat{\sigma}mn + \hat{\Omega}_\Delta n^2 \quad , \quad (3.21)$$

and C is the normalization factor. Finally, evaluation of the binomial coefficients using Stirling's formula results in

$$p_{st}(m,n) \;\tilde{=}\; C' \exp\phi(m,n) \qquad , \qquad (3.22)$$

with

$$\Phi(m,n) = \zeta(m,n) - [(\bar{m}+m)\ln(\bar{m}+m) + (\bar{m}-m)\ln(\bar{m}-m)] -$$

$$[(\bar{n}+n)\ln(\bar{n}+n) + (\bar{n}-n)\ln(\bar{n}-n)] \quad . \quad (3.23)$$

Extrema of the distribution defined by equation (3.22) must satisfy the following conditions:

$$\left.\partial\Phi(m,n)/\partial m\right|_{\hat{m},\hat{n}} = 0 \quad \text{or} \quad \hat{m} = \bar{m}\,\tanh[u(\hat{m},\hat{n})] \quad , \text{ and}$$

$$\left.\partial\zeta(m,n)/\partial n\right|_{\hat{m},\hat{n}} = 0 \quad \text{or} \quad \hat{n} = \bar{n}\,\tanh[v(\hat{m},\hat{n})] \quad . \quad \Big\}\ (3.24)$$

The case in which equation (3.18) does not hold (i.e., detailed balance is not attained) will be considered next. In this case the stationary solution generally available tends to have rather complicated representations. However, the form of the solution with detailed balance defined by equation (3.20) suggests that the following would be an appropriate formulation for the case without detailed balance:

$$p_{st}(m,n) = C \left(\frac{2\bar{m}}{\bar{m}+m}\right)\left(\frac{2\bar{n}}{\bar{n}+n}\right)\exp[\chi(m,n)] \quad , \quad (3.25)$$

where $\chi(m,n)$ is approximated by a polynomial of second order in m and n. Substituting equation (3.25) into the stationary master equation (3.6), expanding the exponential expressions in the appropriate way and then comparing coefficients leads to the determination of $\chi(m,n)$ as

$$\chi(m,n) = 2\pi_\mu m + 2\pi_\Delta n + \hat{\Omega}_\mu m^2 + (\hat{\sigma}_\mu + \hat{\sigma}_\Delta)mn + \hat{\Omega}_\Delta n^2 \quad . \quad (3.26)$$

The form of $p_{st}(m,n)$ given by equation (3.25) together with equation (3.26) extends equation (3.20), as a first approximation, to the case without detailed balance.

3.3. Approximate Closed Equations of Motion

Here a distribution $p(m,n,t)$ will be assumed with only one sharp peak around the mean values $\langle m\rangle$ and $\langle n\rangle$. Hence, the scaled dimensionless time variate $\tau=2\gamma t$ will be used instead of π, and the normalized variables x and y will be used instead of m and n, such that

$$x = m/\bar{m} \quad \text{where} \quad -1 \leq x \leq 1 \text{ , and}$$

$$y = n/\bar{n} \quad \text{where} \quad -1 \leq y \leq 1 \text{ .} \quad (3.27)$$

The approximate mean value equations take the form

$$dx/d\tau = \sinh[u(x,y)] - x \cosh[u(x,y)] = [\cosh(u)][\tanh(u) - x], \text{ and}$$
$$\hspace{11cm} (3.28)$$
$$dy/d\tau = \sinh[v(x,y)] - y \cosh[v(x,y)] = [\cosh(v)][\tanh(v) - y],$$

with

$$u(x,y) = \pi_\mu + \Omega_\mu x + \sigma_\mu y \quad \text{and} \quad v(x,y) = \pi_\mu + \Omega_\Delta y + \sigma_\Delta x . \quad (3.29)$$

The explicit but lengthy variance equation may be obtained in an analogous fashion from equations (3.9), (3.10) and (3.11), but obtaining this result is not necessary for the present purposes. It should be pointed out, however, that for vanishing preference parameters (i.e., for $\pi_\mu = \pi_\Delta = 0$) the equations (3.28) and (3.29) are invariant under the transformations

(a) $\quad x \rightarrow -x$, $\quad\quad y \rightarrow -y$, and

(b) $\quad x \rightarrow x$, $\quad y \rightarrow -y$, $\quad \sigma_\Delta \rightarrow -\sigma_\Delta$, $\quad \sigma_\mu \rightarrow -\sigma_\mu$.

Invariance (a) reflects the equivalence of both parts of the city for $\pi_\mu = \pi_\Delta = 0$. Invariance (b) implies, for instance, that for every solution for which $x(\tau) > 0$ and $y(\tau) > 0$ concerning populations with positive external sympathy parameters, where $\sigma_\mu > 0$ and $\sigma_\mu > 0$ describe a clustering of both populations P_μ and P_Δ in the same area of the city, there exists a corresponding solution $x'(\tau) = x(\tau)$ and $y'(\tau) = -y(\tau)$ for populations with negative external sympathy parameters, where $\sigma'_\mu = -\sigma_\mu$ and $\sigma'_\Delta = -\sigma_\Delta$ describe a separation of these two populations.

The global structure of the change in the mean values of the migration system with time depends decisively on the stationary or singular points $P_j(x_j, y_j)$ of the differential equations (3.28). These are defined by

$$(dx/d\tau)_{P_j} = \cosh(u_j)[\tanh(u_j) - x_j] \equiv 0 , \text{ and}$$
$$\hspace{11cm} \} \ (3.30)$$
$$(dy/d\tau)_{P_j} = \cosh(v_j)[\tanh(v_j) - y_j] \equiv 0 ,$$

or, equivalently, by

$$F_\mu(x_j, y_j) \equiv \tanh[u(x_j, y_j)] - x_j = 0 \quad , \text{ and}$$
$$\hspace{11cm} \} \ (3.31)$$
$$F_\Delta(x_j, y_j) \equiv \tanh[v(x_j, y_j)] - y_j = 0 \quad .$$

The singular points $P_j(x_j, y_j)$ for given propensity parameters π_μ, Ω_μ, σ_μ, π_Δ, Ω_Δ and σ_Δ are intersection points of a graph of $F_\mu(x,y) = 0$ belonging to $(\pi_\mu, \Omega_\mu, \sigma_\mu)$ with a graph of $F_\Delta(x,y) = 0$ belonging to $(\pi_\Delta, \Omega_\Delta, \sigma_\Delta)$. In the symmetrical case (i.e., for $\pi_\mu = \pi_\Delta = 0$) there can exist 1, 3, 5 or 9 singular points $P_j(x_j, y_j)$, depending upon the choice of the trend parameters.

Employing linear stability analysis at this point will help to answer the question of whether or not a point $P[x(\tau),y(\tau)]$ in the vicinity of a singular point $P_j(x_j,y_j)$, moving according to equation (3.28), will approach (or move away from) P_j. In the first (second) case P_j is denoted as a stable (unstable) focus. For this analysis the following small deviations from the singular point $P_j(x_j,y_j)$ are introduced:

$$\xi(\tau) = x(\tau) - x_j \quad \text{and} \quad \eta(\tau) = \mu y(\tau) - y_j \quad , \quad (3.32)$$

and equation (3.28) is linear in $\xi(\tau)$ and $\eta(\tau)$, yielding

$$d\xi/d\tau = -\gamma_{\mu j}\xi + \sigma_{\mu j}\eta \qquad , \text{ and}$$
$$d\eta/d\tau = \sigma_{\Delta j}\xi - \gamma_{\Delta j}\eta \qquad , \qquad \} \qquad (3.33)$$

where

$$\Omega_{\mu j} = \Omega_\mu/\cosh(u_j) \quad \text{and} \quad \Omega_{\Delta j} = \Omega_\Delta/\cosh(v_j) \quad ,$$

$$\gamma_{\mu j} = [\cosh(u_j)-\Omega_{\mu j}] \quad \text{and} \quad \gamma_{\Delta j} = [\cosh(v_j)-\Omega_{\Delta j}] \quad , \} \quad \text{and} \quad (3.34)$$

$$\sigma_{\mu j} = \sigma_\mu/\cosh(u_j) \quad \text{and} \quad \sigma_{\Delta j} = \sigma_\Delta/\cosh(v_j) \quad .$$

The solutions to equation (3.33) are linear combinations of the solutions to the characteristic equations

$$\xi_{\pm}(\tau) = \xi_{0\pm} \exp(\lambda_{j\pm} \tau) \qquad \text{and}$$
$$\eta_{\pm}(\tau) = \eta_{0\pm} \exp(\lambda_{j\pm} \tau) \qquad , \qquad \} \qquad (3.35)$$

with eigenvalues

$$\lambda_{j\pm} = -\{(\gamma_{\mu j} + \gamma_{\Delta j}) \pm [(\gamma_{\mu j}+\gamma_{\Delta j})^2 + 4\varepsilon_j]^{1/2}\}/2 \quad , \qquad (3.36)$$

where $\varepsilon_j \equiv \sigma_{\mu j}\sigma_{\Delta j} - \gamma_{\mu j}\gamma_{\Delta j}$. The eigenvalues λ_{j+} and λ_{j-} are either real or conjugate complex numbers. P_j is a stable focus if $\text{Re }\lambda_{j+}<0$ and $\text{Re }\lambda_{j-}<0$. A case of particular interest arises under the conditions

$$\Omega_\mu + \Omega_\Delta > 2 \quad , \quad \sigma_\mu \sigma_\Delta < 0 \quad \text{and} \quad \pi_\mu = \pi_\Delta = 0 \qquad , \quad (3.37)$$

for which the origin $P(0,0)$ is the only singular point, and hence an unstable focus. Furthermore, the fluxlines of equation (3.28) in the domain $D(-1 \leq x \leq 1, -1 \leq y \leq 1)$ are always directed inward at the boundary because

$$dx/d\tau < 0 \quad \text{for} \quad x = 1 \quad \text{and} \quad dx/d\tau > 0 \quad \text{for} \quad x = -1 \text{ , and}$$
$$\left.\begin{array}{l}\\ \\\end{array}\right\} \ (3.38)$$
$$dy/d\tau < 0 \quad \text{for} \quad y = 1 \quad \text{and} \quad dy/d\tau > 0 \quad \text{for} \quad y = -1 \text{ .}$$

In other words, in the domain D^*, consisting of D but without the origin $P(0,0)$, a limit cycle must exist as a solution to equation (3.28), according to the Poincare-Bendixon theorem (Rosen, 1970). The solution computed and shown in Figure 7a verifies this statement.

3.4. Numerical Illustrations of the Results

Sets of propensity parameters have been chosen in such a way that all possible combinations of critical points associated with the system appear in Figures 1a thru 7a. The resulting paths of the mean values $x(\tau)$ and $y(\tau)$ have been graphed in these figures, too. In addition to these mean values, of particular interest in Figure 7c is the path of the variances $\sigma_{xx}(\tau)$, $\sigma_{xy}(\tau)$ and $\sigma_{yy}(\tau)$ for the limit cycle. Each of these figures results from computed solutions to equations (3.28) and (3.9) for selected values of the trend parameters. Furthermore, the solutions, represented by equations (3.20) and (3.25), to the stationary master equation for the same trend parameters are shown in Figures 1b thru 7b. For illustrative purposes m and n both are given the value of 20 in all calculations, since this numerical value leads to a noticeably large variance for the resulting distributions.

In interpreting these figures it has to be kept in mind that according to the definitions of the variables x and y, the subdomain quadrants of D are defined as

$D_{++}(0 < x < 1$ and $0 < y < 1)$: both populations concentrate in city area 1,

$D_{-+}(-1 < x < 0$ and $0 < y < 1)$: P_μ concentrates in city area 2 and P_Δ concentrates in city area 1,

$D_{+-}(0 < x < 1$ and $-1 < y < 0)$: P_μ concentrates in city area 1 and P_Δ concentrates in city area 2, and

$D_{--}(-1 < x < 0$ and $-1 < y < 0)$: both populations concentrate in city area 2.

For the results shown in Figures 1a and 1b, weak internal and external sympathy parameters are assumed. Figure 1a shows that all mean values tend towards the stable focus $P(0,0)$. This focus represents an equal population density of P_μ and P_Δ over both city areas. Consequently, the stationary distribution shown in Figure 1b concentrates around this stable focus.

This result illustrates that the geographic distribution of both populations over city areas 1 and 2 does not play a prominent role in the migration decisions for members of populations with

38

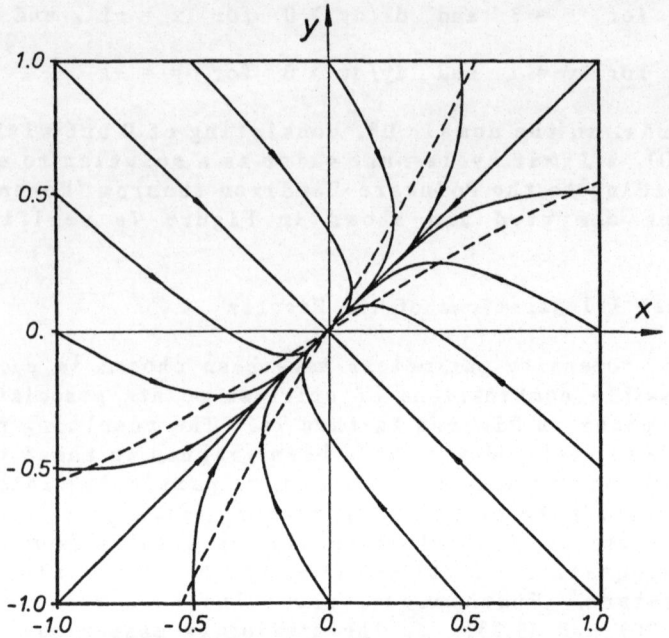

Figure 1a. x-y diagram of fluxlines of mean values using trend parameters $\pi_\mu = \pi_\Delta = 0$, $\Omega_\mu = \Omega_\Delta = .2$ and $\sigma_\mu = \sigma_\Delta = .5$, leading to one stable focus at x=y=0.

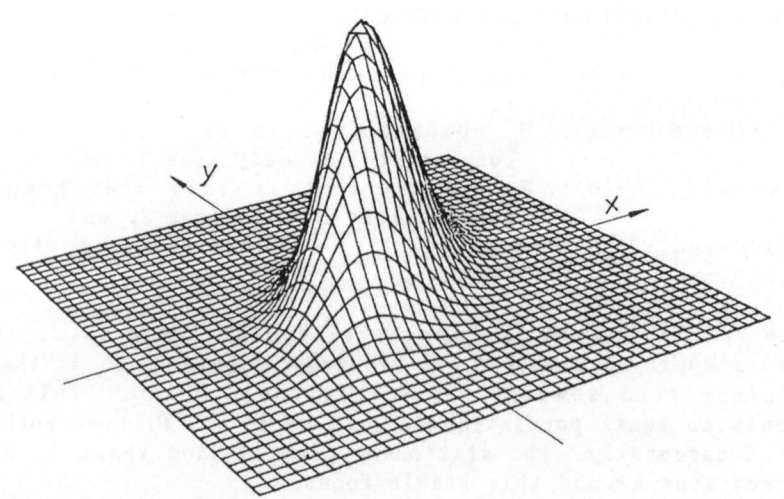

Figure 1b. Stationary solution of the master equation (3.25), where $\bar{m} = \bar{n} = 20$ and $p_{st}(m,n)_{max} = .9937 \times 10^{-2}$.

weak internal and external sympathy parameters. Furthermore, no preference for one of the areas is assumed because $\pi_\mu = \pi_\Delta = 0$. Therefore it is to be expected that any initial non-equilibrium distribution of population finally moves into the state of a homogeneous distribution, as mentioned above.

In Figures 2a and 2b weak internal but stong external sympathy parameters are assumed. Figure 2a shows that, depending upon their initial values, the mean values tend toward one of the stable foci occurring in D_{++} or D_{--}. These two foci represent a concentration of both populations in area 1 or area 2 of the city. On the other hand, the focus $P(0,0)$, which represents a homogeneous population distribution, has become unstable. Thus the stationary distribution of Figure 2b shows symmetrical concentrations around the stable foci. Perhaps, then, the strong external sympathy parameters imply a mutual 'attraction' between the populations, in the sense of implying a strong propensity to live together, and hence a tendency to migrate to the area where the majority of the other population lives. Therefore, a final and stable state is reached in which the majority of both populations settle in the same area of the city. The case described by Figures 2a and 2b is a good example for illustrating the implication of the invariance of equation (3.28) under the transformation (b) of Section 3.3, which holds for $\pi_\mu = \pi_\Delta = 0$.

For the solutions shown in Figures 3a and 3b, extremely strong internal and strong external sympathy parameters have been assumed. Figure 3a indicates that, in addition to the main stable foci in D_{++} and D_{--}, now there exist secondary stable foci in D_{-+} and D_{+-}. These stable foci correspond to a concentration of the populations P_μ and P_Δ. Depending upon their initial values, the mean values move to one of the main or secondary foci. The corresponding stationary distribution shown in Figure 3b exhibits two pronounced maxima around the two main foci.

Figures 4a and 4b show the situation occurring if both populations have a preference for city area 1, in addition to having positive internal and external sympathy parameters. Figure 4a shows that the unstable and the two stable foci of Figure 3 now are shifted. Here the mean value fluxlines tend to end in the main stable focus in D_{++}. The stationary distribution shown in Figure 4b has just one pronounced maximum, which is located around the main stable focus. Moreover, if the sympathy within and between the populations exists along side of a 'natural' or intrinsic preference for area 1, the majority of both populations finally will migrate to and settle in the area. The stable focus in D_{--} represents the much less probable case of both populations remaining settled in the less preferred area 2, because of both the existence of internal and mutual sympathy and the given initial situation.

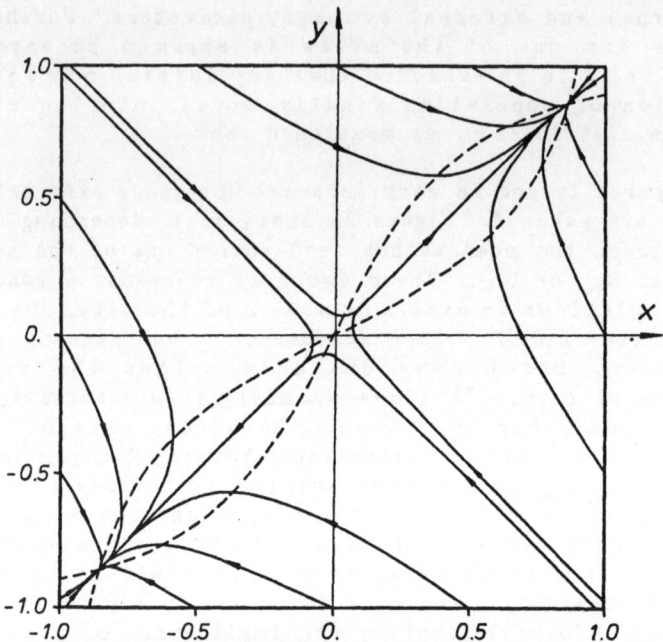

Figure 2a. x-y diagram of fluxlines of mean values using trend
parameters $\pi_\mu=\pi_\Delta=0$, $\Omega_\mu=\Omega_\Delta=.5$ and $\sigma_\mu=\sigma_\Delta=1$, leading to
two stable foci and one unstable focus.

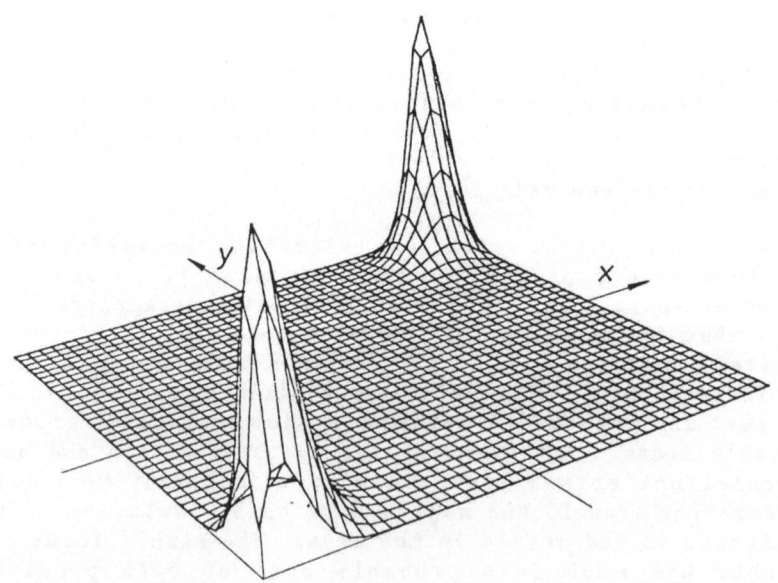

Figure 2b. Stationary solution of the master equation (3.25),
where $\bar{m}=\bar{n}=20$ and $p_{st}(m,n)_{max}=.2615\times10^{-1}$.

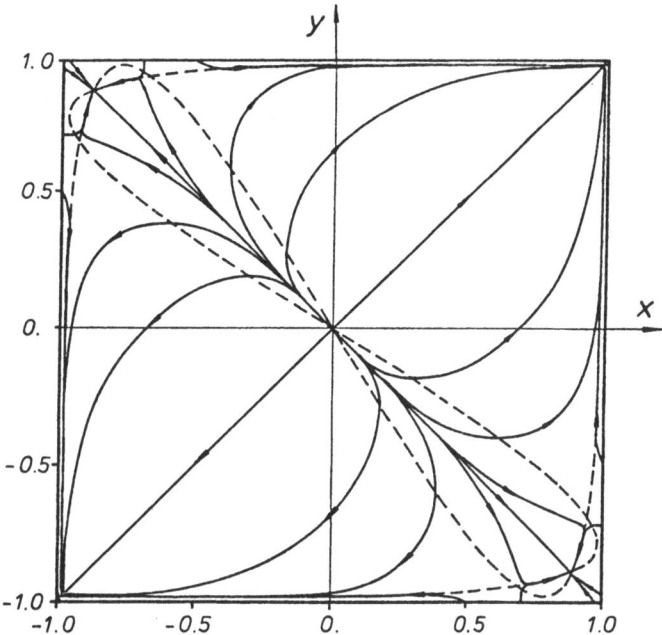

Figure 3a. x–y diagram of fluxlines of mean values using trend parameters $\pi_\mu=\pi_\Delta=0$, $\Omega_\mu=\Omega_\Delta=2.6$ and $\sigma_\mu=\sigma_\Delta=1$, leading to four stable and five unstable foci.

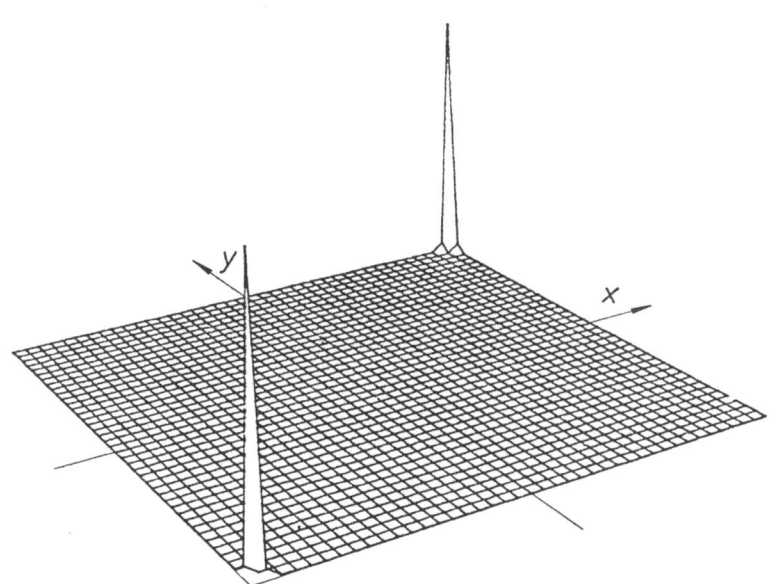

Figure 3b. Stationary solution of the master equation (3.25), where $\bar{m}=\bar{n}=20$ and $p_{st}(m,n)_{max}=.4669$.

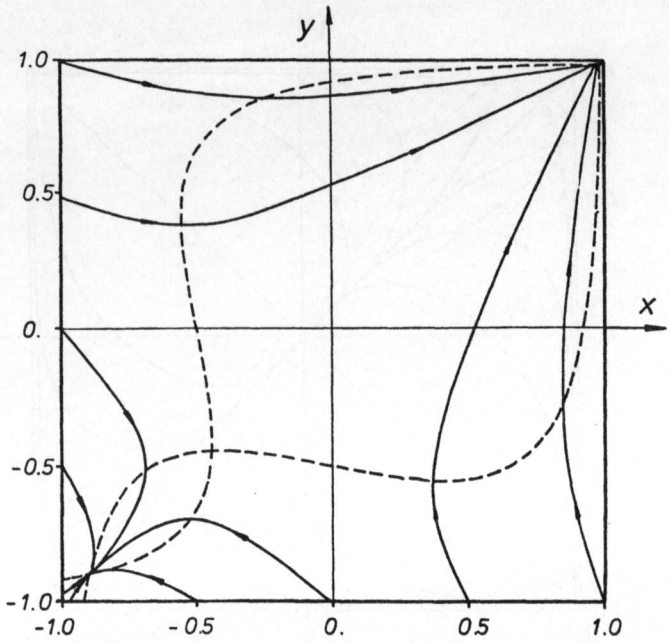

Figure 4a. x-y diagram of fluxlines of mean values using trend parameters $\pi_\mu = \pi_\Delta = .25$, $\Omega_\mu = \Omega_\Delta = 1.2$ and $\sigma_\mu = \sigma_\Delta = 1$, leading to two stable foci and one unstable focus.

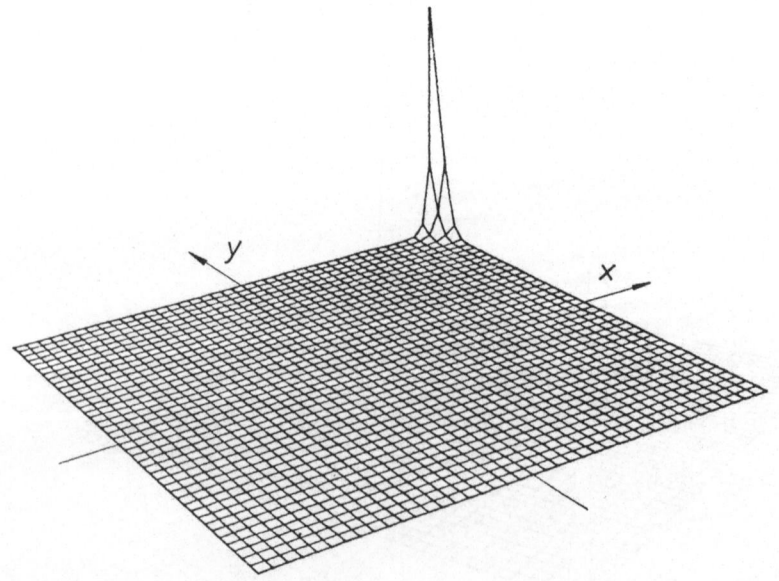

Figure 4b. Stationary solution of the master equation (3.25), where $\bar{m} = \bar{n} = 20$ and $p_{st}(m,n)_{max} = .5197$.

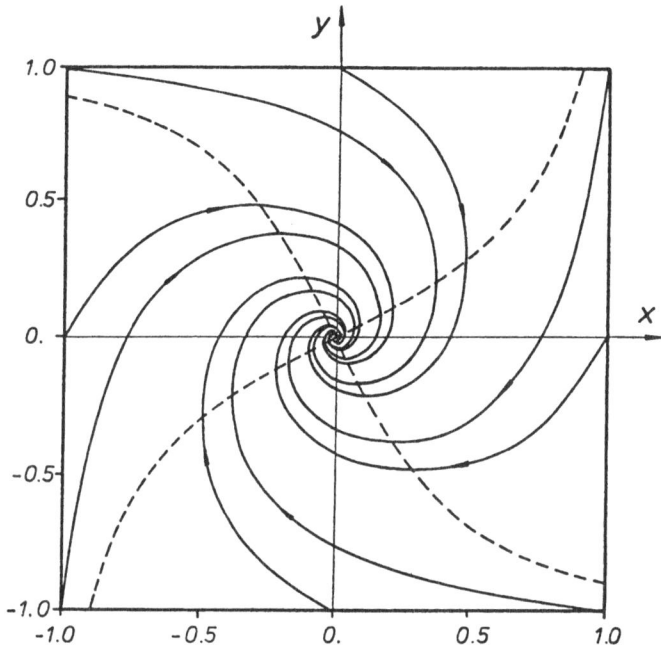

Figure 5a. x-y diagram of fluxlines of mean values using trend
parameters $\pi_\mu=\pi_\Delta=0$, $\Omega_\mu=\Omega_\Delta=.5$ and $\sigma_\mu=-\sigma_\Delta=1$, leading to
one stable focus at x=y=0.

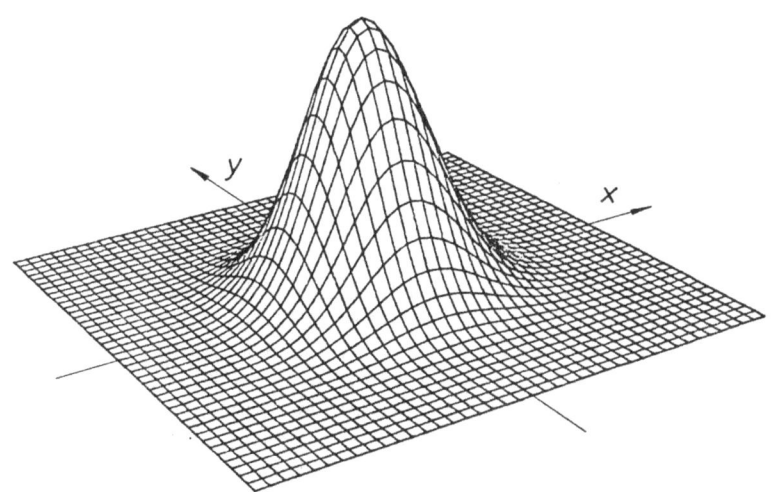

Figure 5b. Stationary solution of the master equation (3.25),
where $\bar{m}=\bar{n}=20$ and $p_{st}(m,n)_{max}=.7952\times10^{-2}$.

For Figures 5a and 5b weak internal sympathy parameters but strong and asymmetrical external sympathy parameters have been assumed. While populations P_μ and P_Λ tend to live together as neighbors of P_Λ ($\sigma_\mu > 0$), P_Λ wants to separate from P_μ ($\sigma_\Lambda < 0$). The fluxlines of Figure 5a then approach the stable focus $P(0,0)$ in spirals that express a new type of migration. The stationary distribution shown in Figure 5b also concentrates around this stable focus. Although the stationary situation, namely the uniform density of both populations over both areas of the city, is almost the same as for the case illustrated in Figures 1a and 1b, the trends behind the migration process in non-equilibrium situations differ from those of Figures 1a and 1b, since population P_Λ tends to avoid close contact with P_μ. However, the weak internal sympathy parameter associated with P_Λ, which indicates only a weak preference for these populations to live together, finally wipes out the difference between this case and that of Figure 1, again leading to the stationary situation of homogeneously distributed populations.

In Figures 6a and 6b strong asymmetrical external sympathy parameters and internal sympathy parameters are shown to lead to a phase transition. Figure 6a shows mean value fluxlines that approach the focus at $x=y=0$ at a very slow rate (i.e., critical slowing down). This phenomenon is due to the fact that the focus becomes unstable for the values assumed for the trend parameters. The stationary distribution shown in Figure 6b covers a large interval around the focus, corresponding to the large 'critical fluctuations' at the phase transition. The transition from Figures 5a and 5b to 6a and 6b is due to the choice of stronger internal sympathy parameters $\Omega_\mu = \Omega_\Lambda = 1$ instead of $\Omega_\mu = \Omega_\Lambda = 0.5$. The solidarity of 'staying together' has been increased for each population. This favors a collective migration of groups of members of P_μ and P_Λ, separately. Consequently, extremely large fluctuations around the homogeneous distribution are observed in Figure 6b.

The assumption of very strong internal sympathy parameters, together with the same strong asymmetrical external sympathy parameters, as assumed in the construction of Figures 5 and 6, leads to the configurations of Figures 7a and 7b. Figure 7a demonstrates that the mean value fluxlines now approach a limit cycle instead of a stable focus. In other words, the migration process does not reach a stable situation in spite of the existence of a stationary solution to the master equation. This type of stationary solution to the master equation appears in Figure 7b. This type of migration has to be interpreted as the expression of the following conflict between the two populations. Starting from a geographic distribution in which both populations are concentrated in city area 1 (i.e., $P[\tau.]\epsilon D_{++}$), since $\sigma_\Lambda < 0$ the population P_Λ tends to avoid P_μ and migrate to and concentrate in city area 2, leaving P_μ behind. This would correspond to a

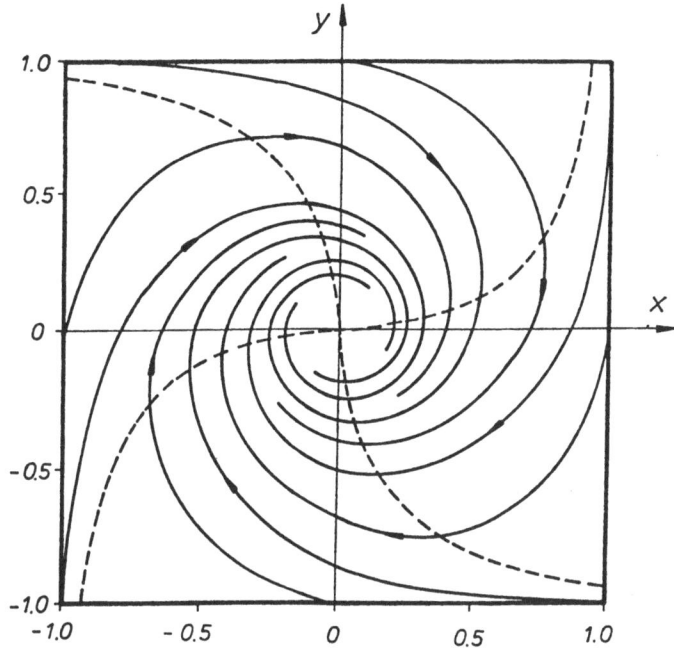

Figure 6a. x-y diagram of fluxlines of mean values using trend parameters $\pi_\mu = \pi_\Delta = 0$, $\Omega_\mu = \Omega_\Delta = 1.0$ and $\sigma_\mu = -\sigma_\Delta = 1$, at the phase transition, where the focus at $x = y = 0$ becomes unstable.

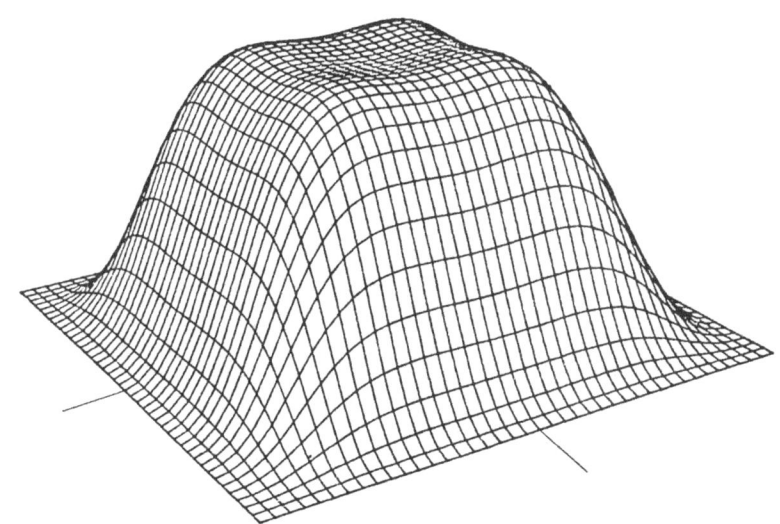

Figure 6b. Stationary solution of the master equation (3.25), where $\bar{m} = \bar{n} = 20$ and $p_{st}(m,n)_{max} = .1323 \times 10^{-2}$.

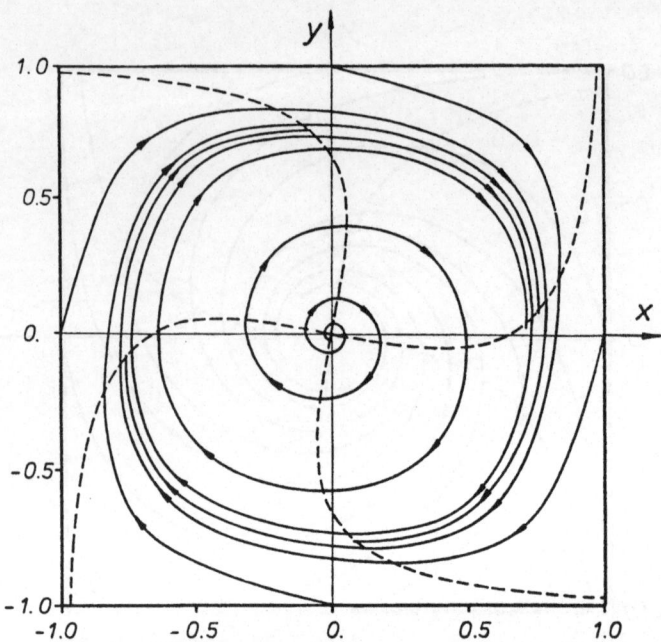

Figure 7a. x—y diagram of fluxlines of mean values using trend
parameters $\pi_\mu=\pi_\Delta=0$, $\Omega_\mu=\Omega_\Delta=1.2$ and $\sigma_\mu=-\sigma_\Delta=1$, showing a
limit cycle.

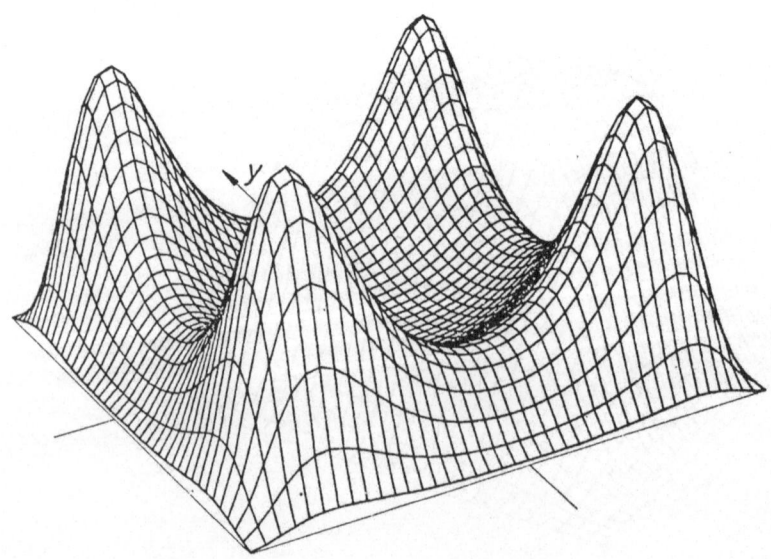

Figure 7b. Stationary solution of the master equation (3.25),
where $\bar{m}=\bar{n}=20$ and $p_{st}(m,n)_{max}=.1957 \times 10^{-2}$.

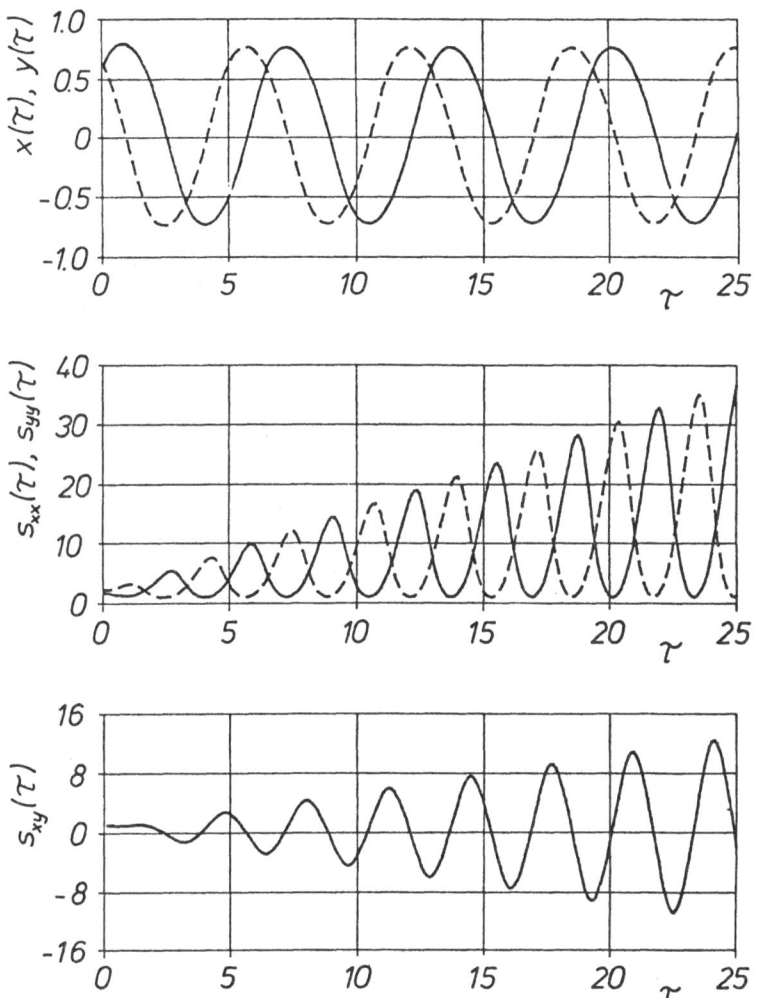

Figure 7c. Mean value x(τ) and variance σ$_{xx}$(τ), denoted by solid
lines, mean value y(τ) and variance σ$_{yy}$(τ), denoted by
broken lines, and variance σ$_{xy}$(τ) plotted as a function
of scaled time τ.

configuration $P(\tau_1)\varepsilon D_{+-}$. Since $\sigma_\mu > 0$ population P_μ will follow
sooner or later, and concentrate in city area 2 as well. The
resulting geographic distribution now has reached that of
$P(\tau_3)\varepsilon D_{--}$. This 'cat and mouse' process is repeated and leads both
populations back to area 1 of the city, at least in this simple

model. In more realistic models with a city split up into more than two areas, a generalization of this type of model may capture the principle lying behind the sequential erosion (i.e., sequence occupance) observed in many big cities of the world, in which occurs migration of populations of different social attitudes who interact in an asymmetrical manner. Figure 7b shows the stationary distribution for this case, which concentrates in the vicinity of the limit cycle. The peaks and connecting ridges of this distribution correspond to the varying speed of traversing the fluxlines around the limit cycle.

To conclude the discussion of the results shown in Figures 1 thru 7, a general caveat regarding the approximate character of equations (3.28) and (3.9) is in order. The mean value equations (3.28) used for calculating the results of Figures 1a thru 7a, in general, will lose their validity for long time horizons. This follows from the fact that the exact and unique mean values belonging to the stationary distributions are $x(\infty)=0$ and $y(\infty)=0$ in all of these cases, with the exception of the case illustrated in Figure 4. But the fluxlines of the approximate equation (3.28), as $\tau \to \infty$, either terminate in one or another stable focus depending upon the initial condition, or approach a limit cycle. These fluxlines approximate the path of the exact mean value only as long as the accompanying distribution is essentially unimodal, as was noted at the outset. This is why this assumption was made in all the models discussed above.

In order to estimate how long a unimodal distribution can survive in the migration process, the approximate mean value and variance equations (3.28) and (3.9) have been solved, as an example, for the parameters chosen for the construction of Figures 7a and 7b, and for a judiciously selected initial variance. Figure 7c shows that the functions $s_{xx}(\tau)=\bar{m}\sigma_{xx}(\tau)$, $s_{xy}(\tau)=\bar{m}\sigma_{xy}(\tau)$ and $s_{yy}(\tau)=\bar{m}\sigma_{yy}(\tau)$, whose equations of motion are independent of the value of $\bar{m}=\bar{n}$, consist of oscillations together with a steady increase of the amplitude as the mean values $x(\tau)$ and $y(\tau)$ traverse the limit cycle. For $\bar{m} \gg 1$ and a set of reasonable initial conditions, $|\sigma_{xx}(\tau)| \ll 1$, $|\sigma_{xy}(\tau)| \ll 1$ and $|\sigma_{yy}(\tau)| \ll 1$ may remain valid for several cycles. This implies that the distribution remains essentially unimodal, and that equations (3.28) and (3.9) still hold. Thus, the shape of the (still unimodal) distribution repeatedly shrinks and widens, alternatingly, in the x and y directions during its motion around the limit cycle.

4. MIGRATION AND PREDATOR-PREY INTERACTION BETWEEN TWO SPECIES

This section is meant to contribute to the old problem of the interaction between two biological species. 'To be or not to be' is the essential question decided by the predator-prey interaction for the members of certain species. The famous Volterra–Lotka model for this problem (Lotka, 1920; Volterra, 1937) has attracted the interest of many researchers who have tried to generalize it in many respects (May, 1973; Getz, 1976; MacDonald, 1976). Here, as above, an attempt is made to shed new light on an old problem. In this case the new twist stems from the combination of non–linear migration with the predator–prey interaction. A useful implication can be anticipated for this combination. The pure Volterra–Lotka model leads to a mathematically marginal case. The singular point, which is equivalent to the stationary state, is a center encircled by closed orbits. These orbits are non–stationary solutions that depend upon the initial conditions, and this is analogous to the solutions of undampened Hamiltonian systems in theoretical mechanics. Therefore, the original model is sensitive to small perturbations that may have different causes (Yoon and Blanch, 1977; Chewning, 1975). On the other hand, Volterra–Lotka cycles have been observed by biologists in nature for several entirely different species (Gause, 1934; Huffaker, 1958; Utida, 1950). These writings stress that the existence of such cycles could depend upon the availability of appropriate facilities for migration (Huffaker, 1958; Chewning, 1975).

The extension of the pure Volterra–Lotka model by adding non–linear migration leads to a complex picture with a relatively large variety of different solution cases. In particular, a result has been obtained for an appropriate 'mixture' of migration and predator–prey interaction in which a limit cycle appears that is similar to the original Volterra–Lotka cycles, but which has a stable solution that is independent of the initial conditions.

4.1. Master and Mean Value Equations for the Special Model

The model proposed in this section is both simple and plausible. Two habitats are assumed, namely an open habitat 1 and a refuge habitat 2. The prey species is able to migrate between both habitats, while for the predatory species only the open habitat 1 is accessible. This situation is illustrated in Figure 8. A simple socio–configuration $\{m_1, m_2, n\}$ can consist of three integers, the numbers m_1 and m_2 of animals that can be preyed upon in the habitats 1 and 2, and the number $n_1 \stackrel{-}{=} n$ of predatory animals in, and restricted, to habitat 1. Denoting the prey and predatory populations by P_δ and P_β, and using the notation of the second section, the following contributions to the configuration transition probabilities are obtained.

Migration of P_δ.

$$w_{21}^\delta(-1,1,0;m_1,m_2,n) = m_1\mu_{21}(m_1,m_2,n) \quad ,$$
$$w_{12}^\delta(1,-1,0;m_1,m_2,n) = m_2\mu_{12}(m_1,m_2,n) \quad ,$$

$$\} \quad (4.1)$$

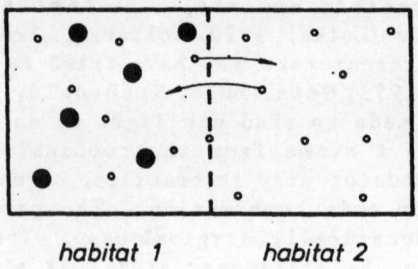

habitat 1   habitat 2

Figure 8.   Preditor population, denoted by solid circles, and prey
           popultion, denoted by open circles, in two habitats.

where the following forms for the individual migration
probabilities are adopted:

$$\mu_{12}(m_1,m_2,n) = \hat{\Delta}\, \exp[u(m_1,m_2,n)] \qquad ,$$
$$\left. \right\} \quad (4.2)$$
$$\mu_{21}(m_1,m_2,n) = \hat{\Delta}\, \exp[-u(m_1,m_2,n)] \qquad ,$$

with $u(m_1,m_2,n) = \hat{\delta} + \hat{\alpha}(m_1-m_2) - \hat{\beta}n$. In accordance with their
substative interpretations, the propensity parameters $\hat{\Delta}$, $\hat{\delta}$, $\hat{\alpha}$ and $\hat{\beta}$
may be labelled as follows:

$\hat{\Delta}$ = the flexibility parameter,

$\hat{\delta}$ = the preference parameter,

$\hat{\alpha}$ = the clustering parameter for population P_δ, and

$\hat{\beta}$ = the fear parameter (i.e., P_δ fears P_β).

Birth-death processes in P_δ and P_β.

$$w_{1+}^{\delta}(1,0,0;m_1,m_2,n) = m_1\beta_1^{\delta} \qquad ,$$
$$\left. \right\} \quad , \quad (4.3a)$$
$$w_{1-}^{\delta}(-1,0,0;m_1,m_2,n) = m_1\delta_1^{\delta} + m_1^2\gamma_1^{\delta}$$

$$w_{2+}^{\delta}(0,1,0;m_1,m_2,n) = m_2\beta_2^{\delta} \qquad ,$$
$$\left. \right\} \quad , \text{ and } (4.3b)$$
$$w_{2-}^{\delta}(0,-1,0;m_1,m_2,n) = m_2\delta_2^{\delta} + m_2^2\gamma_2^{\delta}$$

$$w_{1+}^{\beta}(0,0,1;m_1,m_2,n) = n\beta_1^{\beta} \qquad ,$$
$$\left. \right\} \quad . \quad (4.3c)$$
$$w_{1-}^{\beta}(0,0,-1;m_1,m_2,n) = n\delta_1^{\beta} + n^2\gamma_1^{\beta}$$

Prey-predator interaction between P_δ and P_β.

$$w_1^{\delta\beta}(-1,0,1;m_1,m_2,n) = m_1 n v_{\delta\beta} \quad . \qquad (4.4)$$

For convenience, the parameters are specialized and given a simplified notation for the model to be presented and numerically implemented here. The specific simplications are:

$$\left.\begin{array}{l} \beta_1^\delta = \beta_2^\delta = \Omega_1 \ , \ \gamma_2^\delta = \omega \ , \ \delta_1^\delta = \Omega_2 \ , \ v_{\delta\beta} = \gamma \\[2mm] \delta_1^\delta = \delta_2^\delta = 0 \ , \ \gamma_1^\delta = 0 \ , \ \beta_1^\beta = 0 \ , \ \gamma_1^\beta = 0 \end{array}\right\} \quad . \ (4.5)$$

The total configuration transition probability $w(k_1,k_2,k_3;m_1,m_2,n)$ is the sum of the transition probabilities given by equations (4.1), (4.3) and (4.4) for the individual partial processes [see equation (2.6)]. The master equation for the probability distribution $p(m_1,m_2,n,t)$ over the socio-configuration now has the general form of equation (2.8), namely:

$$\begin{aligned} dp(m_1,m_2,n,t)/dt = \ & (\partial p/\partial t)_M(m_1,m_2,n,t) \ + \\ & (\partial p/\partial t)_{BD}(m_1,m_2,n,t) \ + \\ & (\partial p/\partial t)_{VL}(m_1,m_2,n,t) \quad , \quad (4.6) \end{aligned}$$

where the three terms on the right-hand side have the following explicit form:

$$\begin{aligned} (\partial p/\partial t)_M(m_1,m_2,n,t) = \ & (m_1+1)\mu_{21}(m_1+1,m_2-1,n)p(m_1+1,m_2-1,n) \ - \\ & m_1\mu_{21}(m_1,m_2,n)p(m_1,m_2,n) \ + \\ & (m_2+1)\mu_{12}(m_1-1,m_2+1,n)p(m_1-1,m_2+1,n) \ - \\ & m_2\mu_{12}(m_1,m_2,n)p(m_1,m_2,n) \quad , \qquad (4.7) \end{aligned}$$

$$\begin{aligned} (\partial p/\partial t)_{BD}(m_1,m_2,n,t) = \ & \Omega_1(m_1-1)p(m_1-1,m_2,n) - \Omega_1 p(m_1,m_2,n) \ + \\ & \Omega_1(m_2-1)p(m_1,m_2-1,n) - \Omega_2 m_2 p(m_1,m_2,n) \ + \\ & \omega(m_2+1)^2 p(m_1,m_2+1,n) - \omega m_2^2 p(m_1,m_2,n) \ + \\ & \Omega_2(n+1)p(m_1,m_2,n+1) \ - \\ & \Omega_2 n p(m_1,m_2,n) \quad , \ \text{and} \quad (4.8) \end{aligned}$$

$$\begin{aligned} (\partial p/\partial t)_{VL}(m_1,m_2,n,t) = \ & \gamma(m_1+1)(n-1)p(m_1+1,m_2,n-1) \ - \\ & \gamma m_1 n p(m_1,m_2,n) \quad . \qquad (4.9) \end{aligned}$$

The approximate closed form equations for the mean values $\tilde{x} = \langle m_1 \rangle_t$, $\tilde{y} = \langle m_2 \rangle_t$ and $\tilde{z} = \langle n \rangle_t$ are obtained from the master equation (4.6) associated with concentrated unimodal distributions, and according to equation (2.16) A transformation to scaled variables and scaled time yields

$$x = (\Omega_2/\gamma)\tilde{x} \quad , \quad y = (\Omega_2/\gamma)\tilde{y} \quad , \quad z = (\Omega_1/\gamma)\tilde{z} \quad , \quad \tau = \Omega_1 t \quad , \quad (4.10)$$

whereas a transformation to scaled trend parameters yields

$$\Delta = \hat{\Delta}/\Omega_1 \quad , \quad \delta = \hat{\delta} \quad , \quad \alpha = (\gamma/\Omega_2)\hat{\alpha} \quad ,$$

$$\left. \beta = (\gamma/\Omega_1)\hat{\beta} \quad , \quad y_0 = (\Omega_1 \gamma)/(\Omega_2 \omega) \quad , \quad a = \Omega_2/\Omega_1 \quad , \right\} \text{ and } (4.11)$$

$$u(x,y,z) = \delta + \alpha(x - y) - \beta z \quad . \quad (4.12)$$

Consequently, the final scaled and explicit forms of the mean value equations obtained using equation (4.2) are

$$dx/d\tau = - \Delta[x \exp(-u) - y \exp(u)] + x(1 - z) \quad , \quad (4.13a)$$

$$dy/d\tau = \Delta[x \exp(-u) - y \exp(u)] + y(y_0 - y)/y_0 , \text{ and } (4.13b)$$

$$dz/d\tau = - a(1 - x)z \quad . \quad (4.13c)$$

Equations (4.13) are a foundation for the following analysis.

4.2. Comparison of Predatory-Prey Interaction Without and With Non-Linear Migration

As far as the singular points and their stability are concerned, standard analytical procedures will be applied. It can be shown that the singular points $p_s(x_s, y_s, z_s)$ of the system are solutions to the three transcendental equations

$$0 = - \Delta[x_s \exp(-u_s) - y_s \exp(u_s)] + x_s(1 - z_s) \quad , \quad (4.14a)$$

$$0 = \Delta[x_s \exp(-u_s) - y_s \exp(u_s)] + y_s(y_0 - y_s)/y_0 , \text{ and } (4.14b)$$

$$0 = - a(1 - x_s)z_s \quad . \quad (4.14c)$$

Then the linearized equations of motion (4.13) arround a singular point $p_s(x_s, y_s, z_s)$ are

$$\frac{d}{d\tau} \begin{pmatrix} \xi \\ \eta \\ \zeta \end{pmatrix} = \begin{pmatrix} a_{11} & a_{12} & a_{13} \\ a_{21} & a_{22} & a_{23} \\ a_{31} & a_{32} & a_{33} \end{pmatrix} \begin{pmatrix} \xi \\ \eta \\ \zeta \end{pmatrix} \quad , \quad (4.15)$$

where

$$x(\tau) = x_s + \xi(\tau) \quad ,$$

$$y(\tau) = y_s + \delta(\tau) \quad , \text{ and } \quad \} \quad (4.16)$$

$$z(\tau) = z_s + \zeta(\tau) \quad ,$$

and where the matrix elements in equation (4.15) are defined as follows:

$$a_{11} = 1 - z_s + \Delta(\alpha x_s - 1) \exp(-u_s) + \Delta \alpha y_s \exp(u_s) \; ,$$

$$a_{12} = -[\Delta \alpha x_s \exp(-u_s) + \Delta(\alpha y_s - 1) \exp(u_s)] \; ,$$

$$a_{13} = -[x_s + \Delta \beta x_s \exp(-u_s) + \Delta \beta y_s \exp(u_s)] \; ,$$

$$a_{21} = -[\Delta(\alpha x_s - 1) \exp(-u_s) - \Delta \alpha y_s \exp(u_s)] \; , \qquad \} \quad (4.17)$$

$$a_{22} = (y_0 - 2y_s)/y_0 + \Delta \alpha x_s \exp(-u_s) + \Delta(\alpha y_s - 1) \exp(u_s) \; ,$$

$$a_{23} = \Delta \beta x_s \exp(-u_s) + \Delta \beta y_s \exp(u_s) \; ,$$

$$a_{31} = az_s \; , \quad a_{32} = 0 \; , \text{ and } a_{33} = -a(1 - x_s) \; .$$

For a (linear) stability analysis, equation (4.15) has to be solved using

$$(\xi[\tau] \quad \eta[\tau] \quad \zeta[\tau])^T = (\xi_0 \quad \eta_0 \quad \zeta_0)^T \exp(\lambda \tau) \quad , \quad (4.18)$$

where the eigenvalue λ has to satisfy the characteristic equation

$$P(\lambda) = \det(A - \lambda I) = 0 \quad , \quad (4.19)$$

where matrix A is defined by (4.15). From equation (4.19) it must be determined if all of the eigenvalues λ_j fulfill the condition $\text{Re}(\lambda_j) < 0$, for $j=1,2,3$. $P_s(x_s, y_s, z_s)$ is stable only if this condition holds for all λ_j. The equations (4.14) have a trivial singular point solution $P_0(0,0,0)$, which considering equation (4.13) turns out to be unstable. For non-trivial solutions, equation (4.14c) implies that

$$\text{either (a) } x_s = 1 \; , \text{ or (b) } z_s = 0 \; . \quad (4.20)$$

In case (a) equations (4.14) imply that by elimination of the migration term

$$z_s = 1 + [(y_0 - y_s)/y_0]y_s \quad . \quad (4.21)$$

Then this equation can be inserted into equation (4.14b) to yield the transcendental equation for y_s along:

$$\Delta y_s \exp[v(y_s)] = [(y_0 - y_s)/y_0]y_s + \Delta \exp[-v(y_s)] \quad , \quad (4.22)$$

with $v(y_s) = \rho - \alpha y_s - [\beta(y_0 - y_s)/y_0]y_s$, where $\rho = \delta + \alpha - \beta$. Equation (4.22) can be solved graphically by plotting both of its sides as functions of y.

Until now the discussion of the model has been completely general. However, this equation will be made specific by choosing the following particular but reasonable propensity parameters:

$$\beta = \delta > 0 \qquad \text{and} \qquad y_0 = 1 \qquad . \qquad (4.23)$$

These parameter value choices imply that the fear parameter β of the prey equals its preference parameter δ for the open habitat 1 where the predators are lurking. A plausible explanation for this choice could be, for instance, that food for the prey is better in the dangerous habitat 1. Furthermore, a specific saturation level $y_0 = 1$ has been chosen.

For the case of equation (4.23) the solution of equation (4.14) is particularly simple:

$$x_s = 1 \; , \quad y_s = 1 \; , \quad \text{and} \quad z_s = 1 \qquad (4.24)$$

satisfy equation (4.14) or, alternatively, equations (4.14c), (4.21) and (4.22). For stability analysis the characteristic equation (4.19) assumes the (simple) form:

$$P(\lambda) = C_0\lambda^3 + C_1\lambda^2 + C_2\lambda + C_3 \qquad , \qquad (4.25)$$

with
$$\begin{aligned}
C_0 &= 1 \; , \\
C_1 &= 1 - 2r \; , \\
C_2 &= a(1 + s) - r \; , \text{ and} \\
C_3 &= a(1 + r - r)
\end{aligned}$$

where $r = \Delta(2\alpha - 1)$, and $s = 2\Delta\beta$. $\qquad (4.26)$

A necessary and sufficient condition for the asymptotic stability of the singular point defined by equation (4.24) is that all solutions to equation (4.25) have negative real parts. This condition is fulfilled if the Hurwitz criterion holds (see, for example, Haken, 1977). A straightforward analysis of the Hurwitz criterion for the coefficients of equation (4.25) leads to the simple condition

$$\alpha < 1/2 \qquad , \qquad (4.27)$$

for the asymptotic stability of the singular point $P(1,1,1)$. In Figures 9a, 9b, 10a, 10b, 11a and 11b results are presented for three different values of the clustering parameter α, and for the constant parameters $\beta = \delta = y_0 = \Delta = 1$ and a = 2. Figures 9a, 10a and 11a show the graphical solution of equation (4.22), while

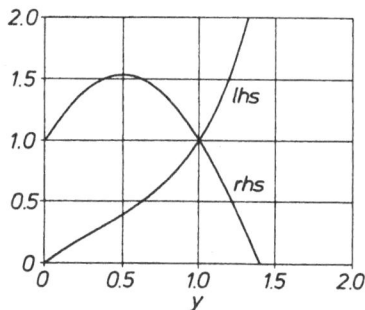

Figure 9a. Values of the left-hand and right-hand sides of equation (4.22), for α=0.

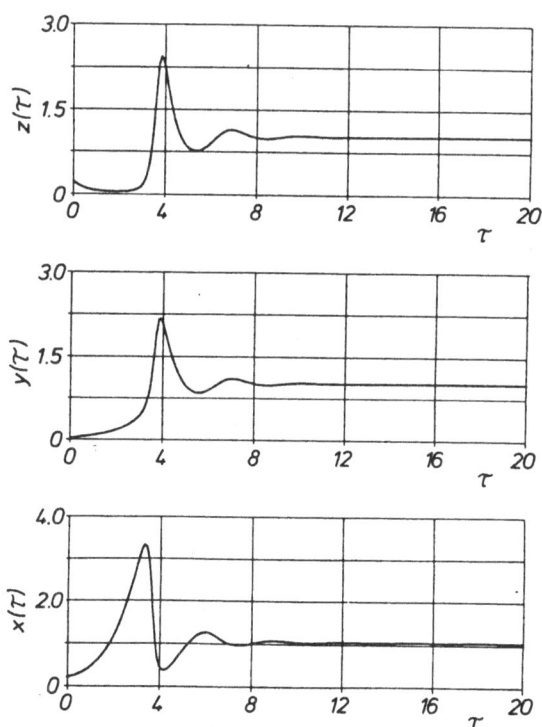

Figure 9b. Mean value of the prey population in habitat 1, x(τ), the prey population in habitat 2, y(τ), and the predator population in habitat 1, z(τ), plotted as functions of scaled time τ.

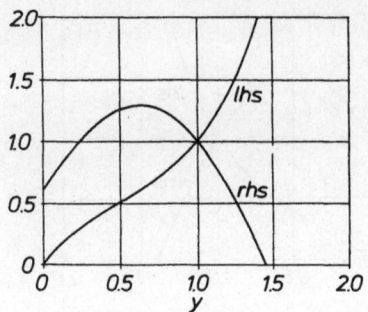

Figure 10a. Values of the left-hand and right-hand sides of equation (4.22), for α=.5.

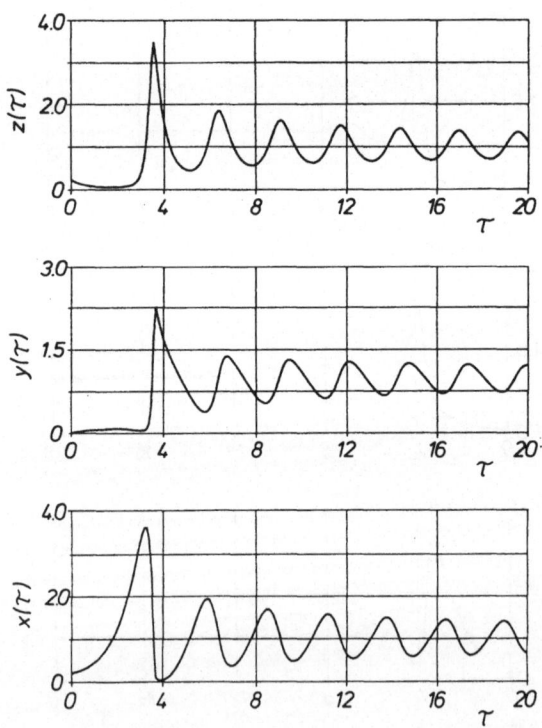

Figure 10b. Mean value of the prey population in habitat 1, x(τ), the prey population in habitat 2, y(τ), and the predator population in habitat 1, z(τ), plotted as functions of scaled time τ.

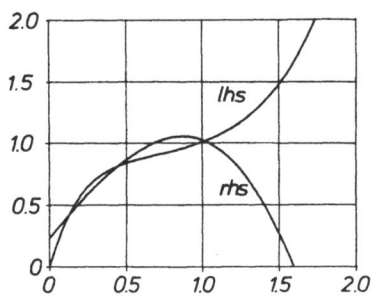

Figure 11a.   Values of the left-hand and right-hand sides of
equation (4.22), for α=1.5.

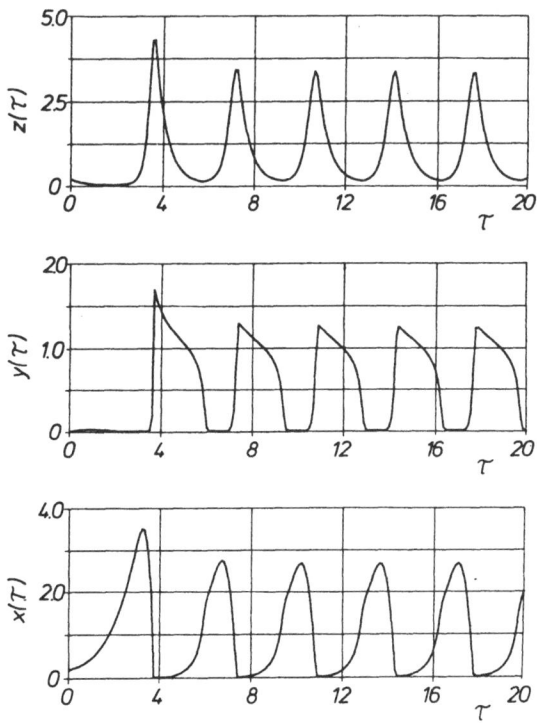

Figure 11b. Mean value of the prey population in habitat 1, $x(\tau)$,
the prey population in habitat 2, $y(\tau)$, and the
predator population in habitat 1, $z(\tau)$, plotted as
functions of scaled time τ.

Figures 9b, 10b and 11b show the time dependence of the variables $x(\tau)$, $y(\tau)$ and $z(\tau)$ for the respective sets of propensity parameters. Figure 9b shows a dampened time dependence of the variables approaching the stable focus $P_1(1,1,1)$ for $\alpha = 0$. Figure 10b depicts the case of a critical 'slowing down' of the oscillatory trajectory, as it approaches the focus $P_1(1,1,1)$, which becomes unstable for $\alpha = 0.5$. Figure 11b shows a limit cycle behavior. The variables $x(\tau)$, $y(\tau)$ and $z(\tau)$ approach a regular pulsatory time dependency, independent of the initial conditions. The singular point $P_1(1,1,1)$ becomes unstable for $\alpha = 1.5$.

This phase transition in the behavior of the predator-prey populations suggests an interesting scenario. For a small clustering trend value of α, for P_δ, an equilibrium situation is reached in which a steady stream of prey from the refuge to the open habitat leads to a stationary number of prey in both habitats, as well as a stationary number of predators living on the prey in the open habitat. For a large clustering trend value of α, for P_δ, however, the migration of the prey population from the refuge habitat 2, where it recovers but finally tends to move into the open habitat 1, takes place periodically in swarms, and thus induces a periodic variation in the predator population P_β.

Case (b) of equation (4.20) also is of interest. For $z_s = 0$, the migration terms in equations (4.14a) and (4.14b) again can be eliminated, yielding

$$x_s = -[(y_0 - y_s)/y_0]y_s \quad , \tag{4.28}$$

which can be inserted into equation (4.14b) in order to obtain the following transcendental equation for y_s alone:

$$[(y_0 - y_s)/y_0] \exp[-w(y_s)] + \exp[w(y_s)] = \Delta^{-1}(y_0 - y_s)/y_0 \ , \tag{4.29}$$

$$\text{with } w(y_s) = \delta - \alpha y_s[1 + (y_0 - y_s)/y_0].$$

Stability analysis then can be used to show that the one stationary point given by equations (4.28), (4.29) and $z_s = 0$ is stable. Figures 12a and 12b show, respectively, the graphical solution of equation (4.29), and the time dependency of $x(\tau)$, $y\tau)$ and $z(\tau)$. The reason for the extinction of the predator population P_β (see Figure 12b), despite finite values for $x(\tau)$ and $y(\tau)$ of the prey, is that the prey has no preference for the open habitat 1 ($\delta = 0$), and therefore freely migrates into the refuge habitat, where its number stabilizes even for values of $y_s > y_0$. The number of prey in the open habitat 1 remains small, and so the effective death rate $a(1 - x)$ of the predator population stays positive until its extinction.

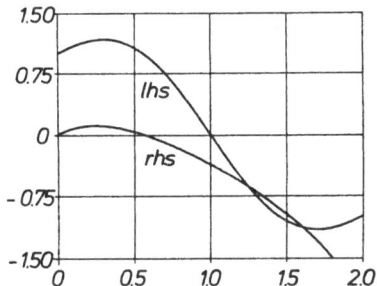

Figure 12a.   Values of the left-hand and right-hand sides of equation (4.29), for $\delta=0$, $\beta=1$, $\Delta=1$, $y_0=1$, $\alpha=1$ and $a=2$.

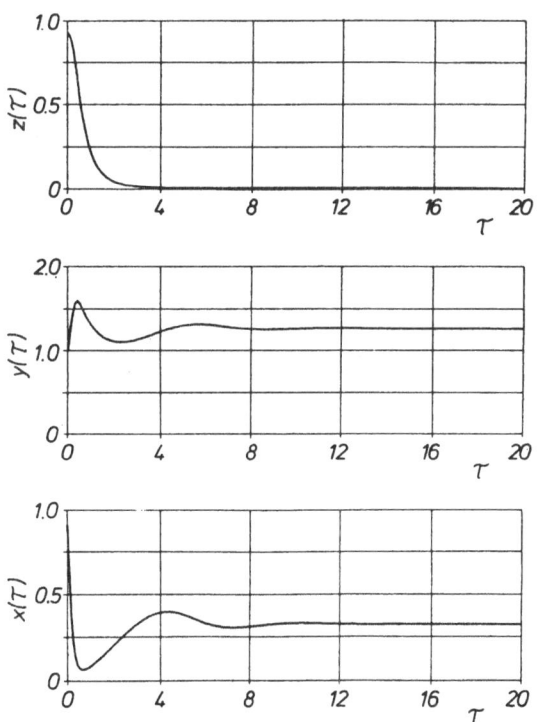

Figure 12b. Mean value of the prey population in habitat 1, $x(\tau)$, the prey population in habitat 2, $y(\tau)$, and the predator population in habitat 1, $z(\tau)$, plotted as functions of scaled time $\tau.5$.

CONCLUSIONS AND OUTLOOK

In the general formulation of the model in Section 2, the problem of human migration and animal migration have been treated simultaneously. The structure of the migration term introduced into this model demonstrates, even in the simplest case of two interacting populations and two zones or habitats, a complex variety of migration patterns. In Section 4 an interesting application of this migration term, for an animal population with predator-prey interaction, has been shown. Several observed stable oscillations in population densities may be explained in this way. Also, the migration of a single human population between a city and its suburban ring, with respect to different economic conditions of these two regions, on the basis of utility optimization, can be treated in a similar way. This treatment will be undertaken in a forthcoming paper. The migration of industrial firms of a certain type, or the migration of a single human population between different zones, also may be treated in this fashion. This kind of investigation permits one to address the question of whether or not he can expect urban concentration or deconcentration, and/or growth pole effects in certain zones. How the model parameters can be estimated, especially in the case of non-linear equations, is one of the next tasks to be tackled.

6. REFERENCES

Chewning, W., 1975, Migratory Effects in Predator-Prey Models, Mathematical Biosciences, 23: 253-262.

Gause, G., 1934, Experimental Demonstration of Volterras' Periodic Oscillations in Numbers of Animals, Journal of Experimental Biology, 12: 44-48.

Getz, W., 1976, Stochastic Equivalents of the Linear and Volterra-Lotka Systems of Equations - A General Birth-and-Death Process Formulation, Mathematical Biosciences, 29: 235-257.

Haken, H., 1977, Synergetics: An Introduction, Berlin: Springer.

Huffaker, C., 1958, Experimental Studies of Predation, Hilgardia, 27: 101-157.

Jorne, J. and S. Carmi, 1977, Liapunov Stability of the Diffusive Lotka-Volterra Equations, Mathematical Biosciences, 37: 51-61.

Lotka, A., 1920, Analytical Note on Certain Rhythmic Relations in Organic Systems, Preceedings of the National Academy of Sciences, United States, 6: 410-415.

MacDonald, N., 1976, Time Delay in Prey-Predator Models, Mathematical Biosciences, 28: 321-330.

May, R., 1973, Stability and Complexity in Model Ecosystems, Princeton: Princeton University Press.

Pielou, E., 1969, An Introduction to Mathematical Ecology, New York: Wiley.

Rosen, R., 1970, Dynamical System Theory in Biology, New York: Wiley.

Utida, S., 1950, On the Equilibrium State of the Interacting Population of an Insect and Its Parasite, Ecology, 31: 165-175.

Volterra, V., 1937, Principes de Biologia Mathematique, Acta Biotheoretica, 3: 1-36.

Weidlich, W. and G. Haag, 1980a, Migration Behaviour of Mixed Population in a Town, Collective Phenomena, 3: 89-102.

------, 1980b, Dynamics of Interacting Groups With Application to the Migration of Population, Systemforschung und Neuerungsmanagement, 11: 114-123.

_____, 1982, Concepts and Models of a Quantitative Sociology: The Dynamics of Interacting Populations, Berlin: Springer, Series in Synergetic No. 14.

Yoon, H. and H. Blanch, 1977, The Stability of Predator-Prey Interactions in Microbial Ecosystems, Mathematical Biosciences, 33: 85-100.

AN OPTIMAL CONTROL REPRESENTATION OF A STOCHASTIC MULTISTAGE-MULTIACTOR CHOICE PROCESS

Giorgio Leonardi

International Institute for Applied Systems Analysis
Austria
(on leave from the Polytechnical Institute of Turin, Italy)

1.   INTRODUCTION

This paper considers a multiactor decision process, where each actor chooses a sequence of actions over time, in order to maximize his utility, subject to a random noise in the utility evaluation. The pool of alternative actions is limited, and all actors compete to use these alternatives, thus generating mutual externalities due to shortages. This process, which has a direct formulation in terms of a nested random utility model, is shown to be equivalent to a suitably built optimal control problem, whose Hamiltonian can be interpreted as a total benefit. A continuous time formulation is outlined, and an application to modelling residential mobility is discussed.

Residential mobility provides a good example of the type of process considered in this paper. Assume that a set of durable and exchangeable goods exists, such as dwellings for rent, and a set of actors, such as households, who make moves between the dwellings according to some choice criteria, but are constrained by current vacancies. Each actor is faced with a sequential decision process, possibly of a stochastic nature, since his knowledge about the future is imperfect. Each actor is influenced by all other actors, since the set of available alternatives, namely vacant dwellings, at each point in time depends upon choices made by households during all previous points in time. The dynamics of such a process can be described in the following two spaces:

(1)   the 'primal' space, related to the time behavior of stocks and flows, such as the number of vacant dwellings and the number of moves, and

(2)   the 'dual' space, related to the time behavior of values guiding the choices, such as the expected utilities for each household at each point in time.

The aim of this paper is to provide a rigorous formulation for the above mentioned duality by embedding the process in an optimal control problem.

2.   A DIRECT FORMULATION

Consider the following multistage choice process. A set of alternatives is partitioned into m subsets, and an actor is said to be in state j at time t if he is currently using an alternative belonging to the jth set at that point in time. The actor makes transitions from state i to state j (j=1,2,...,m) according to the following utility evaluation process:

v_{ij} is the utility (or disutility) associated with a transition from an alternative in the ith subset to an alternative in the jth subset (v_{ij} is assumed to depend upon i and j only, and not on the specific elements of the ith and jth subsets),

$H_i(t)$ is the utility (or disutility) gained per unit time of stay in state i,

$\alpha > 0$ is the discount rate of utilities over time,

Ω_j is a random variable (i.e., the random utility term) accounting for dispersion of preferences in the utility evaluation, and

$V_i(t)$ is the total (discounted) expected utility for a process starting in state i at time t.

Let Δ denote the amount of time that elapses between t and t+1. If at time $t + \Delta$ the actor considers a transition from state i to any alternative in the jth subset, he will attach to this alternative the utility

$$(1 - \alpha\Delta)[v_{ij} + V_j(t + \Delta) + \Omega_j] \quad .$$

If, on the other hand, he considers remaining in state i, he will, starting from time $t + \Delta$, gain the utility

$$(1 - \alpha\Delta)[V_i(t + \Delta) + \Omega_i] \quad .$$

In addition, assuming decisions to move are taken at the end of the interval $(t, t+\Delta)$, he will gain the utility $[h_i(t)\Delta]$ during his stay in i over the time interval $(t,t+\Delta)$.

Suppose the actor collects a sample of size n, out of which M_j are the alternatives sampled from the jth state $(\sum_j M_j = n)$, in order to make a choice. Assuming he chooses by maximizing his utility, the following relationship holds for the functions $V_i(t)$:

$$V_i(t) =$$

$$h_i(t)\Delta + \underset{\Omega}{E}\Big[\max \{V_i(t+\Delta) + \Omega_i, \underset{(M_j)}{\max} [V_j(t+\Delta) + V_{ij} + \Omega_j]\}\Big] , \quad (2.1)$$

where $\underset{(M_j)}{\max}$ means that the maximum, over all M_j elements of the jth set and over all sets $j=1,2,\ldots,m$, is taken, and $\underset{\Omega}{E}$ denotes expectation with respect to the random variables Ω_j. Assuming further that the Ω_j are independent identically distributed random variables with distribution function

$$Pr(\Omega_j < x) = \exp[\exp(-\beta x)] \quad , \beta > 0 .$$

This assumption is unnecessarily restrictive, but simplifies the derivation. Results derived here may be generalized to situations for weaker assumptions (see Leonardi, 1982). After some algebraic manipulation, equation (2.1) becomes

$$V_i(t) = h_i(t)\Delta +$$

$$[(1 - \alpha\Delta)/\beta]\log\Big[\sum_j M_j \exp\{\beta[v_{ij}+V_j(t+\Delta)]\} + \exp[\beta V_i(t+\Delta)]\Big] . \quad (2.2)$$

Next assume that there are constant sampling fractions $w_j > 0$, $\sum_j w_j = 1$, such that

$$M_j = n w_j \quad ,$$

and a sampling intensity λ such that in the interval $(t,t+\Delta)$ $n = \lambda\Delta$. Letting $\Delta \to 0$ in equation (2.2) yields, after some rearrangements

$$\alpha V_i - \dot{V}_i = h_i + (\lambda/\beta) \sum_j w_j f_{ij}\exp[\beta(V_j-V_i)] \quad , \quad (2.3)$$

where $\dot{V}_i = dV_i/dt$ and $f_{ij} = \exp(\beta v_{ij})$.

The differential equation (2.3) was derived (in a simplified form) first by Leonardi and Campisi (1981). The transition rates between two states i and j are derived by observing that equation (2.2) defined above can be transformed into a logit model (Luce, 1959; Block and Marschak, 1960; McFadden, 1973, 1978; De Palma and Ben-Akiva, 1981). Therefore, according to the logit formula, the probability of moving from state i to state j during the time interval $(t, t+\Delta)$ is

$$p_{ij}(t, t+\Delta) =$$

$$M_j \exp\{\beta[v_{ij} + V_j(t + \Delta)]\}/\left[\sum_j M_j \exp\{\beta[v_{ij}+V_j(t+\Delta)]\} + \exp[\beta V_i(t+\Delta)]\right],$$

and using the assumptions introduced above concerning M_j together with the limit as $\Delta \to 0$ yields

$$r_{ij} = \lim_{\Delta \to 0} p_{ij}(t, t+\Delta)/\Delta = \lambda w_j f_{ij} \exp[\beta(V_j - V_i)] . \quad (2.4)$$

If a pure redistribution process is considered (i.e., constant total population, no births, no deaths, and no migrations to and from the rest of the world), then differential equations for the expected population sizes in each state are given by (Leonardi and Campisi, 1981):

$$\dot{P}_j = \sum_i P_i r_{ij} - P_j \sum_i r_{ji} . \quad (2.5)$$

More complex processes allowing for births, deaths, and interactions with the rest of the world basically would leave the structure of equation (2.5) unaltered. Such developments will be discussed in future publications.

Interestingly, in equation (2.4) the transition rates do not seem to depend directly upon the gains $h_j(t)$. These rates do, however, depend implicitly upon these gains, as long as they contribute to the total expected utilities $V_j(t)$ in equation (2.3). Although no explicit assumption has been introduced about the $h_j(t)$ terms, it seems sensible to assume that they might be endogenous, somehow reflecting the externalities due to the interactions among the actors competiting for a limited number of alternatives. The same comment applies to the w_j terms, which intuitively should be proportional to the number of available alternatives, and hence endogenously changing as well. For example, consider again the housing mobility problem. If the $j=1,2,...,m$ label different submarkets of the housing stock, and Q_j is the total number of dwellings in submarket j, then one would expect $w_j \propto (Q_j - P_j)$. In other words, the number of alternatives considered in state j is proportional to the number of actually vacant dwellings in state j, which is not constant and changes over time (even neglecting new constructions and demolitions). Similarly, h_i should depend upon, among other things, the price (e.g., the rent) of the dwellings in submarket i, with this price being neither constant nor independent of market interactions. The next section discusses how these two problems actually are related, and how they can be solved.

3. AN EQUIVALENT OPTIMAL CONTROL PROBLEM

Equations (2.3) and (2.5) respectively provide dual and primal descriptions of the process. While equation (2.5) is stated in terms of the population sizes P_i and the flows r_{ij}, equation (2.3) is stated in terms of the expected utilities V_i. One might conjecture that it is possible to specify this duality relationship more precisely, in the sense that a primal extremal problem exists. This problem would have the flows r_{ij} as control variables, whose dual variables are related to the V_i as given by equation (2.3). This is indeed true, as will now be shown. Consider the function

$$F(r,P) = $$

$$\sum_i P_i \left[a_i + (1/\beta) \sum_j r_{ij}\{1 - \log(r_{ij}) + \log[\lambda f_{ij}(Q_j - P_j)]\} \right], \tag{3.1}$$

where $r = \{r_{ij}\}$, $P = \{P_i\}$, $f_{ij} = \exp(\beta v_{ij})$, and a_i, v_{ij}, Q_j, λ and β all are given non-negative constants. The main result of this section may be summarized as follows:

Proposition 1

The dual variables of the optimal control problem

$$\max_r \, {}_0\!\int^{\infty} \exp(-\alpha t) \, F(r,P) \, dt$$

subject to equation (2.5), satisfy equation (2.3) with

$$w_j = Q_j - P_j \quad , \text{ and}$$

$$h_i = a_i - (1/\beta) \, (\sum_j P_j r_{ij})/(Q_i - P_i) \quad .$$

Proof

The Hamiltonian function for the above problem is given by

$$H(r,P,\mu) = \exp(-\alpha t) \, F(r,P) + \sum_i P_i \sum_j r_{ij}(\mu_j - \mu_i) \quad , \tag{3.2}$$

where μ_i and μ_j are the discounted dual variables associated with equation (2.5). Therefore, according to the Pontryagin maximum principle (see Sengupta and Fox, 1969; Hadley and Kemp, 1971), an optimal (r,μ) pair must satisfy the equations

$$- \dot{\mu}_i = \partial H/\partial P_i \quad , \text{ and} \tag{3.3}$$

$$\partial H/\partial r_{ij} \stackrel{=}{=} 0 \quad , \tag{3.4}$$

where $\dot{\mu}_i$ is the time derivative of μ_i. The relevant derivative of the Hamiltonian of equation (3.2) is found to be

$$\partial H/\partial r_{ij} =$$

$$- \{[\exp(-\alpha t)]/\beta\}P_i \{\log(r_{ij}) - \log[\lambda f_{ij}(Q_j-P_j)] +$$

$$P_i(\mu_j - \mu_i) \, ,$$

and condition (3.4) implies

$$r_{ij} = \lambda f_{ij}\{\exp[\beta(V_j - V_i)]\}(Q_j - P_j) \qquad , \qquad (3.5)$$

having defined

$$V_j = [\exp(\alpha t)] \, \mu_j \qquad\qquad (3.6)$$

(i.e., the dual variable expressed in current value). Taking condition (3.4) into account, for an optimal control, yields

$$\partial H/\partial P_i =$$

$$\exp(-\alpha t) \{a_i + (\sum_j r_{ij})/\beta - [\sum_j P_j r_{ji}]/(Q_i - P_i)]/\beta\} \qquad ,$$

where r_{ij} is defined by equation (3.5). Therefore, using condition (3.3), definition (3.6) and the equality

$$\dot{\mu}_i = \dot{V}_i \exp(-\alpha t) - \alpha V_i \exp(-\alpha t) \qquad ,$$

the following equations are obtained for V_i:

$$\alpha V_i - \dot{V}_i =$$

$$\{a_i - [(\sum_j P_j r_{ij})/(Q_i - P_i)]/\beta\} + (\sum_j r_{ij})/\beta \qquad . \qquad (3.7)$$

Finally, a comparison of equation (3.7) with (2.3) yields the proof of Proposition 1.

In the above result a specific assumption has been introduced for the weights w_j, namely $w_j = (Q_j - P_j)$. This in turn has produced a specific form for h_i, namely

$$h_i = a_i - [(\sum_j P_j r_{ij})/(Q_i - P_i)]/\beta \qquad . \qquad (3.8)$$

This result can be translated into the housing mobility problem terminology. The numbers Q_j can be interpreted as the total housing stock in submarket j. Hence the difference $(Q_j - P_j)$ is the number of vacant dwellings in submarket j. The assumption about w_j means that the probability of choosing a dwelling in submarket j is proportional to the number of vacancies in j. This introduces the shortage effect at the aggregate level, due to competition among households.

The aggregate shortage effect has a disaggregate counterpart—the second term on the right-hand side of equation (3.8). The gain per unit time, according to this equation, is split into an exogenous term a_i, which depends upon the specific qualities of submarket i but not on the households using it, and an endogenous term appearing with a minus sign, which can be interpreted as a disaggregate negative externality due to shortage and competition. Using equation (3.5), this term can be rewritten as

$$(\lambda/\beta) \sum_j P_j f_{ij} \exp[\beta(V_i - V_j)] \qquad , \qquad (3.9)$$

which is proportional to the probability that a vacant dwelling in submarket i will be occupied in a small time interval. In other words, expression (3.9) is a rate at which demand is attracted per vacant dwelling in i.

Making use of well-known geographic concepts concerning mobility, it is helpful to introduce the following definitions:

$$\tau_i = (\sum_j r_{ij})/\lambda = \sum_j f_{ij}\{\exp[\beta(V_j - V_i)]\}(Q_j - P_j) \quad , \quad (3.10)$$

which defines the accessibility to new moves for actors currently in submarket i, and

$$\omega_j = \sum_i P_i f_{ij}\{\exp[\beta(V_i - V_i)] \qquad , \qquad (3.11)$$

which defines the demand potential in submarket j. Substituting equations (3.10) and (3.11) into equation (3.7) yields

$$\alpha V_i - \dot{V}_i = a_i + (\lambda/\beta)(\tau_i - \omega_i) \qquad . \qquad (3.12)$$

This last equation relates the rate of change in expected utility to a positive contribution (i.e., a benefit) due to accessibility, and a negative contribution (i.e., a cost) due to demand potential.

In light of the above discussion, it is natural to think of ω_i as the price and to try to relate it to actual rents. This task together with other economic interpretations are discussed in the next section.

4. ECONOMIC INTERPRETATION

A micro-economic justification for ω_i actually being proportional to a revenue-maximizing market rent now will be given. Consider a landlord owning an occupied dwelling in submarket j and willing to adjust its rent. Due to the assumptions made about the random nature of customer behavior, the bid he can get from the current household occupying the dwelling is a random variable having the distribution function $\exp[-\exp(-\beta x)]$. On the other hand, from transactions occurring in the submarket, the maximum bid he would get from a new household would be a random variable with the following distribution function:

$$\exp\{- \lambda\Delta \sum_i P_i \ f_{ij}\{\exp[\beta(V_j-V_i)] \ \exp(-\beta x)\} \qquad . \qquad (4.1)$$

Straightforward calculations yield a maximum expected bid of

$$[\log(\lambda \ \Delta \ \omega_j + 1)]/\beta \qquad , \qquad (4.2)$$

and by truncating a Taylor's expansion of expression (4.2), approximately equals $\Delta(B/\beta)\omega_j$. If the landlord is revenue-maximizing, he will set the new rent equal to this expected maximum bid. That is, if y_j is the average rent in submarket j, then

$$y_j = (\lambda/\beta) \ \omega_j \qquad , \qquad (4.3)$$

which is exactly the 'externality' part of the right-hand side of equation (3.12). Of course, some implicit assumptions have been made about the landlord's behavior, as well as about the market rules. First, the landlord is driven by the last bids, and has no far-sighted policy, except for his willingness to take the risk of a vacancy because he is sure that he will fill it by new demand in a very short time span. Other things being equal, this is not a terribly unrealistic assumption. Secondly, no external (e.g., public) control is acting on the market, so that rents are free to change and reach any value at any time. This contention is somewhat unrealistic for the specific example of the housing market, but analyzing optimal rent-control policies is beyond the aim of this paper. Again this is a topic that will be reported on in a future paper.

The above micro-economic interpretation is far-reaching, since it shows very clearly the realtionship between a purely macro-level constraint (i.e., the shortage) and the embedded micro-level producer's behavior. One can go one step further and find a macro-economic interpretation of the Hamiltonian function (3.2). In particular, it can be shown that this Hamiltonian may be interpreted as a total benefit (i.e., the sum of consumers' surplus and producers' revenues). This demonstration also is straightforward. By substituting from equations (3.5), (3.6) and (3.12) into equation (3.2), and doing some algebraic manipulations, one gets for the current Hamiltonian the value

$$\exp(\alpha t) \ H = \sum_i P_i \ (\alpha V_i - \dot{V}_i) + \sum_i P_i \ y_i \qquad , \qquad (4.4)$$

where y_i is defined by equation (4.3). The first term in the right-hand side of equation (4.4) is the rate of change in consumers' expected utility (i.e., the consumers' surplus), while the second term is the producers' total revenue.

The exceedingly difficult problem of analyzing the dynamic behavior of equations (2.5) and (3.7) will not be treated here. Rather, assuming an equilibrium exists, the behavior of the system in equilibrium will be considered. Assuming, as $t \to \infty$, that $\dot{P}_j \to 0$ and $\dot{V}_j \to 0$, equation (2.5) yields

$$\lambda \, \omega_j \, (Q_j - P_j) = \lambda \, P_j \, \tau_j \quad , \text{ or}$$

$$P_j/(Q_j - P_j) = \omega_j/\tau_j \quad . \tag{4.5}$$

That is, in equilibrium the ratio of occupied to vacant dwellings is equal to the ratio of demand potential to accessibility. A rearrangement of equation (4.5) yields

$$P_j/Q_j = \omega_j/(\tau_j + \omega_j) \quad .$$

In other words, the density of households per dwelling in submarket j is an increasing function of the demand potential and a decreasing function of accessibility. Equation (4.5) is purely geographic, being built using accessibility and potential measures. It can be translated into purely economic terms by using rents y_j and expected utilities V_j. From equation (3.12), as $t \to \infty$ and after some algebraic manipulation, one gets

$$\tau_j = (\beta/\lambda)(\alpha \, V_j - a_j) + \omega_j \quad . \tag{4.6}$$

Substituting equations (4.3) and (4.6) into (4.5) yields

$$P_j/(Q_j - P_j) = y_j/(\alpha V_j - a_j + y_j) \quad . \tag{4.7}$$

This equation is more meaningfully rearranged by making y_j explicit. Analogous with logistic growth, let the 'carrying capacity' of submarket j be defined as $\bar{P}_j = Q_j/2$. Then equation (4.7) becomes

$$y_i = \{[P_j/(\bar{P}_j - P_j)]/2\}(\alpha V_j - a_j) \quad . \tag{4.8}$$

Thus the rent is asymptotically a linear function of expected utility. The slope of this line is positive when the population P_j is less than the carrying capacity, and negative when the population P_j is greater than the carrying capacity. Since y_j is always nonnegative by construction, this implies that

$$\text{if} \quad P_j < \bar{P}_j \quad \text{then} \quad \alpha V_j \geq a_j \quad , \text{ and}$$

$$\text{if} \quad P_j > \bar{P}_j \quad \text{then} \quad \alpha V_j \leq a_j \quad .$$

Further, assuming that $V_j < \infty$,

$$\text{if} \quad P_j = \bar{P}_j \quad \text{then} \quad \alpha V_j = a_j \quad .$$

In other words, a population less than the carrying capacity is associated with the marginal expected utility needed to move αV_j higher than the exogenous marginal benefit a_j, while a population exceeding the carrying capacity is associated with the marginal expected utility needed to move αV_j lower than the exogenous marginal benefit. This might be losely rephrased by saying that the carrying capacity is exceeded when staying is better than moving.

5. CONCLUDING REMARKS

The main result of this paper has been to embed an essentially disaggregated choice process into an equivalent aggregate extremal principle, the optimal control problem introduced in Section 3. This has been shown to provide further insight into the process, leading to consistent economic interpretations. There is still some theoretical work to be done concerning existence, uniqueness and stability of solutions. Such problems have been set aside in this paper. This latter problem looks like an interesting one strictly from a computational point of view, too.

Although the aims of this paper have been largely theoretical, it is worth mentioning that a discrete-time formulation of the choice process described in Section 2, as well as a discrete-time counterpart of the optimal control embedding problem, can be built (see Leonardi and Campisi, 1981). Since efficient computational versions of the discrete Pontryagin maximum principle are available (Chang, 1961; Fan and Wang, 1964; Sengupta and Fox, 1969), many computational problems are made easier. This topic will be reported on in a subsequent paper.

6. REFERENCES

Block, H. and J. Marschak, 1960, Random Orderings and Stochastic Theories of Responses, in Contributions to Probability and Statistics, edited by I. Olkin, et al., Stanford: Stanford University Press, pp. 97132.

Chang, S., 1961, Synthesis of Optimal Control Systems, New York: McGraw-Hill.

De Palma, A. and M. Ben-Akiva, 1981, An Interactive Dynamic Model of Residential Location Choice, paper presented at the International Conference on Structural Economic Analysis and Planning in Time and Space, Umea, Sweden, June 21-26, 1981.

Fan, L. and C. Wang, 1964, The Discrete Maximum Principle, New York: Wiley.

Hadley, G. and M. Kemp, 1971, Variational Methods in Economics, Amsterdam: North-Holland.

Leonardi, G., 1982, An Extreme Value Approach to Random Utility Search and Choice Models, paper presented at the Course on Transportation Planning Models, International Center for Transportation Studies, Amalfi, Italy, October, 1982.

_____ and M. Campisi, 1981, Dynamic Multistage Random Utility Choice Processes: Models in Discrete and Continuous Time, paper presented at the 2nd meeting of the Regional Science Association, Italian Section, Naples, October 19-21, 1981.

Luce, R., 1959, Individual Choice Behavior: A Theoretical Analysis, New York: Wiley.

McFadden, D., 1973, Conditional Logit Analysis of Qualitative Choice Behavior, in Frontiers in Econometrics, edited by P. Zarembka, New York: Academic Press, pp. 105142.

_____, 1978, Modelling the Choice of Residential Location, in Spatial Interaction Theory and Planning Models, edited by A. Karlqvist et al., Amsterdam: North-Holland, pp. 7596.

Sengupta, J. and K. Fox, 1969, Optimization Techniques in Quantitative Economic Models, Amsterdam: North-Holland.

SECTION 2

MODELS OF INNOVATION DIFFUSION PROCESSES

As mentioned in the Introduction to this book, an evolutionary geographical model must deal with initial conditions, trajectories, and final states. In this section these aspects of innovation diffusion processes are modelled. The first paper, by Nijkamp, discusses the role research and development (RD) investment plays in regional and national growth. Various theories are reviewed that purportedly explain the long term cyclical behavior of Western economies. Several of these theories relate this cyclical behavior to the diffusion of innovations through the space-economy. The model presented is a simple catastrophe-type model of spatio-temporal growth that includes three kinds of capital, namely productive capital, social overhead capital and RD capital, that often are viewed as providing the initial conditions for the appearance and the diffusion of innovations. Dendrinos, in the next section, also makes explicit use of catastrophe theory in this same way. Nijkamp's model is used to illustrate 'bang-bang' switches, bifurcations and perturbations associated with the evolution of a regional system. Based upon this model, Nijkamp is able to set out the conditions under which different growth paths may evolve. The manner in which a production system relates to each type of capital is discussed. Some emphasis is placed on the asymmetric behavior through time displayed by the relationship between production and capital investment, and of course the accompanying catastrophes that may result. In concluding, Nijkamp uses an optimal control model coupled with a social welfare objective function in order to characterize the problem of optimal investment strategies.

The second paper of this section, by Sonis, treats the theory of innovation in an explicitly spatial and dynamic manner. The question addressed asks why certain geographic systems evolve the way they do. The model presented is based upon the Lotka-Volterra model, and focuses on the competitive process operating among innovations over time in a closed system. As such the family of possible innovation trajectories is of utmost interest. This paper is divided into two parts. The first part presents the vectorial differential equation of the temporal diffusion process, and gives results for a local stability analysis involving a linearized Lyapunov approximation of the behavior of solutions to this problem, within the vicinity of equilibrium points. A competitive exclusion principle is formulated for the case of a homogeneous environment with constant competitive interactions between innovations. The Lyapunov approximation generates a non-homogeneous, continuous Markov chain that permits the existence, nature and stability of long-run equilibria to be evaluated. The second part of this paper deals with totally antagonistic innovations, and presents explicit formulae for the solution of the associated non-linear vectorial differential equation. Here the structure of arbitrary competition is ascertained. Sonis concludes by noting that the outcome of his modelling efforts is consistent with empirical regularities, such as the logistic curve. Like Haag and Weidlich in the preceding section, Sonis treats a problem involving predator-prey interactions, and like Dendrinos and Curry, in later sections, he is interested in trajectories that lead to markedly different steady states.

The last paper of this section, by Ralston, is concerned with final states of a geographical system and the dynamic processes leading to these states. The problem he addresses is the supply of information of public consumption and the diffusion of this information over space and through time. Ralston reviews the literture on innovation diffusion and communication that provides (a) some theory and models, and (b) empirical evidence of regularities in information transmission. He follows this with a formulation of a control theoretic analysis of strategies that will enable some agent to induce change in such a way that a particular geographic distribution of knowers is attained. The manner in which various communication 'channels' may be used to optimize selected objectives is treated, as is modelling the optimal supply of information in an urban system. Other contributors to this volume making use of control theory are Leonardi, Nijkamp and Lea. Ralston ends by relating his analysis to Markov random field models. In keeping with the purely temporal models, he attempts to incorporate source and interaction effects into the differential equations describing the rates of change in an adoption process. His ultimate goal is a spatial model, though, which for the time being remains elusive in the form of a Markov random field model.

Consequently, these three papers illustrate three of the ingredients of an evolutionary spatial model of innovation diffusion. Their authors try to take the dynamic modelling of this important process over space one step closer to evolutionary modelling. Considerable insight should be gleaned from these papers.

TECHNOLOGICAL CHANGE, POLICY RESPONSE AND SPATIAL DYNAMICS

Peter Nijkamp

Free University of Amsterdam
The Netherlands

1. LONG-TERM CYCLICAL ECONOMIC DYNAMICS

The eighties seem to be marked by a situation of structural economic change and, in a geographical context, a reorientation of cities, regions and countries all over the world. Such perturbations are not new phenomena in the history of the world; economic cycles (especially long wave patterns) always have drawn a great deal of attention in the history of economics (see, for instance, Adelman, 1965; Schumpeter, 1939). Especially in recent years, many economists have concentrated their efforts on providing contemporary explanations for the emergence of drastic shifts in economic conditions in the Western world. The persistent and deeply-rooted economic recession, the future uncertainties regarding energy and raw materials supply, the divergent development patterns between the developed world and the developing world, and the inability of government policies to control the present unstable economic and technological process have led to a revival of theories and methods aimed at analyzing long-term economic developments. The phenomenon of long waves, which includes such issues as perturbations, balanced growth, stable equilibria, international and geographical discrepancies, and multi-actor conflicts, has been a very popular topic in recent economic literature.

It is no surprise that Kondratieff's theory of long-wave cycles has come to the forefront in recent years (see Clark et al., 1981; Delbeke, 1981; Van Duijn, 1979; Freeman et al., 1982; Mandel, 1980; Rostow, 1978). In his view, the long-run development pattern of a free enterprise economy is normally characterized by cyclical processes composed of five stages: take-off, rapid growth,

maturation, saturation, and decline. However, the existence of
such long-term cyclical patterns is hard to demonstrate because of
a lack of historical data. It is lamentable that, apart from
Schumpeter (1939), most economists have regarded the Kondratieff
cycle mainly as an economic curiosity that was only reflected in
price changes (cf., Mass, 1980). Fortunately, recently many
efforts have been undertaken to provide the long-wave hypothesis
with a more substantial empirical foundation (see, Clark et al.,
1981; Kleinknecht, 1981; Mensch, 1979). A fascinating problem
tackled in this literature asks whether or not a long-term cyclical
pattern is an endogenous phenomenon in Western countries. This
requires a theory explaining the rise of each new stage of a cycle
(i.e., prosperity, recession, depression, and recovery) from the
economic and technological developments during previous stages. In
this respect, one may, for instance, try to answer the question of
whether or not economic recovery would require much technological
progress and innovation during the preceding 'downswing' of the
economy.

A basic problem in research on cyclical economic movements is
evidently the length of the time horizon. In the literature,
several kinds of cycles have been identified: Kondratieff cycles
(40 to 50 years), Kuznets cycles (15 to 25 years), Juglar cycles (5
to 15 years), and business cycles (up to 5 years). Clearly, one
also may observe in reality a superimposition of these different
cycles. At present, following Schumpeter, much attention is being
paid to the Kondratieff cycles, as they may reflect the structural
economic changes in the Western world. This also explains why
almost all authors use Schumpeter as a principal reference, even
though his writings on innovations, market structure and industrial
concentration are not always clear (see Dasgupta, 1982; Futia,
1980; Loury, 1980; Rosenberg, 1976; von Weiszsaecker, 1980)

Various theoretical explanations, not always firmly supported
by empirical evidence, have been given to the presence of long-term
dynamic and cyclical movements of the economy. One type of
explanation is based upon monetary theories. In general these
explanations have referred to the naive quantity theory by assuming
an inverse relationship between price level and gold stock (cf.,
Dupriez, 1947). A second type of explanation is based on bottle-
neck theories. Due to production rigidities in the primary sector
a continuing rise in the industry will be hampered, so that excess
demand—and consequently higher profits—take place in the primary
sector. Then more resources flow to the primary sector, so that
the bottlenecks in this sector are removed, leading to
overproduction and finally reduced profits in the primary sector.
In turn, then, it again is more profitable to invest in this
sector, and so forth (c.f., Delbeke, 1981). A third explanation
may be derived from profit rate theories. In a competitive
situation, profit rates are related to an acceleration and

deceleration of capital accumulation. This will lead, in a free-enterprise economy, to varying profit rates. In a downswing of a cycle, profit rates tend to decline until a depression has been reached. Countermovements, however, may lead to a reverse development, so that a cyclical pattern may emerge (c.f., Mandel, 1980). These countermovements may be caused by a more efficient composition of technological capital, captial saving innovations or a wage decline. Another explanation is based upon investment and capital theories. The demand for productive capital is usually unstable. A rapid expansion during a period of economic growth leading to high capital costs will be followed by a decline in the production of captial goods leading to low capital costs (cf., Graham and Senge, 1980; Heertje, 1981). Various reasons may be mentioned for this cyclical pattern. First, due to indivisibilities in capital stocks an overcapacity may emerge, leading to fluctuations in the rate of use of existing capital. Secondly, it may be argued that the translation of final demand impulses into new productive investments is characterized by threshold effects (e.g., an entrepreneur can decide only whether or not to invest), so that a wave like development may take place. And, thirdly, the long gestation period of productive capacities may cause the emergence of long waves in the economy. By the time investments come into operation, an entirely different economic situation already may exist, so that an unstable and cyclical growth pattern may be induced. In conclusion, the investment and capital theories take for granted the existence of successive stages of over- and underinvestments due to inertia and rigidity in economic behavior. The essence of these processes has been captured in vintage and puttyclay models (cf., Clark, 1980).

A fifth explanation comes from systems dynamic theories. Multiplier and accelerator mechanisms lead to fluctuations throughout the economy. Smooth adjustments are disrupted by discontinuous capital stock adjustments. There is normally too much captial expansion in an upswing stage, with favorable prospects, and too much contraction in a downswing stage, with less favorable prospects (see Forrester, 1977). A sixth explanation derives from resosurce theories. These theories argue that, from a global viewpoint, long-term international fluctuations may emerge due to variations in the supply of food stuff and raw materials, accompanied by corresponding price patterns (see Rostow, 1978). Finally, an explanation can be gleaned from innovation theories. Lack of innovation (or of diffusion of innovation) often is considered to be a source of cyclical economic patterns (see, among others, Clark et al., 1981; Kleinknecht, 1981; Mensch, 1979). Some aspects of innovation theories will be discussed in greater detail in the next section, as these elements will play an important role in the contribution made here to dynamic and geographical aspects of long-term economic growth patterns.

2. INNOVATION AND ECONOMIC DYNAMICS

Innovation will be regarded as a process of research, development, application and exploitation of a technology (see Haustein et al., 1981). A distinction of these stages is meaningful, as very often a certain new invention does not (directly) lead to an application or exploitation of such new findings because of market structures, patent systems, and the such. Also the diffusion of innovation is often hampered by many bottlenecks caused by monopoly situations, lack of information, and so forth (cf., Brown, 1981; Davies, 1979; Rosegger, 1980).

Innovations can be analyzed at two different levels:

(1) the macro level which focuses upon description and explanation of aggregate implications of innovations, such as the impacts of labor-saving and capital-saving technological progress (cf., Kennedy, 1964),

(2) the micro and meso levels which focus on the principal factors stimulating innovations at the level of firms or of the industry (cf., Kamien and Schwartz, 1975).

Research conducted at the micro or meso level certainly is justified, as innovations are not spread uniformly over all sectors of the economy, but usually exist only in a limited number of key sectors (cf., Kleinknecht, 1981; Mahdavi, 1972). The growth pattern of individual economic sectors or firms normally is characterized by a cyclical pattern, too.

At the level of industries (or sectors) and firms it is usual to distinguish between two kinds of innovation: basic innovations that lead to new products or even new industrial sectors, and process innovations that lead to new industrial processes in existing or basic sectors. At present, much attention is being given to basic innovations, as they are assumed to take place periodically and in clusters leading to economic fluctuations. This is in contrast to process innovations. In the literature on basic innovations, it is assumed typically that after a period of growth, a period of saturation may take place, which in turn leads to a recession. Then the struggle for survival will induce entrepreneurs to search for basic changes leading to radical innovations. This so-called 'depression-trigger' hypothesis has been strongly supported by Mensch (1979). Clark et al. (1981) and Freeman et al. (1982), however, have criticized Mensch's hypothesis, as in their view Mensch failed to demonstrate that in an economic downswing investments aimed at innovation are not overly risky. In a further contribution of this debate, Kleinknecht (1981) has claimed that only relative risks--risks related to those faced by competitors--are important, so that in the intensified struggle for survival characteristic of downswings, risk-taking via radical innovation may indeed be a rational behavior. Apparently the 'depression-trigger' hypothesis will be valid only if the products from the related basic innovations can be sold on the market (the demand-pull hypothesis). Furthermore, it should be noted that basic innovations also may be caused by intersectoral linkages (e.g., in an input-output framework).

In any case, innovation processes usually can be described by means of a logistic (S-shaped) curve, implying the following phases: introduction, growth, maturity, saturation (and then eventually decline, which is not shown by the logistic curve). Normally, an economic 'upswing' of a certain sector or firm requires the fulfillment of at least one of the following conditions

(1) a sufficient (potential) demand for the product at hand (Mowery and Rosenberg, 1979),

(2) a technological innovation inducing the demand for the new product,

(3) an availability of sufficient resources to finance new investments, and

(4) a satisfactory endowment of public capital favoring the innovation and investment process.

Thus, some critical combination of research and development (i.e., RD) capital, productive capital and public capital is a necessary condition to fulfill the (potential) demand for new products and to create radical technological changes (cf., Schmookler, 1966). These changes may be regarded as propulsive factors behind the process of structural economic growth. This emphasis on 'supply side economics' (Giersch, 1979) also explains the revival of growth pole theory, as this theory claims that polarization phenomena (scale advantages, intersectoral linkages and technological innovation) shape the necessary conditions for a rapid economic growth process characterized by a diffusion of growth impulses from propulsive sectors to other sectors (cf., Nijkamp and Paelinck, Chapter 7, 1976). There is, however, a basic difference between the Schumpeterian view of innovation and the role it usually is assumed to play in growth pole theory. Schumpeter regards innovation as an endogenous instrument in a profit-maximizing economy, so that cyclical economic patterns may be expected. These cyclical movements are not necessarily smooth and continuous growth processes, since inertia in adopting innovations, rigidities and bottlenecks in exploiting innovations, and indivisibilities at the supply side may cause shocks, perturbations or catastrophes in an economic system. In growth pole theory, innovations are regarded largely as an exogenous instrument that sets the stage for the 'take-off' of less developed areas. Clearly, in a short- or medium-term perspective the resulting economic growth pattern may be the same. In any case, dynamic evolutionary models may be used as meaningful operational tools for describing and analyzing innovation and diffusion processes (cf., Nelson and Winter, 1977).

In both the 'depression-trigger' and the 'demand-pull' hypothesis, innovation plays a crucial role, though it may be induced by different sources. A prerequisite for innovation to take place is sufficient effort in RD sectors. There is a strong positive correlation between RD efforts and innovative output (see Mansfield, 1968). Another necessary condition for innovation is the presence of a satisfactory breeding place, including for example education institutions and facilities (cf., Rosenberg, 1976), relative ease of communication and market entrance, good environmental conditions, and possibilities for agglomeration favoring innovative activities. This also may explain why monopoly situations and industrial concentrations (including patent systems) tend to have a greater potential for technological process and innovation. Thus, in general, the availability of a satisfactory level of public infrastructure capital stock (in its broadest sense) shapes the necessary conditions for innovative capacities in a region (see Nijkamp, 1982a). However, the transmission of innovative efforts to other firms or other sectors of the economy often is hampered by barriers emerging from monopoly situations or patent systems. Although in the short run monopolies and patent systems may protect and incubate new inventions, and even stimulate innovation, in the medium- and long-term they may lead to rigidities precluding new developments (cf., Mansfield et al., 1981).

Two conclusions may be drawn from the aforementioned studies on spatial dynamics. First, innovation may be regarded as a necessary condition (and thus an instrument) for economic growth. Second, RD activities and public infrastructure capital are necessary conditions for innovative activities.

An important aspect of innovation still remains to be discussed, though, namely its precise definition and measurement. Here innovation has been regarded as a process related to structural sectoral change, technological progress in production processes, or adoption of new products or adoption of new marketing strategies. In general, data relating to innovations considered in this way are fugitive and/or of poor quality (e.g., see Terleckyj, 1980). Consequently, a precise measurement of innovation that would allow a cross-sectoral or cross-national comparison is very difficult to obtain. However, it is clear that significant sectoral and national differences in innovative efforts may exist (see van Bochove, 1982; Dasgupta and Stiglitz, 1980). Therefore, often only indirect measurements of innovation are used. Common examples include: relative growth rates of (clusters of) key sectors in relation to other sectors (cf., Mensch, 1979), relative sectoral profit rates in relation to other sectors (cf., Brinner and Alexander, 1979), relative amounts of money spent on RD activities in each sector (cf., Haustein et al., 1981), and the number of (requested or granted) patents on new industrial processes or products (cf., Kleinknecht, 1982; Thomas, 1981). Despite poor data and the necessity of using proxy variables, there is evidence that generally only a limited number of industrial sectors typically account for the major share of expenditures on innovative activities (electronics, petrochemicals and aircraft), although in a few cases small firms may be a source of major innovations [for instance, in the area of micro-processors (cf., Rothwell, 1979; Thomas, 1981)].

3. SPATIAL ASPECTS OF ECONOMIC DYNAMICS AND INNOVATION

In the context of the previous discussion of economic dynamics and innovation, it is interesting to examine some of the spatial aspects of these developments. It is clear that spatial systems also have undergone a dynamic evolution, especially during the last decade. Although spatial systems have never been static, but are always in a state of flux, it is interesting to observe that in recent years several geographers have suggested that a clean break with the past has taken place (see, among others, Berry and Dahmann, 1977; Vining and Kontuly, 1977; Vining and Strauss, 1977). Although this reversal of past spatial trends has been questioned by others (see Gordon, 1982), in many countries waves of urban centralization and decentralization can be observed. It seems as though cities are key factors in generating spatial dynamics. Usually the following phases can be distinguished in urban development patterns (see, for example, Nijkamp and Rietveld, 1981; van Lierop and Nijkamp, 1981; Chatterjee and Nijkamp, 1981):

(1) urbanization - a growth of cities in an economic and demographic respect implying strong agglomeration forces and innovative efforts,

(2) suburbanization - a further economic growth of cities (especially in the tertiary sector) accompanied by a flight of population to the suburbs (in this stage the city is still the heart of innovative opportunities),

(3) desurbanization - a decline of cities from both an economic and demographic point of view, so that the innovative power of cities also may be decreasing

(this may even lead to a decline of whole metropolitan areas; for a more detailed explanation see van den Berg et al., 1982), and

(4) reurbanization – a process of urban revitalization and urban renewal, so that cities again become attractive nuclei for residential and (some) commercial purposes.

Several countries in the Western world (e.g., Germany, The Netherlands, the United States) to a certain extent have demonstrated such patterns of spatial fluctuations in the post-war period. Historically, there is a close connection between innovative activities and spatial dynamics. On the one hand, innovation may alter spatial development processess significantly. For instance, the invention of the steam engine in the last century, and the exploitation of mass transit systems in this era, have had drastic effects on regional and urban growth processes. On the other hand, geographical concentrations of activity and the availability of an efficient spatial communication system may imply better information about new inventions, and may lead to geographical diffusion and adoption of innovations. In turn this can stimulate growth and new types of activities (e.g., in the chemical, aircraft, electronic and micro-processing industry). These observations suggest that spatial dynamics, public infrastructure in large agglomerations, RD activities and invention and innovation diffusion are strongly interrelated, so that product cycles, regional and urban cycles and innovation cycles display parallel patterns and common trends (cf., Nijkamp, 1982b). New invention is most clearly associated with major agglomerations while the actual exploitation of these inventions (e.g., the production of new commodities) may be located in low-wage peripheral areas (especially when standardized products are involved). Therefore, locational conditions, agglomeration economies, infrastructure policy, RD policy and economic developments are closely linked phenomena.

The parallels between dynamics in economic space and geographical space also are reflected in the parallel concepts of a growth pole (as a purely economic concept characterized by propulsive intersectoral growth effects) and a growth center (as a purely geographical concept characterized by centrifugal spatial diffusion processes.) A number of models have emphasized the close connection between economic and spatial developments (see Nijkamp, 1982b). These include base-multiplier models, (inter-)regional input-output models, gravity and income potential models, growth pole models, center-periphery models, unbalanced growth models and development potential models.

The extinsion of the market which follows many innovations has caused the emergence of large scale operations that lead to geographical concentration and specialization, with these in turn inducing innovations, and so forth. Conventional wisdom suggests that innovation potential increases with city size (cf., Alonso, 1971; Nijkamp, 1981; Pred, 1966; Richardson, 1973; Thompson, 1977). The most important reasons for this are that

(1) the geographical concentration of economic activities (implying scale advantages) leads to higher productivity (cf., Kawashima, 1981),

(2) large urban agglomerations demonstrate a high industrial diversification, and a rich social, cultural and educational infrastructure, which all stimulate innovative ability (cf., Nelson and Norman, 1977), and

(3) large agglomerations induce technological progress (cf., Carlino, 1977).

It should be noted, however, that the innovative potential in the United States, which traditionally has been concentrated in urban areas, seems to be declining, especially in the largest urban areas (see Malecki, 1979). Thus there is some evidence that innovative activity may be suffering from diseconomies of size (cf., Sveikauskas, 1979). Such dynamic processes are in agreement with geographical spillover effects known as spread and backwash effects (cf., Myrdal, 1957). Migration, input—output flows, and capital and commodity flows are media through which cumulative spatial processes evolve via multiplier effects. As soon as a critical level of agglomeration diseconomy or congestion effect in the initial center has been reached, innovative activities shift to other areas. Thus, while innovative potential requires a minimum sustainable threshold of infrastructure endowment as a source of regional and/or urban development, beyond a certain critical upper level (negative) congestion effects may take place.

4. A CATASTROPHE MODEL FOR SPATIOTEMPORAL GROWTH PROCESSES

It has been indicated in the previous sections that the evolution of a spatial system may exhibit unbalanced growth with many shocks and perturbations. Several models describing the spatiotemporal dynamics of a system of regions have been developed in the past (see, among others, Allen and Sanglier, 1979; Andersson, 1981; Batten, 1981; Casetti, 1981; Dendrinos, 1981; van Duijn, 1972; Isard and Liossatos, 1979; Nijkamp, 1982b). Such models may be helpful in analyzing the evolution and fluctuations of a spatial system. As current spatial systems demonstrate rather drastic changes, it may be illuminating to build models that feature discontinuous growth paths characterized by 'bang—bang' switches, bifurcations or perturbations.

Now spatiotemporal dynamics will be formalized by means of a model that is able to take into account both threshold and congestion effects. As a result, the phenomenon of spatial waves can be related to the endogenous growth patterns of a spatial system. In this model, social overhead capital and RD capital play a crucial role. A catastrophe-type approach will be employed to construct a discontinuous spatiotemporal model that is able to generate shocks in a dynamic spatial system. This model does not necessarily guarantee a stable equilibrium path, but it sets out the conditions under which various growth paths may evolve. The key relationship in this model is formed by a generalized or quasi-production function (see Biehl, 1980; Nijkamp, 1982a). This production function includes infrastructure capital and RD capital, in addition to traditional production factors. Infrastructure serves as the necessary public capital that complements private productive capital. RD capital (both private and public) serves to generate innovation processes. Thus, the following production function is used here:

$$Y = f(K, I, R) \quad , \quad (4.1)$$

where Y = regional income (or product) per capita,

K = directly productive capital per capita (an aggregate of many production factors),

I = infrastructure capital per capita, and

R = RD capital per capita.

Labor has not been included, as all variables are defined in units per capita. The following important assumption is made regarding this production function:

$$\left. \begin{array}{l} \partial Y / \partial K \geq 0 \quad \text{for} \quad K \leq K^* \quad , \text{ and} \\ \partial Y / \partial K < 0 \quad \text{for} \quad K > K^* \quad . \end{array} \right\} \quad (4.2)$$

These conditions imply a positive marginal product of productive capital during a first stage of economic growth. Beyond a certain critical level K^*, diseconomies of scale (e.g., high density, congestion, environmental decay) may occur, resulting in a negative marginal product. This idea is illustrated in Figure 1. It should be pointed out that the curve in Figure 1 is not necessarily the path taken by a downward movement. Usually production systems are not symmetric in periods of expansion and contraction. A new curve showing typical upward and downward paths appears in Figure 2. This asymmetric behavior may lead to various kinds of so-called catastrophes (see the dashed lines of Figure 2), whose presence precludes a smooth transition.

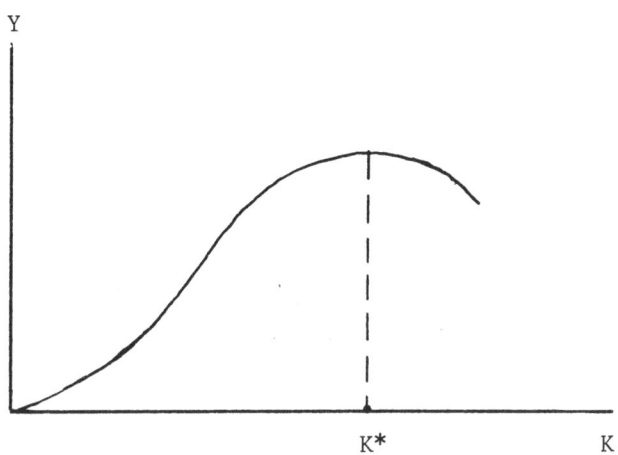

Figure 1. The partial relationship between product and capital.

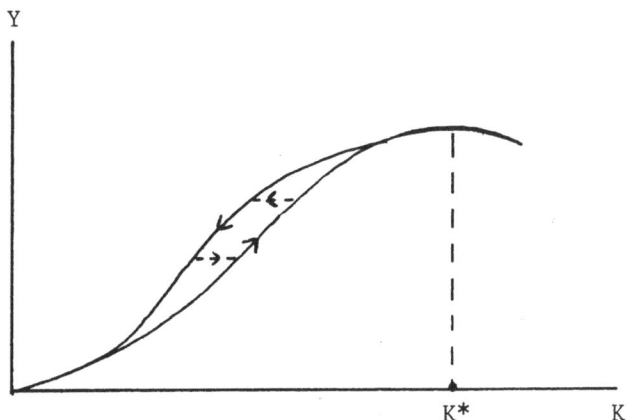

Figure 2. Asymmetric behavior of capital in a production function.

Next, it is assumed that investment in infrastructure and RD can be used to attenuate negative externalities, so that the level of production beyond which external diseconomies emerge can be shifted upwards. Examples of such investments include water sewage plants, communication infrastructure, and energy and environmental research institutions. This implication suggests the following relationship:

$$K^* = f(I,R) \qquad , \qquad (4.3)$$

with the following conditions:

$$\partial K^*/\partial I > 0 \quad \text{and} \quad \partial K^*/\partial R > 0 \qquad . \qquad (4.4)$$

Relationships (4.2) thru (4.4) lead to the three-dimensional surface depicted in Figure 3.

For infrastructure itself, the production function satisfies the following condition:

$$\left.\begin{array}{ll} \partial Y/\partial I \geq 0 & \text{for } I \geq I^* \quad , \text{ and} \\ \partial Y/\partial I = 0 & \text{for } I < I^* \quad . \end{array}\right\} \qquad (4.5)$$

Here I^* denotes a critical level of per capita infrastructural investment. Condition (4.5) states that a city or region needs a minimum endowment of infrastructure in order to reach a selfsustained growth. It is in this sense that infrastructure is a prerequisite for sustained regional development. Thus, the relationship between infrastructure and production may be assumed to look something like that shown in Figure 4. This relationship is built up with a series of logistic growth paths. Because of indivisibilities in infrastructure capital, significant growth effects are only observed following a significant investment in infrastructure. In a period of contraction, an asymmetric pattern may emerge again, due to inertia in infrastructure policy and indivisibilities in infrastructure endowment (see Figure 5). In Figure 5 various kinds of shocks again may be observed for the case of a reversed growth path (leading to various catastrophes). This phenomenon is similar to the one described in Figure 2.

The last argument in the production function, namely RD capital per capita, is assumed to have the following effect on regional income:

$$\partial Y/\partial R \geq 0 \qquad . \qquad (4.6)$$

A typical relationship for this situation is depicted in Figure 6.

Finally, the synergetic effects from interactions among K, I and R can be assessed using the following second-order derivatives:

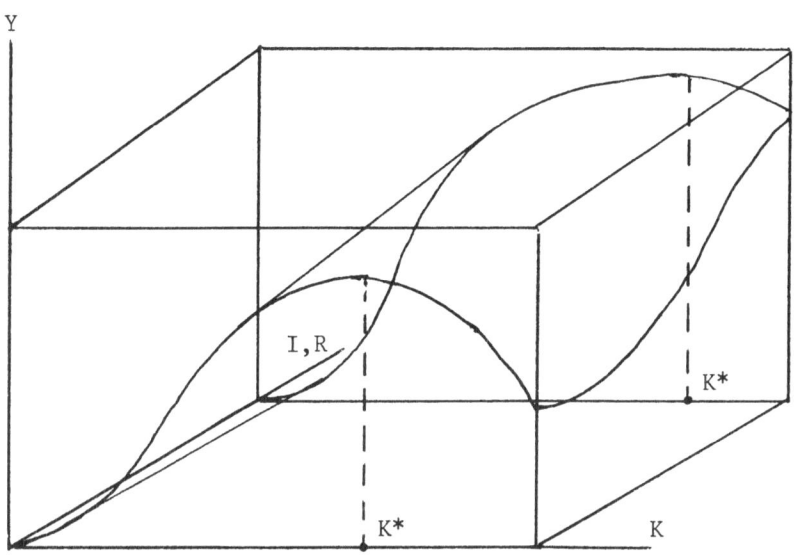

Figure 3. External diseconomies and infrastructure and RD capital
in a production function.

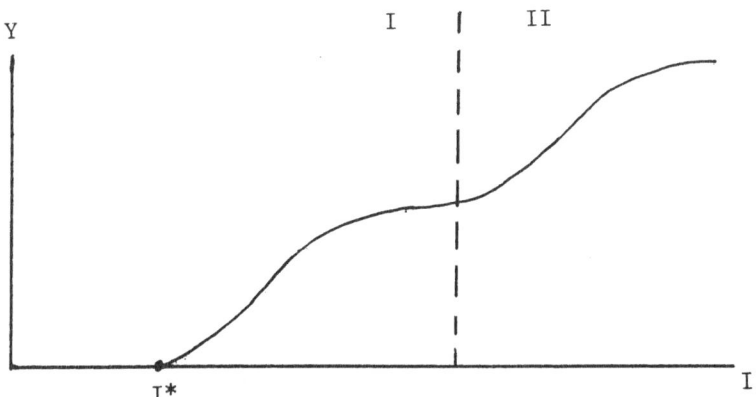

Figure 4. The partial relationship between product and
infrastructure.

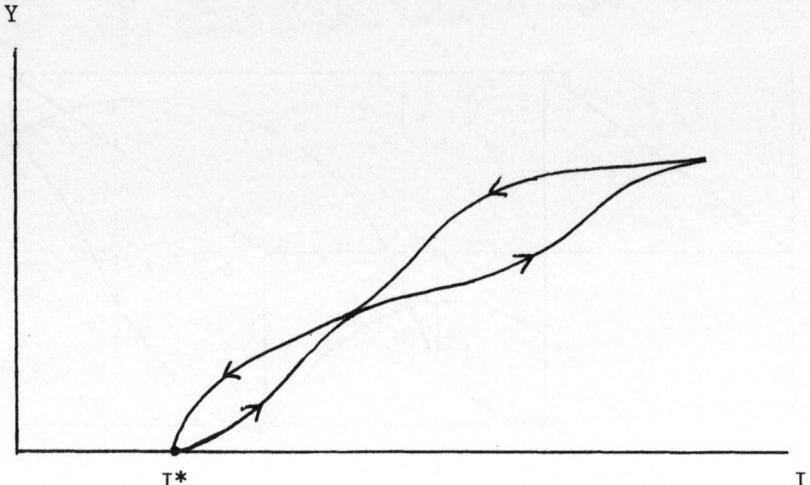

Figure 5. Asymmetric behavior of infrastructure in a production function.

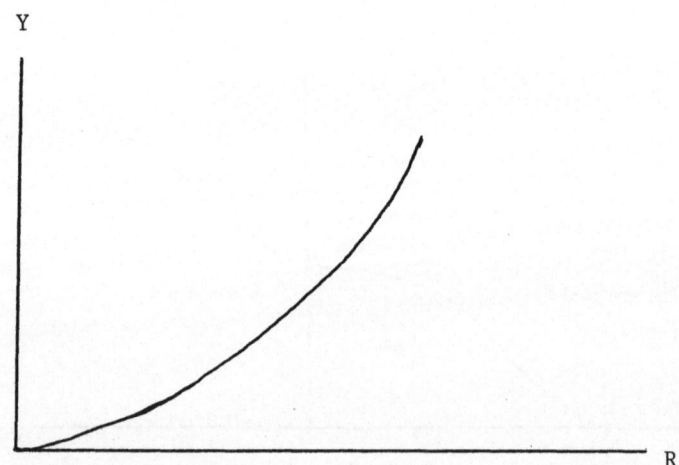

Figure 6. The partial relationship between product and RD capital.

$$\partial^2 Y/(\partial K \partial I) \geq 0 \quad , $$
$$\partial^2 Y/(\partial K \partial I) \geq 0 \quad , \quad \} \qquad (4.7)$$
$$\partial^2 Y/(\partial I \partial R) \geq 0 \quad . $$

These cross impacts once more illustrate the necessity of a fine tuning in planning direct productive capital, infrastructure capital and RD capital. In the case of a lack of coordination, various jumps in the system may occur, leading to various kinds of catastrophes, as can be seen in Figure 3. Such catastrophes also can be depicted in a three-dimensional space as topological singularities, in the form of geometrical projections. In this way one gets more insight into the conditions under which equilibrium states of a system display shocks or smooth transitions (especially if one value of a control variable produces multiple equilibria values of endogenous variables). A smooth change of a control variable may cause a sudden jump of the endogenous variable, which leads to a new value of this endogenous variable across the fold of the equilibrium surface. Depending upon the nature of the equilibrium surface, various kinds of catastrophes may be distinguished, such as cusps, butterflies, or the such (see Nijkamp, 1982a).

Having discussed the relevant aspects of the production system, some motion equations describing the evolution of the system will be presented now. The following equation for productive investments will be assumed:

$$\dot{K} = \eta_1 Y - \delta_1 K \quad , \qquad (4.8)$$

where $\dot{K}$ = the change in capital (i.e., $\partial K/\partial t$),

η_1 = the rate of investment in directly productive activities, and

δ_1 = the depreciation rate for directly productive capital.

A similar equation may be assumed for infrastructure:

$$\dot{I} = \eta_2 Y - \delta_2 I \quad , \qquad (4.9)$$

where η_2 = the rate of investment infrastructure capital, and

δ_2 = the depreciation rate for infrastructure capital.

Finally, the RD investment equation can be written as

$$\dot{R} = \eta_3 Y - \delta_3 R \quad , \qquad (4.10)$$

where η_3 = the rate of investment in RD capital, and

δ_3 = the depreciation rate for RD capital.

The parameter η_3 deserves closer attention, as it may be related to the debate between advocates of the so-called 'demand-pull' hypothesis and those of the so-called 'depression-trigger' hypothesis. If the depression-trigger hypothesis were valid, then η_3 would be higher in the case of a decline in Y than an increase in Y. On the other hand, if the demand-pull hypothesis were valid, then η_3 would be higher in the case of a growth in Y. Suppose for the moment no prior information exists about η_3 (nor on the validity of the 'demand-pull' versus the 'depression-trigger' hypothesis). Then it is more appropriate to consider η_3 as an unknown dynamic control variable whose time path may be assessed on the basis of reasonable assumptions regarding economic behavior of the system in question. Before doing so, however, additional relationships have to be introduced in terms of a consumption equation and its accompanying necessary constraints.

The following consumption model is assumed:

$$C = (1 - \eta_1 - \eta_2 - \eta_3)Y \quad , \quad (4.11)$$

where C = consumption per capita, and

the following conditions must hold:

$$\left.\begin{array}{c} \eta_1 + \eta_2 + \eta_3 \leq 1 \\ \eta_1, \eta_2, \eta_3 \geq 0 \end{array}\right\} \quad . \quad (4.12)$$

The parameters η_1, η_2, and η_3 now will be regarded as control variables. Suppose, for instance, that consumption is receiving a higher priority, then η_1, η_2 and η_3 must be very low. In this case, however, productive capital, infrastructure and RD will be fairly low, so that after some time the productive potential is affected. This, in turn, affects the consumption level. Therefore, a more balanced situation has to be found that guarantees a compromise between short-term desires and a long-term stable growth. For an analysis of a long-term growth strategy for the system at hand, it is helpful to use optimal control theory as a mathematical tool. The use of optimal control theory requires the specification of a multi-temporal objective function. Consider the following social welfare function:

$$\text{MAX:} \quad \omega = {}_0\!\int^T \Omega(C,K) \exp(-rt) \, dt \quad , \quad (4.13)$$

where r is the discount rate for a planning period with time horizon T. The preference function $\Omega=\Omega(C,K)$ reflects a compromise between consumption activities C and production activities K. The

Hamiltonian for this optimal control model can be written directly as follows:

$$H = \Omega \exp(-rt) + \lambda_1 (\eta_1 Y - \delta_1 K) + \lambda_2 (\eta_2 Y - \delta_2 I) + \lambda_3 (\eta_3 Y - \delta_3 R), \quad (4.14)$$

where λ_1, λ_2, and λ_3 are the costate variables (i.e., the Lagrangian multipliers). Recalling that η_1, η_2, and η_3 are considered to be control variables, the following first-order conditions for an interior optimal solution for the motion of the system can be formulated:

$$\partial H / \partial \eta_i = 0 \quad , \quad i=1,2,3 \quad . \quad (4.15)$$

The first-order conditions for the adjoint system are:

$$\dot{\lambda}_1 = - \partial H / \partial K$$

$$\dot{\lambda}_2 = - \partial H / \partial I \quad \} \quad . \quad (4.16)$$

$$\dot{\lambda}_3 = - \partial H / \partial R$$

The conditions for an interior solution of system (4.14) can be written as follows:

$$\begin{aligned} \exp(-rt) \, (\partial \Omega / \partial C) &= \lambda_1 \\ \exp(-rt) \, (\partial \Omega / \partial C) &= \lambda_2 \quad \} \quad . \quad (4.17) \\ \exp(-rt) \, (\partial \Omega / \partial C) &= \lambda_3 \end{aligned}$$

As the λ_is (i=1,2,3) may be regarded as the shadow prices of productive capital, infrastructure capital and RD capital, respectively, condition (4.17) states that the categories of capital have to be utilized in such a way that the shadow prices of all categories are equal. Each of these shadow prices should be equal to the discounted value of the marginal contribution of consumption to social welfare. Consequently this condition guarantees a compromise between productive and consumptive activities.

There is, however, a problem related to the foregoing control model analysis. The control variables are linear in the state space, so that most probably corner solutions will occur (see Nijkamp and Paelinck, 1973). The feasible control space is based on conditions (4.12) and is represented in Figure 7. Suppose, now, that the gradient of H with respect to η_1 is larger than that with respect to η_2, while the latter, in turn, is larger than the gradient of H with respect to η_3. Then

$$\partial H / \partial \eta_1 > \partial H / \partial \eta_2 > \partial H / \partial \eta_3 \quad . \quad (4.18)$$

This result implies that

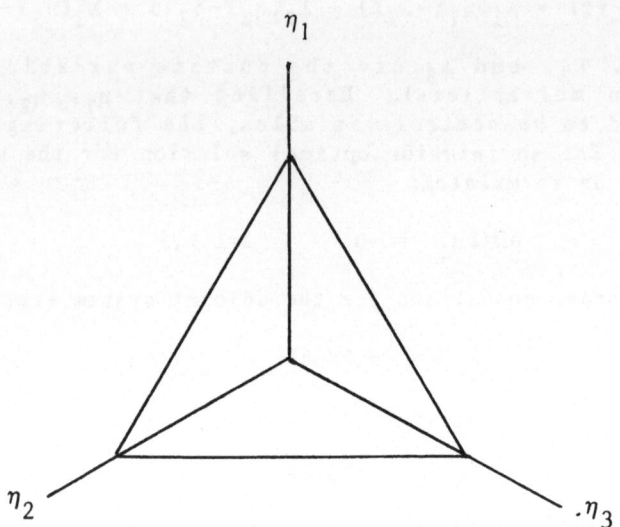

Figure 7. The feasible control space.

$$\lambda_1 \; > \; \lambda_2 \; > \; \lambda_3 \qquad , \qquad\qquad (4.19)$$

so that the dual price of capital is larger than that of infrastructure, which in turn is larger than that of RD capital. Hence, the optimal control solution is:

$$\eta_1 \; = \; 1 \; , \;\; \eta_2 \; = \; 0 \; , \;\; \eta_3 = 0 \qquad . \qquad\qquad (4.20)$$

Such extreme controls eventually will affect the capacity of the system for spending on consumption, so that after a time a shift toward another control is possible, even likely (either another corner solution or an interior solution). All other corner solutions may be analyzed in a similar manner. Apparently, then, the presence of corner solutions may lead to so-called 'bang-bang' strategies that guarantee that there will be continuous shocks to disturb the system.

In conclusion, catastrophes and perturbations in the foregoing system may have two sources:

(1) the asymmetric behavior of the dynamic system reflected by the complex production function, and

(2)   the corner solutions of policy strategies governing the state of the system at hand.

The preceding analysis can be extended either by introducing multiple objective functions (leading to multicriteria optimal control models; see Nijkamp, 1979), or by incorporating spatial interaction (or spillover) effects. The first approach is especially relevant in a policy context in which there is interaction between experts and decision-makers. In such a setting there will be shifts in policy priorities, and these may lead to shocks in the outcomes of the system concerned. The second extension is particularly relevant in cases of diffusion of innovation or of interregional spillovers from infrastructure endowment. By introducing such spatial interaction effects, a fully integrated spatial system may emerge that is capable of describing the spatial dynamics of the interwoven spatial system. Clearly, more analytical and empirical work has to be done before such approaches are operational and suitable for practical policy situations.

5.   CONCLUSION

The analysis presented here has demonstrated various interesting points. First, it turns out that inertia in a dynamic spatial system can be reflected by means of non-linear dynamic models that may generate various fluctuations. Thus, the phenomenon of spatiotemporal waves emerging from recent literature on economic dynamics can be provided with a firm theoretical basis that is in agreement with current economic research in the area of long waves. Second, the subdivision of regional capital equipment into productive capital, social overhead capital, and RD capital appears to yield a meaningful framework for analyzing the differential impact of various capital categories on regional growth phenomena. This framework also offers a possibility for including retardation effects, congestion effects and threshold effects, so that various kinds of catastrophes can be described. Finally, this analysis is of interest because it has been able to study the conditions under which the demand-pull hypothesis and the depression-trigger hypothesis may have some validity. Once again, a close link amongst current economic studies in the area of innovation and economic growth is found to exist.

6.   REFERENCES

Adelman, I., 1965, Long Cycles – Fact or Artifact, American Economic Review, 44: 444-463.

Allen, P. and M. Sanglier, 1979, A Dynamic Model of Growth in a Central Place System, Geographical Analysis, 11: 256-272.

Alonso, W., 1971, The Economics of Urban Size, Papers of the Regional Science Association, 26: 67-83.

Andersson, A., 1981, Structural Change and Technological Development, Regonal Science and Urban Economics, 11: 351-362.

Batten, D., 1981, On the Dynamics of Industrial Evolution,Research Paper, Umea Economic Studies No. 97, University of Umea, Sweden.

Berry, B. and D. Dahmann, 1977, Population Redistribution in the United States in the 1970's, Population Development Review, 3: 443-471.

Biehl, D., 1980, Determinants of Regional Disparities and the Role of Public Finance, Public Finance, 35: 44-71.

Brinner, R. and A. Alexander, 1979, The Role of High Technology Industries in Economic Growth, Cambridge, Mass.: Data Resource.

Brown, L., 1981, Innovation Diffusion, London: Methuen.

Carlino, G., 1977, Economies of Scale in Manufacturing Location, Boston: Kluwer Nijhoff.

Casetti, E., 1981, Technological Progress, Exploitation and Spatial Economic Growth: A Catastrophe Model, in Dynamic Spatial Models, edited by D. Griffith and R. MacKinnon, Alphen aan de Rijn: Sijthoff and Noordhoff, pp. 215-277.

Chatterjee, L. and P. Nijkamp (eds.), 1981, Urban Problems and Economic Development, Alphen aan den Rijn: Sijthoff and Noordhoff.

Clark, J., 1980, A Model of Embodied Technical Change and Employment, Technological Forecasting and Social Change, 16: 47-65.

_____, C. Freeman and L. Soete, 1981, Long Waves and Technological Developments in the 20th Century, in Konjunktur, Krise, Gesellschaft, edited by D. Petzina and G. van Roon, Stuttgart: Klett-Cotta, pp. 132-179.

Dasgupta, P., 1982, The Theory of Technological Competition, paper presented at the IEA Conference on New Developments in the Theory of Market Structures, Ottawa, Canada.

_____ and J. Stiglitz, 1980, Industrial Structure and the Nature of Innovative Activity, Economic Journal, 90: 266-293.

Davies, S., 1979, The Diffusion of Process Innovations, Cambridge: Cambridge University Press.

Delbeke, J., 1981, Recent Long-Wave Theories: A Critical Survey, Futures, 13: 246-257.

Dendrinos, D. (ed.), 1981, Dynamic Non-linear Theory and General Urban/Regional Systems, Lawrence, Kansas: School of Architecture and Urban Design, University of Kansas.

Dupriez, L., 1947, Des Mouvements Economiques Généraux, Louvain: University of Louvain.

Forrester, J., 1977, Growth Cycles, De Economist, 125: 525-543.

Freeman, C., J. Clark and L. Soete, 1982, Unemployment and Technical Innovation, London: Frances Printer.

Futia, C., 1980, Schumpeterian Competition, Quarterly Journal of Economics, 94: 675-695.

Giersch, H., 1979, Aspects of Growth, Structural Change and Employment, Weltwirtschaftliches Archiv, 115: 629-651.

Gordon, P., 1982, Deconcentration Without a 'Clean Break,' in Human Settlement Systems: Spatial Patterns and Trends, edited by T. Kawashima and P. Korcelli, Laxenburg: IIASA, pp. 193-202.

Graham, A. and P. Senge, 1980, A Long Wave Hypothesis of Innovation, Technological Forecasting and Social Change, 16: 283-311.

Haustein, H., H. Maier and L. Uhlmann, 1981, Innovation and Efficiency, Reserch Report RR-81-7, Laxenburg: IIASA.

Heertje, A., 1981, Technical Change and Economics, Man, Environment, Space and Time, 1: 59-76.

Isard, W. and P. Liossatos, 1979, Spatial Dynamics and Optimal Space-Time Development, Amsterdam: North-Holland.

Kamien, M. and N. Schwartz, 1975, Market Structure and Innovation: A Survey, Journal of Economic Literature, 13: 1-37.

Kawashima, T., 1981, Urban Optimality, in Cities in Transition, edited by P. Nijkamp and P. Rietveld, Alphen aan den Rijn: Sijthoff and Noordhoff, pp. 141–156.

Kennedy, C., 1964, Induced Bias in Innovation and the Theory of Distribution, Economic Journal, 74: 541–547.

Kleinknecht, A., 1981, Observations on the Schumpeterian Swarming of Innovations, Futures, 7: 293–307.

_____, 1982, Patenting in the Netherlands, paper presented to the OECD Workshop on Patent and Innovation Statistics, Paris, France.

Loury, G., 1980, Market Structure and Innovation, Quarterly Journal of Economics, 93: 395–410.

Mahdavi, K., 1972, Technological Innovation, Stockholm: Beckmans.

Malecki, E., 1979, Locational Trends in RD by large U. S. Corporations, 1965–1977, Economic Geography, 55: 309–323.

Mandel, E., 1980, Long Waves of Capitalist Development, Cambridge: Cambridge University Press.

Mansfield, E., 1968, Industrial Research and Technological Innovation, New York: Norton.

_____, M. Schwartz and S. Wagner, 1981, Imitation Costs Patents: An Empirical Study, Economic Journal, 91: 907–918.

Mass, N., 1980, Monetary and Real Causes of Investment Booms and Declines, Socio-Economic Planning Sciences, 14: 281–290.

Mensch, G., 1979, Stalemate in Technology, Cambridge, Mass.: Ballinger.

Mowery, D. and N. Rosenberg, 1979, The Influence of Market Demand on Innovation, Research Policy, 8: 102–153.

Myrdal, G., 1957, Economic Theory and Underdeveloped Regions, London: Duckworth.

Nelson, R. and V. Norman, 1977, Technological Change and Factor Mix Over the Product Cycle, Journal of Development Economics, 4: 3–24.

_____ and S. Winter, 1977, In Search of Useful Theory of Innovation, Research Policy, 6: 36–76.

Nijkamp, P., 1979, Multidimensional Spatial Data and Decision Analysis, New York: Wiley.

_____, 1981, Perspectives for Urban Analyses and Policies, in Cities in Transition: Problems and Policies, edited by P. Nijkamp and P. Rietveld, Alphen aan den Rijn: Sijthoff and Noordhoff, pp. 67-98.

_____, 1982a, A Multidimensional Analysis of Regional Infrastructure and Economic Development, in Structural Economic Analysis and Planning in Time and Space, edited by A. Andersson and T. Puu, Amsterdam: North-Holland.

_____, 1982b, Long Waves or Catastrophes in Regional Development, Socio-Eonomic Planning Sciences, forthcoming.

_____ and J. Paelinck, 1973, Some Models for the Economic Evaluation of the Environment, Regional Science and Urban Economics, 3: 33-62.

_____ and J. Paelinck, 1976, Operational Theory and Method in Regional Economics, Aldershot: Gower.

_____ and P. Rietveld (eds.), 1981, Cities in Transition, Alphen aan den Rijn: Sijthoff and Noordhoff.

Pred, A., 1966, The Spatial Dynamics of U. S. Urban-Industrial Growth, 1800-1914, Cambridge, Mass.: MIT Press.

Richardson, H., 1973, The Economics of Urban Size, Lexington: D. C. Heath.

Rosegger, G., 1980, The Economics of Production and Innovation, Oxford: Pergamon Press.

Rosenberg, N., 1976, Perspectives on Technology, Cambridge: Cambridge University Press.

Rostow, W., 1978, The World Economy, London: MacMillan.

Rothwell, R., 1979, Small and Medium Sized Manufacturing Firms and Technological Innovation, Management Decision, 16: 362-370.

Schmookler, J., 1966, Invention and Economic Growth, Cambridge: Cambridge University Press.

Schumpeter, J., 1939, Business Cycles, New York: McGraw-Hill.

Sveikauskas, L., 1979, Interurban Differences in the Innovative Nature of Production, Journal of Urban Economics, 6: 216-227.

Terleckyj, N., 1980, What Do RD Numbers Tell Us About Technological Change?, Papers and Proceedings, American Economic Review, 70: 55-61.

Thomas, M., 1981, Growth and Change in Innovative Manufacturing Industries and Firms, Collaborative Paper CP-81-5, Laxenburg: IIASA.

Thompson, W., 1977, The Urban Development Process, in Small Cities in Transition, edited by H. Bryce, Cambridge, Mass.: Ballinger, pp. 95-112.

van Bochove, C., 1982, Income Elasticities and the Sectoral Distribution of Peroductivity Growth Caused by Innovation, Disussion Paper 8201/G, Institute for Fiscal Studies, Erasmus University, The Netherlands.

van den Berg, L., R. Drewett, L. Klaassen, A. Rossi and C. Vijverberg, Urban Europe, A Study of Growth and Decline, Oxford, England: Pergamon, 1982.

van Duijn, J., 1972, An Interregional Model of Economic Fluctuations, Farnborough: Saxon House.

_____, 1979, De Lange Golf in de Economie, Assen: van Gorcum.

van Lierop, W. and P. Nijkamp (eds.), 1981, Locational Developments and Urban Planning, Alphen aan den Rijn: Sijthoff and Noordhoff.

von Weiszsaecker, C., 1980, Barriers to Entry, Berlin: Springer.

Vining, D. and T. Kontuly, 1977, Population Dispersal from Major Metropolitan Regions, International Regional Science Review, 3: 143-156.

_____, and A. Stauss, 1977, A Demonstration That the Current Deconcentration of Population in the United States is a Clean Break With the Past, Environment and Planning A, 9: 751-758.

COMPETITION AND ENVIRONMENT:
A THEORY OF TEMPORAL INNOVATION DIFFUSION

Michael Sonis

Bar-Ilan University
Israel

1. INTRODUCTION

The following three main empirical regularities form a theoretical basis for an innovation diffusion theory: the neighborhood effect, the hierarchical effect, and the logistic cumulative growth of the percentage of adopters of an innovation. The interconnections between these three effects have been a topic of many studies. This study will be confined to an analytical model of the diffusion process that generates the S-shaped growth only. Methodologically, the choice of only one basic principle enables an evaluation of the importance of this curve to the theory. Ultimately this choice will be shown to be justified, because the underlying interconnections between the proposed model of innovation diffusion process and Volterra-Lotka ecological dynamics are uncovered.

The task set out here will be achieved by dividing the analysis into two parts. In the first part of this study the vectorial differential equation of the temporal diffusion process is presented, and the results of a local stability analysis of the behavior of solutions of the linearized Lyapunov approximation within the vicinity of equilibrium points are given. In the case of a homogeneous environment and constant competition interaction between innovations, a competitive exclusion principle is found to hold. In other words, in the long run only one type of innovation will remain.

The Lyapunov approximation generates a continuous Markov chain. The stochastic matrix, corresponding to this Markov chain, describes the physical and social environment that defines the redistribution of innovations between adopters. This environment includes different conflicting extreme tendencies of behavior of adopters. The decomposition of the stochastic matrix of the Markov chain into a weighted sum of basic matrices leads to a description of the structure of this environment. The conflict between different extreme redistribution tendencies takes on its clearest form within environmental niches supporting and preserving the adoption of definite types of innovations. The minimax algorithm for the attainment of this decomposition, as well as the identification of these places of environmental niches, is employed.

In the second part of this study, totally antagonistic innovations are dealt with, and explicit formulae for the solution of the non-linear vectorial differential equation are presented. These explicit formulae give the generalized relative logistic growth of the proportions of adopters of totally antagonistic innovations. The structure of arbitrary competition is ascertained; each antisymmetric competition matrix is the sum of the antagonistic matrices describing the totally antagonistic competition among different sub-sets of innovations.

2. THE COMPETITIVE EXCLUSION PRINCIPLE, MARKOV CHAIN APPROXIMATIONS AND STRUCTURE OF THE ENVIRONMENT

2.1 · Emitted and Adopted Information Flows Between Adopters

Consider a stable population of P individuals mixing at random, each one of whom adopts some type of innovation from the set of n different innovations. The set of innovations is called the set of competitive innovations if they are mutually exclusive (i.e., if the adoption of one of them excludes the simultaneous adoption of the others). Information flows will be considered between the users and possible adopters of different innovations under the following assumptions (cf., Casetti, 1969): (i) the innovations are competitive, (ii) the individuals may change their opinion and reject the innovation after adopting it, (iii) the individuals become adopters of some kind of innovation under the influence of other individuals (i.e., adopters or non-adopters of this innovation) in the course of direct personal contacts, and (iv) the individuals have different degrees of resistance to and propensity for different innovations that depend upon the type of innovation and on the number of messages received from other individuals.

Let $P_i(t)$ be the number of adopters of the i-th innovation at the point in time t, and let $I_i(t)$ be the information flow emitted by this population. This information flow may be disaggregated into n subflows $I_{ij}(t)$ from the population $P_i(t)$ to the population

$P_j(t)$ such that

$$I_i(t) = I_{i1}(t) + I_{i2}(t) + \ldots + I_{in}(t) \quad .$$

If $s_{ij}(t) = I_{ij}(t)/I_i(t)$, then the matrix $S(t) = [s_{ij}(t)]$ will be a stochastic matrix.

Let $J_j(t)$ be the information flow adopted by the population $P_j(t)$ at the point in time t. Then

$$J_j(t) = I_{1j}(t) + I_{2j}(t) + \ldots + I_{nj}(t)$$

$$= s_{1j}(t) I_1(t) + s_{2j}(t) I_2(t) + \ldots + s_{nj}(t) I_n(t).$$

It is possible to present the influence generated by the interconnections between the set of emitted and adopted information flows, in matrix form, as follows:

$$J(t) = M(t) I(t) \quad ,$$

where $J(t) = [J_i(t)]$ and $I(t) = [I_i(t)]$, i=1,2,...,n, and the matrix $M(t) = [m_{ij}(t)]$ is the Markovian matrix transposed with respect to the matrix $S(t)$ [i.e., $m_{ij}(t) = s_{ji}(t)$]. The vectorial equation $J = M I$ represents a process of redistribution of information emitted by adopters, and, therefore, represents the structure of the social and spatial environment within which the diffusion of innovations occurs. Thus, the structure of a stochastic matrix $S(t)$, or its transposed Markov matrix $M(t)$, depends upon the structure of the social and physical environment (i.e., on the social norms, roles, social controls and cultural values that give context to interpersonal relations).

2.2. A Vectorial Differential Equation of the Temporal Diffusion Process

The per capita rates of change of the emitted information flows $I_i(t)$ are given by

$$d I_i(t)/[I_i(t) d t] = d \ln[I_i(t)]/d t \quad .$$

These rates of change are functions of the amounts of information flows I_1, I_2, ..., I_n, such that

$$d \ln(I_i)/d t = v_i(I_1, I_2, \ldots, I_n) \quad ,$$

where the functions $v_i[I_1(t), I_2(t), \ldots, I_n(t)]$ are the measures of effectiveness of the messages (see Casetti, 1969, p. 104), or measures of the influence of the adoption of each type of innovation on the adoption of every other type of innovation.

Assume that the functions v_i are linear homogeneous functions of I_1, I_2, ..., I_n (the Volterra assumption; see Volterra, 1926), so that

$$d \ln[I_i(t)]/d\, t = b_{i1}(t)\, I_1(t) + b_{i2}(t)\, I_2(t) + ... + b_{in}(t)\, I_n(t),$$

$$i = 1, 2, ..., n.$$

Thus, the vectorial differential equation of information flow s is

$$d \ln[I(t)]/d\, t = B(t)\, I(t) \qquad ,$$

where $B(t) = [b_{ij}(t)]$ is a coinfluence matrix and the operator $d \ln(\cdot)/d\, t$ acts on each component of the vector I.

If the information flows for different innovations are independent, meaning the Markovian matrix $M(t)$ is invertible, then $I(t) = M^{-1}(t)\, J(t)$, and the following vectorial differential equation of information flows is obtained:

$$d \ln(M^{-1}\, J)/d\, t = B\, M^{-1}\, J \qquad . \qquad (2.1)$$

Assume, following Casetti, that the information flows of messages $J_j(t)$ are proportional to the number of adopters of the j-th innovation, such that

$$J_j(t) = \omega\, P_j(t) \qquad , \quad j = 1, 2, ..., n.$$

Let $y_j(t) = P_j(t)/P$ be the portion of adopters of the j-th innovation. Then $J_j(t) = \omega\, P\, y_j(t)$, and the adopted vectorial information flow $J(t)$ is

$$J(t) = \omega\, P\, y(t) \qquad , \qquad (2.2)$$

where $y(t) = [y_i(t)]$, $i = 1, 2, 3, ..., n$, in which

$$\sum_{i=1}^{i=n} y_i(t) = 1 \qquad \text{and} \qquad 0 \leq y_i(t) \leq 1 \qquad ,$$

is the vector of the relative distribution of the set of competitive innovations.

Substituting equation (2.2) into (2.1) yields the following vectorial differential equation of the temporal diffusion process:

$$d \ln[M^{-1}(t)\, y(t)]/d\, t = A(t)\, M^{-1}(t)\, y(t) \qquad , \qquad (2.3)$$

where $y(t)$ is a probability vector and the competition matrix $A(t)$ is defined as $A(t) = \omega\, P\, B(t)$.

The condition $y_1(t) + y_2(t) + \ldots + y_n(t) = 1$ puts strong constraints on the coefficients of the competition matrix $A(t)$. This matrix $A(t)$ must be antisymmetric [i.e., $a_{ij}(t) = -a_{ji}(t)$]. Indeed, the substitution $w(t) = M^{-1}$ transforms the equation

$$d \ln(M^{-1}y)/d t = A M^{-1} y$$

into the form

$$d \ln(w)/d t = A w \qquad , \qquad (2.4)$$

where the vector $w = M^{-1} y$ is a probability vector, too. In other words,

$$w_1(t) + w_2(t) + \ldots + w_n(t) = 1, \qquad \text{and}$$

$$d w_1(t)/d t + d w_2(t)/d t + \ldots + d w_n(t)/d t = 0 \qquad .$$

Since the diffusion process can be continued from an arbitrary initial distribution of adopters, the vector $y(t)$ at each point in time t can be an arbitrary probability vector.

The coordinate form of equation (2.4) is

$$d w_i/(d t) = w_i \sum_{j=1}^{j=n} a_{ij} w_j = f_i(w) , i = 1, 2, \ldots, n. \quad (2.5)$$

Therefore, the condition

$$d w_1/d t + d w_2/d t + \ldots + d w_n/d t = 0$$

means that

$$\sum_{k=1}^{k=n} w_k \left(\sum_{s=1}^{s=n} a_{ks} w_s \right) = 0$$

for each probability vector. The substitution

$$w^T = e_i^T = (0 \ldots 0 \ 1 \ 0 \ldots 0) \qquad ,$$

where unity occurs in the i-th location, and all other elements are zero, immediately implies that $a_{ii} = 0$, and the substitution

$$w^T = (0 \ldots 0 \ 1/2 \ 0 \ldots 0 \ 1/2 \ 0 \ldots 0) \qquad ,$$

where 1/2 occurs in both the i-th and j-th locations implies that

$$(a_{ii}/2 + a_{ij}/2)/2 + (a_{ji}/2 + a_{jj}/2)/2 = 0 \qquad ,$$

or $a_{ij} + a_{ji} = 0$. Thus, mutually exclusive innovations are antagonistic in the sense of the theory of antagonistic games.

The differential equation (2.3) gives a description of the temporal diffusion process at the macro-level and includes the matrix representations $M(t)$ and $A(t)$ of two important parts of the diffusion process, namely (i) the representation of the redistribution of the information between adopters and, thereby, the number of adopters of each type of innovation, and (ii) the representation of the structure of competition (i.e., the influence of adoption of each innovation on the adoption of every other type of innovation). The strength of the interconnections between the i-th and j-th innovations at the point in time t is measured by the coefficient $a_{ij}(t)$ of the antisymmetric matrix $A(t)$.

In the case of a constant competition matrix A [i.e., $A(t)=A$], and an identity matrix version of M [i.e., $M(t) = I$], the vector equation $d \ln(y)/d\ t = A\ y$ resembles the generalized Volterra-Lotka ecological model of competition and predation in animal societies (see MacArthur, 1972, pp. 33-58). The difference between the diffusion model presented here and the generalized ecological model of Volterra and Lotka lies in the fact that the ecological model gives a description of absolute population growth within the neighborhood of the equilibrium, while the model presented here gives a description of relative growth. Moreover, in a special case of totally antagonistic competition, this latter model gives a global description of a diffusion process at each point in time, with the help of explicit formulae (see Section 3).

The simplest case of a constant antisymmetric matrix $A=\begin{bmatrix} 0 & a \\ -a & 0 \end{bmatrix}$ and an identity Markov matrix $M = I$ generates the most commonly noted empirical regularity of the temporal diffusion process, namely the cumulative S-shaped change of the percentage of adopters of an innovation (Brown and Cox, 1971). Indeed, in this case the vectorial differential equation $d \ln(y)/d\ t = A\ y$ is equivalent to Verhulst's equation $d\ y/d\ t = ay[1 - y]$ (Verhulst, 1838). The solution of this equation is a well-known logistic curve $y = 1/[1 + C\exp(-a\ t)]$. In the case of an arbitrary Markovian matrix

$$M = \begin{pmatrix} m_1 & m_2 \\ 1-m_1 & 1-m_2 \end{pmatrix}$$

the Pearl-Reed equation $d\ y/d\ t = [a/(m_1 - m_2)](y - m_2)(m_1 - y)$ with the solution

$$y = [m_1 + m_2\ C\ \exp(-a\ t)]/[1 + C\ \exp(-a\ t)]$$

is obtained, which is the S-shaped curve with a lower limit m_2 and an upper limit m_1. This curve was developed independently by Pearl

and Reed in 1920 (see Pearl, 1925), who have used it to describe
the growth of an albino rat and of a tadpole's tail, the number of
yeast cells in a nutritive solution, the number of fruit flies in a
bottle with a limited food supply, and the number of human beings
in a geographical area. In the case of a non-constant competition
matrix $A(t) = \begin{bmatrix} 0 & a(t) \\ -a(t) & 0 \end{bmatrix}$, and $M = I$, the equation $d\,y/d\,t = a(t)$
$y(1 - y)$ yields

$$y(t) = 1/\{1 + C \exp[-\,_0\!\int^t a(t)\,dt]\} \quad .$$

This curve has been termed the harmonic logistic by Dodd (1956;
also see Hudson, 1972, pp. 64-67). Thus, the diffusion model
$d\,\ln M^{-1}\,y/d\,t = A\,M^{-1}\,y$ describes generalized relative logistic
growth and redistribution. Even in the case of a constant
competition matrix A and a constant Markov matrix M, the proposed
diffusion model includes non-linearities, and usually the explicit
form of the solution for the system of non-linear differential
equations is impossible to determine. Hence local stability
analysis of dynamic equilibria needs to be relied upon in order to
obtain the asymptotic behavior of solutions.

As an aside, the antisymmetry of the competition matrix $A(t)$
gives a possible way for discussing competition between innovations
at the micro-level with the help of the theory of antagonistic
games, in such a way that the element $a_{ij}(t)$ is the value of the
game between the i-th and j-th innovations. From this point of
view a heuristic basis can be provided for competitive exclusion.
If the i(o)-th innovation is a 'winner' in an antagonistic game
against every other innovation, then in the long run only this
i(o)-th innovation will exist under the assumption that the
physical and social environment is indifferent to the diffusion of
innovations.

2.3. The Competitive Exclusion Principle

One of the most important theoretical concepts in general
ecology is the competitive exclusion principle, which forbids the
stable coexistence of two or more species with identical needs and
habits when there exist limited resources (Volterra, 1926; Lotka,
1932; Gause, 1935; Hardin, 1960; Miller, 1967). For the present
model of diffusion of innovations the competitive exclusion
principle gives the conditions of existence in the long run of only
one type of innovation. This principle may be obtained under the
assumption of non-changeable coinfluence (i.e., the antisymmetric
competition matrix A is constant), and of a homogeneous social and
physical environment (i.e., the Markovian matrix M is equal to the
identity matrix). The results of the treatment of a common case
with non-constant matrices $M(t)$ and $A(t)$ is analogous. Such a case
will be presented later, as a special case of an antagonistic
matrix $A(t)$. The analytical tool for the study of the conditions

of existence of competitive exclusion is the Lyapunov neighborhood stability analysis [i.e., the analysis of the local behavior of the solutions of the linearized Lyapunov differential vectorial equation in the vicinity of an equilibrium point (May, 1973)].

From this viewpoint the competitive exclusion principle means that there is an equilibrium orthovector of the type

$$e_{i(o)}^T = (0 \ldots 0 \ 1 \ 0 \ldots 0) \quad ,$$

where unity occurs in the $i(o)$-th location and zero occurs everywhere else, such that this equilibrium is stable with respect to small disturbances of the initial distribution of innovations between adopters. Moreover, this equilibrium is asymptotically stable (i.e., as $t \to \infty$ in the limit, the solution $w(t)$ of the linear Lyapunov approximation equation tends to the equilibrium points $e_{i(o)}$.

The vectorial differential equation $d \ln(w)/d t = A \, w$ has the coordinate form defined by equation (2.5), and the equilibrium points will be the solutions of the system of equations

$$f_i(w) = 0 \quad , \quad i = 1, 2, \ldots, n. \qquad (2.6)$$

The condition of antisymmetry of the competition matrix A means that $a_{ii} = 0$, and, therefore, it is easy to check that each orthovector $e_{i(o)}$, $i(o) = 1, 2, \ldots, n$, is the equilibrium point for the equation $d \ln(w)/d t = A \, w$. For the equilibrium point $e_{i(o)}$ the linear Lyapunov approximating differential equation has the form

$$d \, w/d t = L(w - e_{i(o)}) \quad , \qquad (2.7)$$

where the community matrix L has the coordinates

$$l_{ij} = \partial f_i / \partial w_j \, |_{w=e_{i(o)}}$$

(see, for example, Coppel, 1965; Levins, 1968). Since

$$\partial f_i / \partial w_j = \begin{cases} a_{ij} w_i & , i \neq j, \\ \sum_{k=1}^{k=n} (a_{ik} w_k - a_{ii} w_i) & , i = j, \end{cases}$$

then

$$l_{ii} = \begin{cases} a_{ii(o)} & , i \neq i(o), \\ 0 & , i = i(o). \end{cases}$$

For $i \neq j$,

$$1_{ij} = \left\{ \begin{array}{ll} 0 & , \ i \neq i(o), \\ a_{i(o)j} & , \ i = i(o). \end{array} \right.$$

Thus, L is a matrix of zeros except for the diagonal, whose values are $a_{ii(o)}$, the $i(o)$-th row, whose values are $a_{i(o)i}$, and cell $[i(o),i(o)]=0$. In other words,

$$L = a_{i(o)}^T I + [0 \ \vdots \ -a_{i(o)}^T \ \vdots \ 0]^T \quad , \quad (2.8)$$

since $a_{i\,j} = -a_{j\,i}$, where the second matrix term on the right-hand side of equation (2.8) is a partitioned matrix.

The eigenvalues of this matrix are the elements of its main diagonal (i.e., the coordinates of the $i(o)$-th vector-column $a_{i(o)}$ of competition matrix A. Therefore, the equilibrium vector $e_{i(o)}$ will be stable if the $i(o)$-th column $a_{i(o)}$ of matrix A is non-positive, and will be unstable if the $i(o)$-th column $a_{i(o)}$ includes at least one strongly positive element.

For the study of the asymptotic stability of the equilibrium point $e_{i(o)}$, consider a linearized Lyapunov vectorial differential equation

$$d\ (w - e_{i(o)})/d\ t \ = \ L(w - e_{i(o)}) \qquad \text{or} \qquad d\ w/d\ t \ = \ L\ w \quad ,$$

since $L\ e_{i(o)} = 0$. From the theory of ordinary differential equations the solution of this equation is given by the formula

$$w(t) = [\exp(L\ t)]\ w(0)$$

(see, for example, Gantmacher, 1966, Chapter V, Section 6).

By mathematical induction, since $a_{i\,j} = -a_{j\,i}$, it can be shown that

$$L^k = (a_{i(o)}^T I)^k + [0 \ \vdots \ -(a_{ii(o)}^k)^T \ \vdots \ 0]^T \quad .$$

Therefore, direct calculation gives

$$\exp(L\ t) = I + L\ t + 1/2! - L^2\ t^2 + \ldots + 1/k! - L^k\ t^k + \ldots$$

$$= [\exp(a_{ii(o)}t)]^T I + [0 \ \vdots \ (1 - \exp\{a_{ii(o)}t\})]^T \ \vdots \ 0]^T \quad . \quad (2.9)$$

If the $i(o)$-th column of the matrix A includes only non-positive elements, then for each k and $t > 0$, the elements $\exp(a_{ki(o)}\ t)$ lie between 0 and 1, and the matrix $\exp(L\ t)$ is a Markovian matrix.

Thus, the equation $\mathbf{w} = \exp(L\ t)\ \mathbf{w}(0)$ generates the continuous homogeneous Markov chain with a stochastic matrix $\exp(L^T)$.

If the i(o)-th column includes, besides $a_{i(o)i(o)} = 0$, only negative elements, then the limiting matrix becomes

$$\underset{t \to \infty}{\text{limit}} \exp(L\ t) = [0 : 1^T : 0]^T \quad ,$$

and, therefore,

$$\underset{t \to \infty}{\text{limit}}\ \mathbf{w}(t) = \underset{t \to \infty}{\text{limit}} \exp(L\ t)\ \mathbf{w}(0) = e_{i(o)} \quad .$$

Thus, the equilibrium point $e_{i(o)}$ is the point of asymptotical stability, and hence the competitive exclusion principle may be reformulated as follows. Let the antisymmetric competition matrix $\mathbf{A}$ of the mutually exclusive innovations be constant and the Markovian matrix $\mathbf{M}$ be equal to the identity matrix. If the i(o)-th column of the matrix $\mathbf{A}$ includes, besides element $a_{i(o)i(o)}$, only strongly negative elements [i.e., $a_{ji(o)} < 0,\ j \neq i(o)$], then the stable equilibrium distribution of innovations between adopters has the form of the orthovector

$$e_{i(o)}^T = (0 \ldots 0\ 1\ 0 \ldots 0) \quad ,$$

where unity occupies the i(o)-th location, and all other entries are zero. Moreover, this stable equilibrium is asymptotically stable and coincides with the final probability distribution generated by the continuous regular homogeneous Markov chain with stochastic matrix $\exp(L^T)$ [see equation (2.9) for t=1].

If the non-positive i(o)-th column of the matrix $\mathbf{A}$ includes more than one zero element, for example

$$a_{i(1)i(o)} = a_{i(2)i(o)} = \cdots = a_{i(k)i(o)} = a_{i(o)i(o)} = 0 \quad ,$$

then the limit matrix

$$\underset{t \to \infty}{\text{limit}} \exp(L\ t)$$

includes the unity on its main diagonal in the columns $i_1, i_2, \ldots i_k$, and the i(o)-th row includes zeroes in the same columns and units in every other column. The final distribution of innovations for the linearized system depends on the initial distributon of innovations

$$\underset{t \to \infty}{\text{limit}}\ \mathbf{w}(t) = [\underset{t \to \infty}{\text{limit}} \exp(L\ t)]\ \mathbf{w}(0) =$$

$$
\begin{pmatrix}
0 & \cdots & 0 & 0 & 0 & \cdots & 0 & 0 & 0 & \cdots & 0 & 0 & 0 & \cdots & 0 \\
\cdot & & \cdot & \cdot & \cdot & & \cdot & \cdot & \cdot & & \cdot & \cdot & \cdot & & \cdot \\
\cdot & & \cdot & \cdot & \cdot & & \cdot & \cdot & \cdot & & \cdot & \cdot & \cdot & & \cdot \\
\cdot & & \cdot & \cdot & \cdot & & \cdot & \cdot & \cdot & & \cdot & \cdot & \cdot & & \cdot \\
0 & \cdots & 0 & 0 & 0 & \cdots & 0 & 0 & 0 & \cdots & 0 & 0 & 0 & \cdots & 0 \\
0 & \cdots & 0 & 1 & 0 & \cdots & 0 & 0 & 0 & \cdots & 0 & 0 & 0 & \cdots & 0 \\
0 & \cdots & 0 & 0 & 0 & \cdots & 0 & 0 & 0 & \cdots & 0 & 0 & 0 & \cdots & 0 \\
\cdot & & \cdot & \cdot & \cdot & & \cdot & \cdot & \cdot & & \cdot & \cdot & \cdot & & \cdot \\
\cdot & & \cdot & \cdot & \cdot & & \cdot & \cdot & \cdot & & \cdot & \cdot & \cdot & & \cdot \\
\cdot & & \cdot & \cdot & \cdot & & \cdot & \cdot & \cdot & & \cdot & \cdot & \cdot & & \cdot \\
0 & \cdots & 0 & 0 & 0 & \cdots & 0 & 0 & 0 & \cdots & 0 & 0 & 0 & \cdots & 0 \\
1 & \cdots & 1 & 0 & 1 & \cdots & 1 & 1 & 1 & \cdots & 1 & 0 & 1 & \cdots & 1 \\
0 & \cdots & 0 & 0 & 0 & \cdots & 0 & 0 & 0 & \cdots & 0 & 0 & 0 & \cdots & 0 \\
\cdot & & \cdot & \cdot & \cdot & & \cdot & \cdot & \cdot & & \cdot & \cdot & \cdot & & \cdot \\
\cdot & & \cdot & \cdot & \cdot & & \cdot & \cdot & \cdot & & \cdot & \cdot & \cdot & & \cdot \\
\cdot & & \cdot & \cdot & \cdot & & \cdot & \cdot & \cdot & & \cdot & \cdot & \cdot & & \cdot \\
0 & \cdots & 0 & 0 & 0 & \cdots & 0 & 0 & 0 & \cdots & 0 & 0 & 0 & \cdots & 0 \\
0 & \cdots & 0 & 0 & 0 & \cdots & 0 & 0 & 0 & \cdots & 0 & 1 & 0 & \cdots & 0 \\
0 & \cdots & 0 & 0 & 0 & \cdots & 0 & 0 & 0 & \cdots & 0 & 0 & 0 & \cdots & 0 \\
\cdot & & \cdot & \cdot & \cdot & & \cdot & \cdot & \cdot & & \cdot & \cdot & \cdot & & \cdot \\
\cdot & & \cdot & \cdot & \cdot & & \cdot & \cdot & \cdot & & \cdot & \cdot & \cdot & & \cdot \\
\cdot & & \cdot & \cdot & \cdot & & \cdot & \cdot & \cdot & & \cdot & \cdot & \cdot & & \cdot \\
0 & \cdots & 0 & 0 & 0 & \cdots & 0 & 0 & 0 & \cdots & 0 & 0 & 0 & \cdots & 0
\end{pmatrix}
\begin{pmatrix}
w_1^o \\ \cdot \\ \cdot \\ \cdot \\ w_{i(1)}^o \\ \cdot \\ \cdot \\ \cdot \\ \cdot \\ w_{i(o)}^o \\ \cdot \\ \cdot \\ \cdot \\ \cdot \\ w_{i(k)}^o \\ \cdot \\ \cdot \\ \cdot \\ \cdot \\ w_n^o
\end{pmatrix}
=
\begin{pmatrix}
0 \\ \cdot \\ \cdot \\ \cdot \\ w_{i(1)}^o \\ \cdot \\ \cdot \\ \cdot \\ 0 \\ 1 - \sum_k w_{i(k)}^o \\ 0 \\ \cdot \\ \cdot \\ \cdot \\ 0 \\ w_{i(k)}^o \\ 0 \\ \cdot \\ \cdot \\ 0
\end{pmatrix}.
$$

Thus, in this case the equilibrium is stable, but not asymptotically so.

2.4. Asymptotical Stability of the Multi-innovation Equilibrium

It is important to stress that, methodologically, in the case of a non-homogeneous environment (i.e., in the case of the non-identity Markovian matrix M), the competitive exclusion principle implies the existence of an asymptotically stable equilibrium that includes a set of different innovations. Consider the diffusion model with a constant antisymmetric competition matrix A and a constant Markovian matrix M. The vectorial differential equation $d \ln(M^{-1} y)/dt = AM^{-1} y$ can be transformed to a simpler form when $M = I$, namely $d \ln(w)/(dt) = Aw$, by making the substitution $w = M^{-1} y$. This substitution determines a one-to-one correspondence between the equilibrium points and the linear Lyapunov approximations for the differential equation and its simple form. Therefore, the equilibrium points as well as the conditions of its asymptotical stability can be obtained from the competitive exclusion principle.

The substitution $w = M^{-1} y$ means that $Mw = y$, and therefore the equilibrium point for the equation $d \ln(M^{-1} y)/dt = AM^{-1} y$ is $Me_{i(o)} = m_{i(o)}$, where the vector $m_{i(o)}$ is the $i(o)$-th vector-column of the Markovian matrix M. Furthermore, the simplified Lyapunov approximation equation $dw/dt = L(w - e_{i(o)})$, after making the substitution $w = M^{-1}y$, takes on the form

$$dw/dt = d\ M^{-1}y/dt = M^{-1}\ dy/dt = L(M^{-1}y - e_{i(o)}) \quad .$$

Therefore, the linearized equation for the non-identity constant Markovian matrix M will be

$$dy/dt = ML(M^{-1}y - e_{i(o)}) = MLM^{-1}y - m_{i(o)} \quad .$$

In other words, the community matrix for this linearized differential equation will be $C = MLM^{-1}$. The solution of this equation also can be obtained from the solution $w(t) = [\exp(Lt)]w(0)$ in the form $y(t) = [\exp(Ct)]y(0)$.

Since the matrix $\exp(Lt)$ is Markovian, then the matrix $\exp(Ct) = \exp(MLM^{-1}t) = M[\exp(Lt)]M^{-1}$ is Markovian also, and generates the continuous Markov chain. The eigenvalues of matrix $\exp(C)$ coincide with the eigenvalues of matrix $\exp(L)$, and therefore the following conditions of a stable equilibrium are obtained. Consider the vectorial differential equation

$$d\ \ln[M^{-1}y(t)]/dt = AM^{-1}y(t) \quad ,$$

with a constant Markovian invertible matrix M and a constant antisymmetric competition matrix A. This equation has vectorial equilibrium points that are the vector-columns m_i of the matrix M. Moreover, if the $i(o)$-th column of the matrix A includes at least one positive element, then the equilibrium $m_{i(o)}$ is unstable. If the $i(o)$-th column of matrix A is non-positive with the unique zero element $a_{i(o),i(o)} = 0$, then the equilibrium distribution of adopters $m_{i(o)}$ is asymptotically stable, and coincides with the final probability vector for the continuous Markov chain with a stochastic matrix $Q = \exp(C^T) = M^{-1}[\exp(L^T)]M$. If the non-positive $i(o)$-th column includes more than one zero element, then the equilibrium $m_{i(o)}$ is stable, but not asymptotically so.

The simplest constant antisymmetric matrix $A = \begin{pmatrix} 0 & a \\ -a & 0 \end{pmatrix}$ always includes a non-positive column, and, therefore, the corresponding diffusion differential equation has one stable equilibrium. But the next more complicated case of a 3-by-3 antisymmetric matrix A yields examples of competitive innovations without any stable equilibrium. Indeed, in this case there are only eight different possibilities for the distribution of signs for an antisymmetric matrix. They are:

$$\begin{pmatrix} 0 & + & + \\ - & 0 & + \\ - & - & 0 \end{pmatrix} \begin{pmatrix} 0 & - & - \\ + & 0 & - \\ + & + & 0 \end{pmatrix} \begin{pmatrix} 0 & - & + \\ + & 0 & + \\ - & - & 0 \end{pmatrix} \begin{pmatrix} 0 & + & - \\ - & 0 & - \\ + & + & 0 \end{pmatrix} \begin{pmatrix} 0 & + & - \\ - & 0 & - \\ + & + & 0 \end{pmatrix} \begin{pmatrix} 0 & - & + \\ + & 0 & + \\ - & - & 0 \end{pmatrix} \begin{pmatrix} 0 & + & - \\ - & 0 & + \\ + & - & 0 \end{pmatrix} \begin{pmatrix} 0 & - & + \\ + & 0 & - \\ - & + & 0 \end{pmatrix} \quad .$$

The first six combinations of signs give the possibility of existence for an asymptotically stable equilibrium. The last two sign combinations exclude this possibility.

The interesting fact is that if there is an asymptotically stable equilibrium, then it must be unique, since the existence of a column with strongly negative non-diagonal elements implies, by antisymmetry, the existence of strongly positive elements in every other column of the matrix A (i.e., every other equilibrium must be unstable).

2.5. Redistribution Processes, the Structure of Environment and Environmental Niches

The main assumption, which leads to the vectorial differential equation of a diffusion innovation process, is that of the ratio of information flows between adopters to the number of adopters themselves; therefore, the distribution of information between adopters implies the redistribution of different types of innovations between adopters. The process of redistribution of innovations between adopters has its most distinct form in the vicinity of the asymptotically stable equilibrium distribution of innovations. The stochastic matrix $Q = M^{-1}[\exp(L^T)]M = [q_{ij}]$ describes this redistribution process in such a way that the component q_{ij} of matrix Q is the portion of adopters of the i-th innovation who reject this type of innovation and adopt the j-th innovation. In probabilistic terms, q_{ij} is the probability that an adopter of the i-th innovation will switch to the j-th innovation.

The matrix Q gives only the simplified form of the redistribution process. The real redistribution process is more complicated. Associated with it are the following properties:

(a) the possibility of a real switch of each adopter from each type of innovation,

(b) non-proportionate (or, as a special case, proportionate) entry or exit of adopters to different types of innovations from the world outside of a given region,

(c) non-proportionate internal growth or decline of the number of adopters of different types of innovations, and

(d) the ability to conceptually reclassify the adopters of the region in question into any m (or, as a special case, n) types of innovations.

As can be seen, not all of these properties are associated with actual switches of adopters from one type of innovation to another. Property (a) clearly is connected with actual transfer. Property

(b) is purely spatial in nature. Property (c), which refers to natural growth, is demographic in nature. Finally, property (d), which is reclassification, results in a conditional redistribution, but no actual transfer. The simplest form of redistribution process corresponds to the orthostochastic matrix [i.e., to a matrix whose rows have the form of orthovectors with coordinates ⟨0, ..., 0, 1, 0, ..., 0⟩]. Each orthostochastic matrix represents one of the extreme tendencies of innovation redistribution that acts according to the following redistribution principle: 'everything or nothing.' In other words, an orthostochastic matrix represents an extreme redistribution of innovations that generates the transfer of all adopters of one type of innovations to only one other type.

In the actual environment there are different conflicting extreme tendencies, and each of these becomes only partially expressed in reality. The aggregation effect of the totality of extreme tendencies will be presented with the help of the decomposition of the stochastic redistribution matrix Q into the additive weighted sum of the orthostochastic matrices Q_k, with the weights p_k, such that

$$Q = \sum_{k=1}^{k=m} p_k Q_k \quad ,$$

where the weights p_k have the properties of

$$0 \leq p_k \leq 1 \qquad \text{and} \qquad \sum_{k=1}^{k=m} p_k = 1 \quad .$$

The conflict between the different extreme redistribution tendencies is expressed with the help of interdictions (i.e., prohibition) on transfer from one type of innovations to another. This interdiction has its sharpest form within an environmental niche, supporting and preserving the adoption and usage of a definite type of innovation—only one extreme tendency acts within the niche.

The algebraic algorithm for the construction of extreme tendencies Q_k and the places of environmental niches within the social and physical environment now will be outlined. The construction of the orthostochastic matrix Q_1, which represents the main extreme tendency, will be given with the help of a minimax assignment algorithm (Sonis, 1980, 1981b) based upon the 'everything or nothing' principle. This algorithm may be summarized as follows:

In each row of the stochastic matrix Q the maximal

component $a_{ij(i)}$ is chosen, and in its place is put unity, whereas zeroes are put in the other places in the row. The weight p_1 of this extreme tendency in the actual redistribution process is equal to the minimal column element from the elements $a_{ij(i)}$, such that

$$p_1 = \min_i \{a_{ij(i)}\} = \min_i \max_j \{a_{ij}\} = a_{i(o),j(o)} \quad .$$

The difference $Q - p_1Q_1 = (1 - p_1)Q'$ gives the unexplored remainder Q', and so the actual decomposition is

$$Q = p_1Q_1 + (1 - p_1)Q' \quad .$$

The conflict between the main extreme tendency (given by Q_1) and other extreme tendencies (given by the remainder Q') achieves a complete form in that the remainder Q' will include a zero coordinate in the place of the choice of p_1 (i.e., in the place of minimax). This minimax represents the place [i(o),j(o)] of maximal counteraction to the extreme tendency Q_1 [i.e., it represents the strong tendency to retain the i(o)-th innovation rather than abandon it for the j(o)-th innovation]. Hence the location [i(o),j(o)] represents the place of an environmental niche that protects and supports the adoption of the i(o)-th innovation against the attraction forces of the j(o)-th innovation.

The unexplored remainder Q' can be calculated by the formula

$$Q' = [1/(1 - p_1)]Q - [p_1/(1 - p_1)]Q_1 \quad .$$

Now the problem becomes one of choosing the next extreme tendency Q_2 acting within Q'. To solve this problem the previous steps of the algorithm need to be repeated, relative to Q', which will give the decomposition

$$Q' = q_1Q_2 + (1 - q_1)Q'' \quad ;$$

therefore,

$$Q = p_1Q_1 + (1 - p_1)[q_1Q_2 + (1-q_1)Q''] =$$

$$p_1Q_1 + p_2Q_2 + (1 - p_1 - p_2)q'' \quad ,$$

where $p_2 = (1 - p_1)q_1$. The stochastic matrix Q'' has two zero components in the places [i(o),j(o)] and [i(1),j(1)]. In other words, the location has been found of the additional environmental niche corresponding to the prohibition on transfering from the i(1)-th innovation to the j(1)-th, and so forth. After a finite number of steps, m, the set of extreme tendencies acting in the environment will be obtained, together with their orthostochastic matrices $\mathbf{Q}_k$ with the weights p_k and the series of the places of environmental niches preserving the existence of some innovations (the stopping rule is that the number of steps must not exceed $n^2 -$ n, which is the dimension of the space for all stochastic matrices.

The same type of analysis can be repeated for the matrix $\mathbf{Q}^t = \mathbf{M}^{-1}[\exp(\mathbf{C}t)]\mathbf{M}$ at each point in time, t. In this way one may observe the dynamics of the main forces acting within the social and physical environment, which lead eventually to the asymptotically stable distribution of innovations between adopters.

If, for some time interval, the decomposition includes the same set of extreme tendencies, so that

$$\mathbf{Q}^t = p_1(t)\mathbf{Q}_1 + p_2(t)\mathbf{Q}_2 + \ldots + p_m(t)\mathbf{Q}_m \quad ,$$

with the constant orthostochastic matrices $\mathbf{Q}_k$ and the variable weights $p_k(t)$, then a structurally stable innovation redistribution process will be obtained. Structural change would result from the substitution of one or a few orthostochastic matrices for other ones. Thus, the innovation redistribution at each point in time is the dynamic resultant, or the center of gravity, of the extreme redistribution tendencies acting within the environment, and forming the dynamic environmental niches.

3. TOTALLY ANTAGONISTIC COMPETITION BETWEEN INNOVATIONS, GENERALIZED RELATIVE LOGISTIC GROWTH AND ADDITIVE STRUCTURE OF COMPETITION

3.1. Totally Antagonistic Innovations

In the first part of this paper the vectorial differential equation of the temporal diffusion process was presented with equation (2.3). The essential deficiency of previous considerations relates to the absence of explicit formulae for the solution of this vectorial non-linear differential equation. For the explicit treatment of this equation one needs a notion of totally antagonistic innovations, which is a specialization of the notion of mutually exclusive competitive innovations (see Sonis, 1981b). In the case of two innovations the competitive pair of innovations is always totally antagonistic.

Mutually exclusive innovations are called totally antagonistic innovations if for each closed chain of innovations i(1), i(2), coinfluence matrix $A = [a_{ij}(t)]$ is equal to zero for each point in time t, such that

$$a_{i(1),i(2)} + a_{i(2),i(3)} + \ldots + a_{i(m),i(1)} = 0 \quad . \quad (3.1)$$

For instance, if the chain is i,i then $a_{ii}(t) = 0$, and if the chain is i,j,i then $a_{ij}(t) + a_{ji}(t) = 0$ (i.e., the totally antagonistic competition is a special case of mutually exclusive competition). Moreover, the requirement (3.1) is equivalent to the property

$$a_{ij}(t) = a_{ik}(t) - a_{jk}(t) \quad , \text{ for each i, j and k. (3.2)}$$

Indeed, the chain i,j,k,i gives

$$a_{ij} + a_{jk} + a_{ki} = 0$$

$$(\text{i.e., } a_{ij} = -a_{jk} - a_{ki} = a_{ik} - a_{jk}).$$

Conversely, if condition (3.2) holds, then for the chain i(1), i(2), ..., i(m)

$$a_{i(1),i(2)} + a_{i(2),i(3)} + a_{i(m-1),i(m)} + a_{i(m),i(1)} =$$

$$a_{i(1),1} - a_{i(2),1} + a_{i(2),1} - a_{i(3),1} + \cdots + a_{i(m),1}$$

$$- a_{i(m-1),1} + a_{i(m),1} - a_{i(1),1} = 0 \quad .$$

The condition (3.2) means that each column of the competition matrix **A** uniquely defines all the components of the matrix **A**, and the competition matrix **A** has the form $A(t) =$

$$\begin{bmatrix} 0 & -a_{21}(t) & -a_{31}(t) & \cdots & -a_{n1}(t) \\ a_{21}(t) & 0 & a_{21}(t)-a_{31}(t) & \cdots & a_{21}(t)-a_{n1}(t) \\ \cdot & \cdot & \cdot & \cdots & \cdot \\ \cdot & \cdot & \cdot & \cdots & \cdot \\ \cdot & \cdot & \cdot & \cdots & \cdot \\ a_{n1}(t) & a_{n1}(t)-a_{21}(t) & a_{n1}(t)-a_{31}(t) & \cdots & 0 \end{bmatrix} .$$

This type of matrix will be referred to here as the totally antagonistic matrix.

Each totally antagonistic matrix $A(t)$ includes a non-negative and a non-positive column, both of which define the asymptotical behavior of the model for $t \to \pm\infty$. More precisely, if for a fixed t

$$\max_i\{a_{i1}(t)\} = a_{i(o),1}(t) \quad , \text{ and}$$

$$\min_i\{a_{i1}(t)\} = a_{i(1),1}(t) \quad ,$$

then the i(o)-th column includes only non-positive components, and the i(1)-th column includes only non-negative components. As such, the structure of the i-th column of the totally antagonistic matrix A is $(-a_{i1}, a_{21}-a_{i1}, a_{31}-a_{i1}, \ldots, a_{n1}-a_{i1})$. Since $a_{11}=0$, then

$$a_{i(o),1} = \max_i\{a_{i1}\} \geq a_{ii} = 0 \text{ and } -a_{i(o),1} \leq 0.$$

Furthermore, $a_{i(o),1} \geq a_{i1}$ and, $a_{i,i(o)} = a_{i1} - a_{i(o),1} \leq 0$. Analogously, the i(1)-th column is non-negative.

3.2. The Explicit Solution of the Vectorial Differential Equation of the Temporal Innovation Diffusion Process for Totally Antagonistic Competition

Once again consider equation (2.3). Let $u = M^{-1}y$. Since y is a probabilistic vector and M^{-1} is a Markovian matrix, then the vector u also will be a probabilistic vector, and

$$d \ln[u(t)]/dt = A(t) u(t) \quad ;$$

or, in the coordinate form,

$$d \ln[u_i(t)]/dt = \sum_{j=1}^{j=n} a_{ij}(t) u_j(t) \quad , i=1, 2, \ldots, n, \quad (3.3)$$

$$\sum_{j=1}^{j=n} u_j = 1 \quad .$$

Since $a_{ij} - a_{1j} = a_{i1}$, then

$$d \ln[u_i(t)/u_1(t)]/dt = a_{i1}(t) \quad .$$

Therefore,

$$\ln[u_i(t)/u_1(t)] = \int_{t_0}^{t} a_{i1}(t) dt + \ln[u_i(t_0)/u_1(t_0)] \quad .$$

In other words,

$$u_i(t)/u_1(t) = [u_i(t_0)/u_1(t_0)] \exp[\int_{t_0}^{t} a_{i1}(t)] \quad .$$

Since

$$u_i(t) = u_1(t)[u_i(t_0)/u_1(t_0)] \exp[\int_{t_0}^{t} a_{i1}(t)] \quad ,$$

then

$$1 = \sum_{j=1}^{j=n} u_j(t) = [u_1(t)/u_1(t_0)] \sum_{j=1}^{j=n} u_j(t_0) \exp[\int_{t_0}^{t} a_{j1}(t)] \quad .$$

Thus,

$$u_1(t) = u_1(t_o)/\{ \sum_{j=1}^{j=n} u_j(t_o) \exp[_{t_o}\!\int^t a_{j1}(t)]\} \quad ,$$

and

$$u_i(t) = u_1(t_o) \exp[_{t_o}\!\int^t a_{i1}(t)]/\{ \sum_{j=1}^{j=n} u_j(t_o) \exp[_{t_o}\!\int^t a_{j1}(t)]\} .$$

Let $C_i = u_i(t_o)$. Then $0 \leq C_i \leq 1$ and $\sum_{j=1}^{j=n} C_j = 1$. Since

$$u(t) = M^{-1}(t)u(t) , \quad \text{then } y = Mu , \quad \text{and}$$

$$y_i(t) = \sum_{k=1}^{k=n} m_{ik}(t) u_k(t) =$$

$$\sum_{k=1}^{k=n} C_k m_{ik}(t) \exp[_{t_o}\!\int^t a_{k1}(t)]/\{ \sum_{j=1}^{j=n} C_j \exp[_{t_o}\!\int^t a_{j1}(t)]\} .$$

These formulae include the components of the first column of matrix A(t). It also is possible, with the help of the condition $a_{ij} = a_{i1} - a_{j1}$, to present these formulae in a form that depends upon the componets of every other column. For example, the i(o)-th column may be included in the formulae as

$$y_i(t) =$$

$$\sum_{k=1}^{k=n} C_k m_{ik}(t) \exp[_{t_o}\!\int^t a_{k,i(o)}(t)]/\{ \sum_{j=1}^{j=n} C_j \exp[_{t_o}\!\int^t a_{j,i(o)}(t)]\} , \quad (3.4)$$
$$i = 1, 2, \ldots, n,$$

where the vector of constants $C^T = (C_1,C_2,\ldots,C_n)$ is given by $C = M^{-1}(t_o)y(t_o)$.

In the case of constant matrices M and A the following solution is obtained, which is a generalization of the logistic curve:

$$y_i(t) = \sum_{k=1}^{k=n} C_k m_{ik} \exp[a_{k,i(o)}(t)]/\{ \sum_{j=1}^{j=n} C_j \exp[a_{j,i(o)}(t)]\} \quad (3.5)$$
$$i = 1, 2, \ldots, n.$$

When matrix M is the identity matrix I, these formulae are similar to the multinomial logit model in random utility choice theory (see Hensher and Johnson, 1981). This similarity illustrates the duality between individual choice of alternative innovations, and the behavior of the innovations themselves. The explicit formulae (3.4) make possible to study the asymptotic behavior of solutions, and to construct an approximating non-homogeneous Markov chain model directly, without using the Lyapunov stability analysis. The methodological importance of this approximation lies in the fact that if the model achieves logistic growth in the long run, then this model can be simplified to a continuous Markov chain model that generates asymptotic exponential growth (cf., Webbeer and Joseph, 1976; Allen and Sanglier, 1981).

Assume that for the i(o)-th column of the antagonistic matrix $A(t) = [a_{ij}(t)]$, the following intergrals are divergent to $-\infty$:

$$\lim_{t \to +\infty} {_{t_0}}\!\int^{t} a_{i,i(o)}(t) \, dt = {_{t_0}}\!\int^{\infty} a_{i,i(o)}(t) \, dt = -\infty \quad , \quad (3.6)$$

$$i = 1, 2, \ldots, n.$$

Additionally, let there be for each coordinate of the Markovian matrix $M(t) = [m_{ij}(t)]$ the limit

$$\lim_{t \to +\infty} m_{ij}(t) = m_{ij}^{+} \quad , \text{ or } \quad \lim_{t \to +\infty} M(t) = M_{+} \quad . \quad (3.7)$$

Further, assume initially that the vector $C = M^{-1}(t_0)y(t_0)$ includes only non-zero components. If the i(o)-th column of the matrix $A(t)$ includes only one zero element $a_{i(o),i(o)}(t) = 0$, then

$$y_i(t) =$$

$$C_{i(o)} \, m_{i,i(o)}(t) + \sum_{\substack{k=1 \\ k \neq i(o)}}^{k=n} C_k \, m_{ik}(t) \, \exp[{_{t_0}}\!\int^{t} a_{k,i(o)}(t)] / \{C_{i(o)} +$$

$$\sum_{\substack{j=1 \\ j \neq i(o)}}^{j=n} C_j \, \exp[{_{t_0}}\!\int^{t} a_{j,i(o)}(t)]\}, \quad i = 1, 2, \ldots, n,$$

and, since for each $k \neq i(o)$

$$\lim_{t \to +\infty} {_{t_0}}\!\int^{t} a_{k,i(o)}(t) \, dt = -\infty \quad ,$$

then

$$\lim_{t \to +\infty} y_i(t) = m_{i,i(o)}^{+}$$

[i.e., the final (for $t \to +\infty$) equilibrium distribution of adopters of the totally antagonistic innovations--the 'ceiling' distributon-- coincides with the i(o)-th vector-column $\mathbf{m}_{i(o)}^{+}$ of the limiting Markovian matrix $\mathbf{M}_{+}$. This equilibrium distribution of adopters is asymptotically stable, since it is not dependent on the initial distribution of adopters. This fact is referred to as equifinal behavior, and constitutes an important characteristic of open systems (von Bertalanffy, 1962).

Analogously, if for the j(o)-th column of the competition matrix $\mathbf{A}$

$$\lim_{t \to -\infty} {}_{t}\int{}^{t}_{o} \, a_{k,j(o)}(t) \, dt = +\infty \quad \text{and} \quad \lim_{t \to -\infty} M(t) = \mathbf{M}_{-} = [m_{ij}^{-}] \quad ,$$

then

$$\lim_{t \to -\infty} y_i(t) = m_{i,j(o)}^{-} \quad ,$$

and the lowest asymptotically stable distribution of adopters is located in the j(o)-th column of the limiting matrix $\mathbf{M}_{-}$ (cf., Rapoport, 1963, pp. 495-579).

If the conditions (3.6) and (3.7) hold for the i(o)-th column, which includes the following zero components:

$$a_{i(o),i(o)}(t) = a_{i,i(o)}(t) = \ldots = a_{i(s),i(o)}(t) = 0 \quad ,$$

then

$$\lim_{t \to +\infty} y_i(t) =$$

$$[C_{i(o)} \, m_{i,i(o)}^{+} + C_{i(1)} \, m_{i,i(1)}^{+} + \ldots + C_{i(s)} \, m_{i,i(s)}^{+}]/[C_{i(o)} +$$

$$C_{i(1)} + \ldots + C_{i(s)}] =$$

$$p_o \, m_{i,i(o)}^{+} + p_1 \, m_{i,i(1)}^{+} + \ldots + p_s \, m_{i,i(s)}^{+} \quad ,$$

where $p_r = C_{i(r)}/[C_{i(o)} + C_{i(1)} + \ldots + C_{i(s)}]$, r=1,2,...,s [i.e., the 'ceiling' distribution of adopters is the convex combination- weighted sum of the columns $\langle i(o),i(1),\ldots,i(s)\rangle$ of the Markovian matrix $\mathbf{M}_{+}$]. This ceiling distribution is stable, but not asymptotically stable, since it depends on the initial distribution. It is possible to obtain an analogous description for the stable lowest distribution of adopters.

If the matrices $\mathbf{A}(t)$ and $\mathbf{M}(t)$ are constant, then $\mathbf{A}(t) = \mathbf{A}$, $\mathbf{M}(t) = \mathbf{M}$, and if the i(o)-th column is non-positive and the j(o)-th column is non-negative, then the ceiling distribution of adopters is located in the i(o)-th vector-column $\mathbf{m}_{i(o)}$ of matrix $\mathbf{M}$, whereas the bottom distribution of adopters is in the j(o)-th vector-column

$\blacksquare_{j(o)}$ of matrix **M** (see Figure 1). This asymptotic behavior is analogous to the coexistence or sole survivor asymptotic behavior of competitive populations in Slobodkin's model of competitive growth of two animal populations (Slobodkin, 1961).

3.3. The Non-homogeneous Continuous Markov Chain Approximation of the Asymptotic Behavior of Adopters of Totally Antagonistic Innovations

The existence of the explicit formulae (3.4) immediately gives an approximate description of the asymptotical behavior of the totally antagonistic innovations, when time t tends to $\pm\infty$. This result has been obtained in the first section, with the help of stability analysis.

Consider, for the sake of simplicity, the case of a constant vector **C** with non-zero components, and let the elements $a_{j,i(o)}(t)$ of the i(o)-th column of the antagonistic matrix **A** have the following property:

$$\lim_{t \to +\infty} {}_t\!\int_0^t a_{j,i(o)}(t)\, dt = -\infty \qquad , \; j \neq i(o), \text{ and}$$

$$a_{i(o),i(o)} = 0 \qquad \text{otherwise.}$$

The generalized logistic growth is given by the formulae

$$y_i(t) =$$
$$\{m_{i,i(o)}(t) + \sum_{\substack{k=1 \\ k \neq i(o)}}^{k=n} (C_k/C_{i(o)})\, m_{ik}(t)\, \exp[{}_t\!\int_0^t a_{k,i(o)}(t)\, dt]\}/\{1 +$$

$$\sum_{\substack{j=1 \\ j \neq i(o)}}^{j=n} (C_j/C_{i(o)})\, \exp[{}_t\!\int_0^t a_{j,i(o)}(t)\, dt]\} \qquad , \qquad (3.8)$$
$$i = 1, 2, \ldots, n.$$

If t is sufficiently large, then the sum

$$\sum_{\substack{j=1 \\ j \neq i(o)}}^{j=n} (C_j/C_{i(o)})\, \exp[{}_t\!\int_0^t a_{j,i(o)}(t)\, dt] < 1 \qquad ,$$

since

$${}_t\!\int_0^{+\infty} a_{j,i(o)}(t)\, dt = -\infty \qquad \text{for } j \neq i(o).$$

Next the following approximation formula may be used:

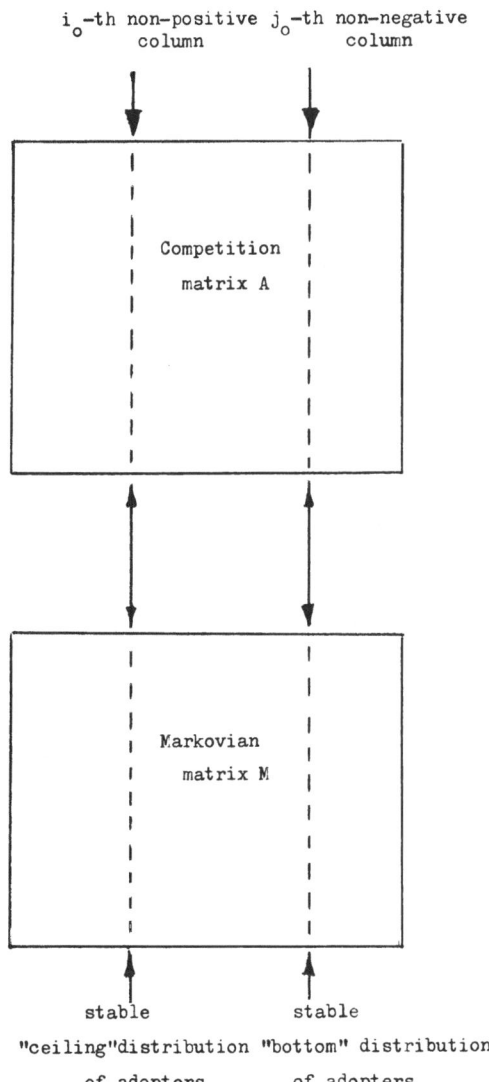

Figure 1. Asymptotical behavior of adopters—the final equilibrium distribution—for the constant competition and Markovian matrices **A** and **M.**

$$1/(1 + x) \; \widetilde{} \; 1 - x \qquad \text{for } |x| < 1.$$

Consequently,

$$y_i(t) \; \widetilde{}$$

$$m_{i,i(o)}(t) + \sum_{\substack{k=1 \\ k \neq i(o)}}^{k=n} (C_k/C_{i(o)}) \, m_{ik}(t) \, \exp[{}_{t_o}\!\!\int^t a_{k,i(o)}(t) \, dt]$$

$$- m_{ii(o)}(t) \sum_{\substack{j=1 \\ j \neq i(o)}}^{j=n} (C_j/C_{i(o)}) \, \exp[{}_{t_o}\!\!\int^t a_{j,i(o)}(t) \, dt].$$

Thus, if t is sufficiently large,

$$y_i(t) \; \widetilde{} \; w_i(t) = m_{i,i(o)}(t) +$$

$$\sum_{k=1}^{k=n} [m_{ik}(t) - m_{i,i(o)}(t)](C_k/C_{i(o)}) \, \exp[{}_{t_o}\!\!\int^t a_{k,i(o)}(t) \, dt] \, . \tag{3.9}$$

Thus, $w_i(t) \geq 0$, and hence

$$\sum_{i=1}^{i=n} w_i(t) = \sum_{i=1}^{i=n} m_{i,i(o)}(t) +$$

$$\{ \sum_{\substack{k=1 \\ k \neq i(o)}}^{k=n} (C_k/C_{i(o)}) \exp[{}_{t_o}\!\!\int^t a_{k,i(o)}(t) \, dt]\} \{ \sum_{i=1}^{i=n} [m_{ik}(t) - m_{i,i(o)}(t)]\}$$

$$= 1 \quad ,$$

since

$$\sum_{i=1}^{i=n} m_{i,i(o)}(t) = 1 \quad \text{and} \quad \sum_{i=1}^{i=n} [m_{ik}(t) - m_{i,i(o)}(t)] = 0 \quad .$$

Thus, the vector $w(t)$ with the components $w_i(t)$ is a probabilistic vector. Accordingly,

$$w_i(t_o) = m_{i,i(o)}(t_o) +$$

$$\sum_{\substack{k=1 \\ k \neq i(o)}}^{k=n} [m_{ik}(t_o) - m_{i,i(o)}(t_o)] (C_k/C_{i(o)}) \qquad . \qquad (3.10)$$

The probabilistic vector $w(t)$ can be represented in the form

$$w(t) = m_{i(o)}(t) +$$

$$\{ \sum_{\substack{k=1 \\ k \neq i(o)}}^{k=n} (C_k/C_{i(o)}) \exp[{}_{t_o}\!\int^t a_{k,i(o)}(t) \, dt] \}[m_k(t) - m_{i(o)}(t)] , \qquad (3.11)$$

where the vectors $m_k(t)$ are the vector-columns of the Markovian matrix $M(t)$.

Now the non-homogeneous Markov chain will be constructed that generates the behavior given by equaiton (3.11). As a set of infinitesimal generators (i.e., generators that approach 0 in the limit) of the temporally non-homogeneous continuous Markov chain, consider the matrices $M(t)L(t)M^{-1}(t_o)$, where $L(t)$ is defined the same way that matrix L was defined in equation (2.7). According to the definition of an integral of the matrix function (see Gantmacher, 1966, Chapter V):

$$_{t_o}\!\int^t L(t) \, dt =$$

$$\begin{pmatrix}
{}_{t_o}\!\int^t a_{1,i(o)}(t)dt & 0 & \dots 0 \dots & 0 \\
0 & {}_{t_o}\!\int^t a_{2,i(o)}(t)dt & \dots 0 \dots & 0 \\
\vdots & \vdots & \vdots & \vdots \\
-{}_{t_o}\!\int^t a_{1,i(o)}(t)dt & -{}_{t_o}\!\int^t a_{2,i(o)}(t)dt & \dots 0 \dots & -{}_{t_o}\!\int^t a_{n,i(o)}(t)dt \\
\vdots & \vdots & \vdots & \vdots \\
0 & 0 & \dots 0 \dots & {}_{t_o}\!\int^t a_{n,i(o)}(t)dt
\end{pmatrix} .$$

Therefore, it is possible to calculate $R_+(t)$ using the series:

$$R_+(t) = \exp[{}_{t_o}\!\!\int^t L(t)\ dt] =$$

$$I + {}_{t_o}\!\!\int^t L(t)\ dt + (1/2!)[{}_{t_o}\!\!\int^t L(t)\ dt]^2 + \ldots$$

$$(1/k!)[{}_{t_o}\!\!\int^t L(t)\ dt]^k + \ldots =$$

$$\begin{pmatrix} a_1 & 0 & \ldots\ 0\ \ldots & 0 \\ 0 & a_2 & \ldots\ 0\ \ldots & 0 \\ \vdots & \vdots & \vdots & \vdots \\ 1 - a_1 & 1 - a_2 & \ldots\ 1\ \ldots & 1 - a_n \\ \vdots & \vdots & \vdots & \vdots \\ 0 & 0 & \ldots\ 0\ \ldots & a_n \end{pmatrix},$$

where

$$a_j = \exp[{}_{t_o}\!\!\int^t a_{j,i(o)}(t)\ dt] \qquad , \quad j = 1, 2, \ldots, n.$$

The function $R_+(t)$ is the infinite product of the exponents of the infinitesimal generators. This type of product–integral was first introduced by Volterra in 1887 (see Gantmacher, 1966, Chapter XV).

The Markovian matrix–function $R_+(t)$ preserves the orthovector $e_{i(o)}$ such that $R_+(t)\ e_{i(o)} = e_{i(o)}$, and for $k \neq i(o)$

$$[R_+(t)\ e_k]^T = \langle 0,\ldots,0,a_k,0,\ldots,0,1-a_k,0,\ldots,0\rangle$$

$$= a_k\ e_k + (1 - a_k)e_{i(o)} \quad .$$

It can be shown that the infinitesimal generators $M(t)L(t)M^{-1}(t_o)$ generates the non–homogeneous Markov chain according to the formula

$$w(t) = M(t)\ R_+(t)\ M^{-1}(t_o)\ w(t_o) \quad . \tag{3.12}$$

Indeed, the Markovian matrix $M(t)$ transforms the orthovector e_k into its own vector–column $m_k(t)$, and therefore the inverse matrix $M^{-1}(t_o)$ transforms the vectors $m_k(t_o)$ into e_k. Let

$$Q_+(t) = M(t)\ R_+(t)\ M^{-1}(t_o) \quad .$$

With the help of formulae (3.10) and (3.11), the following result is obtained:

$$Q_+(t) \; w(t_o) \;\; = \;\; m_{i(o)}(t) \;\; +$$

$$\{ \sum_{k=1}^{k=n} (C_k/C_{i(o)}) \; \exp[{}_{t_o}\!\!\int^t a_{k,i(o)}(t) \; dt]\}[m_k(t) - m_{i(o)}(t)]$$

$$= \;\; w(t) \qquad .$$

When information about the past of the innovation diffusion process is needed, the time-direction can be changed. If for the $j(o)$-th column of matrix A the following integrals are divergent

$$\underset{t \to -\infty}{\text{limit}} \; {}_t\!\!\int^{t_o} a_{k,j(o)}(t) \; dt \;\; = \;\; {}_{-\infty}\!\!\int^{t_o} a_{k,j(o)}(t) \; dt \;\; = \;\; \infty \quad ,$$

$$\text{for } k \neq j(o) \quad ,$$

then for $t \to -\infty$ the same type of approximation can be obtained, with the help of the Markovian matrix-function

$$Q_-(t) \;\; = \;\; M(t) \; R_-(t) \; M^{-1}(t_o) \qquad .$$

The restoration of the time-direction gives the generalized exponential growth equation corresponding to the inverse matrix-function $Q_-^{-1}(t) = M(t_o)R_-^{-1}(t)M^{-1}(t)$. Thus, the combination of Q_-^{-1} and $Q_+(t)$ of generalized exponential growth and asymptotical exponential behavior is a substitute for generalized logistic growth (cf., Goodman, 1974).

If the $i(o)$-th column includes the following zero elements:

$$a_{i(o),i(o)}(t) = a_{i(1),i(o)}(t) = \ldots = a_{i(s),i(o)}(t) = 0 \quad ,$$

and for $k \neq i(o), i(1), \ldots, i(s)$,

$$_t\!\!\int_0^\infty a_{k,i(o)}(t) \; dt \;\; = \;\; -\infty \quad ,$$

then the Markov chain approximation will be the same as that given by the formulae (3.8), (3.9) and (3.11). But, the probability vectors $w(t)$ will have the form

$$w(t) \;\; = \;\; p_0 m_{i(o)}(t) + p_1 m_{i(1)}(t) + \ldots + p_s m_{i(s)}(t) +$$

$$\sum_{\substack{k=1 \\ k \neq i(j)}}^{k=n} [C_k/(C_{i(o)} + C_{i(1)} + \ldots + C_{i(s)})][m_k(t) -$$

$$p_0 m_{i(0)}(t) - \ldots - p_s m_{i(s)}(t)] \exp[_{t_0}\int^t a_{k,i(0)}(t) \, dt] \quad ,$$

$$j = 1, 2, \ldots, s,$$

where p_r is defined as before.

In the case of a constant competition matrix **A** and a constant Markovian matrix **M**, the homogeneous continuous Markov chain is obtained, which describes asymptotic behavior of a distribution of adopters when $t \to \pm\infty$. The constant competition antagonistic matrix **A** includes at least one non-positive $i(0)$-th column and at least one non-negative $j(0)$-th column. The non-positive column $a_{i(0)}$ generates the unique infinitesimal constant generator MLM^{-1}, as well as the Markov matrix $\exp[MLM^{-1}] = M[\exp(L)]M^{-1}$. In turn, this generates the homogeneous Markov chain by the formula

$$w(t) = M[\exp(L)]M^{-1} w(0) \quad ,$$

where $Q_+(t) = \{M[\exp(L)]M^{-1}\}^T$ and $t_0 = 0$. Analogous formulae hold for the non-positive column of the matrix **A**.

It is important to emphasize that for the constant matrices **A** and **M** the approximating Markov matrix $M[\exp(L)]M^{-1}$ includes complete information about the antagonistic matrix **A**, since the eigenvalues of the matrix $M[\exp(L)]M^{-1}$ are the values $\exp(a_{k,i(0)})$, $k = 1, 2, \ldots, n$. In this way the $i(0)$-th column of matrix **A** is obtained, which generates every other column of matrix **A**. The matrix $M[\exp(L)]M^{-1}$ includes in its $i(0)$-th column either the $i(0)$-th column of matrix **M** (see Figure 1), which gives the final ceiling distribution of adopters, or the convex combination of a few non-positive columns of matrix **A**. It is possible to use this information for the statistical evaluation of the coefficients of matrices **A** and **M**.

4. CONCLUDING REMARKS AND UNSOLVED PROBLEMS

This paper has presented a model of the temporal innovation diffusion process that generates the main empirical regularity of such a process, namely the S-shaped growth of the cumulative percentage of users of an innovation. The underlying idea of this study has been that of competition between innovations. This idea of competition permits introducing into the analysis the main concepts of the Volterra-Lotka ecological dynamics of competition and predation in animal societies. Briefly speaking, at a micro-level each innovation plays the antagonistic zero-value game against every other innovation, while at the macro-level the competitive exclusion principle (i.e., if some innovation beats all others, then in the long run only the winning innovation will remain in space) is obtained.

The simplest case of antagonistic competition is equivalent to the logit model of the theory of random utility choice. The equivalence between these two constructs raises the question of the establishment of duality between the behavior of alternative competitive innovations and the behavior of the individual who chooses those alternatives. Presumably this duality must serve as the basis for developing the generalizations of random utility choice models.

A discrete model paradox also may be presented. For the discrete model of the innovation diffusion process, the main problem is the choice of an adequate discrete analogue of the continuous model. The discrete model paradox lies in that the discrete analogue of the logistic curve does not fit the logistic difference equation. The correct difference equation, corresponding to the logistic curve, is the Volterra-Lotka type non-linear difference equation, which has an explicit solution. The natural discrete logistic difference equation generates very complicated behavior without an ability to represent it in an explicit manner. A key question that remains unanswered asks which type of difference equation system, the correct one or the natural one, will provide a better description of reality.

5. REFERENCES

Allen, P. and M. Sanglier, 1981, A Dynamic Model of Growth of a Central Place System - II, Geographical Analysis, 13: 149-164.

Brown, L. and K. Cox, 1971, Empirical Regularities in the Diffusion of Innovation, Annals of the Association of American Geographers, 61: 551-559.

Casetti, E., 1969, Why Do Diffusion Processes Conform to Logistic Trends?, Geographical Analysis, 1: 101-105.

Coppell, W., 1965, Stability and Asymptotic Behaviour of Differential Equations, Boston: D. C. Heath.

Dodd, S., 1956, Testing Message Diffusion in Harmonic Logistic Curves, Psychometrica, 21: 191-205.

Gantmacher, F., 1966, Theory of Matrices, 2nd ed., Moscow: Nauka.

Gause, G., 1935, La Théorie Mathématique de la Lutte Pour la Vie, Paris: Hermanne et Cie.

Goodman, M., 1974, Study Notes in System Dynamics, Cambridge, Mass.: Wright Allen Press.

128

Hardin, G., 1960, The Competitive Exclusion Principle, _Science_, 131: 1292-1298.

Henscher, D. and L. Johnson, 1981, _Applied Discrete Choice Modeling_, London: Croom Helm.

Hudson, J., 1972, _Geographical Diffusion Theory_, Evanston, Ill.: Northwestern University Press, Studies in Geography No. 19.

Levins, R., 1968, _Evolution in Changing Environments_, Princeton, New Jersey: Princeton University Press.

Lotka, A., 1932, The Growth of Mixed Populations, Two Species Competing for a Common Food Supply, _Journal of the Washington Academy of Science_, 22: 461-469.

MacArthur, R., 1972, _Geographical Ecology: Patterns in the Distribution of Species_, New York: Harper and Row.

May, R., 1973, _Stability and Complexity in Model Ecosystems_, Princeton, New Jersey: Princeton University Press.

_____, 1976, Simple Mathematical Models with Very Complicated Dynamics, _Nature_, 261: 459-467.

Miller, R., 1967, Pattern and Process in Competition, in _Advances in Ecological Research_, Vol. 4, edited by J. Gragg, New York: Academic Press, pp. 1-74.

Pearl, R., 1925, _The Biology of Population Growth_, New York, Knopf.

Rapoport, A., 1963, Mathematical Models of Social Interaction, in _Handbook of Mathematical Psychology_, Vol. 2, edited by R. Luce, E. Galanter and R. Bush, New York: Wiley, pp. 495-579.

Slobodkin, L., 1961, _Growth and Regulation of Animal Populations_, New York: Holt, Rinehart and Winston.

Sonis, M., 1980, Push-pull Analysis of Migration Streams, _Geographical Analysis_, 12: 80-97.

_____, 1981a, Diffusion of Competitive Innovations, in _Proceedings_ of the 12th Annual Pittsburgh Conference on Modeling and Simulation, Vol. 12, part 13: Socio-Economics and Biomedical, edited by W. Vogt and M. Mickle, Instrument Society of America: University of Pittsburgh, School of Engineering, pp. 1037-1041.

_____, 1981b, The Decomposition Principle Versus Optimization in Regional Analysis--The Inverted Problem of Multiobjective Programming, _Proceedings_, Regional Science Association (Athens Colloque), in press.

Verhulst, P., 1838, Notice sur la loi que la Population Suit Dans son Accroissement, _Correspodence Mathématique et Physique_, 10: 113-121.

Volterra, V., 1926, Variazioni e Fluttuazioni del Numero d'Individui in Specie Animali Conviventi, _Memorie Accademia Nazionale Lincei_, 2 (series 6): 31-113.

von Bertalanffy, L., 1962, General Systems Theory--A Critical Review, _General Systems_, 7: 1-20.

Weber, M. and A. Joseph, 1976, Diffusion of Innovations and the Growth of Cities, _Discussion Paper_ No. 8, Hamilton, Ontario: McMaster University, Department of Geography.

Wilson, A., 1979, Some New Sources of Instability and Oscillation in Dynamic Models of Shopping Centers and Other Urban Structures, _Working Paper_ No. 267, Leeds: University of Leeds, School of Geography.

THE DYNAMICS OF COMMUNICATION

Bruce Ralston

University of Tennessee
United States

1. INTRODUCTION

Communication is studied here from a supply side point of view. That is, emphasis is placed on supplying information to the public, rather than on the public's use and evaluation of information. The models used to study the communication channels, however, are based upon well known empirical regularities from the communication and diffusion of innovation literature. The main body of the paper begins with a description of, and the derivation of models for, differing communication channels. The characteristics of changes in agency's optimal use of these channels are discussed, and accompanied by numerical examples. In the final two sections communication channels are incorporated into two spatial-temporal models. The former, diffusion through an urban system, is implicitly spatial, while the latter, random field models of diffusion, is an explicit spatial-temporal model.

2. COMMUNICATION CHANNELS

There are several perspectives from which to view communication processes. The approach used in this paper views them as occurring through two distinct channels: source, or media, channels, and interaction channels. Each of these modes of communication will be studied separately, and then together.

2.1. Source Communication

The source, or media, communication channels have two basic characteristics. First, they are impersonal; and second, they

produce a one way flow of knowledge. Examples of source channels
are radio, television, film, leaflets, newspapers, and magazines.
These avenues of communication often have the ability to reach
large audiences rapidly, allowing a single change agency to contact
many potential adopters. Through these contacts the agency can
spread information concerning innovation, which possibly will lead
to changes in attitude and adoption.

Studies of source channel effects occurred mostly after 1920.
This interest came about because,

> in the 1920's it was widely held that the newspapers and
> their propaganda 'got us into the war', while in the
> 1930's many saw in the Roosevelt campaign 'proof' that a
> 'golden voice' on the radio could sway men in any
> direction (Katz and Lazarsfeld, 1955).

This belief that source channels had led to the manipulation of
personal opinion gave rise to the early communication theory model
of its effects. In this model the audience consisted of
undifferentiated individuals with sponge-like minds who quickly
took in all media inputs, and these individuals never discussed the
information they had learned (see Blumer, 1946). The policy
implication of this model is that when change agencies use source
channels '... the appeal has to be addressed to the anonymous
individual,' because 'the purchasers are a heterogeneous group
coming from all walks of life', as members of the mass, however,
because of their anonyminity, they are homogeneous or essentially
alike.'.

Fear of the power of source channels to reach and inform is
still discussed today, but the idea of the masses as empty headed,
isolated individuals waiting to be informed fell out of favor in
1955 with the publication of Katz and Lazarsfeld's _Personal
Influence_. Source effects since have been found to be a very
important part of diffusion, but not the only part.

The use of impersonal channels of communication and their
effects has been the subject of several studies. Three of the
major research themes in this area are discussed here, namely media
use in underdeveloped countries, source channels and the spread of
news stories, and measuring the effect of source channels through
leaflet experiments. Studies of media use in underdeveloped
countries have been directed toward using source communication to
educate and affect change, and measuring media use and comparing it
to changes in development. The first approach is policy oriented
and somewhat abstract, while the second is empirical and looks for
statistical measures of source effects.

The policy oriented literature considers source communication channels as tools in the development process. The need is to develop general rules for using these channels, and this literature stresses the responsibilities of the communication industry in not abusing its power. Schramm (1967) maintains that the principal mission of communications research is to determine 'how society can use its mass media to its greatest good.' Schramm (1963) defined six essential functions for source communication to play in national development. One of the most important roles is that 'the media must be the voice of national planning.'

The empirical literature on source communication and underdeveloped societies attempts to measure the correlation between media use and indices of development, such as the adoption of agricultural innovations. In these studies it has been found that exposure to impersonal communication channels is positively correlated with modernization, and leads to early adoption of innovations. Deutschmann (1963) states that media use in Colombia is an important part of the development process because the media carried 'modernizing' information. He also found that farmers with higher media exposure adopted new agricultural innovations sooner than people who had low media exposure. Similarly, Rogers (1965) found '... a positive, significant relationship' between media exposure and two measures of innovativeness. In general, empirical studies of media use in underdeveloped societies have found that persons with high media exposure scored well on the researchers' measures of development and innovativeness, while those persons dependent on interpersonal communication lagged behind. Unfortunately, many of these studies overlooked the fact that exposure to media may be a surrogate for ability to adopt rather than willingness to adopt.

Studies of the diffusion of news events are concerned with how fast news diffuses through various communication channels. The results are conflicting as to whether source or interaction is more important in the diffusion of news events. This is because several of the studies deal with the assassination of President Kennedy. It is fair to say that a news event of this magnitude does not reflect the normal, day-to-day effects of the media. Greenberg (1964) first raised this point, and he has shown that the importance of the message may alter the workings of the communication channels (Greenberg, et. al., 1965).

The temporal characteristics of the diffusion of news events are difficult to assess, since the focus of many reports is on which communication channel was used first rather than when and on which channel the message was heard. Deutschmann and Danielson (1960) did consider the temporal spread of knowledge and found that for major news stories initially there was a rapid rise in the number informed, with this increase soon leveling off, as in an exponential growth curve.

This same behavior has found in leaflet studies. These papers
stemmed from Project Revere, an Air Force sponsored study testing
the effectiveness of airborne leaflets. Several towns were
'bombed' with yellow 5-by-8 cards containing a civil defense
message. The results of these experiments lead to two major
findings. First, the intensity of the message (leaflets per
capita) increased, the number of knowers increased, but at a
decreasing rate. Second, 'every additional repetition (drop)
increased the learning that took place, but with smaller and
smaller increments' (Defluer and Rainboth, 1952). The plotting of
the growth in adoption produced an exponential curve like the one
appearing in Figure 1 (where n is the number of adopters, N is the
total population and t is time).

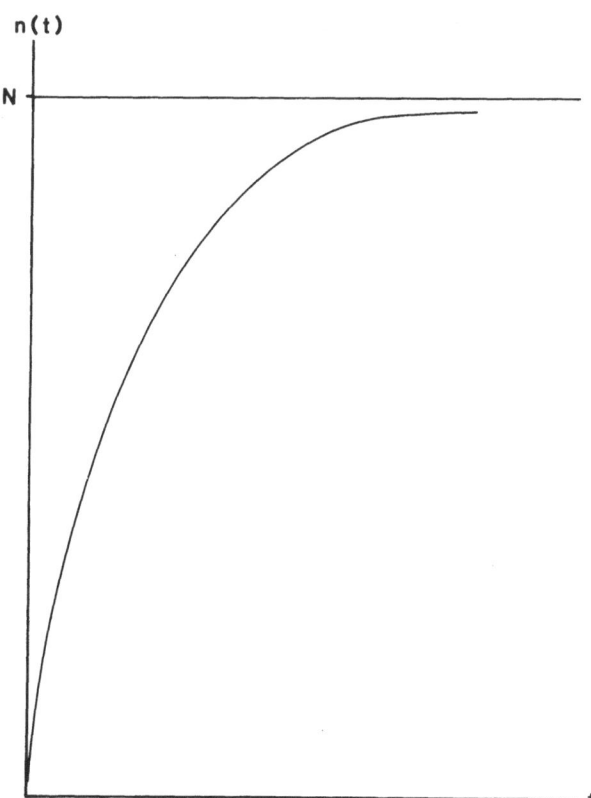

Figure 1. Growth curve for a pure source process.

The modeling of the effects of source channels dates back to Dodd, who in 1955 stated that if a population is '... distributed with somewhat equal opportunity ...' to receive messages '... at a steady rate throughout a period, ...' then a '... convex upwards decaying exponential curve ...' will result. Katz (1961) speculated that an empirically derived exponential growth curve for the number of doctors using a new drug '... might result, for example, from some constant stimulus–say, advertising–operating each month so as to influence a constant proportion of those who have not yet adopted.'

Each of these formulations is based upon a constant source rate reaching the uninformed (nonadopting) population. Mathematically this can be expressed as

$$dn/dt = a[N - n(t)] \qquad , \qquad (2.1)$$

where a is the source rate, $n(\cdot)$ is the number of knowers, and N is the total population. Solving this differential equation yields

$$n(t) = N[1 - n_0 \exp(-a\,t)] \qquad , \qquad (2.2)$$

where n_0 is the initial level of knowledge. Equation (2.1) will be referred to as a pure source diffusion model, and it generates the type of growth curve shown in Figure 1.

2.2. Interpersonal Communication

Interpersonal communication of innovation is a result of interaction between knowers and nonknowers, and it produces a personal, usually face–to–face, two–way flow of knowledge, as opposed to the impersonal, one–way flow associated with source communication. This means of communication does not have the potential for reaching large audiences rapidly, as in the source case, but it does have the advantage of reaching individuals, rather than amorphous masses. Interpersonal communication initially produces a slower rise in the spread of knowledge than does the source channels, but parts of its growth curve can be extremely steep.

The early studies of diffusion as interpersonal interaction were anthropologically oriented. The interaction of cultures and the spread of traits were studied by several scholars. The hypothesis that cultural traits diffused from society to society as a result of interaction was by no means universally accepted. Competing hypotheses included independent invention and convergence in development (see Wissler, 1923). The diffusionists' model was that an innovation would be invented, and then spread like a wave outward from the inventing culture. 'Successive elaborations thus arise in the nuclear area and spread, as it were, like successive

lava flows, of which the latest covers the smallest area ...'
(Dixon, 1928). Given this theory it is possible to map the various
changes in certain cultural traits and '... determine where the
trait arose, and unravel something of its history.' The importance
of space in this model stems from its ability to indicate the
center of the diffusion process and underpin the determination of
the amount of time that has passed since the innovation was
introduced.

The mathematical formalization of this model has been studied
by several researchers. As Dodd (1955) found in 'Operation
Coffee,' a spinoff of Project Revere, '... the diffusion waned with
distance. The further the message travelled, the fewer the people
who knew it.' He went on to posit the following mathematical
model:

Briefly, suppose a particle has a constant amount of
energy (or social force of any sort) which is equally
likely to diffuse in any direction. Draw concentric
zones of equal width around the particle ... The energy
per unit space at distance L will then be inversely
proportional to L ... For the simplest social
diffusion, one assumes that the potency is unity and
that people are uniformly distributed in an area ...
Then the proportion of people influenced in each zone
will necessarily be inverse to that zone's origin point.

More recent studies of interaction diffusion are either
geographically or temporally oriented. One body of literature is
concerned with interaction networks and potentials over space,
while the other is concerned with the growth in adoption over time.
Not surprisingly, the former body of literature derives from work
by geographers, while the latter is an amalgam of works by other
social scientists. Potential models will be dealt with in a later
section. The focus now will turn to temporally oriented studies.

Temporal analysis of interaction diffusion is concerned with
the flow of information over time and the subsequent growth in
adoption. In this communication setting, the only way a nonadopter
can learn of the innovation is by contact with an adopter. As Dodd
(1955) states:

We define the diffusion variables as involving solely
three dimensions of actors acting in time. In the case
of diffusing messages, the actors are persons, the acts
are tellings and hearings whose end product is the
attribute 'knowing the message', and the time is the
period of diffusing.

The interaction hypothesis arose because of the empirical

regularity of the S-shaped, or logistic, growth in adoption. Pemberton (1936) was the first to postulate a mathematical model to explain this phenomena. After seeing Chapin's (1928) success in using the S curve to plot the growth in the number of municipalities adopting the city manager and city commission forms of givernment, Pemberton found the same curve to hold for three other traits, namely the number of countries using postage stamps, the number of states placing limits on municipal taxation, and the number of states with compulsory education. He was motivated to seek a mathematical model to explain this regularity. Pemberton's model, however, had very little to do with 'cultural interaction.' He believed that the cultural forces working for early adoption tended to balance those working for late adoption, so that '... the most probable time of trait acceptance [would be] in the middle or average time and the probabilities of frequencies in other time periods [would] follow the normal distribution.'

Again, it was Dodd (1955) who first postulated that the logistic growth curve was the result of the interaction of adopters with nonadopters:

The simplest logistic curve is generated albegraically as the cumulation over t periods of the joint probability of knowers meeting nonknowers in each period. It is socially generated by mixing knowers and nonknowers fully and freely.

The analogy of this description with the description for the spread of contagious diseases is obvious. In fact, interaction diffusion often is called contagious diffusion, and diffusionists have turned to epidemiologists for additional mathematical models.

Mathematically the pure interaction model can be expressed as

$$dn/dt = b\, n(t)[N - n(t)] \quad , \quad (2.3)$$

where N and $n(\cdot)$ are defined as in equation (2.1) and b is the rate of random mixing. The solution to this differential equation is:

$$n(t) = N/\{1 - [n - n_0][\exp(-b\, t)]/n_0\} \quad , \quad (2.4)$$

where n_0 is the initial number of knowers. This is a logistic curve and is pictured in Figure 2.

2.3. Source and Interaction Communication Together

The communication of ideas and technologies is most likely a process composed of both source and interaction communication. People read and hear about innovations from various source channels and talk to others about these inputs. Often professional

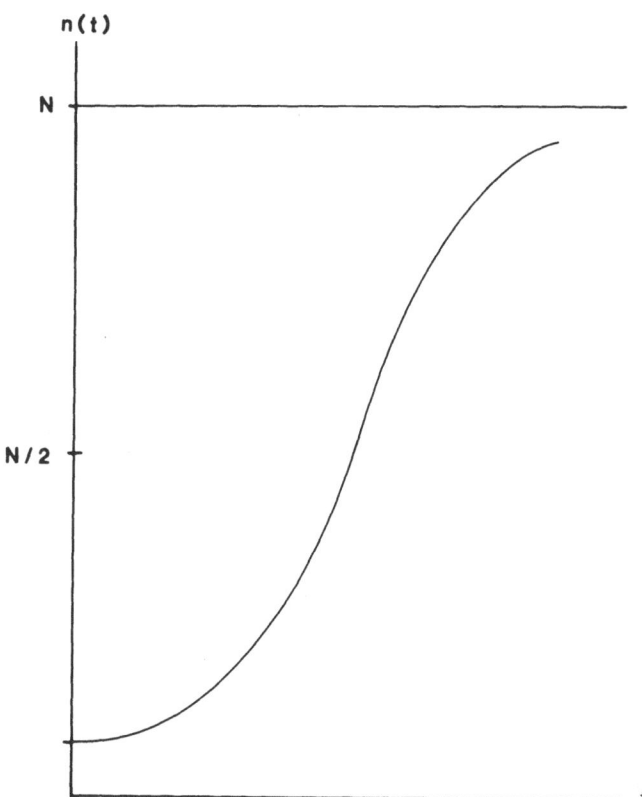

Figure 2.   Growth curve for a pure interaction process.

interactors or change agents seek out people to tell them of new innovations, making the overall process of communication of innovation a result of both source and interaction communication. The use of both channels creates a communication process that is both personal and impersonal, with one-way and two-way flows of knowledge that have the potential for reaching large audiences rapidly, while still having personal communication.

The model of the media being able to shape the minds of the atomistic, noninteracting masses was laid to rest with the publication of Katz and Lazarsfeld's Personal Influence (1955). In this text the role of interaction was rediscovered to be an important part of the communication process. In their study of voting behavior in Decatur, Illinois, the authors found that the adoption (voting) decision was a result of two steps, one based upon source communication and the other upon interpersonal interaction. First of all, source communication influences the opinion leaders in a society, who come from all walks of life and social strata. Then, the public discusses these inputs, and in this way the majority of people form their opinions (i.e., decide whether or not to adopt). The temporal role of the two communication channnels is crucial. The source channels are important first, followed by the interaction channels.

This temporal arrangement of the importance of source and interaction channels was found some years earlier in Ryan and Gross' (1943) classic paper, 'The Diffusion of Hybrid Seed Corn in Two Iowa Communities.' Amongst the findings of this study was that the typical farmer heard of the hybrid seed from impersonal sources, but adopted only after consulting his neighbors. However, the typical farmer was not the only type found. Early adopters credited source channels with their decision to adopt, while late adopters credited their neighbors with helping them to decide. The role of the two communication channels is summed up by the authors:

'The spread of knowledge and the spread of 'conviction'
are, analytically at least, distinct processes, and in
this case have appeared to operate in part through
different although complementary channels.'

These two works have spawned a vast number of studies which have, by and large, supported their findings. The two-step flow hypothesis of Katz and Lazarsfeld has undergone some modifications, and now a multi-step flow model is in vogue. This rather nebulous model reflects the findings that people are influenced by different communication channels at different times, and there are no distinct steps. However, the communication process is at all times made up of source and interaction flows of differing intensity, with either one coming first (see Rogers and Shoemaker, 1971).

The Ryan and Gross study also has had a strong impact on the study of source and interaction effects. Their paper was the first to suggest that there are stages to the adoption process, and that each communication channel played certain roles at each stage. These findings have been replicated in several studies, and the notion of stages in the adoption process has become accepted dogma, as has the classification of adopters. In one of the finest studies on the roles of communication channels, Coleman, Katz and Menzel (1966) find that the growth in the number of doctors using the drug gammanym is a result of both a source and an interaction process. Earlier Katz (1961) had noted that the growth of adoption curve for integrated doctors was a result of interaction and resembles the S-shaped curve found for hybrid corn, whereas the curve for the isolated doctors is a result of a pure source diffusion process. These findings indicate that the growth in adoption can be separated, theoretically at least, into source and interaction processes, resulting in a right-hand skewed S-shaped adoption curve (i.e., the point of inflection is shifted to the left). Mathematically this entails combining equations (2.1) and (2.3), resulting in the following differential equation capturing both source and interaction diffusion:

$$dn/dt = a[N - n(t)] + b\,n(t)[N - n(t)] \quad . \quad (2.5)$$

The solution of equation (2.5) is

$$n(t) =$$

$$\{N - \exp[-(a+b)t]Na[N-n_0]/[bn_0+aN]\}\{\exp[-(a+b)t]b[N-n_0]/[bn_0+aN]\}^{-1} \quad (2.6)$$

Defining the relative strength of source and interaction diffusion as

$$r = a/(a + b) \quad ,$$

the effect of varying the relative sizes of a and b can be found. The growth in adoption for a selection of values of r, ranging from 0 to 1, is illustrated in Figure 3. One should note in this figure that as r increases, source rates become more important, and the inflection point shifts to the left.

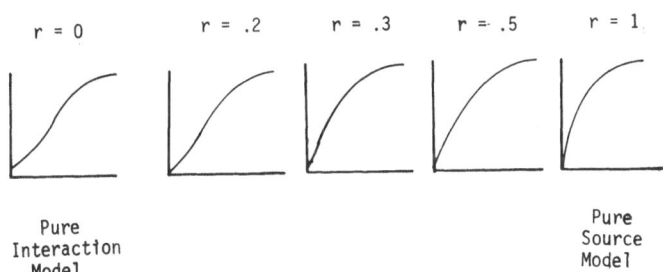

Figure 3. Growth curves for varying strengths of source and interaction processes.

(From Lekvall and Wahlbin, A study of some assumptions underlying innovation diffusion functions.)

3. CHARACTERISTICS OF OPTIMAL COMMUNICATION POLICIES

Each of these source, interaction, and combined source and interaction models can be viewed as a dynamic production density function that leads to an aggregate production function (Solow, 1957). The types of communication one chooses to stress, namely media, interpersonal or both, will determine the mix of communication resources. How does one choose this mix? In practice there is media advertising. One also can imagine a situation for affecting, if not controlling, interaction. For example, many book clubs give members free books if these members get others to join the club. By manipulating the flow of information through these channels, change agencies can direct the spatial and temporal spread of innovation. The optimal use of these channels within a single population has been studied by several authors. The two goals considered thus far are profit maximization, and reaching a given level of adoption at a minimal cost. This former goal has been studied in an infinite planning horizon framework, for instance, by Gould (1970) and Sethi (1974), whereas this latter goal has been studied for a fixed planning period, for example, by Ralston (1977) and Sethi (1974).

In any given city an increase in adoption can result from nonadopters interacting with adopters in either that city or another city. A change agency can influence the rate of adoption by controlling the amount of source inputs, $u(t)$, and by influencing the amount of interaction with some interaction control, $v(t)$. It may not always seem reasonable to assume that a change agency can fully control the rate at which people in a society interact. Most likely, people interact at some intrinsic rate β that a change agency can influence, but not completely control. With this assumption, the effect of interaction communication is given by

$$dn/dt = \beta[1 + v(t)] n(t) [N - n(t)] \qquad ,$$

where $v(t)$ is the rate of interaction influence. The term $v(t)$ may be scaled to reflect the ability to influence but not control interaction, giving

$$dn/dt = \beta[1 + v(t)/M] n(t) [N - n(t)] \qquad .$$

Controlling the rate of source communication and influencing the rate of interaction cost money, and there is probably some upper limit, M, at which a change agency can spend during any given time period. Accordingly, the most general form of an objective function for a change agency is:

$$\text{MIN:} \quad {}_{t_0}\!\int^{T} f(n,u,v,t) \, dt \qquad ,$$

S.T.: $n(t_0) = n_0$,   (3.1)

$u + v \leq M$,   (3.2)

$u \geq 0$,   (3.3)

$v \geq 0$, and   (3.4)

$dn/dt = u(t)[N-n(t)] + \beta[1 + v(t)/M]n(t)[N-n(t)]$,(3.5)

where the following assumptions are made:

(1) f is continuous and twice differentiable,

(2) dn/dt is continuous and twice differentiable, and

(3) u and v are piecewise continuous.

The first-order conditions for this problem are

$- d\lambda/dt = f_n - \lambda u + \lambda\beta[(1+v)/M](N-2n)$, (3.6)

$0 = f_u + \lambda(N-n) - \mu_1 + \mu_2$, and   (3.7)

$0 = f_v + \lambda\beta n(N-n)/M - \mu_2 + \mu_3$,   (3.8)

where λ is the standard Euler-Lagrange multiplier,

μ_1 is a constraint multiplier that is nonzero only if u=0,

μ_2 is a constraint multiplier that is nonzero only if v=0, and

μ_3 is a constraint multiplier that is nonzero only if u+v=M.

The constraint multipliers are always non-negative.

It is convenient to break the study of the first-order conditions into two cases. The first case is where none of the constraints are binding, and the second case is where some of the constraints are binding. For the first case, the change agency operates in both communication channels such that

$$f_u/(N-n) = M f_v/[\beta n(N-n)] \quad .$$ (3.9)

Equation (3.9) states that, in both the cost minimization and profit maximization cases, the ratio of dynamic marginal costs (DMC) to dynamic marginal productivities (DMP) must be the same for both communication channels. These DMCs and DMPs are the partial derivatives of dn/dt with respect to the control used (i.e., u or v). Moreover, the agency's rate of expenditures on the two communication channels is based upon the return per dollar spent. In the early stages of information spread, $n(t)$ will tend to be small, and since the DMP of source communication will be much greater than that for interaction communication, the agency will stress, in a cost sense, source communiction. As time passes the DMP for source communication decreases, while interactions will increase at first, and then decrease. Early on, then, source communication should dominate an agency's communication policy, with interaction becoming more important, in a cost sense, as time passes. These findings complement those found in the empirical literature by researchers such as Rogers and Shoemaker (1971).

The second case is when some constraints are binding, and thus the corresponding multipliers may be positive. The possible configurations of binding constraints may be summarized in tabular form as follows:

Possible Constraint Cases
1 = binding   and   0 = nonbinding

	Configuration					
Constraint	1	2	3	4	5	6
$u = 0$	1	0	0	1	0	1
$v = 0$	0	1	0	0	1	1
$u + v = M$	0	0	1	1	1	0

It is asserted that configurations 1 and 2 are impossible. Configuration 3 follows the same rule given in equation (3.9). Configurations 4 and 5 imply the ratio of DMP to DMC is higher for the channel used. And configuration 6 will not occur at the initial time, if f is a profit or cost function that treats knowers as positive goods and costs as negative goods. The proofs of these assertions are straightforward when done by contradiction.

A study of the actual behavior of u and v and a test for sufficiency require an articulation of the exact form of $f(\cdot,\cdot,\cdot,\cdot)$. Two objectives will be considered. The first is the achievement of a given level of adoption at a minimal cost within a fixed planning period, and is studied for both a pure source communication process and a source and interaction communication process. The second is the minimization of a weighted sum of costs and deviations from a desired growth path, and is studied only with the latter dynamics. In all cases, cost functions are assumed to be quadratic in form.

3.1. Objective 1: Reaching a Desired Level of Adoption at a Minimal Cost

The first case to be treated here is that of a pure source diffusion process. If the only influence on the number of adopters is the source rate chosen by the change agency, then

$$dn/dt = u(t) [N - n(t)] \quad ,$$

$$n(0) = n_0 \quad , \quad n(T) = n_T \quad , \quad u \geq 0 \quad ,$$

where $u(t)$ is the source rate chosen, $n(T)$ is the number of adopters at time t, and n_0 and n_t are the given initial and final adoption levels. For the agency to achieve the target of n_T at a minimal cost, it must minimize

$$\int_0^T A\, u(t)^2 \, dt \quad .$$

Since only one diffusion control is involved, it can be assumed without loss of generality that A=1.

The Hamiltonian for this problem is

$$H[u(t),n(t),\lambda(t),t] = u^2(t) + \lambda(t)\, u(t)\, [N-n(t)] \quad . \quad (3.10)$$

This equation results in the following Euler-Lagrange equations:

$$(\partial H/\partial n) = -\dot{\lambda} = -\lambda u \quad , \text{ and}$$

$$(\partial H/\partial u) = 0 = 2u + \lambda (N-n) + \mu(-1) \quad .$$

When the source rate is positive

$$u = -\lambda (N-n)/2 \quad , \text{ and}$$

$$-\lambda = 2u/(N-n) \quad , \quad (3.11)$$

which is the ratio of dynamic marginal costs to dynamic marginal producivity of u.

Since the Hamiltonian is not an explicit function of time, it may be assumed to be equal to a constant, say c_0. The source rate must be positive for any diffusion to take place. Therefore, combining equations (3.10) and (3.11) yields

$$-\lambda^2 (N-n)/4 = c_0 \quad ,$$

which implies that

$$u(t) = (-c_0)^{1/2} \quad .$$

The goal of the agency is to reach n_T at time T. Thus, the optimal source rate is given by

$$u* = \{ \ln[(N-n_0)/N-n_T)]\}/T \qquad . \qquad (3.12)$$

To prove that $u*$ in equation (3.12) is sufficient, the following theorem is needed (Leitmann, 1968). Consider the following problem

$$\text{MIN:} \quad {}_{t_0}\!\int^{t_1} L[x(t),u(t)] \, dt \qquad\qquad (3.13)$$

$$\text{S.T.:} \quad dx/dt = f[x(t),u(t)] \qquad , \text{ and}$$

$$x(t_0) = x_0 \qquad ,$$

where $x(t_1)\varepsilon\Omega_1$, the set of all admissible final states, and (t_1-t_0) is not specified. The function $u(t)$ is an element of a bounded set U, and is assumed to be piecewise continuous. Let X be an open set in E^n. The control $u*(t)$ with corresponding trajectory $x*(t)$, $t_0 \leq t \leq t_1^*$, is optimal with respect to all admissible controls $u(t)\varepsilon U$, with corresponding solutions $x(t)\varepsilon X$, for all $t\varepsilon(t_0,t_1)$, if there exists a scalar function $V(x)$ with continuous first derivatives on X such that

(1) $\lim\limits_{x \to x^1} V(x) = 0$, for all $x^1 \varepsilon\Omega_1$ and $x\varepsilon n$ contained in X,

where n is a solution curve of equation (3.13) generated by an admissible control,

(2) $L[x*(t),u*(t)] + \text{grad}[V(x*) \, f(x*,u*)] = 0$ for all $[t_0,t_1^*)$, and

(3) $L(x,u) + \text{grad}[V(x) \, n(x,u)] \geq 0$ for all $u\varepsilon U$ and for all $x\varepsilon X$.

To apply this theorem in the present case, the following definitions are necessary. Let

$$n_1 = n(t) < N \qquad , \text{ and}$$

$$dn/dt = u(t) [N - n(t)] \qquad .$$

Since the final time period T is specified, a second state variable must be defined. Let $n_2=t$, $n_2(0)=0$, and $n(t_1)=T$. The scalar function $V(n_1,n_2)$ is defined as

$$(T - n_2)^{-1} \ln^2[(N-n_1)/(n-n_T)] \qquad ,$$

$$\Omega_1 = \{(n_1,n_2) \, | n_T - n_1(t_1) = 0 \quad \text{and} \quad n_2 - T = 0\} \qquad .$$

To find the limit of $V(n_1,n_2)$ it is necessary to use l'Hopital's rule for several variables. Suppose

$$f(n_1,n_2) = \ln^2[(N-n_1)/(N-n_T)] \quad , \text{ and}$$

$$g(n_1,n_2) = T - n_2 \quad .$$

Then

$$f_{n_1} = (-2/-n) \ln[(N-n_1)/(N-n_T)] \quad ,$$

$$f_{n_2} = 0 \quad , \qquad g_{n_1} = 0 \quad , \text{ and} \qquad g_{n_2} = -1 \quad .$$

Therefore,

$$\lim_{n \to n^1 \varepsilon \Omega_1} V(n_1,n_2) = 0/1 = 0 \quad ,$$

satisfying the first condition of the theorem. Further, defining V^* as $V(n_1^*, n_2^*)$ yields

$$(\partial V^*/\partial n_2) = (T-n_2)^{-2} \ln^2[(N-n_1)/(N-n_T)] \quad , \text{ and} \quad (3.14)$$

$$(\partial V^*/\partial n_1) = \{-2/(T-n_2) \ln[(N-n_1)/(N-n_T)]\}/(N-n_1) \quad . (3.15)$$

Substituting equations (3.14) and (3.15) into the equation of parts (2) and (3) of the theorem yields

$$u^2 - [2n/(T-n_2)] \ln[(N-n_1)/(N-n_T)] +$$

$$[1/(T-n_2)^2] \ln^2[(N-n_1)/(N-n_T)] =$$

$$\{u - (T-n_2)^{-1} \ln[(N-n_1)/(N-n_T)]\}^2 \quad , \qquad (3.16)$$

which is non-negative for all $u \neq u^*$. For $u=u^*$ and $n_1=n^*$ equation (3.16) reduces to

$$\{u^* - [1/(T-t)] \ln[(N-n_0)/(N-n_T)]^{(T-t)/T}\}^2 = 0 \quad .$$

Hence, u^* is both necessary and sufficient for minimizing costs.

The second case to be treated here is that for source and interaction control. When a change agency can choose between communication channels, its optimization problem becomes

$$\text{MIN:} \quad {_0\int^T} (A u^2 + B v^2) \, dt$$

$$\text{S.T.:} \quad dn/dt = u(N-n) + \beta(1 + v/M) n(N-n) \quad ,$$

$$u, v \geq 0 \quad ,$$

$$u + v \leq M \qquad , \text{ and}$$

$$n(T) = n_T \qquad .$$

The following properties, which are derived from first-order conditions, namely equatins (3.6) thru (3.8), can be shown to hold for this problem:

(1) if both u and v are positive, meaning both types of communication are used, then they are chosen so that

$$n = (M/\beta) \, DMC_v/DMC_u = (M/\beta)(2Bv/2Au) \qquad , \qquad (3.17)$$

(2) if both u and v are positive, then the source rate is decreasing in time so that

$$du/dt = -(A\beta/BM) \, u^2 \, n(N-n) < 0 \qquad ,$$

(3) if both u and v are positive, then the interaction rate is increasing in time so that

$$dn/dt = (Bv^2)/(An^2) \, (N-n) > 0 \qquad ,$$

(4) if both u and v are positive, then the difference in the percentage change in communication controls equals the percentage change in adoption so that

$$\dot{n}/n = \dot{v}/v - \dot{u}/u \qquad , \text{ and}$$

(5) if one control equals zero and the other is positive, then the positive control must be at its maximum level (i.e., if v=0 and u≠0, then u=M, whereas if u=0 and v≠0, then v=M).

Some exemplary optimal control solutions to this problem are shown in Figures 4 and 5. One should note in these figures that when source rates are costly, an agency relies more on interaction. In each case the source rate monotonically decreases, while the interaction rate monotonically increases.

3.2. Objective 2: Minimizing Costs and a Weighted Sum of Squared Deviations From a Desired Growth Path

In this instance, the objective function becomes

$$\text{MIN:} \quad {}_0\!\int^T [A(n-\hat{n})^2 + Bu^2 + Cv^2] \, dt \qquad ,$$

where $\hat{n}$ is the desired growth path. The constraints to this problem are given by equations (3.1) thru (3.5). Many of the properties found in the previous example no longer hold in this

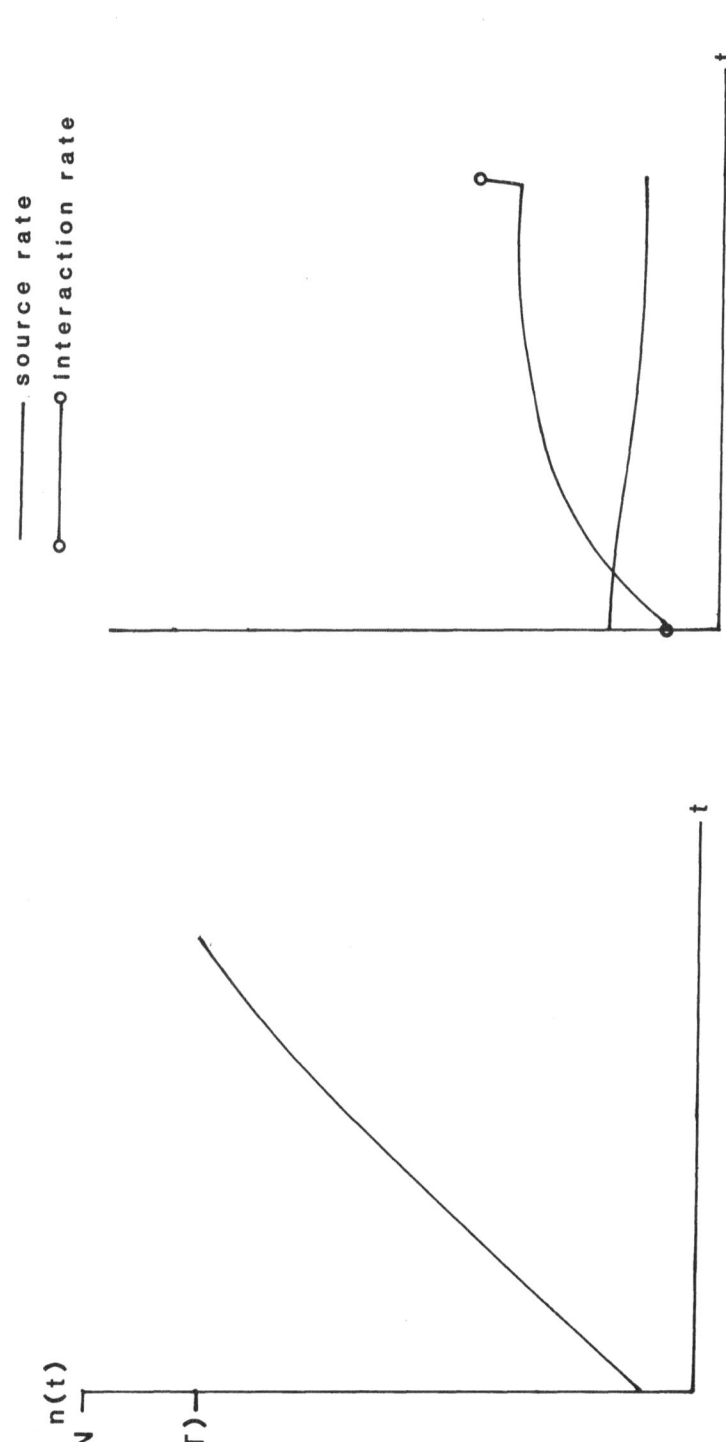

Figure 4. Exemplary optimal control solution.

148

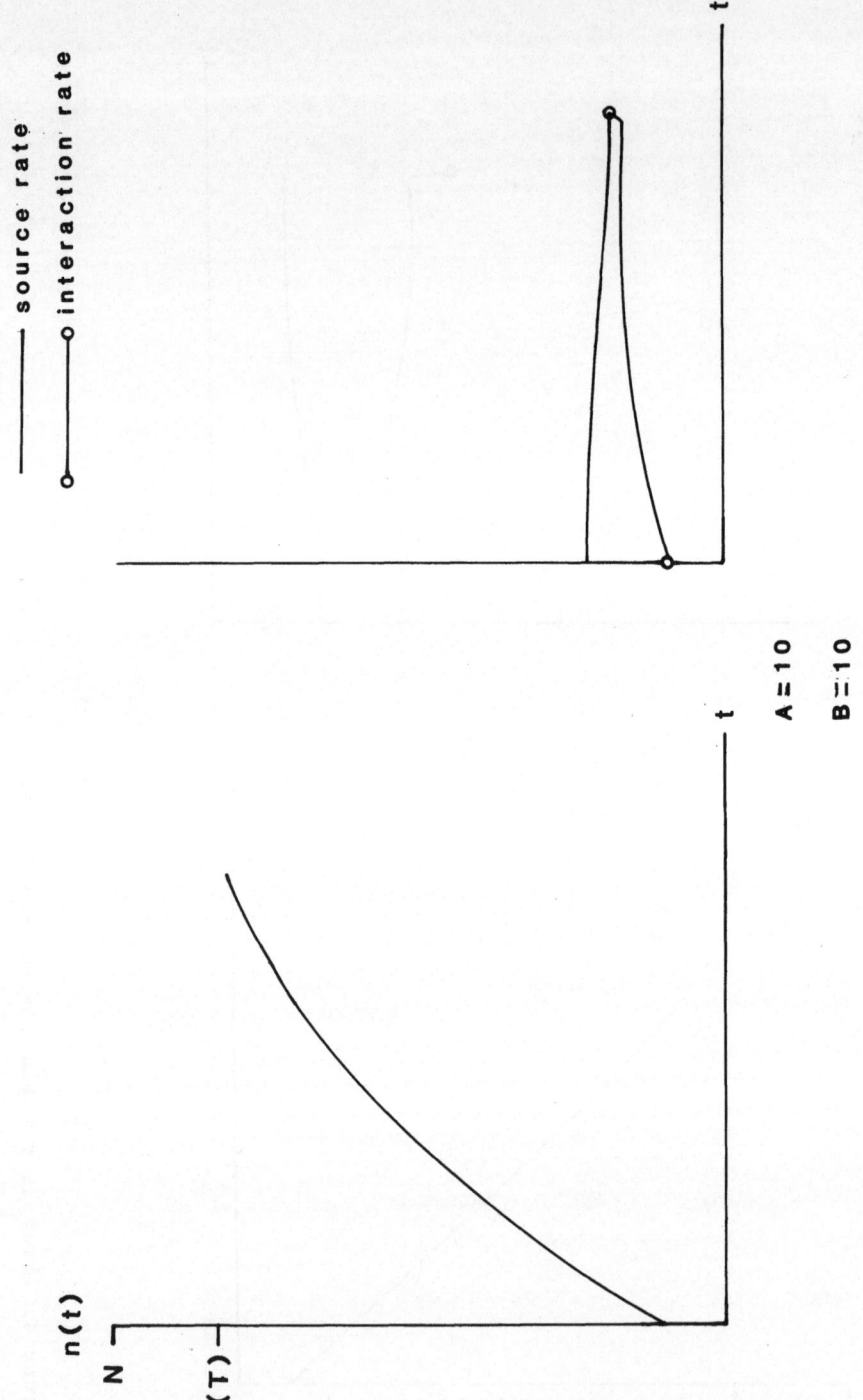

A=10

B=10

Figure 5. Exemplary optimal control solution.

source rate

interaction rate

case. In fact, only the fifth property remains valid. The monotonicity of the controls no longer holds because their values depend upon whether one is above or below the desired growth path. This property is captured by:

$$du/dt = (A/B)(N-n)(n-\hat{n}) - v u (N-n) \quad , \text{ and}$$

$$dv/dt = (A/C)(N-n)(n-\hat{n}) + (v^2/n^2)(N-n) \quad .$$

The behavior of the optimal control solutions for specific illustrative cases is shown in Figures 6 and 7. In Figure 6 both diffusion controls eventually increase rapidly, in order to induce a growth curve that follows the desired path. Not surprisingly, the source control reaches its peak before the interaction rate does, since the former's DMP more quickly achieves a higher value. One should note that in Figure 7 the constraints u=M and v=0 are binding up to t_1, and u+v=M is binding from t_1 to t_2.

4. URBAN SYSTEMS DIFFUSION

Extension of the previous results to urban systems diffusion will be undertaken in this section. Studies of diffusion through an urban system primarily are concerned with the effects of city size and distance on the diffusion process. These studies attempt to measure whether the diffusion process is directed down through the urban hierarchy (i.e., hierarchical diffusion), or whether it is based upon spatial separation (i.e., neighborhood diffusion). The realization of the importance of city size and distance on the diffusion process dates back to Bowers (1937), who studied the diffusion of ham radio sets in the United States. He found both regional trends and a city size effect. His findings were supported by Crain (1966) in his study of the diffusion of flouridation. Both of these studies found a prominent hierarchical bias to diffusion. In his study of the diffusion of Rotary Clubs, Hagerstrand (1966) found an initial hierarchical effect that eventually was replaced by neighborhood diffusion. These results are supported by Pyle (1969), who found that as the transportation system improved, the hierarchical effects became stronger.

This association between development and hierarchical diffusion has led several authors to the study of diffusion and urban development. Berry (1971) has stated that

... the role played by growth centers in regional development is a particular case of the general process of innovation diffusion, and therefore that the sadly deficient 'theory' of growth centers can be enriched by turning to the better developed general case.

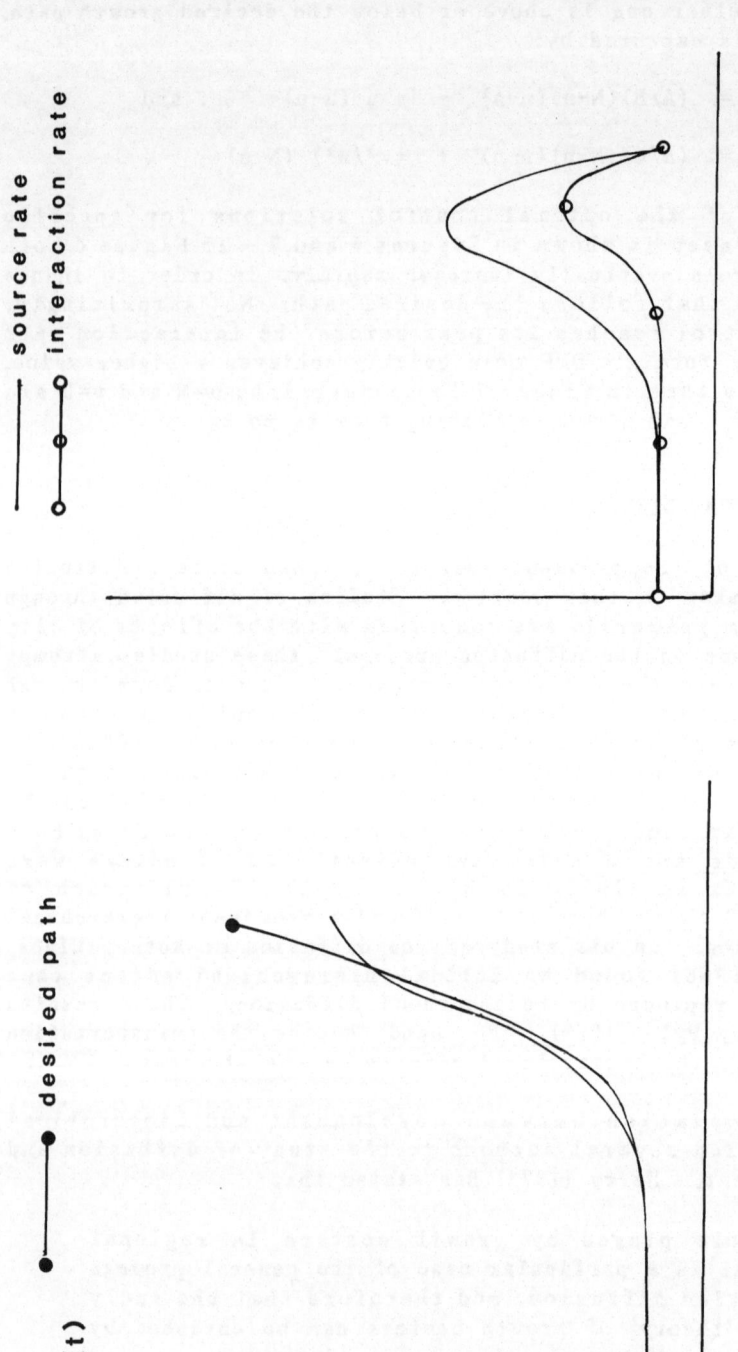

n(t)

• • desired path

source rate
• • interaction rate

A = 100

B = 10

C = 10

Figure 6. Behavior of the illustrative optimal control solutions.

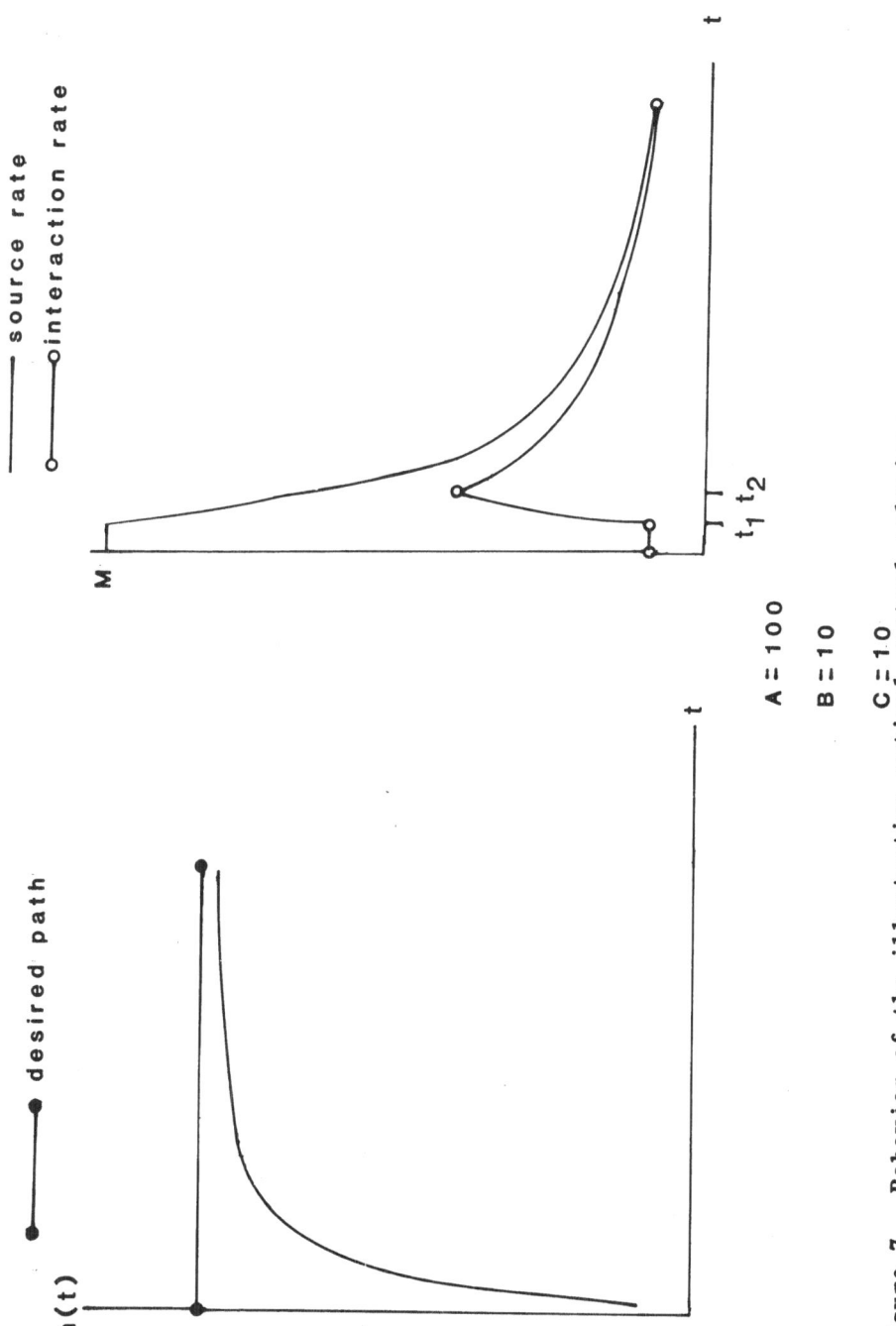

Figure 7. Behavior of the illustrative optimal control solutions.

The theory to which Berry alludes is the theory of diffusion through an urban system. A rigorous model of such processes has been developed by Hudson (1969, 1972). This model is based upon a central place distribution of cities:

> in a hierarchical system of influence flows it is assumed that the adopters of a change successively influence those at lower levels in the hierarchy. Although there can be other communication flows in the system, these do not lead to adoption when there is an ordering of influence according to hierarchy

Hudson's strict dominance assumption limits his model to a purely descriptive level. Others have taken this ability to get a good fit from a descriptive model as a cue for using central place diffusion as a basis for determining developmental policies. Models based on pure central place dominance tell much about diffusion in a central place system, but often little about anything else (see Pred, 1973). There is a clear need to develop a diffusion model whose goal is not just a good fit, but one designed to reflect the communication process in an urban system. In other words, it is necessary to include both channels of communication in the urban system framework, and to see how the communication process within and between cities of all ranks induces the empirically derived diffusion trends.

To adapt the channel models to an urban system, a system of communicatin dynamics must be specified. An urban system can be depicted as a group of multiple populations, such that each city i is treated as a population, denoted n_i. These populations, of course, may interact. A standard approach to adapting single population dynamics to multiple populations is to index each popualtion. Thus, a system of differential equations results, each with appropriate subscripts and locationally based parameters (Bailey, 1968; Ralston, 1977). A drawback of this approach is that it often leads to complex systems of equations, even for small numbers of locations. Despite this drawback, this approach can offer insights, and shall be pursued here.

Consider a region consisting of m cities whose populations are denoted by N_1, N_2, ..., N_m, and ordered so that

$$N_1 > N_2 > \ldots > N_m \quad .$$

At time t the number of adopters in each city is given by $n_1(t)$, $n_2(t)$, ..., $n_m(t)$, and the number of nonadopters by $\bar{N}_1(t)$, $\bar{N}_2(t)$, tries to control the path of diffusion both by controlling the source rate, and by using change agents to influence the interaction rate. Since the diffusion process is to take place within and between several cities, a distinction must be made between within-city interaction and between-city interaction. β_{ii} refers to within-city interction, and β_{ij} refers to between-city interaction. It is a well-documented empirical regularity that interaction is a decreasing function of distance. Therefore, an ordering of interaction level is based upon distance, so that

$$\beta_{ii} > \beta_{ij} \qquad \text{for all i and j } (i \neq j) \qquad .$$

If the j-th city is closer to the i-th city than is the k-th city, then $\beta_{ij} > \beta_{ik}$. The total interaction effect is an implicit function of distance, as well as an explicit function of the adopting and nonadopting populations. For example, it could be assumed that

$$\beta_{ij} = \beta/d_{ij}^2 \qquad ,$$

where d_{ij} is the distance from city i to city j, and β is a scaling parameter. Then the total interaction between adopters in city j and nonadopters in city i is given by

$$(\beta/d_{ij}^2) \ \bar{N}_i(t) \ n_j(t) \qquad ,$$

the standard gravity model. Consequently, the diffusion dynamics for the urban system may be characterized by the following set of simultaneous equations:

$$n_i(t) = u_i(t) \ \bar{N}_i(t) + \sum_{j=1}^{j=m} \beta_{ij} \ [1 + v_{ij}(t)/M] \ \bar{N}_i(t) \ n_j(t) , \qquad (4.1)$$
$$i = 1, 2, \ldots, m \qquad .$$

The choice of the source and interaction rates depends upon the agency's goals, resources, and costs. For every unit of source control the agency incurs some cost, and the same is true for controlling interaction. The goal considered here is that of reaching a given level of adoption, n_T, at a minimal cost within a fixed planning period [0,T]. It is assumed that the agency's resources are limited, constraining the amount of information control the agency can exert at any given time. Symbolically, then, the agency's task is to

$$\text{MIN:} \quad {}_0\!\int^T W(\mathbf{U}) \ dt \qquad ,$$

$$\text{S.T.:} \quad \mathbf{U} \ 1 \ \leq \ M \qquad \text{and}$$

$$\text{equations (4.1)} \qquad ,$$

where $\mathbf{U}$ is a vector of communication controls such that

$$\mathbf{U}^T(t) = [u_1(t), \ u_2(t), \ \ldots, \ u_m, \ v_{11}(t),$$

$$v_{12}(t), \ldots, v_{1m}(t), \ldots, v_{mm}(t)] \quad ,$$

and W(U) is the cost function that will be assumed to be convex.

4.1. Case 1: Source Control Only

For a two-city system the agency will choose a strategy to meet its adoption goal, and hence will face the following optimization problem:

$$\text{MIN:} \quad _0\!\!\int^T W(u_1, u_2)\, dt \quad ,$$

$$\text{S.T.:} \quad n_1(0) = n_{10} \quad \text{and} \quad n_2(0) = n_{20} \quad ,$$

$$n(T) = n_T = n_1(T) + n_2(t) \quad ,$$

$$\dot{n}_1 = u_1 \bar{N}_1 + \beta_{11}\bar{N}_1 n_1 + \beta_{12}\bar{N}_1 n_2 \quad ,$$

$$\dot{n}_2 = u_2\bar{N}_2 + \beta_{22}\bar{N}_2 n_2 + \beta_{21}\bar{N}_2 n_1 \quad ,$$

$$u_1(t) \geq 0,\ u_2(t) \geq 0,\ \text{and}$$

$$u_1(t) + u_2(t) \leq M.$$

The corresponding first-order conditions are

$$- \dot{\lambda}_1 = - \lambda_1 u_1 + \lambda_1\beta_{11}(N_1 - 2n_1) + \lambda_1\beta_{12}(-n_2) + \lambda_2\beta_{21}\bar{N}_2 \quad ,$$

$$- \dot{\lambda}_2 = - \lambda_2 u_2 + \lambda_2\beta_{22}(N_2 - 2n_2) + \lambda_2\beta_{21}(-n_1) + \lambda_1\beta_{12}\bar{N}_1 \quad ,$$

$$0 = DMC(u_1) + \lambda_1\, DMP(u_1) - \mu_1 + \mu_3 \quad ,\ \text{and}$$

$$\lambda_1(T) = \lambda_2(T) \quad .$$

If the costs for controlling the source rate are dependent only on the level of source inputs, the agency should concentrate its efforts in cities where it receives the highest return per dollar spent. More succinctly,

$$u_1 > u_2 \quad \text{implies} \quad \bar{N}_1 > \bar{N}_2 \quad .$$

Assuming that there are initially few adopters, then the largest city, or cities, would receive the greatest amount of source inputs, and experience a faster growth in adoption than in smaller cities. If the level of adoption rises to the point at which the largest cities have the same number of nonadopters as the smaller cities, the focus of source rate control would shift to the smaller cities. This shift is attributable to the expectation that large amounts of within-city interaction would take place in the larger cities. This indicates that pure source diffusion processes would give rise to hierarchical filtering.

If the constraint on the agency's actions is binding, then the agency must decide which cities are to receive source inputs and which cities are to be ignored. Again, if the agency is to get the highest increase in adoption per dollar spent, it will concentrate its efforts in those cities with the largest market potential. In other words,

$$\text{if} \quad u_1 > 0 \quad \text{and} \quad u_j = 0 \quad , \quad \text{then} \quad \bar{N}_i > \bar{N}_j \quad .$$

Unfortunately a rigorous derivation of this scenario is rather difficult. In order to derive these properties in an m- city system it is necessary to compute the whole set of $\lambda_i(T)$ to $\lambda_i(0)$, for each i. Interaction effects mean that each λ_i depends on other λs, and their dependence on the adoption levels in each city further complicates matters to the point of intractability. Nevertheless, standard marginal analysis supports the above argument.

4.2. Case 2:   Source and Within-city Interaction Control

When a change agency can control the source rate and influence the interaction rate within cities, it will choose those rates in accordance with the following optimization problem:

$$\text{MIN:} \quad {}_0\!\int^T \dot{W}(U) \, dt \quad ,$$

which in a two-city system this objective function reduces to

$$\text{MIN:} \quad {}_0\!\int^T W(u_1, u_2, v_{11}, v_{22}) \, dt \quad .$$

The accompanying constraints, for this two-city case, would be written as

$$\text{S.T.:} \quad \dot{n}_1 = u_1\bar{N}_1 + \beta_{11}(1 + v_{11}/M)\bar{N}_1 n_1 + \beta_{12}\bar{N}_1 n_2 \quad ,$$

$$\dot{n}_2 = u_2\bar{N}_2 + \beta_{22}(1 + v_{22}/M)\bar{N}_2 n_2 + \beta_{21}\bar{N}_2 n_1 \quad ,$$

$$u_1 \geq 0 \quad , \quad u_2 \geq 0 \quad , \quad v_{11} \geq 0 \quad , \quad v_{22} \geq 0 \quad ,$$

$$u_1 + u_2 + v_{11} + v_{22} \leq M \quad ,$$

$$n_1(0) = n_{10} \quad , \quad n_2(0) = n_{20} \quad , \text{ and}$$

$$n(T) = n_T = n_1(T) + n_2(T) \quad .$$

The dynamic marginal productivity of v_{ii} s given by

$$(\beta_{ii}/M) \, \bar{N}_i \, n_i \quad .$$

When the marginal cost of influencing within-city interaction is dependent only on the level of interaction control, the amount of natural interaction becomes crucial. If within-city interaction is greatest in the first city, then that city will receive the greatest share of within-city interaction control, since the agency concentrates its efforts in the areas with the highest return per dollar spent. If there are initially few adopters, the source rates will dominate because of their higher productivities. As shown in the previous section, adoption in the larger city will grow rapidly. This will result in an increse in $\bar{N}_1 n_1$. Hence, in the initial stages of diffusion, the largest city, or cities, will receive the majority of source inputs. This, in turn, results in these cities receiving the largest amounts of within-city interaction control. This behavior reinforces the importance of large cities as centers of early adoption.

Two of the necessary conditions the agency must meet are

$$0 = DMC(u_1) + \lambda_1 \bar{N}_1 - \mu_1 + \mu_5 \qquad , \text{ and} \qquad (4.2)$$

$$0 = DMC(v_{11}) + \lambda_1 (\beta_{11}/M) n\bar{N}_1 - \mu_3 + \mu_5 \qquad . \qquad (4.3)$$

When the source and interaction controls are both present, meaning u_1 and $v_{11} > 0$, equations (4.2) and (4.3) can be combined to yield

$$n_1 = (M/\beta) \ DMC(v_{11})/DMC(u_1) \qquad ,$$

which is analogous to equation (3.17). This implies that a necessary condition for the number of adopters in a city to be increasing is that the capital outlays for within-city interaction are increasing relative to those for source inputs.

4.3. Case 3: Between-city Interaction Control

The agency's ability to influence interaction between cities allows for a more direct assessment of distance effects. The agency will select diffusion controls on the basis of the marginal productivities and costs of these controls. The marginal productivity of the source rate depends upon the size of the nonadopting population, whereas the marginal productivity of the interaction inputs depends upon the sizes of the adopting and nonadopting populations in each city, and on the interaction rates β_{ij}. These latter rates are distance dependent, and any decision rule concerning the use of between-city interaction inputs also must be distance dependent. Therefore, in this most general case city size and distance both affect the choice of diffusion controls. The distance effect is reinforced by the costs associated with the v_{ij}. Since interaction takes place over space, it is reasonable to assume that costs are incurred in overcoming this spatial separation.

In a two-city system, the agency should choose its policy in keeping with the following optimization problem:

MIN: $\quad _0\!\int^T W(u_1,\ u_2,\ v_{11},\ v_{22},\ v_{12},\ v_{21})\ dt$

S.T.: $\quad \dot{n}_1 = u_1\bar{N}_1 + \beta_{11}(1 + v_{11}/M)\bar{N}_1 n_1 + \beta_{12}(1 + v_{12}/M)\bar{N}_1 n_2$,

$\quad\quad\ \dot{n}_2 = u_2\bar{N}_2 + \beta_{22}(1 + v_{22}/M)\bar{N}_2 n_2 + \beta_{21}(1 + v_{21}/M)\bar{N}_2 n_1$,

$\quad\quad\ \beta_{11} > \beta_{12}\quad,\quad\quad \beta_{22} > \beta_{21}$,

$\quad\quad\ u_1 \geq 0\ ,\ u_2 \geq 0\ ,\ v_{11} \geq 0\ ,\ v_{12} \geq 0\ ,\ v_{21} \geq 0$,

$\quad\quad\ u_1 + u_2 + v_{11} + v_{12} + v_{21} \leq M$,

$\quad\quad\ n_1(0) = n_{10}\quad,\quad n_2(0) = n_{20}\quad,\ \text{and}$

$\quad\quad\ n(T) = n_T = n_1(T) + n_2(T)$.

If the constraints are nonbinding, then the rules governing the use of source and within-city interaction controls, derived above in the second case, still hold. Further, it can be shown that

$$n_2/n_1\ =\ (\beta_{21}/\beta_{22})\ \mathrm{DMC}(v_{22})/\mathrm{DMC}(v_{21})\quad.\qquad (4.4)$$

For any given level of interaction control v, $\mathrm{DMC}(v_{ii})$ is less than $\mathrm{DMC}(v_{ij})$. The reverse is true for the interaction coefficients. Therefore, the number of adopters in the first city must be much greater than the number of adopters in the second city for it to be expedient for the agency to stimulate more interaction between nonadopters in the second and first cities, than with adopters in the second city. That is

$$v_{21} > v_{22}\quad \text{implies that}\ n_1 \gg n_2\quad.$$

Since the largest city receives the greatest initial source and within-city interaction control, it is likely that this is true only in the early stages of diffusion (since $N_1 > N_2$). This reinforces the role of large cities as initial centers of adoption and sources of information concerning innovation.

Equation (4.4) implies that as the diffusion process evolves, the spatial sphere of influence of large cities, measured in terms of $\mathrm{DMC}(v_{ii})/\mathrm{DMC}(v_{ij})$, will increase or decrease as

$$\dot{n}_i/n_i \underset{>}{<} \dot{n}_j/n_j\quad,$$

when it is assumed that the j-th city is dominant. Moreover, the change in the spatial sphere of influence of large cities depends upon the percentage growth in adoption in all cities. If the influence of large cities over smaller cities is to increase, the percentage growth in adoption in those cities must be greater than that in smaller cities. In the early stages of diffusion the growth in large cities is rapid, owing to the source and within-city interaction controls, and it is likely that the spatial sphere of influence of larger cities will increase. As the diffusion process evolves and the number of adopters in large cities becomes great, it is likely that $(\dot{n}_i/\dot{n}_i)>(n_i/n_j)$, and hence the spatial sphere of influence for large cities will diminish. In other words, as the diffusion process continues, the agency puts less emphasis on interaction over long distances, between nonadopters in small cities with adopters in large cities, and more emphasis on interaction over short distances. Therefore, although the initial process may be hierarchical, in the latter stages of diffusion there will be an increase in the importance of neighborhood diffusion.

These general results indicate that the urban systems diffusion pattern most often observed may be a result of an economically sound use of the communication channels. However, specific diffusion policies will depend upon specific cost structures and city distributions. Given all the information needed for calibration, it is likely that the problem of deriving specific policies will be intractable. Thus only the general nature, based upon marginal costs and productivities, can be found.

The empirical observations that as transport-communication systems develop, hierarchical effects become stronger, now can be viewed from an economic perspective. Any improvement in these systems has the effect of reducing the marginal costs of source diffusion. Owing to this lowering of costs, source inputs play a more important role in the diffusion process. As a result, the diffusion process exhibits a strong hierarchical path in the early stages of diffusion.

5.   INTERACTION FIELD MODELS

Of the attempts to model spatial diffusion via interaction fields, the most influential has been the work of Hagestrand (1967). His famous model was based upon a spatial pure birth process. The resulting simulation model has received much attention from geographers, with Gale (1972) and Yapa (1975) studying formal properties of this model. The concept of spatial diffusion as a birth-death process remains quite undeveloped, even though this is an obvious extension of the spatial pure birth model. The reason for this seems to be complexity. The pure birth approach seems difficult enough (hence the reliance on simulation) without adding more complexity. Models of diffusion as a temporal birth-death process have been developed, but the spatial aspects of such a process have proved difficult to assess.

Hagerstrand used the mean information field (MIF), a table giving the probability that any two persons will interact, as means of studying the impact of location on communication. He postulated that the probability of interaction was a symmetric function of distance, and then built a simulation model based upon the following five assumptions:

(A1) the initial distribution of knowers is given,

(A2) acceptance of the message occurs immediately upon contact with a knower,

(A3) knowledge of the innovation is passed only via interaction communication,

(A4) information concerning the innovation is forwarded at constant time intervals by knowers to all other persons, and

(A5) the probability that any two persons interact is given by the mean information field.

Because of assumption A3 the model is of pure interaction and this rules out any source input in the diffusion process. This assumtpion may become less tenable as societies advance in literacy and technical knowledge. Since the probability of interaction between adopters and nonadopters is based solely on location, and these interactions take place at discrete time intervals, this characterization of diffusion may be viewed as a first-order, Makovian, autoregressive process. However, the long-run behavior of this model is trivial. Since this is a pure birth model, as t tends to infinity so does $N(t)$. Thus, in the long-run there is an adopter at every location.

Permanent acceptance of innovation is rarely the only possibility in a diffusion process. People may accept an innovation at first and then reject it later, or they may never accept the innovation at all, or they may always use the innovation. If these aspects of diffusion are incorporated into the spatial process, then the long-run behavior of the population is no longer trivial. This extension of the model, while substantively pleasing, seems to make the problem less tractable. Under such conditions, however, it is possible to obtain certain properties of the diffusion process, and to introduce explicitly the workings of the communication channels present in temporal diffusion models.

In the simulation model the entire population eventually

accepts the innovation. In practice, however, the 'final' state acheived is the result of the researcher halting the simulation process, rather than being determined endogenously by the communication process itself. The asymptotic distribution of adopters is clearly a function of the communication process. Further research should be directed towards developing models of spatial diffusion including facets of the adoption process that allow for the determination of the long-run behavior. In these improved models it could be possible to make long-run behavior a function of communication, as well as the importance of opposition in the diffusion process could be captured. A model incorporating these aspects of diffusion in its strucure would be based upon the following new assumptions:

(B1) the region under study, say R, is of finite area, and the population P is distributed on a uniform lattice (not necessarily square), and

(B2) information encouraging acceptance of an innovation is forwarded either by a source that disseminates information uniformly over the region and/or by private information flows.

If each person communicates only with those persons in the MIF concerning the innovation, then the following assumpitons need to be made:

(B3) information encouraging rejection of the innovation is forwarded through the same communication channels as in assumption (B2),

(B4) information flows continually over time, and

(B5) the probability that an adopter becomes a nonadopter is a function of the amount of information he receives encouraging rejection of the innovation, symmetrically, the probability that a nonadopter becomes an adopter is a function of the amount of information he receives encouraging acceptance.

Since adopters may become nonadopters, the number of adopters at time t may be less than at previous times. All that can be said is that the number of adopters is between 0 and P. The task of further analysis is to determine the circumstances under which an asymptotic distribution exists and is unique (and given these circumstances, what knowledge can be gained).

Going beyond the Hagerstrand model, it is assumed here that the probability that an individual will change from an adopter to a nonadopter, or vice versa, depends upon space as well as time. In an arbitrarily short period of time, Δt, an individual may receive some source messages that encourage acceptance of an innovation, and some that encourage rejection of this innovation. It is further assumed that in this time interval Δt an individual interacts with the individuals in his MIF. Suppose that at time t i is a nonadopter, and A is the configuration of adopters in region R. The probability that i becomes an adopter in the time interval $(t, t+\Delta t)$ is given by

$$P\{(i,A), t+\Delta t\} = P\{(i,A),t\} + \beta(i,A)\Delta t + O(\Delta t^2) \quad ,$$

where $\beta(i,A)$ is a function both of the amount of information received encouraging acceptance and the amount encouraging rejection. This formulation leads to the following conditon:

$$dP(i,A)/dt = \beta(i,A) \quad .$$

A further articulation of the functioning of the communication channels is given by

$$\beta(i,A) = \beta_0 + \beta_1 |A_i| - \delta(N - |A_i|) \quad , \quad (5.1)$$

where β_0 is the source rate for messages encouraging adoption,

β_1 is the rate at which i mixes with adopters,

δ_1 is the rate at which i mixes with nonadopters,

$|A_i|$ is the cardinality of $A \cap MIF_i$, and

N is the number of persons in an MIF.

If i is in the set of adopters at time t, the probability that i becomes a nonadopter in the interval $(t, t+\Delta t)$ is given by

$$P\{(i,A\backslash i),t+\Delta t\} = P\{(i,A\backslash i),t\} + \delta(i,A)\Delta t + O(At^2) \quad ,$$

where the elements of $\delta(\cdot,\cdot)$ are defined analogously to those for $\beta(\cdot,\cdot)$, and $A\backslash i$ means excluding i from A. This result leads to the differential equation

$$dP(i,A\backslash i)/dt = \delta(i,A) = \delta_0 + \delta_1(N - |A_i|) - \beta_1 |A_i| \quad .$$

It is assumed that in this short time period, Δt, the only other possibility for i is to remain in the state it was in at time t. This second situation gives rise to the following set of differential equations:

$$dP(B,A)/dt = \begin{cases} \beta(i,A) & \text{if } B = A \cup i, \ i \notin A \\ \delta(i,A) & \text{if } B = A\backslash i, \ i \varepsilon A \\ -\sum_{j \notin a} \beta(j,A) - \sum_{k \varepsilon A} \delta(k,A) & \text{if } B=A \\ 0 & \text{otherwise.} \end{cases} \qquad (5.3)$$

If both $\beta(\cdot,\cdot)$ and $\delta(\cdot,\cdot)$ are positive, then the matrix of all the differential equations, called the generator and denoted as $G(A,B)$, is said to be irreducible. When the generator of a semigroup, in this case the transition probabilities, is irreducible, an asymptotic distribution exists. Questions still remain, however, concerning uniqueness and calculability.

As with most spatial models, it is easiest to start with an example in one-dimensional space. Suppose each person living along a line interacts with his two geographic neighbors. The interaction rate is independent of which states the neighbors are in (i.e., it does not matter if a person's neighbors are knowers or nonknowers, the interaction rate is still the same). Let the interaction rate be defined as $\alpha/2 = \beta_1 = \delta_1$, $\alpha < 1$. The source rates are assumed to be equal, and will be given a value of 1. The resulting generator has the following form:

card$\|A \cap MIF_i\|$	$\beta(i,A)$	$\delta(i,A\backslash i)$
0	$1 + \alpha(0-2)/2$	$1 + \alpha(2-0)/2$
1	$1 + \alpha(1-1)/2$	$1 + \alpha(1-1)/2$
2	$1 + \alpha(2-0)/2$	$1 + \alpha(0-2)/2$

The values taken on by card$\|A \cap MIF_i\|$ represent the number of knowners in A that also are in the information field of i. Only three possibilities exist here. Either none of i's neighbors are knowers, or one neighbor is a knower, or both are knowers. This tabular information desribes a communication process that meets the requirements for an MIF with asymptotic distribution, say π (Spitzer, 1971).

A strong result is implied by this finding. Not only is π a Markov random field with an asymptotic distribution, but also it is equivalent to a grand canonical Gibbs state with a given interaction potential, $N(\cdot,\cdot)$. In such a case, the entire distribution, π, can be characterized by two values, $N(x,x)$ and $N(x,y)$. Once these values are known, the probability of any configuration of adopters A occurring is given by

$$\pi(A) = \exp[-N(A,A)/2]/Z \qquad ,$$

where Z is a normalizing constant and

$$N(A,A) \quad = \quad \sum_{x \varepsilon A} \sum_{y \varepsilon A} N(x,y)$$

Although the diffusion dynamics described by equations (5.1) and (5.2) are easily extended to two dimensions in symbols, the calculability property is lost. The reason for this is that the Gibbs state equivalence requires the process to be time reversible, an extremely limiting assumption (Haining, 1982). Unfortunately here calculability has a very high price, namely relevance. If the goal of the modeling process is to capture source and interaction communication within explicitly spatial models, then use of Gibbs states as limiting processes to random fields seems limited. On the other hand, Markov random fields are very alluring conceptualizations. They capture the autoregressive nature of space-time processes. In addition, it is possible to build the important source and interaction channels directly into the generator, as Dodd did some years ago in purely temporal models. Finally, the notion of a boundary to the range of interaction is isomorphic to the MIF concept popoularized by Hagerstrand. It would appear that the further use of random fields in this area lies in a two-step modelling process. First, it is necessary to use less restrictive ergodic theorems to prove that a unique equilibrium exists. Some progress in this area has been made by Holley (1972a, 1972b). Once existence of a unique equilibrium state is shown, it then may be possible to simulate the process to see how it evolves over time. It is in this direction that random fields seem to have the most to offer students of communication.

6. CONCLUSION

Various strategies for modeling communication channels have been discussed. The relevance of source and interaction channels, it is hoped, has been made clear. The roles these two channels play in determining communication policies are clearly a function of a change agency's goals and the workings of the channels themselves. In the early sections of this paper, it was shown that in a single spaceless population the dynamic analogues to marginal cost and marginal productivity of the channels play a key role in deciding their optimal use.

Attempts to embed source and interaction channels into spatial models followed two distinct paths. The first was to study channel use wihin an urban system. While explicit communication policies could not be derived, qualitative insights did result. Specifically, pure source processes lead to hierarchical filtering. Source and interaction between all cities in an urban system gives rise to a process that is initially hierarchical, with neighborhood interaction becoming more important as time passes. One could study other forms of communication organization within an urban system. An obvious extension would be regional or national media channels (Lin and Burt, 1975). The second approach to imbedding source and interaction models in a spatial model was via random fields. As in the purely temporal models, an attempt was made to incorporate these channels into the differential equations describing the rates of change in the adoption process (i.e., in the generator). While a plausible model can be constructed for one-dimensional space, the assumptions that must be made to use standard ergodic theorems in two dimensions proved too restrictive. If random fields are to be helpful conceptualizations in which to study communiction channels, less restrictive ergodic theorems need to be found.

7. REFERENCES

Bailey, N., 1968, Stochastic Birth, Death, and Migration Processes for Spatially Distributed Populations, Biometrika, 55: 189–198.

Berry, B., 1971, Hierarchical Diffusion: The Basis of Developmental Filtering and Spread in a System of Growth Centers, in Growth Centers and Regional Economic Development, edited by N. Hansen, New York: The Free Press, pp. 108–138.

Blummer, H., 1946, Collective Behavior, in New Outline of the Principles of Sociology, edited by A. Lee, New York: Barnes and Noble, pp. 167–221.

Bowers, R., 1937, The Direction of Intra-Societal Diffusion, American Sociolgical Review, 2: 826–836.

Chapin, F., 1928, Cultural Change, New York: The Century Company.

Coleman, J., E. Katz and H. Menzel, 1966, Medical Innovation, New York: Bobbs–Merrill.

Crain, R., 1966, Flouridation: The Diffusion of Innovation Among Cities, Social Forces, 44: 467–476.

DeFleur, M. and E. Rainboth, 1952, Testing Message Diffusion in Four Communities, American Sociological Review, 17: 734–737.

Deutschmann, P., 1963, The Mass Media in an Underdeveloped Village, Jouralism Quarterly, 40: 27–35.

_____ and W. Danielson, 1960, Diffusion of Knowledge of a Major New Story, Journalism Quarterly, 37: 345–355.

Dixon, R., 1928, The Building of Culture, New York: Charles Scribner's.

Dodd, S., 1955, Diffusion is Predictable: Testing Probability Models for Laws of Interaction, American Sociological Review, 20: 392–401.

Gale, S., 1972, Some Formal Properties of Hagerstrand's Model of Spatial Interaction, Journal of Regional Science, 12: 199–217.

Gould, J., 1970, Diffusion Processes and Optimal Advertising Policies, in Microeconomic Foundations of Employment and Inflation Theory, edited by E. Phelps, New York: Norton, pp. 338–368.

Greenberg, B., 1964, Person to Person Communciation in the Diffusion of News Events, Journalism Quarterly, 41: 489–494.

_____, J. Brinton and R. Farr, 1965, Diffusion of News About Anticipated Major News Events, Journal of Broadcasting, 9: 129–141.

Hagerstrand, T., 1966, Aspects of the Spatial Structure of Social Communciation and the Diffusion of Innovation, Papers of the Regional Science Association, 15: 27–42.

_____, 1967, Innovation Diffusion as a Spatial Process, Chicago: University of Chicago Press.

Haining, R., 1982, Interaction Models and Spatial Diffusion Processes, Geographical Analysis, 14: 95–108.

Holley, R., 1972a, An Ergodic Theorem for Interacting Systems with Attractive Interactions, Zeitschrift fur Wahrscheinlichkeits-theorie und Verwandte Gebiete, 24: 325–334.

_____, 1972b, Markovian Interaction Processes with Finite Range Interactions, The Annals of Mathematical Statistics, 43: 1961–1967.

Hudson, J., 1969, Diffusion in a Central Place System, Geographical Analysis, 1: 45–58.

_____, 1972, Geographical Diffusion Theory, Evanston, Ill.: Northwestern University Press, Studies in Geography No. 19.

Katz, E., 1961, The Social Itinerary of Technical Change, Human Organization, 20: 70–72.

_____ and P. Lazarsfeld, 1955, Personal Influence, Glencoe: The Free Press.

Leitmann, G., 1968, Sufficiency Theorems for Optimal Control, Journal of Optimization Theory and Applications, 2: 285–292.

Lekvall, P. and C. Wahlbin, 1973, A Study of Some Assumptions Undelying Innovation Diffusion Functions, Swedish Journal of Economics, 75: 362–377.

Pemberton, H., 1936, The Curve of Cultural Diffusion, American Sociological Review, 1: 547–556.

Pred, A., 1973, The Growth and Development of Systems of Cities in Advanced Economies, in System of Cities and Information Flows, edited by A. Pred and G. Tornqvist, Lund: The Royal University of Lund Studies in Geography, Series B, No. 38, pp. 9–82.

Pyle, G., 1969, The Diffusion of Cholera in the United States in the Nineteenth Century, Geographcal Analysis 1: 59–75.

Ralston, B., 1977, A Neoclassical Approach to Urban Systems Diffusion, Environment and Planning A, 10: 267–273.

Rogers, E., 1965, Mass Media Exposure and Modernization Among Colombian Peasants, Public Opinion Quarterly, 29: 614–625.

_____ and F. Shoemaker, 1971, Communication of Innovation, New York: The Free Press.

Ryan, B. and N. Gross, 1943, The Diffusion of Hybrid Seed Corn in Two Iowa Communities, Rural Sociology, 8: 15–24.

Schramm, W., 1963, Communication Research in the United States, in The Science of Human Communication, edited by W. Schumm, New York: Basic Books, pp. 1–16.

Sethi, S., 1974, Optimal Control Problems in Advertising, in Lecture Notes in Mathematics and Economics, vol. 106, edited by M. Beckmann and H. Kunzi, New York: Springer–Verlag, pp. 301–337.

Solow, R., 1957, Technical Change and the Aggregate Production Function, Review of Economics and Statistics, 39: 312–320.

Spitzer, I., 1971, Random Fields and Interacting Particle Systems, lecture notes from the 1971 Mathematical Associaton of America summer seminar at Williams College, Williamstown, Massachusetts.

Wissler, C., 1923, Man and Culture, New York: Thomas Y. Crowell.

Yapa, L., 1975, Analytic Alternatives to Monte Carlo Simulation of Spatial Diffusion, <u>Annals</u>, Association of American Geographers, 65: 163-176.

SECTION 3

URBAN AND REGIONAL CHANGE: GROWTH, DEVELOPMENT AND DECLINE

A prelude to evolutionary modelling is dynamic modelling. The
evolutionary model emphasizes asymmetries of temporal trajectories,
and the existence of a family of geographical distribution time
paths, relating not only to initial conditions but also to the
history of the system. In contrast, the dynamic model is concerned
with a single trajectory and its mathematical description, feedback
mechanisms, and lagged relationships. As was mentioned in the
Introduction to this volume, one fundamental difference between
these two types of models is that the dynamic ones embrace time
reversibility. The papers of this section highlight these points.
In the first paper, Dendrinos and Mullally discuss the use of
simple Volterra-Lotka models to describe the dynamics of cities.
Using population and income as key variables, they are able to show
the time paths associated with individual urban economies derived
from a catastrophe theoretic model. Then each city, from a sample
of United States urban areas, is classified in accordance with
these different trajectories. The three urban areas of Pittsburgh,
Miami and Seattle-Everett are analyzed in greater detail. Thus,
the empirical existence of multiple trajectories is demonstrated.
The authors conclud by noting that regional location and age of a
city are not strongly related to the parameters of the Volterra-
Lotka model. Like Nijkamp, in the preceding section, Dendrinos and
Mullally's model makes explicit use of catastrophe theory.

In the second paper of this section, by Griffith, spatial as
well as classical statistical theory is used to construct a space-
time data cube that will permit analysis of the trajectory for a
contracting geographical system. The empirical problem addressed
here concerns the phasing-out of the Puerto Rican sugar industry.
The observed trajectory appears to be unrelated to production
parameters, and strongly related to the given spatial configuration
of sugar mills. Haining, in a later paper in this volume, also is
concerned about the role played by spatial configuration. This
Puerto Rican trajectory also does not seem to reflect unconstrained
distance minimizing behavior. Census data are available for the
years 1959, 1964, 1969, 1974 and 1978. Spatial statistical theory
is used to estimate data for the missing years during 1959 thru
1978. Then, classical statistical theory is employed to evaluate
the reasonableness of these estimates. Griffith concludes by
noting that the asymmetries of the trajectory prevent the
utilization of dynamic location-allocation models that have been
formulated for expanding geographical systems.

The third paper of this section, by Ancot and Paelinck,

presents an intersectoral-interregional dynamic model for the countries of the European Community. this model emphasizes the two notions of a threshold value for growth to take place, and the attractiveness of a region. A threshold value is determined by combining a discriminant-analytic approach with a modal-split type model. Attractiveness is defined in terms of an exponential distance decay function. Dynamic features of this model are captured predominantly in a lagged (temporal) specification, in terms of a linear difference equation, of the equilibrium values of the number of regions classified as growers and non-growers. Ancot and Paelinck conclude by discussing estimation problems and refinements for their model. The other paper in this volume that utilizes spatio-temporal econometrics is by Bennett, and appears in the last section.

The final paper of this section, by Shoenebeck, reports on a simulation modelling of regional demographic and economic development in North Rhine-Westphalia, Federal Republic of Germany. The purpose of this model is to investigate the trajectory of economic, technical and social change, given certain migration patterns and investment decisions. The demo-economic components that are focused on are regional labor markets, market potentials, and housing, while the purely demographic components that are focused on are the aging and migration of population. Complexities of the relationships amongst these variables necessitate the use of an iterative calibration technique for modelling. Essentially the study is one in comparative statics. The model specifies some temporal relationships in terms of a forward shift operator on (t,t+1), and others in terms of a backward shift operator on (t,t-1). Shoenebeck concludes with a discussion of iterative calibration techniques appropriate for his model.

These four papers illuminate the central role played by trajectories in dynamic and evolutionary models. These papers also exemplify the conceptual gap that must be transcended in order to move from the world of dynamic modelling into that of evolutionary modelling.

EMPIRICAL EVIDENCE OF VOLTERRA-LOTKA DYNAMICS
IN UNITED STATES METROPOLITAN AREAS: 1940-1977

Dimitrios Dendrinos
University of Kansas

Henry Mullally
University of Missouri at St. Louis

United States

1.   INTRODUCTION

The subject of non-linear urban ecology is emerging as a distinct area of study in the field of urban and regional science and geography. Extending the previous work of the Chicago School on Urban Ecology, the new line of research is grounded in empirical studies and the mathematical theory of ecology and population dynamics. Recent work on United States cities by Berry and Kasarda (1977) drew from ecological and economic theory and made use of metropolitan data, at both the intra- and the inter-urban levels. As early as 1971, Samuelson (1971) had proposed the integration of economics and ecology, a suggestion that seems as promising today as it looked then.

Following an earlier formulation by Dendrinos(1979) and Dendrinos and Mullally (1981) presented a theory of metropolitan dynamics based on a Volterra-Lotka type ecology. These authors proposed a model of single city aggregate urban ecological dynamics operating within a prespecified environment, which depicts the competitive nature of an open urban system in the context of a national economy containing a large number of cities. In accordance with a Volterra-Lotka predator-prey model with limits to population growth, an urban population-reward model was proposed. According to this model, each city was expected to exhibit

---

Part of this work was done under contract with the U. S. Department of Transportation. We would like to thank Kirk Suther and Mell Henderson, research assistants from the University of Kansas, for their help in doing the computer simulations.

dynamically stable oscillations, in relative population size and in the ratio of its average per capita income to the nationally prevailing one. Normalized population counts (NPS) and the ratio of metropolitan per capita income (PCI) to the prevailing national average were suggested in the proposed model as the two state variables. The rates of change in these two variables (i.e., relative population and resource accumulation) were formulated in a Volterra–Lotka type system of kinematic ordinary differential equations.

It was demonstrated that following this formulation a city's dynamic path in the population/per capita income space should be expected to obey, in general a spiral–sink type dynamic equilibrium. Under very particular conditions it was demonstrated that a dynamically unstable orbital motion was possible. The damping force is the presence of urban friction (i.e., diseconomies of agglomeration). Since absence of friction is associated with orbital motion, the frictionless city must be a very rare occurrence. When this occurs, slight variations in the factors affecting friction must induce a sudden shift from a center to a sink following a truncated Hopf-type bifurcation in the neighborhood of the frictionless condition. The full Hopf bifurcation (shifting sinks to stable limit cycles) is not feasible in the case of the urban Volterra–Lotka model with linear isoclines (Dendrinos, 1979). Although this particular structural instability ought to be very rare in metropolitan dynamics, finding an illustrative case would greatly enhance the confidence in the validity of the model.

Empirical investigation was initiated by a desire to test the ecological model. A first test had been conducted with data from the Pittsburgh, Pennsylvania, Standard Metropolitan Statistical Area (SMSA) using total number of families and average family income. The test revealed a dynamic behavior in the population/income space that is anomalous with respect to standard economic theory (Dendrinos, 1979). Specifically, the presence of an aggregate urban population cycle was detected in contrast to unimodal motion expected from urban economic theory. In a paper by Dendrinos and Mullally (1981), a more extensive data analysis involving United States SMSAs was performed, where the phenomenon of oscillations in metropolitan population dynamics was found to be abundant. This finding induced development of the original thesis along two directions. First, it established a preliminary classification of 90 selected United States SMSAs according to the typology provided from the single city Volterra–Lotka population–reward model. Second, it employed a Runga–Kutta technique to estimate the parameters of the single city population–reward ecological model for one United States SMSA. This technique is discussed in Moursund and Davis (1967). It approximates the Taylor series expansion of a function. Subroutine DVERK, developed by

Hull, Enright and Jackson (1976), was used in the simulation reported on here, and employed Runge-Kutta formulae of orders five and six. A more detailed description of this technique appears in Dendrinos and Mullally (1981). The Tacoma, Washington, SMSA was selected as the first metropolitan area to carry out such analysis because it exhibited pronounced oscillations during the study period (1940-1977), demonstrating a 22-year cycle, and thus fell well inside the observation time span. Since the completion of this last study, the remaining 175 United States SMSAs (categorized by the United States Bureau of the Census as of 1977) have been classified on a preliminary basis. This classification is presented in Table A3 of the Appendix of this paper. Following this taxonomy, 28 SMSAs have been closely examined using the Runga-Kutte method, and the study of these SMSAs is the subject of this paper.

As mentioned earlier, the truncated Hopf bifurcation is a condition predicted by the specific model proposed. Success in capturing a city during a time period in which it undergoes this event would provide significant evidence in support of the validity of the model. One candidate was found in the 265 SMSAs which could possibly exhibit part of this event. The Rochester, New York, SMSA currently oscillates with an amplitude of 14.6 percent of its minimum population size, with a period of 34 years (see Table 1). This is the only SMSA found so far by numerical analysis to show a simulated PCI below 1.00 during certain years, a requirement for orbital motion. Given the short time span (i.e., 37 years being only a snapshot in a metropolitan life cycle), and given the relatively few metropolitan areas examined (i.e., 265 being a rather small number for such a rare occurrence), the event is extraordinary. If subsequent observations confirm a damping motion towards the center, then Rochester could provide the rather persuasive demonstration sought after.

Next, a theoretical foundation will be outlined, and then empirical evidence will be discussed.

2. A REVIEW OF THE SINGLE CITY ECOLOGICAL MODEL

The changes in a city's normalized population size x, and its per capita income y, are given by the following two kinetic simultaneous ordinary differential equations:

$$\dot{x} = \alpha(y - \bar{y})x - \beta x^2 \quad , \text{ and} \qquad (2.1)$$

$$\dot{y} = \gamma(\bar{x} - x)y \quad , \qquad (2.2)$$

where α and γ are parameters indicating speed of adjustment along the x- and y-axis, respectively, β is a coefficient of friction

Table 1. The Rochester, N.Y. SMSA: a candidate for orbital motion.

Year	NPS[a] $\times 10^{-5}$ actual	NPS $\times 10^{-5}$ simulated	PCI
1940	462.30	462.00	.9950
1942		452.12	.9936
1944		440.84	.9931
1946		429.71	.9934
1948		420.20	.9947
1950	443.30	413.39	.9967
1952		409.99	.9990
1954		410.45	1.0015
1956		414.64	1.0039
1958		422.14	1.0057
1960	441.90	432.20	1.0068
1962		443.54	1.0069
1964		454.69	1.0062
1966	455.00	463.98	1.0045
1968		469.98	1.0023
1970	469.50	471.72	.9997
1971	466.50	470.80	.9984
1972	464.50	468.28	.9971
1973	461.00	465.81	.9960
1974	456.30	461.85	.9950
1975	455.10	457.18	.9942
1976	452.70	451.98	.9936
1977	450.00	446.39	.9932
1980		429.57	.9934

[a]Sources: U.S. Department of Commerce, Bureau of the Census, P-25 series; various issues.

(the term βx^2 represents urban friction or net external disceconomies of agglomeration), $\bar{x}$ is the intrinsic carrying capacity the city enjoys relative to a prespecified environment, and $\bar{y}$ is a nationally prevailing average per capita income. The meaning of β can be captured by its magnitude at the equilibrium point,

$$\beta = \alpha(\tilde{y} - \bar{y})/\bar{x} \quad , \tag{2.3}$$

where $\tilde{y}$ is the equilibrium per capita urban income. Thus, β represents the per capita income equivalent congestion effect per urban resident. Environment in this study is taken to be the national economy of the United States. Although not necessarily so for all urban areas, on the basis of empirical evidence provided, the nation seems to act as the environment for the metropolitan areas examined, and therefore is used for this purpose. The environment here is captured by $\bar{x}$, $\bar{y}$, and the total population over which the metropolitan population is normalized. The magnitude of the adjustment parameters also is tied to the prespecified environment.

According to the formulation in equations (2.1) and (2.2), a city's NPS adjusts exponentially fast at a rate proportional to the difference between the urban area's per capita income and the average nationally prevailing level. The growth is not unrestricted, but rather subject to limits imposed by relative friction resulting from relative agglomeration of popualtion in the metropolitan area, within the context of the prespecified environment. This equation depicts a demand type function in population growth. Per capita income also grows exponentially, subject to no limits, as a result of constant growth in resource exploitation, investment and technological progress. Its rate of change, however, depends upon the intrinsic carrying capacity the city enjoys. For levels of urban population exceeding its relative carrying capacity, imposed by the environment as well as its endowed resources, per capita income declines. In contradistinction, per capita income increases for NPS levels below its relative carrying capacity. This proposition reflects a supply side view of income generating activities: income producing activities are attracted to an urban area if it has not reached its carrying capacity. If its carrying capacity is exceeded then income producing activities leave, for other cities become more attractive.

According to equations (2.1) and (2.2), relative population accumulation is affected by current relative income differentials, and in turn relative income accumulation is affected by current relative population size. This condition implies myopic behavior from both population (demand for) and suppliers of income generating activities. The contention is that within an open system, such as an urban economy, only current levels of NPS and PCI count, as if one cannot foresee the dynamics of the environment. Subsequent empirical tests bear this out.

The left-hand side of equation (2.1) is a function of the ratio of the urban to the national growth rate (k). It is positive when k is greater than one; otherwise it is negative. Similarly, the left-hand side of equation (2.2) is a function of the ratio of urban average income growth to urban population growth (h). If w is the metropolitan population, W is the national count, and z is the city's total income, then

$$x = w/W \qquad , \qquad (2.4)$$

$$\dot{x} = (k - 1) \, w \, (\partial W/\partial t)/W^2 \qquad , \qquad (2.5)$$

$$k = [(\partial w/\partial t)/w]/[(\partial W/\partial t)/W] \qquad , \qquad (2.6)$$

$$y = z/w \qquad , \qquad (2.7)$$

$$\dot{y} = (h - 1) \, z \, [(\partial w/\partial t)/w^2] \qquad , \text{ and} \qquad (2.8)$$

$$h = [(\partial z/\partial t)/z]/[(\partial w/\partial t)/w] \qquad . \qquad (2.9)$$

This seemingly simple system of simultaneous equations (2.1) and (2.2) is proposed as the parsimonious descriptor of the involved internal as well as external ecological interactions of a city within a region, or set of regions, or a nation.

Accepting for a moment that the system in equations (2.1) and (2.2) describes single city aggregate NPS/PCI dynamics, there are a number of implications that can be drawn, the validity of which one can empirically test. First, one must expect a clockwise motion in the NPS/PCI space for all SMSAs during the observation period. Second, oscillations of higher amplitudes must always be followed by lower amplitude motion. Third, since the probability that the value of β is very close to zero is extremely small, the chance of observing in aggregate urban dynamics an orbital motion (a dynamic neutrally stable movement) must be very small. Fourth, the per capita income in which each SMSA will settle, operating under friction, must be higher than the national average per capita income. These conditions are associated with the Volterra-Lotka system of equations as stated in equations (2.1) and (2.2). All four conditions are met in the empirical test of the model. These, together with certain other findings that relate to broader aspects of metropolitan evolution, have been discussed extensively by Dendrinos and Mullally (1981). Here the focus will be shifted to the main subject of this paper, namely a numerical analysis of 28 SMSAs.

3. THE EMPIRICAL EVIDENCE

In Dendrinos and Mullally (1981) the first 90 United States SMSAs were classified on the basis of their apparent dynamic behavior in their average PCI and NPS, with the nation acting as the relevant environment. However, only one SMSA (Tacoma, Washington) was analyzed in detail. The remaining SMSAs (those designated by the 1977 Bureau of the Census) were classified on the basis of the same scheme, and are summarized in Tables A1 and A2 of the Appendix, with the individual SMSA classification appearing in Table A3. Table A3 demonstrates the dominance of spiral type sinks in the dynamics of urban areas in the United States, together with certain particular events of a discontinuous nature. A review of these phenomena provides strong evidence that the dynamics of United States SMSAs exhibits stability, and that the urban sector of the United States has been experiencing shocks over the last 37 years. The nature and effects of these shocks are the subject of this paper, too. These findings were reported in Dendrinos (1981), and contained no numerical simulation. In this paper 28 SMSAs are analyzed and the findings reported.

It should be noted that the classification of each SMSA is tentative, especially for those SMSAs which have not yet been numerically simulated. Independent confirmation from other researchers is needed before these results attain scientific acceptance. With this caveat in mind, the findings for the 28 SMSAs studied in detail now will be reported. In certain instances these findings suggest the need for changes in the originally reported classification appearing in Table A3 of the Appendix, and thus the summaries of Tables A1 and A2 of this Appendix.

Parameter values for the 28 SMSAs are shown in Tables 2 and 3. Simulation results are presented in these tables in order to furnish necessary information for initiating simulations of these as well as other SMSAs comparable in size by interested researchers. These parameters correspond to a slightly different formulation of the system than do equations (2.1) and (2.2), namely

$$dx/dt = (\alpha y/\bar{y} - \alpha - \beta x)x \quad , \text{ and} \quad (3.1)$$

$$d(y/\bar{y})/dt = \gamma(\bar{x} - x)y \quad . \quad (3.2)$$

Results were obtained by using the Runge–Kutta method as reported in Dendrinos and Mullally (1981). The approximately ten percent (non–random) sample of selected United States metropolitan areas contains urban settings from all four major regions of the United States, from all age groups, and (steady state and starting population) sizes. The simulated values of the three parameters, the speeds of population (α) and income (γ) adjustment, and the

Table 2. Parameters of the Volterra-Lotka Model for 28 U.S. Metropolitan Areas: (1940 - 77)

SMSA[1]	β	γ	β'	It/y[2]	$\tilde{\alpha}$[3]	$\tilde{\beta}$[3]	$\tilde{\gamma}$[3]
Atlanta	1.080	.0200	.0015	.0185	1.0	1.080	.0185	.0015
Birmingham (AL)	1.120	.0200	.0018	.0179	1.2	1.344	.0215	.0022
Chicago	1.250	.0090	.0008	.0072	2.5	3.125	.0180	.0020
Denver	1.123	.0263	.0020	.0234	.5	.562	.0117	.0010
Detroit	1.120	.0100	.0030	.0089	.3	.336	.0027	.0009
Houston	1.130	.0200	.0019	.0177	.4	.452	.0071	.0008
Kansas City	1.120	.0183	.0020	.0163	1.2	1.344	.0196	.0024
Los Angeles	1.210	.0155	.0025	.0128	.5	.605	.0064	.0013
Miami[4]	1.120	.0150	.0007	.0134	1.3	1.456	.0174	.0009
Minneapolis	1.120	.0200	.0022	.0179	.8	.896	.0143	.0018
New Haven	1.120	.0200	.0020	.0179	2.6	2.912	.0465	.0052
New Orleans	1.140	.0200	.0022	.0175	1.0	1.140	.0175	.0022
New York	1.250	.0050	.0001	.0040	2.0	2.500	.0080	.0001
Philadelphia	1.270	.0100	.0010	.0079	2.0	2.540	.0158	.0020
Phoenix	1.120	.0315	.0020	.0281	.5	.560	.0141	.0010
Pittsburgh	1.150	.0071	.0001	.0062	2.0	2.300	.0124	.0002
Rochester	1.120	.0000	.0025	.0000	1.6	1.792	.0000	.0040
San Francisco	1.140	.0170	.0015	.0149	1.0	1.140	.0149	.0015
Seattle[4]	1.120	.0180	.0017	.0161	1.4	1.568	.0225	.0024
St. Cloud	1.120	.0200	.0020	.0179	2.5	2.800	.0448	.0050
St. Louis	1.120	.0200	.0020	.0179	2.8	3.136	.0501	.0056
Tacoma	1.120	.0330	.0030	.0295	2.0	2.240	.0590	.0060
Terre Haute	1.118	.0200	.0026	.0179	4.0	4.472	.0716	.0104
Toledo	1.119	.0180	.0017	.0161	3.0	3.357	.0483	.0051
Tulsa	1.120	.0200	.0020	.0179	4.0	4.480	.0716	.0080
Tuscon	1.120	.0270	.0015	.0241	.8	.896	.0193	.0012
Wichita	1.120	.0164	.0020	.0146	6.4	7.168	.0934	.0128
York	1.120	.0200	.0020	.0179	5.6	6.272	.1002	.0112

[1] Only the first central city is identified.

[2] This is the number of iterations per calendar year using the Runge-Kutta method.

[3] The product of the corresponding coefficient with the number of iterations per year.

[4] Sink spiral motion perturbed around 1963 to another sink spiral; the values of the second spiral are shown.

Table 3. Period, carrying capacity, steady state PCI, starting values, and classification of 28 SMSAs: (1940 - 77)

SMSA	Period[1]	$\bar{x}$[2]	$\bar{y}$[3]	x_o[2]	y_o[3]	Type	Description
Atlanta	34	8.90	1.163	5.260	1.100	A6ciii	Sink Spiral
Birmingham (AL)	35	3.85	1.068	4.370	1.090	A1b	Sink Spiral
Chicago	9	34.00	1.242	34.470	1.270	C2b	Sink Spiral[a]
Denver	67	5.85	1.135	3.380	1.115	A5ci	Sink Spiral
Detroit	40	20.60	1.184	18.910	1.200	A6b	Sink Spiral[a]
Houston	60	11.00	1.193	4.880	1.120	A6ciii	Sink Spiral
Kansas City	25	6.07	1.099	5.570	1.085	A6b	Sink Spiral
Los Angeles	24	32.50	1.416	21.010	1.260	A6b	Sink Spiral
Miami[4]	40	5.90	1.078	2.020	1.040	C1b	Sink Spiral[d]
Minneapolis	34	9.40	1.163	8.040	1.130	A1b	Sink Spiral
New Haven	14	3.62	1.063	3.650	1.064	C2b	Sink Spiral[a]
New Orleans	30	4.34	1.090	5.130	1.080	A6b	Sink Spiral
New York	100	35.00	1.140	64.320	1.270	A5bii	Sink Spiral
Philadelphia	11	23.65	1.186	23.360	1.100	C2b	Sink Spiral[a]
Phoenix	80	4.55	1.128	1.403	1.113	A5ci	Sink Spiral
Pittsburgh	90	8.75	1.054	15.710	1.100	A5bii	Sink Spiral
Rochester	17	4.40	1.000	4.620	.995	A3	Orbital
San Francisco	25	14.75	1.220	10.660	1.120	A6b	Sink Spiral
Seattle[4]	23	6.60	1.106	5.750	1.050	A6b	Sink Spiral
St. Cloud	17	.69	1.011	.710	1.007	A6a	Sink Spiral
St. Louis	9	11.70	1.209	11.570	1.180	C2b	Sink Spiral[a]
Tacoma	11	1.96	1.057	1.370	1.000	A1b	Sink Spiral
Terre Haute	17	.82	1.014	.752	1.000	A1b	Sink Spiral
Toledo	13	3.65	1.058	3.780	1.045	A1b	Sink Spiral
Tulsa	12	2.61	1.046	2.670	1.030	C2b	Sink Spiral[b]
Tuscon	45	1.20	1.033	.550	1.070	A5ci	Sink Spiral
Wichita	8	1.93	1.029	1.320	1.060	A1b	Sink Spiral[c]
York	10	1.62	1.029	1.640	1.015	A1b	Sink Spiral[c]

[1]Ellapsed calendar time between a local max and min in NPS.

[2]NPS is in 10^5.

[3]PCI is the ratio to the U.S.

[4]Second spiral is recorded.

[a]Drastic decline starting in 1970.

[b]Drastic increase starting in 1970.

[c]With steady state.

[d]Switch to a lower energy trajectory: 1965.

coefficient of friction ($\beta' = \beta/\alpha$) are indicated in Table 2. Their period of motion (i.e., the average time elapsed between a local minimum and the next local maximum), steady state, and starting NPS and PCI values together with the SMSA's classification type are provided in Table 3. In reading the notes that follow, one should keep in mind that these remarks are extracted from a sample covering a 37-year time period of observations only, and that these comments represent in this time span relatively static relationships. Over longer time spans these relationships evolve, forming an urban landscape. In order to identify the relatively slow dynamics, more data are required. Also, it should be noted once again that with the possible exception of the Rochester, New York, SMSA, no other metropolitan area was identified as a candidate for orbital motion (frictionless city). The calibrated values for β were quite a bit higher than zero for the remaining 27 SMSAs.

From a preliminary inspection of the α and γ values one can conclude that NPS is a relatively fast adjusting variable, whereas PCI adjusts relatively slowly, α being about 100 times larger than γ. The steady state equilibrium factor reward $\tilde{y}$ has been plotted against the equilibrium size $\bar{x}$ (the urban carrying capacity), and appears in Figure 1. It is shown that $\tilde{y}$ is a fourth-degree

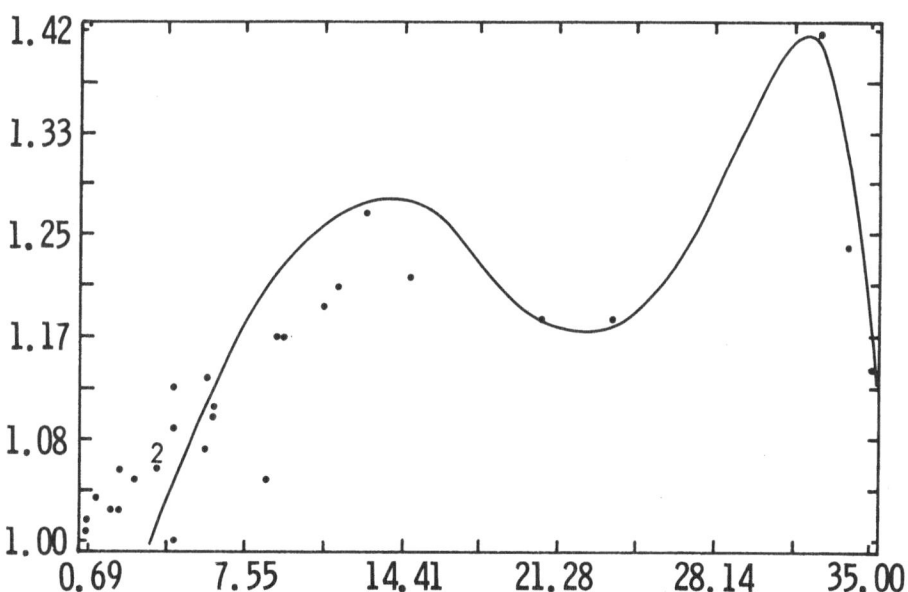

Figure 1.   Per capita income ratio (vertical axis) plotted against NPS at their steady state.

equation in $\bar{x}$. This result supplies strong evidence in support of
the theoretical claim that factor reward and metropolitan size are
connected in a non-convex fashion. Such claims have been made
earlier by Casetti (1980), Dendrinos (1980), and Papageorgiou
(1980). In Dendrinos (1980) it is argued that the existence of
different optimum population sizes for various urban sectors might
result in a reward/size relationship exhibiting multiple local
maxima, and a catastrophe theoretic framework was provided to model
the local dynamics. In Figure 1 the two local maxima are shown to
exist around NPS of 14.40 and 31.50 (x 10^{-4}) in metropolitan areas
in the United States, with a local minimum around 21.20 (x 10^{-4}).
Both β and γ seem to be negative exponential functions of $\bar{x}$;
further, β is linearly related to γ. The steady state equilibrium
size $\bar{x}$ and the period of motion (where period here indicates the
time ellapsed between a local minimum and the subsequent maximum,
in either NPS or PCI, in the NPS/PCI space) do note exhibit any
particular relationship.

The U-shaped curve of Figure 2 shows that the speed of
population adjustment, α, is a second-degree equation of the period
of oscillation in the NPS/PCI space. The minimum in α occurs when
the period of motion is approximately 60 years. High velocity in
the cyclical motion is associated with a relatively fast damping

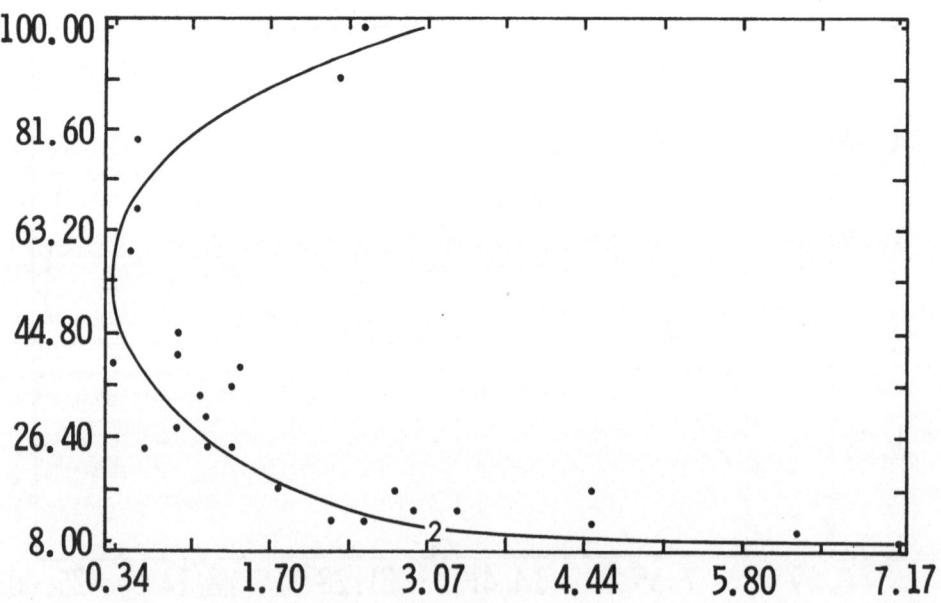

Figure 2. Period of motion (vertical axis) plotted against the
speed of NPS α.

movement towards the steady state. There is evidence that for the majority of the metropolitan areas a smaller period of motion in the NSP/PCI space is associated with a higher speed of population adjustment. A further finding is that the coefficient of urban friction, β, is a negative exponential function of the period of motion. Further, the period of metropolitan population is found to be a second-degree equation of the velocity of income adjustment, γ (neither graph has been included here). Since the magnitude of the urban friction coefficient is linked to the occurrence of the Hopf bifurcation phenomenon, it is concluded that metropolitan areas with longer periods in their NPS/PCI motion must be more likely candidates for such a bifurcation. The maximum period occurs at a γ equal to 0.001.

In order to identify statistically significant relationships among the parameters of the Volterra-Lotka urban dynamic model, disaggregated by age and region of SMSA, it would be necessary to increase the sample size substantially. This is the single most important element of future research. From an initial inspection of the well-identified relationships among period of motion, $\bar{x}$, α, β, and γ it appears to be a reasonable hypothesis that region is not as important as age in metropolitan dynamics. Younger SMSAs seem to lie on the left-hand side of the U-shaped curve relating $\tilde{\alpha}$ and period of motion in Figure 2, indicating a negative relationship for newer metropolitan areas. Middle-aged cities seem to lie at the bottom part, whereas older urban settings tend to line up along the right-hand side of the U-shaped curve (positive correlation between $\tilde{\alpha}$ and period of motion). A need for more data also characterizes the quest for establishing connections between type of dynamic equilibrium and age, region or size of SMSAs.

4. THREE CASE STUDIES

Three specific SMSAs are examined in more detail in this section. The Pittsburgh, Pennsylvania, SMSA is analyzed because data on per capita income as well as population are available for selected years. Miami, Florida, and Seattle-Everett, Washington, are closely studied because these two SMSAs have experienced events linked to effects upon the dynamic stability due to perturbations of their dynamic equilibria. The analysis of these events provides further support for the overall validity of the model.

4.1. The Pittsburgh, Pennsylvania, SMSA

This particular SMSA was selected for a closer look for a number of reasons. First, there are eighteen observations of the SMSA's actual population counts in the period 1940 thru 1977. Second, there are two years, namely 1969 and 1977, for which the real per capita income counts are either directly available or possible to compute from existing data. Both counts, together with their simulated values, are supplied in Table 4. Findings for this SMSA are critical to the validity of the ecological model presented in this paper. For NPS, the average absolute deviation from the actual values of a typical simulated count is less than one percent. This provides strongly supportive evidence for the validity of the model, especially in view of the accuracy obtained from simualtions of other SMSAs' NPS as well.

Table 4. Pittsburgh, SMSA. Simulated and actual NPS and PCI counts: 1940-77.

Year	Simulated NPS $X10^{-5}$	Actual NPS $X10^{-5}$	Simulated PCI ratio to U.S.	Actual PCI ratio deflated[a]
1940	1571.0	1571.0	1.1000	NA
1950	1471.6	1453.3	1.0854	–
1960	1293.3	1326.8	1.0745	–
1961	1277.6	1285.1	1.0736	–
1962	1262.3	1257.9	1.0727	–
1963	1247.5	1244.3	1.0719	–
1964	1233.1	1230.4	1.0711	–
1965	1219.2	1225.3	1.0704	–
1966	1205.7	1204.8	1.0697	–
1967	1192.8	1196.4	1.0690	–
1969	1168.0	–	1.0677	1.0188[1]
1970	1156.3	1171.4	1.0671	–
1971	1145.0	1161.1	1.0665	–
1972	1132.1	1142.5	1.0659	–
1973	1123.6	1121.1	1.0654	–
1974	1113.4	1101.5	1.0648	–
1975	1103.6	1089.2	1.0644	–
1976	1094.2	1072.7	1.0639	–
1977	1085.1	1059.1	1.0634	1.0398[2]

[1]Sources: U.S. Department of Commerce, Bureau of the Census, P-25/882 Report, and Table 124, 1970 Census, 40-678 Pennsylvania; and computation by authors.

[2]Sources: P-25/882 and computations by authors.

[a]Sources: U.S. Department of Commerce, Bureau of the Census, U.S. Statistical Abstract, various years.

Actual PCI values are scarce and less reliable than NPS counts. The two PCI values provided in Table 4 have been deflated by applying appropriate consumer price deflators for both the United States and Pittsburgh for these years such that real incomes have been obtained. For the simulated runs, it should be noted, there is no need to provide the actual starting value for the 1940 PCI; only a guess is necessary. The model is calibrated by varying the starting (1940) value of PCI. Thus, it is rather comforting that the actual values for 1969 and 1977 are no more than 4.6% and 2.2% different from their corresponding simulated values. Also interesting is that NPS declined (over the 1940–1977 horizon) even though the Pittsburgh real per capita income remained above the national average. This qualitative result is an expected outcome from the ecological model, although not from the purely economic one. What is not reflected in the 1969 and 1977 PCI levels is the real income decline that occurred during this period in the simulated series. It is attributable to a possible overestimation of the observed Pittsburgh deflator for the 1969 year.

4.2. The Miami, Florida, SMSA

This SMSA underwent a switch in trajectories around 1963. A perturbation is recorded in the NPS count during the period 1963–1967, which is reflected in the two simulation runs shown in Figure 3. This figure was constructed by differing the speed of PCI adjustments, and depicts two different possible paths for the Miami SMSA. The outer trajectory corresponds to a PCI speed of 0.0008, and best fits the period 1940–1963, whereas the inner, lower energy, trajectory corresponds to $\gamma=0.0007$ and best fits the span 1967–1977. This 12.5% decline in PCI speed γ shifted Miami to an inner path, but towards the same carrying capacity $\bar{x}$ and steady state $\tilde{y}$. In Figure 3 these two paths belong to two different starting values, and to two different values of γ. The γ dimension is been supressed in this figure. This event also is very supportive of the existence of sink-type dynamics in metropolitan time series, since this particular dynamic equilibrium is stable, and slight perturbation are expected to produce the event recorded, namely the shift to another path leading to the same equilibrium point.

Apparently the shift to a lower energy path represents some change in the economic base of the Miami SMSA. In physical terms, an inner path represents energy loss. What forms the energy emission process takes in the case of a city are still an unanswered research question. It should be noted, however, that this shift occurred close to a maximum of PCI. In Figure 3 the perturbation (indicated by the arrow leaving the outer trajectory around 1963 and reaching the inner trajectory about 1967) seems to be directed toward the steady state point. The perturbation apparently was of a small scale; the Miami incident lasted for

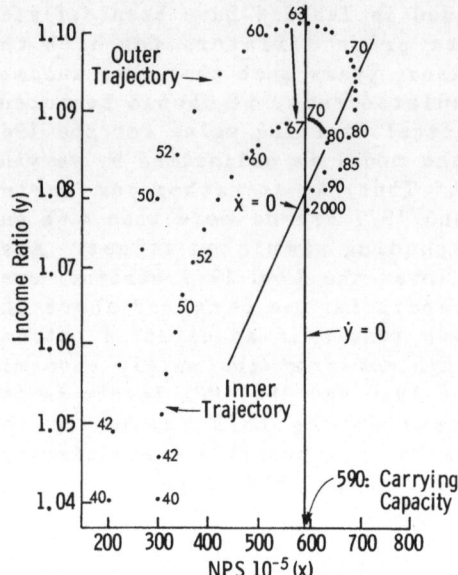

Figure 3. The perturbation of Miami's spiral sink path from an outer to an inner one.

about six calendar years. Whether or not this perturbation is typical of the shocks experienced by United States metropolitan areas over longer time periods is an open question. More data are needed before a definitive answer to this question can be rendered. Longer observed time periods also are needed to determine whether the results obtained here constitute fluctuations in time paths or shifts in dynamic trajectories.

4.3. The Seattle–Everett, Washington, SMSA

In the sample of 28 SMSAs analyzed, the case of a metropolitan area being shifted from an initial spiral sink trajectory to another totally different sink also was recorded. This phenomenon applies to the Seattle–Everett, Washington, SMSA, where an abrupt change in the early sixties, which lasted until 1970, moved Seattle from path (a) to path (b), as shown in Figure 4. The new path corresponds to a higher carrying capacity $\bar{x}$ (from 6.00 to 6.60 NPS), a lower equilibrium $\tilde{y}$, and a higher β (from 0.018 to 0.025). Seattle and Miami, when shifted either to a different or to a neighboring path, experienced drastic declines in expected factor reward during the event. Both shifts were of relatively short duration (six years), and both occurred during the same time period

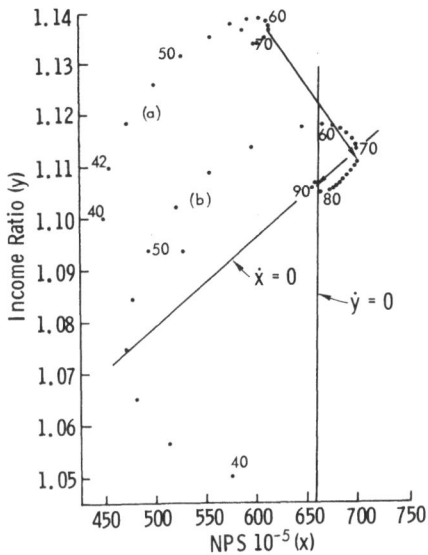

Figure 4. Seattle's shift of trajectory from (a) to (b).

(mid-60s). It is of interest to note that both of these SMSAs were of the same NPS size at the time of the event.

5. CHANGING THE ENVIRONMENT

The analysis summarized in the preceding section was based on the use of total population counts for the United States as the normalizing factor. By normalizing urban population counts on the basis of the national total, one eliminates background noise, such as nationally prevailing demographic events (for example, changes in fertility rates or international migration). However, ecological factors operating between urban and rural settings within a nation (the phenomenon of urbanization, for instance) are not neutralized. If one wishes to examine the ecological competition among cities, then one must normalize, for each time period, using total population residing in the metropolitan areas of a nation. The Volterra-Lotka type model described earlier can be used with the total United States urban population as the normalizing factor. Only those years should be considered in this case for which data exist for all (1977 composition) counties in all SMSAs. As a result, only the years of 1940, 1950, 1960, 1970 and 1977 have been used. Simulations were carried out for a ten percent sample from the 265 SMSAs.

The qualitative properties of the Volterra-Lotka type model were found to be present in the NPS/PCI profiles of this new series. In Table A4 of the Appendix a classification is provided for 27 metropolitan areas that had been randomly selected from the total of 265 SMSAs. Summary totals by type are close to those appearing in Table A2, where total national population is employed. The relationships between the behavior of the various SMSAs under the two environments are the subject of future research efforts.

6. CONCLUSIONS

Using the Runge-Kutta method 28 United States metropolitan areas were simulated according to a Volterra-Lotka type single city aggregate urban dynamic model. The coefficients of individual city dynamics were estimated and relationships among them and the period of motion were established. It was concluded, on a preliminary basis, that geographical region and age of SMSA are not strongly related to the parameters of the Volterra-Lotka model.

Although the research results reported here are preliminary in nature, and need to be confirmed and amplified by more exhaustive empirical studies, some speculation as to the emerging picture of urban evolutionary processes may be justified. What seems to take shape is a picture of an urban landscape in which individual cities, while different in internal structure, trace out qualitatively similar paths in the income/population space within the national environment. What possibly is being observed here is a long-run tendency of cities to adjust to each other, as well as to underlying social, technological and physical circumstances. The individual urban area is an open system within a larger environment, and follows a deterministic evolutionary path toward a steady state. Steady states for all urban areas in the environment considered seem to form a particular structure, for example a fourth-degree equation in the income/population space, when the national economy is the relevant environment.

Here each city's adjustment process was found to be quite predictable, rather than random, following almost a code of urban dynamics. This code was identified by observing Volterra-Lotka type dynamics in the income/population space. Further empirical investigation covering more cities and longer time spans may provide even stronger evidence for such an evolutionary scheme. This work, in turn, may significantly affect perceptions of urban dynamic processes and the role of urban policy making.

7. REFERENCES

Berry, B. and J. Kasarda, 1977, Contemporary Urban Ecology, New York: Macmillan.

Casetti, E., 1980, Equilibrium Population Partitions Between Urban and Agricultural Occupations, Geographical Analysis, 12: 47–54.

Dendrinos, D., 1979, A Basic Model of Urban Dynamics Expressed as a Set of Volterra–Lotka Equations, in Catastrophe Theory in Urban and Transport Analysis, edited by D. Dendrinos, Washington, D. C.: Report # DOT/RSPA/DPB-25/80/20, U. S. Department of Transportation, pp. 79–103.

_____, 1980, Dynamics of City Size and Structural Stability: The Case of a Single City, Geographical Analysis, 12: 236–244.

_____, 1981, Population Growth Patterns for all U. S. Standard Metropolitan Statistical Areas (SMSA), preliminary report to the U. S. Department of Transportation, unpublished paper, Center for Research Incorporated, University of Kansas.

----- and H. Mullally, 1981, Evolutionary Patterns of Metropolitan Populations, Geographical Analysis, 13: 328–344.

Hull, T., W. Enright and K. Jackson, 1976, User's Guide to DVERK-A Subroutine for Solving Non-Stiff ODE's, Technical Report TR100, Toronto: Department of Computer Science, University of Toronto.

Moursund, D. and C. Davis, 1967, Elementary Theory and Applications of Numerical Analysis, New York: McGraw-Hill.

Papageorgiou, G., 1980, On Sudden Urban Growth, Environmnt and Planning A, 12: 1035–1050.

Samuelson, P., 1971, Generalized Predator-Prey Oscillations in Ecological and Economic Equilibrium, Proceedings of the National Academy of Sciences, 68: 980–983.

8. APPENDIX: CLASSIFICATION OF SMSAS

Table A1

Dynamic Patterns of Metropolitan Normalized Population Histories for the Period 1940-77; 175 U.S. SMSAs.

Type of History			Number of Observations
A. Oscillatory Behavior			131
1. Spiral Sink		35	
a. with steady state	7		
b. towards steady state	28		
2. Spiral Source		--	
3. Center		--	
4. Limit Cycles		--	
5. Long-term Oscillatory		48	
6. Medium-term Oscillatory		48	
B. Steady State			2
C. Perturbations			36
1. Structurally Stable		14	
a. switch of long-term motion	10		
b. switch of medium-term motion	4		
2. Change of State		22	
a. naked discontinuity	--		
b. one mode plus discontinuity	. 22		
c. Hopf Bifurcation	--		
D. Unclassified			3
E. Not Enough Data			3

Table A2

Various Long- and Medium-Term Oscillations; their incidence in the Period 1940-77 for 96 SMSAs.

Type	Shape	Occurrence
A.5 Long-term Oscillatory		48
a. Linear*		36
i $\rightarrow \dot{x} > 0 \quad \ddot{x} = 0$		26
ii $\rightarrow \dot{x} < 0 \quad \ddot{x} = 0$		10
b. Concave		10
i $\rightarrow \dot{x} > 0 \quad \ddot{x} > 0$		6
ii $\rightarrow \dot{x} < 0 \quad \ddot{x} < 0$		4
c. Convex		2
i $\rightarrow \dot{x} > 0 \quad \ddot{x} < 0$		1
ii $\rightarrow \dot{x} < 0 \quad \ddot{x} > 0$		1
A.6 Medium-term Oscillatory		48
a. Concave		13
b. Convex		18
c. With inflection point		17
i $\rightarrow$		1
ii $\rightarrow$		--
iii $\rightarrow$		15
iv $\rightarrow$		1

---

*Although in actual NPS counts a straight line may be present, in the simulated runs there are no straight lines but only curves, thus this category does not exist in the classification scheme of numerically analyzed SMSAs.

Table A3

Classification of 175 U.S. SMSAs*: 1940-77

SMSA**	Size§ X10⁻⁵	Description	Type of Event
Columbia, SC	172.9	Long-term linear up	A5ai
Columbus, GA	105.2	Sink spiral	A1b
Columbus, OH	485.0	Sink spiral	A1b
Dallas, TX	1232.5	Long-term linear up	A5ai
Davenport, IA	172.0	Medium-term oscillatory	A6b
Daytona, FL	98.7	Long-term linear up	A5ai
Decator, IL	58.6	Perturbed steady state with discontinuity	C2b
Denver, CO	676.0	Long-term linear up	A5ai
Debuque, IA	43.3	Sink spiral	A1b
Eau-Clair, WI	57.0	Sink Spiral	A1b
Elkhart, IN	61.8	Medium-term oscillatory with inflection point	A6ciii
Elmira, NY	95.7	Medium-term oscillatory	A6b
Eugene, OR	115.3	Long-term linear up	A5ai
Evansville, IN	133.7	Perturbed steady state with discontinuity	C2b
Fargo, ND	60.0	Sink spiral	A1b
Fayetteville, NC	106.0	Medium-term oscillatory	A6b
Fayetteville, AR	71.9	Perturbed steady state with discontinuity	C2b
Florence, AL	58.2	Perturbed steady state with discontinuity	C2b
Fort Collins, CO	58.1	Perturbed steady state with discontinuity	C2b
Ft. Lauderdale, FL	398.8	Medium-term oscillatory	A6ciii
Fort Meyers, FL	80.23	Medium-term oscillatory	A6ciii
Fort Smith, AR	85.65	Medium-term oscillatory	A6a
Gadsden, AL	44.70	Sink spiral with steady state	A1a
Gainsville, FL	58.99	Perturbed medium-term spiral towards steady state	C1b
Galveston, TX	89.91	Perturbed long-term oscillatory	C1a
Gary, IN	298.00	Spiral sink	A1b
Grand Forks, ND	47.00	Medium-term oscillatory	A6a
Great Falls, MT	39.30	Medium-term oscillatory	A6a
Greeley, CO	50.00	Medium-term oscillatory	A6a
Green Bay, WI	80.09	Medium-term oscillatory	A6ciii
Greensboro, NC	357.00	Sink spiral	A1b
Greensville, SC	246.00	Perturbed medium-term oscillation with discontinuity	C2b

SMSA**	Size$\times 10^{-5}$	Description	Type of Event
Honolulu, HI	331.00	Long-term oscillatory	A5ci
Houston, TX	1154.10	Long-term, linear up	A5ai
Huntington, WV	140.00	Medium-term oscillatory with inflection point	A6civ
Huntsville, AL	134.00	Perturbed steady state with discontinuity	C2b
Jackson, MI	68.50	Medium-term oscillatory	A6b
Jackson, MS	134.50	Sink spiral	A1b
Jacksonville, FL	318.00	Medium-term oscillatory	A6b
Janesville, WI	61.65	Medium-term oscillatory	A6b
Jersey City, NJ	259.59	Long-term, linear down	A5aii
Johnson City, TN	187.70	Not enough data	E
Kalamazoo, MI	123.50	Sink spiral with steady state	A1a
Kankakee, IL	44.26	Medium-term oscillatory	A6b
Kenosha, WI	56.60	Discontinuity and steady state	C2b
Killeen, TX	96.10	Long-term, linear up	A5ai
Kokomo, IN	48.00	Spiral sink	A1b
La Crosse, WI	40.50	Medium-term oscillatory	A6a
La Fayette, LA	60.86	Long-term, linear up	A5ai
La Fayette, IN	52.30	Medium-term oscillatory	A6b
Lake Charles, LA	74.00	Spiral sink	A1b
Lakeland, FL	128.00	Long-term, linear up	A5ai
Lansing, MI	206.00	Medium-term oscillatory with inflection point	A6ciii
Laredo, TX	39.20	Unclassified	D
Las Vegas, NV	166.45	Medium-term oscillatory	A6ciii
Lawrence, KS	30.00	Medium-term oscillatory	A6b
Lawton, OK	53.95	Spiral sink	A1b
Lewiston, ME	44.40	Long-term oscillatory	A5bii
Lexington, KY	136.02	Medium-term oscillatory	A6ciii
Lincoln, NB	84.60	Spiral sink	A1b
Little Rock, AR	173.50	Long-term, linear up	A5ai
Long Beach, NY	228.50	Medium-term oscillatory	A6ciii
Long View, TX	61.50	Medium-term oscillatory	A6a
Los Angeles, CA	3232.00	Perturbed medium-term oscillation A6b	C1b
Louisville, KY	482.00	Medium-term oscillatory	A6b
Lynchburg, VA	64.80	Perturbed long-term A5bii oscillation	C1a
Madison, WI	144.80	Medium-term oscillatory	A6b
Manchester, WA	175.50	Medium-term oscillatory	A6a
Mansfield, OH	59.60	Medium-term oscillatory	A6b
McAllen, TX	108.00	Perturbed long-term A5ai oscillation	C1a
Melbourne, FL	105.00	Discontinuity from one steady state to another	C2b
Memphis, TN	405.50	Medium-term oscillatory	A6a

SMSA**	Size§ X10^{-5}	Description	Type of Event
Miami, FL	662.50	Long-term, linear up	A5ai
Midland, TX	33.60	Medium-term oscillatory	A6b
Milwaukee, WI	658.60	Perturbed long-term A6b oscillation	C1a
Minneapolis, MN	943.00	Medium-term oscillation with inflection point	A6ciii
Mobile, AL	195.70	Spiral sink	A1a
Modesto, CA	110.50	Long-term, linear up	A5ai
Monroe, LA	61.00	Perturbed, long-term A5ai oscillatory	C1a
Montgomery, AL		Unclassified	D
Muncie, IN	59.48	Discontinuity following a long-term A5ai motion	C2b
Muskegon, MI	83.17	Medium-term oscillation	A6b
Nashville, TN	355.60	Long-term oscillatory	A5bi
Nassau, NY	1237.50	Medium-term oscillation with inflection point	A6ciii
Newbedford, MA	211.00	Unclassified	D
New Brunswick, NJ	271.00	Medium-term oscillation with inflection point	**A6ciii**
New Haven, CT	349.50	Steady state with discontinuity	C2b
New London, CT	112.50	Medium-term oscillation with inflection point	A6ciii
New Orleans, LA	521.00	Spiral sink	A1b
New York, NY	4350.00	Long-term, linear down	A5aii
Newark, NJ	907.50	Steady state with a discontinuity	C2b
Newport, VA	166.00	Not enough data	E
Norfolk, VA	365.50	Not enough data	E
Northeast, PA	290.00	Long-term oscillatory	A5bii
Odessa, TX	47.80	Spiral sink	A1b
Oklahoma City, OK	356.75	Long-term oscillatory	A5bi
Omaha, NB	267.75	Spiral sink	A1b
Orlando, FL	275.00	Long-term, linear up	A5ai
Owensboro, FL	37.35	Spiral sink plus discontinuity	C2b
Ozmond, CA	214.50	Medium-term oscillatory with inflection point	A6ciii
Panama City, FL	42.00	Spiraling sink	A1b
Parkersburg, WV	71.00	Sink with steady state	A1a
Pascagolo, MS	52.75	Long-term, linear up	A5ai
Patterson, NJ	211.60	Sink spiral with a discontinuity	C2b
Pensacola, FL	127.20	Long-term, linear up	A5ai
Peoria, IL	164.50	Spiral sink	A1b
Petersburg, VA	58.50	Perturbed spiral sink (A1b)	C1b
Philadelphia, PA	2210.00	Sink spiral with a discontinuity	C2b

SMSA**	Size $\times 10^{-5}$	Description	Type of Event
Phoenix, AZ	580.00	Long-term, linear up	A5ai
Pine Bluff, AR	39.20	Steady state with discontinuity	C2b
Pittsburgh, PA	1050.00	Long-term, linear down	A5aii
Pittsfield, MA	66.75	Long-term, linear down	A5aii
Portland, ME	107.00	Steady state succeeded by a discontinuity	C2b
Portland, OR	515.50	Perturbed long-term A5ai motion	C1a
Poughkeepsie, NY	107.85	Medium-term oscillatory with inflection point	A6ciii
Providence, RI	394.50	Long-term, linear down	A5aii
Provo, UT	83.00	Long-term oscillatory	A5bi
Pueblo, CO	56.70	Spiral sink	A1a
Racine, WI	81.15	Steady state with discontinuity	C2b
Raleigh, NC	223.20	Perturbed long-term oscillatory, A5ai	C1a
Rapid City, SD	41.00	Medium-term oscillatory	A6b
Reading, PA	139.70	Long-term oscillatory	A5bii
Reno, NV	72.80	Long-term oscillatory	A5bi
Richland, WA	54.90	Perturbed long-term oscillatory, A5ai	C1a
Richmond, VA	278.00	Medium-term oscillatory with inflection point	A6ci
Riverside, CA	572.00	Long-term, linear up	**A5ai**
Rochester, MN	42.00	Medium-term oscillatory	A6a
Sacramento, CA	427.90	Long-term, linear up	A5ai
St. Cloud, MN	72.40	Medium-term oscillatory	A6a
St. Joseph, MO	46.25	Long-term, linear down	A5aii
Salem, OR	101.00	Perturbed, long-term oscillatory A5ai	C1a
Salinas, CA	127.70	Long-term, linear up	A5ai
Salt Lake City, UT	378.00	Long-term, linear up	A5ai
San Angelo, TX	35.45	Spiral sink	A1b
Santa Barbara, CA	133.20	Medium-term oscillatory with inflection point	A6ciii
Santa Cruz, CA	78.40	Long-term oscillatory	A5bi
Santa Rosa, CA	118.50	Long-term, linear up	A5ai
Sarasota, FL	77.50	Long-term, linear up	A5ai
Savannah, GA	100.00	Spiral sink	A1b
Seattle, WA	665.00	Spiral sink with steady state	A1a
Sherman, TX	38.60	Long-term oscillatory	A5bii
Sioux City, IA	55.40	Long-term, linear down	A5aii
Sioux Falls, SD	47.00	Perturbed, long-term oscillatory A5ai	C1a
Springfield, IL	85.30	Medium-term oscillatory	A6a
Springfield, MO	88.50	Medium-term oscillatory	A6a

SMSA**	Size§ X10^{-5}	Description	Type of Event
Springfield, OH	84.35	Steady state with a discontinuity	C2b
Springfield, MA	271.50	Unclassified	D
Steubenville, WV	75.50	Steady state with a discontinuity	C2b
Tallahassee, FL	63.00	Long-term, linear up	A5ai
Tampa, FL	633.00	Long-term oscillatory	A5bi
Terre Haute, IN	80.00	Sink spiral	A1b
Texarcoma, TX	55.80	Sink spiral with steady state	A1a
Toledo, OH	356.00	Sink spiral	A1b
Topeka, KS	84.50	Sink spiral	A1b
Trenton, NJ	146.10	Steady state	B
Tuscon, AZ	210.00	Long-term, linear up	A5ai
Tulsa, OK	282.50	Medium-term oscillatory	A6a
Tuscoloosa, AL	57.25	Steady state	B
Tyler, TX	50.00	Steady state with a discontinuity	C2b
Utica, NY	152.00	Long-term motion	A5cii
Vallejo, CA	148.00	Long-term, linear up	A5ai
Vineland, NJ	61.00	Spiral sink	A1b
Waco, TX	73.50	Spiral sink	A1b
Washington, DC	1315.00	Medium-term oscillatory	A6b
Waterloo, IA	63.20	Spiral sink	A1b
West Palm Beach, FL	220.00	Long-term, linear up	A5ai
Wheeling, WV	83.00	Long-term oscillatory	A5bii
Wichita, KS	180.20	Spiral sink	A1b
Wichita Falls, TX	60.50	Medium-term oscillatory	A6b
Williamsport, PA	52.50	Long-term, linear down	A5aii
Wilmington, DE	238.00	Medium-term oscillatory	A6b
Wilmington, NC	60.00	Perturbed medium-term oscillatory, A6b	C1b
Worcester, MA	295.00	Long-term, linear down	A5aii
Yakima, WA	73.00	Sink spiral	A1b
York, PA	162.50	Sink spiral	A1b
Youngstown, OH	250.00	Steady state succeeded by a discontinuity	C2b

*175 SMSAs are included in here; 90 SMSAs were included in [1]. Three SMSAs do not have sufficient data to classify (E).

**Only the first Central City and first state are listed in the name.

§The size is as of July 1, 1977.

Table A4

The list of 27 SMSAs* analyzed with NPS employing as basis the U.S.
total urban[1] population count.

Name of SMSA**	Size§ X10⁻⁵	Description	Type of Event
Abilene, TX	82.32	Discontinuity	C2b
Anderson, IN	87.06	Long-term, linear down	A5ai
Baltimore, MD	1353.00	Medium-term oscillatory	A6b
Bloomington, IN	578.10	Unclassified	D
Canton, OH	253.30	Long-term oscillatory	A5cii
Cleveland, OH	1227.00	Long-term oscillatory	A5cii
Daytona Beach, FL	134.80	Long-term oscillatory	A5bi
Elmira, NY	62.47	Long-term, linear down	A5aii
Ft. Lauderdale, FL	545.40	Medium-term oscillatory	A6a
Grand Rapids, MI	364.10	Medium-term oscillatory	A6a
Houston, TX	1578.00	Long-term, linear up	A5ai
Johnstown, PA	168.90	Long-term oscillatory	A5bii
Lafayette, IN	72.11	Spiral sink	A1a
Lexington, KY	185.80	Medium-term oscillatory	A6a
Lynchburg, VA	91.97	Medium-term oscillatory	A6a
Milwaukee, WI	894.10	Long-term, linear down	A5aii
Nassau, NY .	295.70	Long-term oscillatory	A5bii
Odessa, TX	64.80	Spiral sink	A1a
Pensacola, FL	174.10	Long-term, linear up	A5ai
Poughkeepsie, NY	147.80	Spiral sink	A1a
Richmond, VA	380.39	Medium-term oscillatory	A6a
St. Louis, MO	1503.00	Long-term, linear down	A5aii
Santa Cruz, CA	107.00	Long-term oscillatory	A5bi
Spokane, WA	197.30	Spiral sink	A1b
Tampa, FL	866.00	Long-term, linear up	A5ai
Utica, NY	207.60	Long-term, linear down	A5aii
Williamsport, PA	71.86	Long-term, linear down	A5aii

*27 randomly selected SMSAs from the sample of all (265) U.S. SMSAs:
every tenth from an alphabetical order.

**Only first Central City and State mentioned.

§as of July 1, 1977.

[1]Total urban population is the sum of all counties' population for all
265 SMSAs according to their 1977 composition.

PHASING-OUT OF THE SUGAR INDUSTRY IN PUERTO RICO

Daniel A. Griffith

State University of New York at Buffalo
United States

1. INTRODUCTION

This is the second in a series of papers dealing with the rise and fall of the commercial sugar industry in Puerto Rico (see Griffith, 1982). This paper has three major goals. Its first objective is to conduct a preliminary analysis of the space-time trajectory followed by an unfolding and then contraction of the Puerto Rican sugar industry. A more indepth analysis of this trajectory will be reported on subsequently. The second objective, which is a methodological one, is to construct a space-time data cube that will permit the exploration of the phasing-out of the Puerto Rican sugar industry. The third objective is to explore, in a preliminary fashion, features essential to the sketching of an optimal dynamic location-allocation trajectory for the mill closure sequence.

Research results to be presented here will be organized into four sections. In the first section two simple, coarse models of firm closures are described. One model attempts to predict physical plant closures on the basis of production parameter values. The other model considers the importance of an 85-90% weight loss in sugarcane processing, and attempts to predict physical plant closures on the basis of the changing geographical configuration of plants coupled with their locations relative to major urban areas.

This research was funded through a Faculty Fellowship from the Baldy Center for Law and Social Policy, State University of New York at Buffalo.

In the second and third sections a space-time data cube is constructed, and evaluated. United States Department of Agriculture census figures furnish observed data for selected years by municipio, for this cube. The Orchard and Woodbury (1972) missing information principle is used as the basis for constructing the unobserved data for this cube. Various statistical criteria are employed in order to try to distinguish between the constructed and observed portions of the data cube. Also, a STAR (i.e., space-time autoregressive) model is fitted to these data.

The fourth section of this paper views the closure sequence as a dynamic location-allocation optimization problem. Preliminary results are reported, and the necessary computer software is discussed. Results summarized in this section clearly show a need to employ an algorithm that uses capacity constraints.

2. TWO SIMPLE CLOSURE MODELS

The first model of closures to be discussed here relates physical plant production parameters to the closure sequence. The following Cobb–Douglas type production function was assumed:

$$O = M D^\alpha T^\beta R^\gamma \quad , \quad (2.1)$$

where O = the output percentage of sugar extracted from a ton of cane,

D = the annual number of days a mill was in operation,

T = the total tonnage of sugarcane processed,

R = the tons of cane ground per hour, and

M, α, β, γ = parameters.

For this production process sugarcane is a raw input analogous to iron ore in the production of steel, whereas the amount of sugar extracted is analogous to the amout of steel produced. Equation (2.1) was converted to its log–log counterpart, and calibrated using ordinary least squares regression on a 1938–1950 time series for each of 41 physical plants (United States Department of Agriculture, 1938–1950). This calibration was done first ignoring serial correlation, denoted as the unadjusted results in Table 1, and then adjusting for it, denoted as the adjusted results in Table 1. Adjustments were made in the standard econometric fashion, such that equation (2.1) was rewritten as

$$\ln(O_t) = \ln(M) + \alpha[\ln(D_t) - \ln(D_{t-1})] + \beta[\ln(T_t) - \ln(T_{t-1})]$$
$$+ \gamma[\ln(R_t) - \ln(R_{t-1})] + \rho\ln(O_{t-1}) \quad ,$$

where t denotes a year in the time series,

t-1 denotes the preceding year in this time series, and

ρ denotes the serial correlation latent in the time series of percentage of sugar extracted.

The results of these analyses appear in Table 1. Whereas the scale parameter M seems to be the only significant production parameter, these findings nevertheless demonstrate spatial variation in the production process. Further, the model fits tend to be reasonably good, which largely may be attributable to the small sample sizes involved.

These numerical results permit sugar mills to be grouped according to the signs of their production function parameter values. Table 2 summarizes these groups, and gives their corresponding sugar mill frequencies for both the adjusted and unadjusted OLS calibrations. The first group to be identified has been labelled long-term waste minimizers. Sugar mills belonging to this group are characterized by the following three tendencies: as the percentage of sugar extracted from a ton of cane increases,

(a) the annual number of operating days tends to increase,

(b) the total tonnage of sugarcane ground tends to decrease, and

(c) the grinding rate tends to increase.

In other words, these sugar mills benefit from an increased operating season, from techniques that extract a comparatively high percentage of sugar from cane, and from relatively fast handling of small volumes of sugarcane.

The second group of mills that has been identified is characterized by the following three tendencies: as the percentage of sugar extracted from a ton of cane increases,

(a) the annual number of operating days tends to decrease,

(b) the total tonnage of sugarcane ground tends to increase, and

(c) the grinding rate tends to increase.

TABLE 1

PRODUCTION FUNCTION PARAMETERS[+] BY
SUGAR MILL FOR THE PERIOD 1938-1950

Mill	Model Type[++]	$\hat{M}$	$\hat{\alpha}$	$\hat{\beta}$	$\hat{\gamma}$	R	n	Durbin-Watson Statistic	$\hat{\rho}$
AGUIRRE	A	3.73278	.01281	-.11158	.03917	.53261	13	1.37609	.41729
	U	3.38883	.09307	-.15831	.22979	.48891	8	2.38205	
BOCA CHICA	A	2.65088*	.04751	-.07102	.09734	.89651	13	2.54037	-.74172
	U	3.96071	.10362	-.12994	-.11424				
CAÑABALANCHE	A	4.94740***	-.04932	-.17650	-.00147	.70467	13	1.49916	-.48664
	U	4.66015	-.01335	-.16758	-.00206				
CANOVANAS	A	3.13312***	-.04727	-.09326	.15657	.81296	9	1.82843	.06291
	U	3.43264	.03022	-.15391	.17462				
CARIBE	A	.99977	.05252	-.29366*	1.36307*	.95321	8	1.56760	.35541
	U	2.69510	.01432	-.20304	.61924				
CARMEN	A	3.40581	-.08781	-.19119	.41901	.71898	13	2.00026	-.07621
	U	3.46037	-.04087	-.26056	.54615				
CAYEY	A	2.40937**	-.27789	.17424	-.16207	.84299	13	2.02014	-.07106
	U	1.79623	-.45060	.32575	-.23945				
COLOSO	A	3.45447***	-.21049**	.05526	-.12112*	.82384	13	1.24986	-.28807
	U	3.66873	-.18302	.01078	-.07704				
CONSTANCIA-PONCE	A	3.06046***	-.05648	-.11061	.28571**	.80929	13	1.61578	-.03385
	U	3.14704	-.06378	-.14088	.37017				
CONSTANCIA-TOA	A	2.39033***	-.51218**	.32059*	-.31639*	.74476	13	2.51800	.58986
	U	3.11755	-.30911	.05668	.01953				
CORTADA	A	-2.69312	.25967	.21150	1.53601	--	9	--	-.44073
	U	-1.52031	.17349	-.10589	1.05584				
DEFENSA	--	--	--	--	--	.43603	1	2.27048	-.16606
EL EJEMPLO	A	2.23932*	-.29076	.22949	-.26729		13		
	U	1.93811	-.31071	.20627	-.10123				
EUREKA	A	4.14264	-.19319	-.04552	-.02337	.79522	13	1.35124	.60029
	U	3.46249	-.21493	-.00742	.04639				

TABLE 1 (Continued)

PRODUCTION FUNCTION PARAMETERS† BY
SUGAR MILL FOR THE PERIOD 1938-1950

Mill	Model Type††	M̂	α̂	β̂	γ̂	R	n	Durbin-Watson Statistic	ρ̂
FAJARDO	A	3.24605**	-.13252	-.04074	.07234	.80411	13	1.36441	.18270
	U	2.68041	-.01862	-.20018	.46024				
GUAMANI	A	3.18843	.01483	-.08747	.68467	.45956	13	1.14163	.34201
	U	3.45009	.05022	-.14789	.13280				
GUANICA	A	3.14679	-.09309	-.00792	-.76769	.48017	13	.90645	.59312
	U	2.84202	-.14652	.03752	-.01552				
HERMINIA	A	3.98054***	.07801	-.21126	.07828	.86566	10	1.64520	.11581
	U	3.97520	.09571	-.21780	.07423				
IGUALDAD	A	3.30024***	-.14079	.02700	-.09346	.91601	13	2.52929	-.50137
	U	3.02946	-.20465	.11078	-.19353				
JUANITA	A	3.73259***	.06554	-.17593	.12201	.73644	13	1.33597	.79503
	U	4.55815	.09181	-.27740	.18765				
JUNCOS	A	3.23942***	-.11459	.03642	-.13645	.59058	13	1.50930	.31407
	U	3.00996	-.23448	.14238	-.24261				
LAFAYETTE	A	3.69516***	-.08614	-.17300	.29647*	.80606	13	1.33284	.16908
	U	3.66429	-.06631	-.19303	.33388				
LOS CANOS	A	3.69743***	-.17634	-.01214	-.05254	.81792	13	2.03724	-.03923
	U	3.51819	-.38153	.10968	-.12329				
MACHETE	A	3.49004***	.13145	-.36790***	.66270*	.72610	13	1.29635	.09348
	U	3.85340	.16233	-.42894	.71273				
MERCEDITA	A	2.52728***	-.16093	.06010	.00490	.53695	13	1.38792	.33941
	U	2.38312	-.22908	.11699	-.04398				
MONSERRATE	A	2.86259***	-.38154	.20045	-.22723	.58066	13	1.67050	.58891
	U	3.85236	-.06816	-.16404	.20013				
PASTO VIEJO	A	2.69907	-.39705*	.22776	-.25354	.66590	13	1.43315	.15494
	U	1.88224	-.22951	-.04147	.43027				
PELLEJAS	A	2.94073***	-.19855	-.06277	-.05540	.63835	12	2.09549	-.29729
	U	3.08612	-.06027	-.01538	-.05615				

TABLE 1 (Continued)

PRODUCTION FUNCTION PARAMETERS† BY
SUGAR MILL FOR THE PERIOD 1938–1950

Mill	Model Type††	$\hat{M}$	$\hat{\alpha}$	$\hat{\beta}$	$\hat{\gamma}$	R	n	Durbin-Watson Statistic	$\hat{\rho}$
PLATA	A	3.39719***	-.32372***	.12767	-.18948	.94978	13	1.87330	.21649
	U	3.46056	-.28246	.09485	-.15874				
PLAYA GRANDE	A	4.67485	-.23873	.19710	-.86323	.96405	5	1.68461	.98878
	U	7.38446	-.29217	.17987	-1.14369				
PLAZUELA	A	4.01411***	-.20918*	.06014	.15101	.74391	13	1.41765	-.04408
	U	4.59427	.12289	-.32969	.27958				
RIO LLANO	A	2.26417***	-.68207	.37911	-.23182	.72290	12	2.14812	.54084
	U	4.89128	-.01287	-.18373	-.05405				
ROCHELAISE	A	2.49101***	-.25365	.08288	.07762	.70805	13	1.09861	.80071
	U	1.77391	-.10349	-.02532	.40256				
ROIG	A	5.77417***	-.25059**	.07498	-.63780****	.91176	13	2.00078	-.16979
	U	5.92701	-.24935	.07212	-.66290				
RUFINA	A	3.31498*	-.19017	.04740	-.10541	.52009	13	.98787	.56095
	U	2.94959	-.19711	.03573	.01337				
SAN FRANCISCO	A	2.88603	-.15307	-.02665	.21518	.75421	13	2.74469	-.68415
	U	2.47833	-.08403	-.06335	.37017				
SAN JOSE (VANNINA)	A	3.91776***	-.03732	-.07296	-.09270	.79882	13	1.58156	.65928
	U	4.31820	-.07520	-.15356	.07956				
SAN VICENTE	A	3.30539***	-.33041	.17531	-.29160	.82249	13	1.78417	.50080
	U	4.21386	-.10489	-.10204	.01350				
SANTA BARBARA	A	3.35109***	.16181	-.14535	-.06489	.59595	11	1.28306	.58908
	U	3.38609	.30582	-.25378	.06295				
SANTA JUANA	A	4.58339***	.16952	-.34934*	.29535	.88440	13	1.22911	.10202
	U	4.43824	.18055	-.38250	.40015				

TABLE 1 (Continued)

PRODUCTION FUNCTION PARAMETERS[+] BY
SUGAR MILL FOR THE PERIOD 1938–1950

Mill	Model Type[++]	$\hat{M}$	$\hat{\alpha}$	$\hat{\beta}$	$\hat{\gamma}$	R	n	Durbin-Watson Statistic	$\hat{\rho}$
SOLLER	A	3.94454	-.18649*	-.05618	.02312	.82236	13	1.24930	
	U	3.94202	-.20216	-.05172	.03072				.33431
VICTORIA	A	4.23444***	-.02623	-.11345	-.06031	.82794	13	1.37557	
	U	4.01194	-.02157	-.12842	.02512				.14636

* Denotes a significant difference from zero at the 0.10 level.

** Denotes a significant difference from zero at the 0.05 level.

*** Denotes a significant difference from zero at the 0.01 level.

[+] $\hat{M}$ is the scale parameter; $\hat{\alpha}$ is the exponent for annual number of days a mill was in operation; $\hat{\beta}$ is the exponent for tons of sugar cane ground; and $\hat{\gamma}$ is the exponent for the tons/hour grinding rate.

[++] A indicates no correction for serial correlation, while U indicates a correction was made for serial correlation.

TABLE 2

GENERIC PRODUCTION PARAMETER CLASSES:
A TAXONOMY AND INTERPRETATION

Pattern of Signs			Frequency of Occurrence		Interpretation
$\hat{\alpha}$	$\hat{\beta}$	$\hat{\gamma}$	Autocorrelated Parameters	Unautocorrelated Parameters	
+	+	+	0	0	------------
+	+	−	0	0	------------
+	−	+	16	11	Sugar mills that benefit from a long operating season, experience diseconomies of scale in the handling of sugarcane, and are efficient in their grinding: long-term waste minimizers.
−	+	+	3	2	Sugar mills that benefit from a short operating season, can realize economies of scale in the handling of sugar-cane, and are efficient in their grinding: prudent firms.
+	−	−	1	1	Sugar mills that benefit from a long operating season, but experience diseconomies of scale as well as inefficiencies in their grinding: slack firms.
−	+	−	8	11	Sugar mills that benefit from a short operating season, experience economies of scale in the handling of sugar-cane, but are inefficient in their grinding: volume/efficiency trade-off firms.
−	−	+	7	13	Sugar mills that benefit from a short operating season, experience diseconomies of scale in the handling of sugarcane, but are efficient in their grinding: short-term waste minimizers.
−	−	−	6	3	Sugar mills that benefit from a short operating season, experience diseconomies of scale in the handling of sugarcane, and are inefficient in their grinding: marginal firms.

Sugar mills belonging to this group have been labelled prudent firms. They benefit from a short season in which relatively large quantities of cane can be ground at a fast rate. In other words, these mills would benefit by taking advantage of annual peaks in the sugar industry.

The third group of mills has been labelled slack firms. Sugar mills belonging to this group are characterized by the following three tendencies: as the percentage of sugar extracted from a ton of cane increases,

(a) the annual number of operating days tends to increase,

(b) the total tonnage of sugarcane ground tends to decrease, and

(c) the grinding rate tends to decrease.

In other words, these firms are better off when they process small amounts of sugarcane, over long periods of time. The increased cost of this type of strategy would be partly offset by increased extraction due to more thorough grinding.

The fourth group of mills is characterized by the following three tendencies: as the percentage of sugar extracted from a ton of cane increases,

(a) the annual number of operating days tends to decrease,

(b) the total tonnage of sugarcane ground tends to increase, and

(c) the grinding rate tends to decrease.

Sugar mills belonging to this group benefit from high volume and fast processing. Hence, they should trade off volume against efficiency. This group of mills could be referred to as quick and sloppy.

The fifth group of mills has been labelled short-term waste minimizers. Sugar mills belonging to this group, in contradistinction with those in the first group, are characterized by the following three tendencies: as the percentage of sugar extracted from a ton of cane increases,

(a) the annual number of operating days tends to decrease,

(b) the total tonnage of sugarcane ground tends to decrease, and

(c) the grinding rate tends to increase.

Moreover, these mills benefit from short but efficient seasons in which relatively small amounts of cane are thoroughly ground. Regardless of whether or not parameter estimates have been adjusted for the presence of serial correlation, approximately 50% of all mills fall into the two categories of short- and long-term waste minimizers.

The final group of sugar mills is characterized by the following three tendencies: as the percentage of sugar extracted from a ton of cane increases,

(a) the annual number of operating days tends to decrease,

(b) the total tonnage of sugarcane ground tends to decrease, and

(c) the grinding rate tends to decrease.

These mills have been labelled marginal, for their best production situation occurs as they approach an inactive state. Diseconomies of scale exist for volume of cane processed as well as number of days these mills operate. These deterrents are coupled with inefficient grinding.

Once these clusters of sugar mills were identified and interpreted, the following null hypothesis was formulated:

H_0:   the sequence of mill closures is a function of the individual mill production parameters.

This hypothesis was evaluated by ranking mills in their order of closure, and then regressing this ranking on the matrix of production parameters. Some problems arise when doing this, since ordinal data are being treated as interval/ratio data, and little is known about the correlation between variates for these two scales. However, the purpose here is an exploratory one, and to some degree the ordinal scale has been modified because of mill ties in the closure sequence. Results of this regression appear in Table 3. Both autocorrelated and unautocorrelated production parameter regression models were estimated. Each of these data sets was analyzed in order to see whether or not their respective underlying structures implied different conclusions. Very similar results would suggest that the presence of serial correlation in the time series of percentage of sugar extracted did not obscure relationships between the mill closure sequence and mill production parameters. Different results would suggest that the presence of such serial correlation did mask these relationships.

TABLE 3

REGRESSION MODELS OF THE CLOSING
SEQUENCE FOR PUERTO RICAN SUGAR MILLS

Model Type	Variable	Regression Coefficient	Multiple R	Durbin-Watson Statistic
Autocorrelated	B_0	24.98877***	.27901	.12821
Production Parameters[†]	$\hat{M}$	-1.66647		
	$\hat{\alpha}$	-21.90382		
	$\hat{\beta}$	-30.08423		
	$\hat{\gamma}$	-11.55069		
Unautocorrelated	B_0	-9.28656	.40183	.26661
Production Parameters[†]	$\hat{M}$	12.69400		
	$\hat{\alpha}$	47.56907		
	$\hat{\beta}$	156.46190		
	$\hat{\gamma}$	49.79336		
Geometric	B_0	-8.15584		
Configuration of Mills	average distance to active mills	-.09484	.83913	.94716
	distance to closest, most recently closed mill	.15885		
	distance to closest active mill	.99632***		
	distance to Arecibo	-.25752*		
	distance to Caguas	-.31971		
	distance to Mayaguez	.15734		
	distance to Ponce	.21297		
	distance to San Juan	.74483**		

* Denotes a significant difference from zero at the 0.10 level.

** Denotes a significant difference from zero at the 0.05 level

*** Denotes a significant difference from zero at the 0.01 level

[†] See Table 1 for production parameter definitions.

The regression equation for autocorrelated production parameters suggests that sugar mills that benefit the most from scale economies, efficient grinding, and long operating seasons tend to close first. However, roughly only 8% of the variance in the closure ranking is accounted for by these production parameters. Furthermore, only the intercept term b_0 is found to be significant. On the other hand, the regression equation for the unautocorrelated production parameters suggests the completely opposite conclusion. Sugar mills that benefit the most from scale economies, efficient grinding, and long operating seasons tend to close last. Roughly 16% of the variance in the closure ranking is accounted for here. However none of the regression coefficients are significantly different from zero in this model. Neither of these regression equations is very powerful, for both are poor predictors of the closure sequence, and both display serial correlation in their residuals. Nevertheless, if serial correlation is not corrected for in the mills' timer series, counter-intuitive results are obtained. One can conclude in general that sugar mill closings appear to be related to a dimension other than the production process.

As was mentioned earlier, sugar processing results in 85-90% weight loss. The classical Weberian industrial location model maintains that in this kind of industry the distances from raw material sources will tend to play an important role in location decisions. With more than one mill, the Weberian problem becomes a multi-locational firm type of problem. Accordingly, a set of n physical plants should be located in such a way that interplant distances are maximized, and assembly and distribution costs are minimized. The optimal locational configuration is complicated by a changing geographical distribution of sugarcane harvest with the passing of time. As such, locationally advantageous sites at one point in time become marginal sites at later points in time, and vice versa. Meanwhile, the major urban centers serve as market locations, since most Puerto Rican sugar is exported, usually in the form of either refined sugar or rum. Therefore, the following null hypothesis related to locational configuration was posited:

H_0: the sequence of mill closures is a function of individual mill locations within the configuration of mills and major urban areas.

The map appearing in Figure 1 indentifies sugar mill locations, as well as the municipio partitioning of Puerto Rico. This hypothesis was evaluated by considering nearest neighbor distances to both active and most recently closed mills, as well as average distance to all active mills. In addition, distance to each of the five major urban centers (i.e., Arecibo, Caguas, Mayaguez, Ponce and San Juan) was included in the analysis. Results for regressing the closure ranking on this matrix of distances appear in Table 3, too.

Distance to the closest active mill was found to be highly significant, and positively related to the ranking. This distance variate tends to measure spatial competition. As time passes, distance between nearest neighbor active mills and mills that close tends to increase, in an attempt to minimize interplant competition for the sugarcane input. Meanwhile, distance to Arecibo displays an inverse relationship to the closure ranking, whereas distance to San Juan displays a direct relationship. This finding is not surprising, since Arecibo is the site of much rum production, while

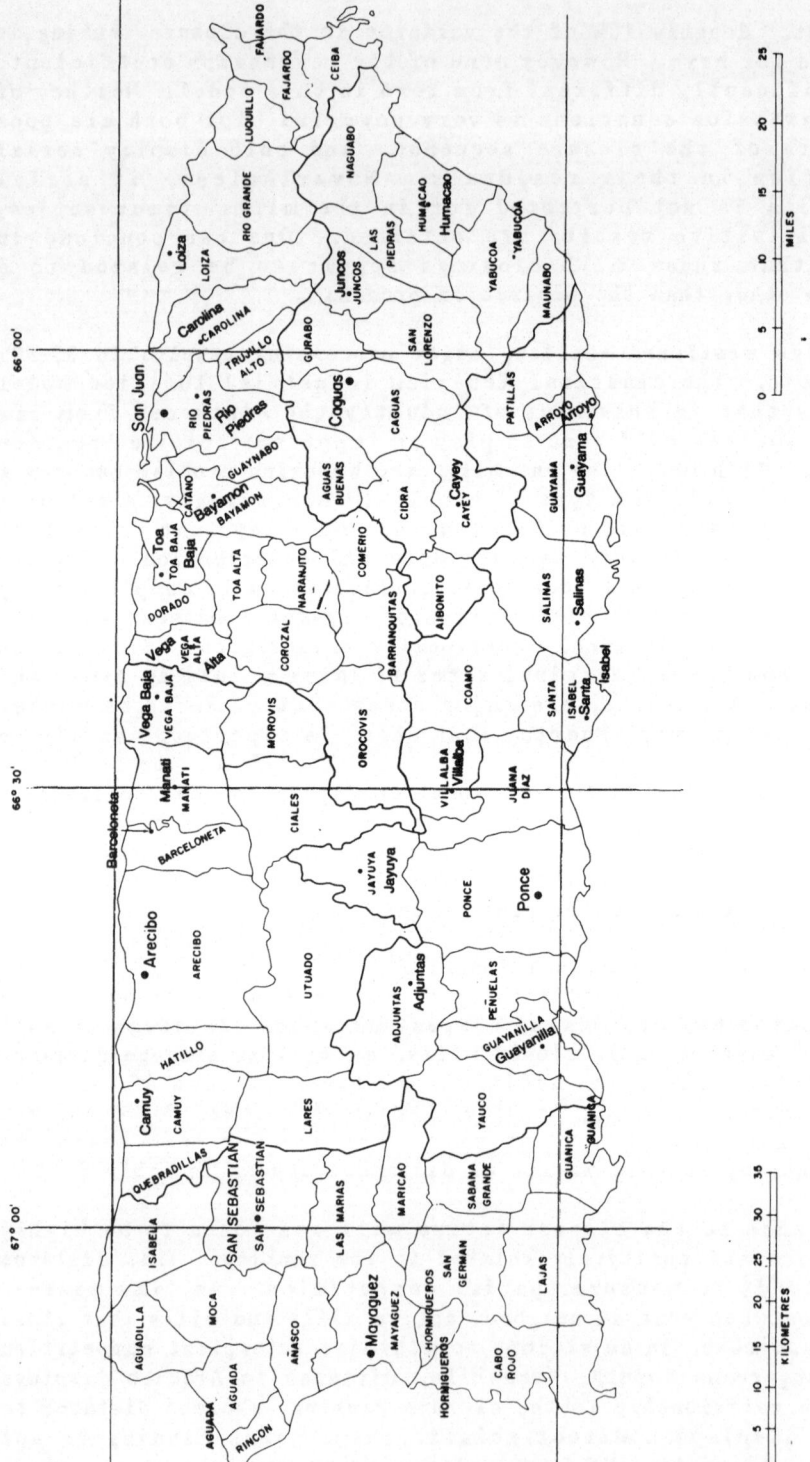

Figure 1. Map showing the partitioning of Puerto Rico into municipios, and the location of sugar mills.

urbanization associated with San Juan is displacing most agricultural and older industrial activities located in its hinterland. Moreover, physical plants supplying extracted sugar for rum production will tend to have a locational orientation towards Arecibo. Areas experiencing extensive urbanization will have their mill sites changed to retail or residential land use, because these two kinds of land use will command a higher location rent than will agricultural activities as the urban area expands outward from its core. All in all, these configuration proxies account for roughly 70% of the variance in the closure ranking.

This analysis suggests that geographical/accessibility variables play a much more important role in explaining mill closures over time than do variables related to production processes. The geometric model of the configuration hypothesis that results also displays negligible serial correlation in its residuals, having a Durbin–Watson statistic value of nearly 1, whereas both of the production parameter regression models have values close to zero.

3.   CONSTRUCTION OF A SPACE–TIME DATA CUBE

Findings discussed so far imply that sugar mill closures are based upon geographic factors. In the context of Weberian analysis, these factors will include assembly costs, production costs, and distribution costs. The distribution of physical plants over the classes, in the previous section, indicate that production costs will tend to be spatially variant. With an 85–90% weight loss during the production process, the transport costs of collecting the raw materials will be sizeable, relative to the transport costs of distributing the final product. In order to undertake a more detailed analysis of the closure sequence, then, data must be collected for the spatial distribution of sugarcane harvest, and the cost of processing cane for each mill. This section will focus on sugarcane harvest data.

The United States Bureau of the Census has published sugarcane harvest data by municipio for the years 1959, 1964, 1969, 1974 and 1978. Unfortunately, the 1974 and 1978 data are incomplete. In addition, this government agency has published total sugarcane harvest on the island for these five years as well as for the years 1909, 1919, 1929, 1935, 1939, 1949 and 1954. The Caribbean Area Agricultural Stabilization and Conservation Service has published annual total tons of sugarcane processed for Puerto Rico, by mill, from 1937 thru 1980. These three data sets constitute the known information for constructing a space–time data cube.

The problem at this point is one of estimating any unknown data. Since spatially distributed data are available only beginning in 1959, the data cube will be for the period 1959–1978, covering 20 years. Estimation of these annual data needs to be constrained by total harvest figures for each year, since the estimated total harvest should equal the observed total harvest. Furthermore, censored municipio harvest figures in the 1974 and 1978 census data need to be estimated. Finally, spatial distributions for the inter–census years need to be estimated, since closures are occurring on a yearly basis. Each of these estimation problems will be dealt with in turn in this section. Then these estimates will be evaluated.

3.1. Construction of the Data Cube

A survey of the two time series of total sugarcane harvest and total sugarcane processed for Puerto Rico illustrate that shortly after the Spanish–American War (ca. 1900) the sugar industry began a commercial expansion, which then peaked in the mid–1950s, and finally returned to early 1900s levels by 1980. This time series may be described by a parabolic curve, which is quadratic in time. Accordingly, the following equation was fitted to the 1937 thru 1978 processed sugarcane time series, for which none of the yearly data were missing:

$$0 = \alpha + \beta T + \gamma T^2 \quad , \quad (3.1)$$

where 0 = the number of tons of sugarcane processed,

T = the year of the 20th century, and

α, β, γ = parameters.

Results obtained from an OLS calibration of equation (3.1), adjusting for the presence of serial correlation, appear in Table 4 under the heading of 'Mill Tonnage: 1938–1980.' Approximately 72% of the variance is accounted for, with virtually no serial correlation being observed in the adjusted time series. The linear trend appears to dominate the second–order trend. Therefore, data for processed sugarcane display a great deal of similarity in adjacent years, as well as a trend over the 50 years in question.

Unfortunately the time series of sugarcane harvest figures was incomplete. So next, since harvest tonnage and processed tonnage should be very similar, the Orchard and Woodbury missing information principle was used to estimate missing values in the 1937 thru 1980 harvest time series. The operationalization of this principle starts with a calibration of the following equation:

$$H_t = \alpha + \beta P_t \quad , \quad (3.2)$$

TABLE 4

REGRESSION MODELS OF TONS OF SUGAR CANE HARVEST
AND MILL INPUT (in millions of tons) IN PUERTO RICO: 1909-1980

Model Name	Variable	Statistic	Percent of Variance Accounted
Mill Tonnage: 1938-1980	Serial Correlation	.41249[1]	
	Constant	-18.04977***	
	Time	1.50888***	26.1
	Time^2	- .02369***	46.1
	Durbin-Watson Statistic	2.09039	
Missing Harvest[2] Data	Constant	.35287	
	Mill Output	.95242	99.9
Harvest: 1909-1980	Serial Correlation	.52681[3]	
	Constant	-10.59280***	
	Time	1.22552***	29.0
	Time^2	- .02448***	34.1
	Durbin-Watson Statistic	2.36751	

*** Denotes a significant difference from zero at the 1% level.

[1] Non-linear maximization of the likelihood function was achieved in 409 iterations using IMSL subroutine ZXMIN.

[2] The number of missing data values was 35; convergence was achieved with two iterations after initialization with the regression of the 8 known data pairs.

[3] Non-linear maximization of the likelihood function was achieved in 527 iterations using IMLS subroutine ZXMIN.

where H_t = the total tons of sugarcane harvested in year t,

P_t = the total tons of sugarcane processed in year t, and

α, β = parameters.

This calibration involved estimating α and β for the twelve known data points, estimating H_t for the 32 unknown data points, and then entering these estimates as observed data in a new regression of H_t on P_t. This procedure must be repeated until estimates of α and β converge to stable values. Equation (3.1) was fitted for this resulting time series of harvest data coupled with the four harvest values that pre-date 1937 (i.e., 1909, 1919, 1929 and 1935). The description rendered by this second calibration of equation (3.1) is consistent with that obtained for the processed sugarcane time series, as shown in Table 4. The estimated harvest figures subsequently will be used to constrain the estimation of annual spatial distributions.

A second task was to estimate censored municipio harvest figures in the 1974 and 1978 census data. Six of these missing values occurred in 1974, while 21 of them occurred in 1978. A modified version of the Orchard and Woodbury missing information principle was utilized here. After Haining, Griffith and Bennett (1982), the following equation was used for estimation purposes:

$$H_{m,t} = \alpha + \beta H_{m,t-1} + \rho(I_{m,m} - \rho W_{m,m})^{-1} W_{m,o}(H_{o,t} - \bar{H}_{o,t}) , \tag{3.3}$$

where W = a stochastic version of the binary connectivity matrix depicting juxtapositions of Puerto Rican municipios,

H_t = the vector of harvest figures for time t,

I = the identity matrix,

o, m = subscripts respectively denoting observed and missing data,

$\bar{H}_t$ = a vector whose elements all are the average harvest figure for time t, and

α, β, ρ = parameters.

Equation (3.3) was estimated separately for 1974 and 1978 annual harvest by municipio, subject to a total annual harvest constraint. Both results for equation (3.3) appear in Table 5, together with the estimated missing data values. A comparison of the relative sizes of the statistics suggest that temporal dependence is more pronounced than spatial dependence, in both calibrations for the tons of sugarcane harvested.

Before undertaking the third estimation task, a preliminary attempt was made to evaluate the quality of the estimates of the missing data. If the predicted values constitute good missing data estimates, then they should blend into patterns latent in the observed data. These estimates also should reflect the same type of covariations with other variates displayed by their observed

TABLE 5

REGRESSION MODELS[1] FOR ESTIMATING MISSING
CENSUS DATA FOR TONS OF SUGARCANE HARVEST IN
PUERTO RICO BY MUNICIPIO: 1974 & 1978

Year	Variable	Statistic	Municipio	Estimate
1974[2]	1974 Spatial Autocorrelation	.18562	Dorado	50251
			Guayama	129286
	Constant	52848.082	Loiza	28175
	1969 Yield	.49850	Rio Grande	12234
			Salinas	207567
			San Lorenzo	151608
1978[3]	1978 Spatial Autocorrelation	.14176	Caguas	0
			Cayey	2769
	Constant	40093.014	Ceiba	14890
	1974 Yield	.71554	Cidra	0
			Guayama	78539
			Guaynabo	0
			Humacao	80507
			Las Marias	5767
			Luquillo	11361
			Manati	27138
			Ponce	106222
			Rincon	14746
			Salinas	124317
			San Juan	0
			San Lorenzo	13313
			Toa Baja	48400
			Trujillo Alto	0
			Vega Alta	17909
			Vega Baja	68756
			Villalba	4297
			Yabucoa	90508

[1] Non-linear maximization of the likelihood function was achieved using IMSL subroutine ZXMIN subject to the constraints of (1) all estimates of missing values must be non-negative, and (2) the estimates plus the known values must sum to their respective regional totals.

[2] Convergence was achieved in 6 supraiterations, involving a total of 354 IMSL iterations.

[3] Convergence was achieved in 7 supraiterations, involving a total of 441 IMSL iterations.

value counterparts. One means of addressing these two concerns is to find out whether or not the underlying data dimensions are a function of the missing value estimates. The five census years were subjected to a principal components analysis, whose results appear in Table 6. Although 1974 and 1978 figures load onto a separate simple structure component, the principal component suggests that all spatial distributions are very similar. Furthermore, 1969 fails to align with either the first two or the

214

TABLE 6

DIMENSIONS OF THE COMPLETE SPACE-TIME TONS OF
SUGARCANE HARVEST CENSUS DATA PARALLELEPIPED FOR PUERTO RICO: 1959-1978

Variable	Unrotated . Principal Components Solution Loadings PCI	Varimax Rotated Principal Components Solution Loadings PCI	PCII	PCIII
1959	.88613	.90096	.30168	.27358
1964	.96074	.74532	.41642	.41197
1969	.91545	.37668	.41886	.80644
1974	.94499	.40575	.61936	.44917
1978	.93665	.40481	.80518	.40641
Eigenvalue	4.31662	1.83762	1.47178	1.26183
Percent of Variance Accounted	86.33	36.75	29.44	25.24

NOTE: Prominent component loadings are underlined.

last two years. These three dimensions uncovered here may be given generic labels for descriptive purposes. PCI uncovers a trend of declining similarity through time, and may be called an increasing dissimilarity trajectory component. PCII uncovers a trend of developing similarity through time, and may be called an increasing similarity trajectory component. PCIII uncovers a trend of developing similariy followed by declining similarity with the passing of time, and may be called a quadratic trajectory component. Unfortunately these findings are inconclusive, though.

The third task was to estimate the entire set of spatial distributions of the harvest data for the inter-census years. The undertaking of this task was motivated by a need for annual geographic distributions of sugarcane harvest, so that the annual mill closures could be properly evaluated. After considerable experimentation with constant and variable coefficient STAR models, the best estimates were found to be based upon differencing methods. In other words, $\Delta = (H_t - H_{t-k})/k$, and $H_t = H_{t-1} + \Delta$. The value for Δ was specific to consecutive pairs of census figures, with the resulting space-time series retaining the five years of observed census values. Once again, these estimates were constrained so that they summed to the estimated annual total tons of sugarcane harvested. Calibration results for the following STAR model appear in Table 7:

$$H_{i,t} = \alpha + (\beta + \gamma T + \delta T^2)H_{i,t-1} + \rho \sum_{j=1}^{j=73} w_{ij} H_{j,t-1} \quad , \quad (3.4)$$

where $H_{i,t}$ = the tons of sugarcane harvest in areal unit i at time t,

T = the year of the 20th century,

TABLE 7

VARIABLE COEFFICIENT STAR MODELS FOR SPACE-TIME TONS OF
SUGARCANE HARVEST DATA FOR PUERTO RICO: 1959-1978

Model	Variable	Statistic	Percent of Variance Accounted
Census Data[1]	Space-time Covariation	.13163***	.4
	Constant	.00001***	
	Lagged Harvest	1.6595 ***	76.0
	Time × Lagged Harvest	− .14414***	.9
	Time^2× Lagged Harvest	.00469***	1.5
Differencing Interpolation[2]	Space-time Covariation	.05970	----
	Constant	.00003	----
	Lagged Harvest	1.1096	----
	Time × Lagged Harvest	− .03534	----
	Time^2× Lagged Harvest	.00129	----
STAR Expectations for Census Years	Constant	−2559.45936***	98.9
	STAR Prediction	1.03176 ***	

** Denotes a significant difference from zero at the 5% level.
*** Denotes a significant difference from zero at the 1% level.

[1] Non-linear maximization of the likelihood function was achieved in 214 iterations using IMSL subroutine ZXMIN; "Percent of Variance Accounted" and significance levels were approximated by regressing synthetic variates from the MLE results on the original data.

[2] Convergence was achieved in 16 supraiterations, involving a total of 2700 IMSL iterations.

$$w_{ij} = \text{the element of matrix } \mathbf{W} \text{ for areal units i and j,}$$
and

$$\alpha, \beta, \gamma, \delta, \rho = \text{parameters.}$$

In order to evaluate the goodness-of-fit of equation (3.4), $H_{i,t}$ was compared with $H_{i,t}$ for the five census time periods only. The remaining 15 time period harvest yields were not used in this test because their values were estimated by the aforementioned differencing procedure. Hence, only data for the five recorded years could be viewed as being free of measurement error. Approximately 99% of the variance of the known census data was accounted for by equation (3.4). Very similar results were obtained for the estimated space-time cube, with predicted census year values being almost perfectly correlated with observed census year values (see Figure 2). Consequently, the spatio-temporal data set constructed here should be representative of the actual but unknown data set.

3.2. Statistical Evaluation of the Estimated Space-time Data Cube

Estimates calculated in the preceding section need to be scrutinized carefully before they can be labelled acceptable as inputs into subsequent research. A number of statistical tests for assessing the suitability of the data will be reported on before moving on. One goal of these tests is to determine whether or not successful discrimination can be achieved between the observed data and the estimated data. Another goal is to determine whether or not selected estimated spatial distributions are consistent with existing piecemeal data.

This first goal was attained by subjecting the entire space-time data cube to a principal components analysis, where years constituted the variables and municipios constituted the observations. Results from this analysis appear in Table 8. As in the census data, simple structure components identified an early period, a middle period, and a later period. Here these components respectively reveal a descending trend (PCI), an ascending trend (PCII), and a quadratic type trend (PCIII). The dominant unrotated principal component suggests that all spatial distributions are very similar. But the crucial finding is that the estimated values have not been isolated from the known census data. Nor have strictly intervening census periods been extracted. Therefore, the estimated data not only exhibit patterns similar to those appearing in Table 6, but they also provide a consistent space-time picture.

A second data set has been compiled by the Department of Agriculture of the Commonwealth of Puerto Rico (1970). These data provide annual figures for the period 1965 thru 1970 on area, measured in cuerdas, of sugarcane harvested, by municipio. Results

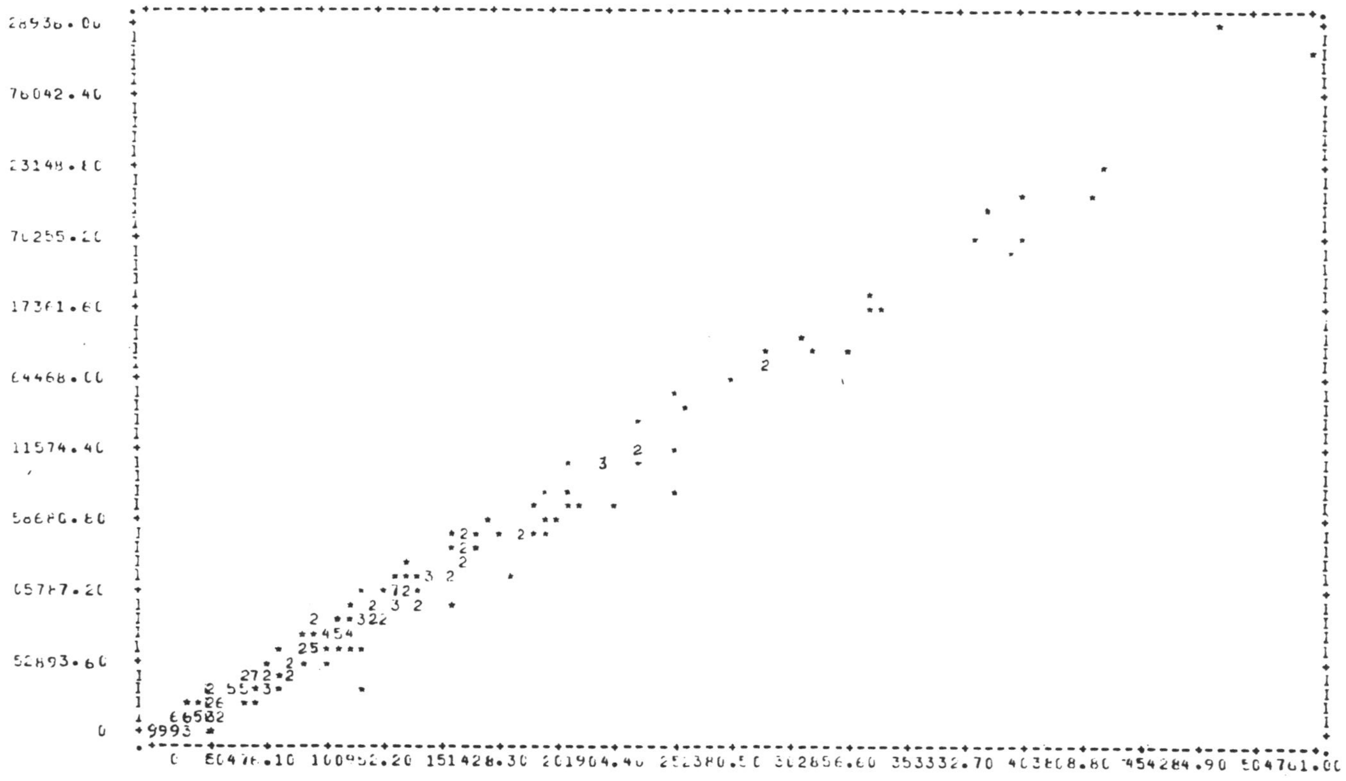

Figure 2. Scatter diagram for the predicted $H_{i,t}$ from equation (3.4) versus the actual tons of sugarcane harvested for the combined set of years 1964, 1969, 1974 and 1978.

TABLE 8

DIMENSIONS OF THE INTERPOLATED SPACE-TIME
TONS OF SUGARCANE HARVEST DATA PARALLELEPIPED FOR PUERTO RICO:
1959-1978

Variable	Unrotated Principal Components Solution Loadings	Varimax Rotated Principal Components Solution Loadings		
	PCI	PCI	PCII	PCIII
1959	.87819	.90242	.30472	.27822
1960	.90567	.88741	.33357	.31180
1961	.92787	.86718	.36049	.34358
1962	.94447	.84215	.38504	.37305
1963	.95547	.81298	.40692	.39980
1964	.96110	.78048	.42597	.42361
1965	.97963	.73031	.43826	.50326
1966	.98895	.66621	.44688	.58590
1967	.98549	.58728	.45010	.66781
1968	.96637	.49434	.44647	.74424
1969	.93042	.39040	.43522	.81032
1970	.94714	.40216	.48562	.77609
1971	.96157	.41397	.54393	.72597
1972	.97027	.42455	.61025	.66582
1973	.96734	.43157	.68285	.57851
1974	.94389	.43117	.75655	.46108
1975	.95312	.43827	.76930	.45650
1976	.95508	.44239	.77685	.44735
1977	.94685	.44213	.77675	.43231
1978	.92554	.43607	.76650	.41031
Eigenvalue	18.05303	7.71096	6.14140	5.90785
Percent of Variance Accounted	90.27	38.55	30.71	29.54

NOTE: Prominent component loadings are underlined.

of a principal components analysis for these six years is presented
in Table 9. Even in this short time series the previously
identified trends seem to be evident. These data also permit a
more fundamental check to be performed. Now the question that may
be raised asks whether or not common dimensions run across both
these area harvested spatial distributions and the corresponding
1965 thru 1970 estimated (and in the case of 1969, observed) tons
of sugarcane harvested spatial distributions. This question may be
answered with the help of the canonical correlation technique.

TABLE 9

DIMENSIONS OF THE SPACE-TIME CUERDAS OF SUGARCANE
HARVESTED DATA PARALLELEPIPED FOR PUERTO RICO:
1965-1970

Variable	Unrotated Principal Components Solution Loadings	Varimax Rotated Principal Components Solution Loadings	
	PCI	PCI	PCII
1965	.99028	.80229	.59477
1966	.99343	.79088	.60933
1967	.99661	.75835	.64486
1968	.99699	.71546	.68880
1969	.98518	.61206	.78810
1970	.98569	.61818	.77608
Eigenvalue	5.89692	3.11290	2.83873
Percent of Variance Accounted	98.28	51.88	47.31

NOTE: Prominent component loadings are underlined

Three significant dimensions were found to underlie both of
these data sets (see Table 10). Two were significant at the 1%
level, while one was significant at the 10% level. Before the
first linear combination of yearly tons and cuerdas harvested was
extracted, the chi-square statistic calculated for the generalized
variance of these two data matrices was found to be 243.66120.
Once this pair of linear combinations was factored out, new
generalized variances were calculated for the residual data values,
and then the new chi-square statistic was found to be 61.76048. A
second linear combination was extracted from these residual data,
with the constraint that this second pair must be orthogonal to the
first pair. This procedure was repeated six (i.e., the number of
variables) times, with each extracted linear combination being
forced to be orthogonal to all preceding pairs. Hence, 'before
removal' in Table 10 refers to the generalized variance of the two
data matrices prior to a canonical variate extraction. Only the
principal canonical dimension will be discussed here, because it
alone accounts for roughly 90% of the variance displayed in each of
these data sets. This principal pair is highly colinear, with a
canonical correlation of 0.968. In other words, given the six

TABLE 10

SIGNIFICANCE OF CANONICAL ROOTS FOR
INTERPOLATED TONS OF AND ACTUAL CUERDAS OF SUGARCANE HARVESTED
IN PUERTO RICO: 1965-1970

Canonical Pair	Eigenvalue	Wilk's Lambda Before Removal	Approximate χ^2 Before Removal	Degrees of Freedom
1	.93646	.02493	243.66120***	36
2	.41812	.39229	61.76048***	25
3	.32583	.67417	26.02180*	16
4	0	1.00000	0	9
5	0	1.00000	0	4
6	0	1.00000	0	1

* Denotes a significant difference from zero at the 10 % level.

*** Denotes a significant difference from zero at the 1% level.

spatial distributions of cuerdas of sugarcane harvested, one would know what the relative geographic distribution of sugarcane harvest would be like. Since total annual production also is known, this relative distribution can be easily converted to an absolute distribution. Canonical variate loadings for it appear in Table 11. All twelve spatial distributions have very high loadings on this canonical variate. Consequently, convincing evidence exists to suggest that the estimated space—time data cube is a good approximation to the actual but unknown cube. Because this data cube appears to be a good approximation of the unknown data cube, it will be used in the next section, which will discuss normative aspects of the sugar mill closures.

4. DYNAMIC LOCATION–ALLOCATION MODELLING OF MILL CLOSURES

The third major objective of this research was the exploration of features that would enable an optimal dynamic location-allocation trajectory to be described for the sequence of mill closures, and a comparison to be made between the observed and the optimal time paths. This exploration required the digitizing of the municipio centroids and the mill locations. An algorithm described in Olson (1976) was used to estimate municipio centroids. The output from this algorithm included centroids as well as municipio areas. Since each municipio was being approximated by an

TABLE 11

CANONICAL VARIATE FOR SUGARCANE
CULTIVATION IN PUERTO RICO: 1965-1970

Data Set	Variable	Correlation With Canonical Variate
Tons of Sugarcane Harvested	1965	.99998
	1966	.99371
	1967	.97321
	1968	.93621
	1969	.88239
	1970	.89576
Cuerdas of Sugarcane Harvested	1965	.95530
	1966	.96169
	1967	.95604
	1968	.94844
	1969	.87637
	1970	.89783

Canonical Correlation	= .96771
Proportion of generalized variance for interpolated tons of sugarcane harvested accounted for by canonical variate	= 89.87
Proportion of generalized variance for actual cuerdas of sugarcane harvested accounted for by canonical variate	= 88.39

n-sided polygon, how well this polygon represented its corresponding digitized municipio can be assessed by comparing the areas of the two. A close correspondence means that the centroid of the polygon lies very close to the centroid of the municipio. Although many different polygon shapes will have the same area, one should keep in mind that those being discussed here were obtained through a digitizing of the actual municipios, which should highly restrict the range of areas that can result. Hence, these areas then were correlated with actual municipio areas provided by the Commonwealth of Puerto Rico (1970). The scatter diagram presented in Figure 3 indicates that the predicted areas are almost identical to the observed areas, implying that the estimated municipio centroids should be reasonably accurate.

The location of active sugar mills is fixed during the 20-year time horizon. One question that is of interest asks whether or not the processing of sugarcane by mills during the 20-year horizon represents a distance minimizing solution. A simple distance

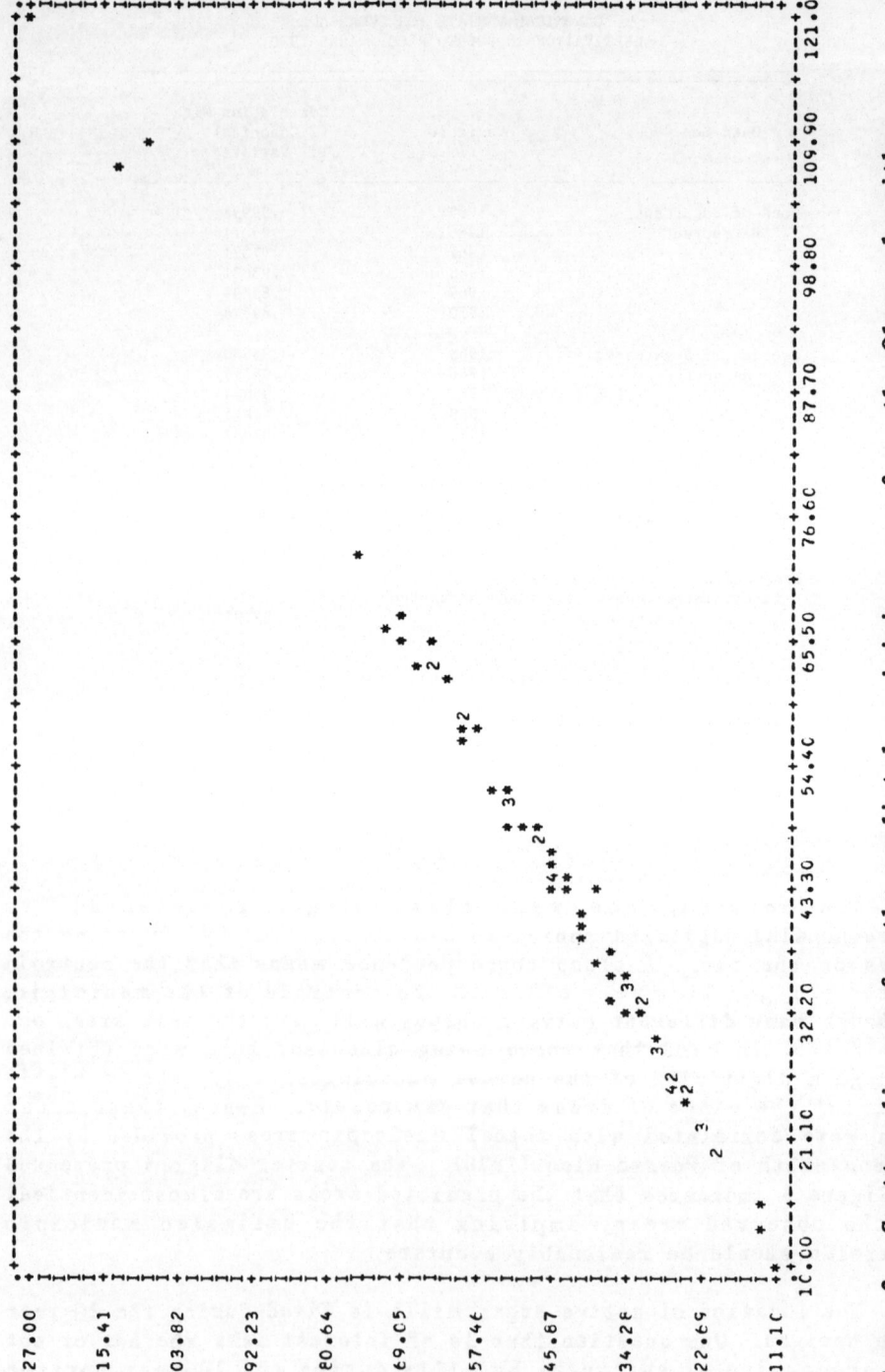

Figure 3. Scatter diagram for the predicted municipio areas from the Olson algorithm versus the actual municipio areas.

minimizing solution was determined for each year, so that municipio harvests were allocated to the closest active mill sites. Results of these allocations appear in Table 12. Not surprisingly, the average number of miles a ton of sugarcane had to be shipped from farm to mill tends to increase as the number of mills decreases.

The correlations between the tons allocated to mills based upon minimum transport costs and actual tons of sugarcane processed at mills also is given in Table 12. Obviously, since virtually none of these correlation coefficients are significant, no relationship exists between these two variables. The single significant coefficient is what one would expect simply because of

TABLE 12

ANNUAL MINIMUM TRANSPORT COST SOLUTION
FOR MOVING SUGARCANE TO SUGAR MILLS:
1959-1978

Year	Number of Operating Mill Locations	Average Miles Shipped Per Ton	Correlation Between Minimum Transport Cost Solution and Actual Mill Tonnage[1]
1959	25	4.477	.12651
1960	25	4.465	.21473
1961	25	4.452	.24056
1962	24	4.709	.23682
1963	23	4.743	.24824
1964	21	4.870	.16517
1965	21	4.823	.24713
1966	20	4.897	.26994
1967	19	4.887	.20955
1968	16	5.191	.03864
1969	16	5.013	.27372
1970	15	5.156	.09753
1971	15	5.178	.16348
1972	14	6.041	.19217
1973	13	6.479	.23201
1974	11	6.872	.46093
1975	11	6.825	.54452*
1976	11	6.772	.51472
1977	11	6.710	.40700
1978	7	9.003	-.36294

*Denotes a significant difference from zero at the .10 level.

[1]The overall correlation, for n = 343, is .27548, which is significantly different from zero at the .01 level.

the presence of sampling error. This result indicates that pure distance minimization has not occurred during the assemblying of sugarcane by mills, and thus a dynamic location-allocation solution should include a capacity constraint. Such a constraint will increase the average number of miles a ton of sugarcane needed to be shipped, but it will allocate sugarcane harvest to mills so that the amount received will equal the amount processed.

This particular finding is an extremely important one, since few dynamic location-allocation algorithms include capacity constraints. In part, this type of constraint is overlooked because in expansion the problem is to determine new locations for physical plants as well as by how much to increase in situ capacity of existing physical plants, and which physical plant locations should experience such expansion. In contraction, however, there will be little incentive to make physical plant investments if an entrepreneur senses that his plant may close sometime in the next few years. This fundamental difference between the situations of expansion and contraction means a capacity constraint is far more important when devising a dynamic location-allocation algorithm for contraction.

5. CONCLUSIONS

In conclusion, the phasing out of the commercial sugar industry in Puerto Rico seems to have a prominent geographical component. Strong evidence exists to suggest that conditional spatial and temporal estimation procedures yield a space-time data cube that closely mirrors reality. This data cube is presented here in Appendix A. Finally, dynamic location-allocation modelling of the spatio-temporal trajectory followed by mill closures must involve capacity constraints for the mills. Current algorithms that are available fail to include these capacity constraints (see Erlenkotter, 1975, 1981). These algorithms also fail to acknowledge the asymmetry of expansion and contraction trajectories.

6. REFERENCES

Commonwealth of Puerto Rico, Department of Agriculture, 1970, Facts and Figures on Puerto Rico's Agriculture 1969/70, Santurce: Office of Agricultural Statistics.

Erlenkotter, D., 1975, Bibliography on Dynamic Location Models, Discusson Paper No. 55, UCLA, Management Science Study Center.

_____, 1981, A Comparative Study of Approaches to Dynamic Location Problems, European Journal of Operational Research, 6: 133–143.

Griffith, D., Contraction Dynamics of the Puerto Rican Sugar Industry: A Preliminary Analysis, paper presented to the annual Conference of Latin Americanist Geographers, SUNY/Buffalo, October 15–18.

Haining, R., D. Griffith and R. Bennett, 1982, A Statistical Approach to the Problem of Missing Spatial Data, Professional Geographer, forthcoming.

Olson, J., 1976, Noncontiguous Area Cartograms, Professional Geographer, 28: 371–380.

Orchard, R. and M. Woodbury, 1972, A Missing Information Principle: Theory and Applications, in Proceedings of the Sixth Berkeley Symposium on Mathematical Statistics and Probability, Vol. 1, edited by L. le Cam, J. Neyman and E. Scott, Berkeley: University of California Press, pp. 697–715.

United States Bureau of the Census, Census of Agriculture, 1964, 1969, 1974, 1978, Puerto Rico, Vol. 1, Part 52, Washington, D. C.: U. S. Government Printing Office.

_____. 1954, Special Reports, Vol. 3, Part 3, Washington, D. C.: U. S. Government Printing Office.

United States Department of Agriculture, 1938–1978, Annual Report of the Caribbean Area Office, San Juan: Agricultural Stabilization and Conservation Service.

7. APPENDIX: CONSTRUCTED SPACE–TIME DATA CUBE FOR PUERTO RICO

AREAL UNIT (MUNICIPIO)	1959	1960	1961	1962	1963	1964	1965
ADJUNTAS	22504	17768	14506	8853	4700	0	0
ACUADA	82709	97569	123562	129013	153849	163376	152111
ACUADILLA	152294	148300	159254	143808	150573	142072	137080
ACUAS BUENAS	22129	18107	15646	10605	7310	3215	2601
AIBONITO	20805	17462	15666	11283	8731	5246	4724
ANASCO	200515	181479	179683	148151	139935	117299	117411
ARECIBO	530374	520669	563765	513392	542176	516063	472413
ARROYO	79640	80973	90730	85434	93224	91621	96907
BARCELONETA	229707	213712	218497	187097	184909	163806	146614
BARRANOUITAS	2880	2401	2134	1514	1141	645	717
BAYAMON/CATANO	126964	110308	103755	80049	69129	50974	40890
CABO ROJO	294222	307504	353382	340607	379771	380788	385333
CAGUAS	190106	170779	167572	136637	127259	104736	86187
CANUY	230428	238099	270814	258590	285869	284397	259744
CAROLINA	171515	149830	141937	110576	96836	72988	60555
CAYEY	123153	111801	111099	92009	87386	73766	65732
CEIBA	61395	63935	73235	70382	78267	78287	66594
CIALES	18305	15798	14728	11224	9518	6813	5464
CIDRA	160026	144598	142887	117530	110680	92419	79760
COAMO	1499	1215	1034	682	444	157	126
COMERIO	11645	9752	8722	6250	4795	2827	3051
COROZAL	88402	73964	66057	47243	36108	21108	17017
DORADO	53610	51112	53683	47359	48383	44484	53926
FAJARDO	129646	126715	136591	123823	130164	123316	107261
GUANICA	76540	82814	98067	97035	110727	113333	120843
GUAYAMA	277665	284830	321805	305386	335673	332172	286408
GUAYANILLA	129343	129594	143186	133030	143303	139110	134048
GUAYNABO	18440	15312	13526	9507	7040	3814	3059
GURABO	130368	124195	130330	114870	117238	107673	103901
HATILLO	143060	139803	150676	136570	143540	135966	111705
HORMIGUEROS	116989	112350	118915	105774	109018	101183	97694
HUMACAO	430736	413648	437812	389422	401358	372504	305306
.ISABELA	157785	165812	191480	185364	207491	208787	193917
JAYUYA	19593	17257	16522	13054	11659	9052	8241
JUAN DIAZ	441400	420699	441706	389525	397792	365575	318458
JUNCOS	112762	105633	108834	94008	93834	84075	76149
LAJAS	207887	217062	249231	240034	267445	267989	274370
LARES	27554	27098	29395	26817	28374	27057	22248
LAS MARIAS	13812	14417	16548	15933	17748	17780	15184
LAS PIEDRAS	78858	76357	81520	73170	76135	71373	67334
LOIZA	189158	168811	164307	132620	121902	98565	88247
LUQUILLO	90581	77794	72054	54406	45491	31787	28621
MANATI	182941	181197	197950	181879	193800	186126	169765
MARICAO	9126	7580	6699	4712	3493	1899	1623
MAUNABO	90228	89283	97445	89449	95222	91366	81960
MAYAGUEZ	177627	181180	203622	192282	210377	207283	202948
MOCA	201215	201750	223063	207380	223540	217136	225707
MOROVIS	81203	74081	74045	61752	59155	50476	41972
NAGUABO	330882	295287	287403	231971	213218	172393	149054
NARANJITO	5876	4816	4172	2840	1975	894	717
OROCOVIS	0	0	0	0	0	0	0
PATILLAS	89007	99632	121289	122794	142853	148649	128854
PENUELAS	83122	87321	100808	97562	109182	109841	112778
PONCE	426563	421009	458319	419635	445583	426450	448599
QUEBRADILLAS	79870	84494	98146	95506	107403	108525	100581
RINCON	115255	105825	106572	89676	86833	75077	65341
RIO GRANDE	100451	94672	98195	85452	86006	77787	64838
SABANA GRANDE	91467	96426	111665	108368	121574	122579	112355
SALINAS	248959	255692	289210	274740	302281	299400	275786
SAN GERMAN	256309	257035	284237	264295	284936	276815	266184
SAN JUAN/RIO PIEDRAS	14328	11489	9621	6167	3743	897	770
SAN LORENZO	96870	94255	101135	91250	95461	89992	76590
SAN SEBASTIAN	315830	329203	377409	362978	403920	404277	382475
SANTA ISABEL	437050	426353	458688	414985	435349	411588	396140
TOA ALTA	113250	102318	101092	83137	78273	65339	52404
TOA BAJA	97214	87765	86636	71170	66914	55758	52641
TRUJILLO ALTO	28219	24651	23351	18191	15930	12005	9723
UTUADO	28916	26584	26811	22600	21928	19007	15372
VEGA ALTA	75129	67233	65662	53227	49201	40086	32468
VEGA BAJA	244137	224245	225926	190203	184283	159451	149815
VILLALBA	64508	66858	76255	72997	80885	80641	67188
YABUCOA	287416	291642	326162	306573	333959	327687	298580
YAUCO	115160	120582	138801	133983	149590	150174	139656

AREAL UNIT (MUNICIPIO)	1966	1967	1968	1969	1970	1971	1972
ADJUNTAS	0	0	0	0	0	0	0
AGUADA	163621	142537	116921	104622	125189	116754	132038
AGUADILLA	153478	139905	120866	114824	121978	101795	103687
AGUAS BUENAS	2305	1501	732	110	132	123	139
AIBONITO	4881	4046	3115	2565	2244	1451	1029
AÑASCO	136581	129569	116702	115815	126238	108153	113157
ARECIBO	498036	423443	337094	290428	286090	219146	202280
ARROYO	118796	118470	111922	116265	109560	79240	67693
BARCELONETA	150308	123323	93667	75621	82017	69918	72790
BARRANQUITAS	917	949	926	990	862	554	387
BAYAMON/CATANO	35742	22653	10235	33	33	29	13
CABO ROJO	453133	434516	395565	396739	391499	300538	278161
CAGUAS	78496	53754	29700	10841	9426	6040	4201
CAMUY	273068	231368	183379	157082	143903	100019	80440
CAROLINA	55852	39099	22669	10005	16749	19327	25399
CAYEY	67007	54566	41014	32605	28350	18164	12634
CEIBA	63736	47396	30880	18889	18893	14742	13922
CIALES	4775	3025	1364	0	179	306	480
CIDRA	77907	59750	41024	27976	24599	16053	11572
COAMO	110	70	31	0	0	0	0
COMERIO	3807	3857	3696	3889	3419	2230	1605
COROZAL	14999	9663	4578	436	379	243	169
DORADO	73700	80356	81809	90579	89682	69126	64306
FAJARDO	105900	82501	58078	41477	49649	46318	52395
GUANICA	149213	149774	142330	148644	145024	109769	99785
GUAYAMA	279397	213875	146391	99237	114391	103306	113601
GUAYANILLA	149873	136410	117651	111567	111036	86125	80742
GUAYNABO	2673	1693	764	0	0	0	0
GURABO	116344	106068	91646	87078	83112	61151	53521
HATILLO	101481	69182	37835	13177	13310	10505	10058
HORMIGUEROS	109461	99859	86345	82106	84810	68670	67696
HUMACAO	276321	187108	100732	32470	48252	52294	66110
ISABELA	207995	180581	147526	131349	128179	97048	88255
JAYUYA	8630	7276	5731	4868	4410	3015	2360
JUAN DIAZ	315064	246172	174091	125342	142981	127914	139468
JUNCOS	79240	66280	51663	43271	37816	24434	17279
LAJAS	326318	316369	291100	295008	290884	223085	206227
LARES	20237	13828	7602	2712	2920	2471	2554
LAS MARIAS	14613	10961	7250	4583	4727	3822	3760
LAS PIEDRAS	73536	65200	54603	50085	44802	30045	22760
LOIZA	90506	74297	56470	45638	45807	35890	34061
LUQUILLO	29565	24497	18857	15521	15465	12012	11280
MANATI	178185	150671	119112	101683	94595	67211	55931
MARICAO	1563	1174	779	495	430	276	192
MAUNABO	84263	69394	52975	43087	52305	49361	56378
MAYAGUEZ	230797	213915	188124	182166	177260	133723	121036
MOCA	272322	267631	249449	255910	256042	199856	188811
MOROVIS	38846	27354	16056	7390	6454	4166	2940
NAGUABO	145965	112368	77623	53550	55498	45101	44643
NARANJITO	627	397	179	0	0	0	0
OROCOVIS	0	0	0	0	5670	9688	15161
PATILLAS	126626	97982	68244	47805	44669	31934	26821
PEÑUELAS	134498	130740	120600	122514	123624	97468	93192
PONCE	547221	543276	511148	528988	499649	362527	311116
QUEBRADILLAS	107613	93151	75824	67208	68241	54193	52257
RINCON	64564	50356	35513	25443	26757	22093	22251
RIO GRANDE	60226	42669	25362	12156	13229	11316	11821
SABANA GRANDE	118633	101057	80644	69699	68060	51571	46947
SALINAS	292925	251342	202397	176979	199006	175692	189227
SAN GERMAN	296932	269588	231883	219234	223581	178447	173072
SAN JUAN/RIO PIEDRAS	746	566	381	250	264	219	222
SAN LORENZO	73357	54611	35653	21906	22412	17954	17486
SAN SEBASTIAN	419034	372880	313585	289033	304975	252708	255478
SANTA ISABEL	442338	402037	346217	327759	329564	258763	246197
TOA ALTA	45794	29009	13085	0	0	0	0
TOA BAJA	57536	51062	42810	39317	43057	37062	38956
TRUJILLO ALTO	8633	5641	2779	465	404	259	180
UTUADO	13619	8862	4316	633	712	629	678
VEGA ALTA	28836	18853	9298	1578	4675	6522	9442
VEGA BAJA	162855	143624	119539	108852	113003	92007	91262
VILLALBA	62368	44139	26178	12460	12085	9080	8174
YABUCOA	313001	264259	208493	177514	177405	138289	130434
YAUCO	150018	130476	106820	95356	90742	66501	57884

AREAL UNIT (MUNICIPIO)	1973	1974	1975	1976	1977	1978
ADJUNTAS	0	0	0	0	0	0
AGUADA	130893	157415	135473	118906	87076	56111
AGUADILLA	93083	101839	94890	92664	78998	64563
AGUAS BUENAS	137	165	127	92	44	0
AIBONITO	500	60	46	33	16	0
ANASCO	104405	117470	111289	110870	96860	81626
ARECIBO	161861	154398	155867	166558	157312	144700
ARROYO	48501	38955	40193	43908	42405	39893
BARCELONETA	66827	74817	85312	101958	106829	108275
BARRANQUITAS	181	6	5	3	2	0
BAYAMON/CATANO	6	0	0	0	0	0
CABO ROJO	223363	214076	225257	250799	246720	236297
CAGUAS	1932	0	0	0	0	0
CAMUY	52012	33676	38618	46365	48766	49585
CAROLINA	28185	37027	29274	22341	12377	3092
CAYEY	5812	0	710	1542	2221	2769
CEIBA	11467	11360	12556	14618	14977	14890
CIALES	588	825	635	459	221	0
CIDRA	5885	1260	969	702	337	0
COAMO	0	0	0	0	0	0
COMERIO	814	169	294	449	557	638
COROZAL	78	0	0	0	0	0
DORADO	51977	50251	38654	27983	13433	0
FAJARDO	51952	62491	59540	59713	52582	44739
GUANICA	78249	72581	76714	85775	84719	81450
GUAYAMA	109876	129286	119586	115731	97543	78539
GUAYANILLA	65904	64537	63024	65001	59104	52189
GUAYNABO	0	0	0	0	0	0
GURABO	39786	34029	28526	24054	16448	9167
HATILLO	8423	8519	6758	5189	2919	800
HORMIGUEROS	58652	61730	73912	91775	99161	103075
HUMACAO	71452	92089	91478	96113	89178	80507
ISABELA	69243	64274	69041	78362	78485	76445
JAYUYA	1449	815	627	454	218	0
JUAN DIAZ	133851	156384	143908	138374	115663	92102
JUNCOS	8342	882	2001	3363	4371	5157
LAJAS	165342	158136	153041	156258	140481	122465
LARES	2327	2585	2297	2109	1656	1203
LAS MARIAS	3251	3414	4105	5113	5537	5767
LAS PIEDRAS	13041	5769	7374	9589	10725	11451
LOIZA	28225	28175	23911	20551	14532	8730
LUQUILLO	9227	9060	9882	11372	11533	11361
MANATI	38404	28441	28835	30950	29366	27138
MARICAO	88	0	0	0	0	0
MAUNABO	56359	68269	60724	55849	43929	32023
MAYAGUEZ	94360	86796	77383	71396	56413	41414
MOCA	155593	154237	150364	154786	140445	123721
MOROVIS	1411	132	102	74	35	0
NAGUABO	38856	41106	39431	39856	35420	30466
NARANJITO	0	0	0	0	0	0
OROCOVIS	18598	26083	20063	14525	6972	0
PATILLAS	18700	14275	19395	26223	30132	32816
PENUELAS	77924	78657	63102	49443	29150	10131
PONCE	224502	182614	167705	160844	133998	106222
QUEBRADILLAS	44139	45101	39558	35684	27275	18978
RINCON	19737	21319	20180	20083	17524	14746
RIO GRANDE	10890	12234	9413	6818	3278	9
SABANA GRANDE	36885	34305	36586	41252	41065	39773
SALINAS	179551	207567	191538	184816	155179	124317
SAN GERMAN	147181	151608	139094	133239	110822	87657
SAN JUAN/RIO PIEDRAS	198	215	165	120	57	0
SAN LORENZO	14943	15480	15321	16034	14814	13313
SAN SEBASTIAN	227535	246850	242026	250716	229079	203374
SANTA ISABEL	204644	205067	202480	211363	194746	174490
TOA ALTA	0	0	0	0	0	0
TOA BAJA	36108	40810	43802	49678	49723	48400
TRUJILLO ALTO	83	0	0	0	0	0
UTUADO	643	744	572	414	199	0
VEGA ALTA	11113	15191	16277	18432	18423	17909
VEGA BAJA	79616	84445	82585	85313	77711	68756
VILLALBA	6325	5756	5529	5598	4985	4297
YABUCOA	107271	106065	104793	109465	100934	90508
YAUCO	42679	36018	35076	36066	32681	28747

THE SPATIAL ECONOMETRICS OF THE EUROPEAN FLEUR-MODEL

Jean Paul Ancot
Jean Paelinck
Netherlands Economic Institute, Rotterdam
The Netherlands

1. INTRODUCTION

In 1972 a large project was started to develop an intersectoral-interregional dynamic model for the countries of the European Community (Netherlands Economic Institute, 1981). This paper relates to the spatial econometric aspects of the project, and will successively develop the specification of the model, its dynamics, the spatial interaction inside the model, and some estimation problems. For the empirical description of the variables used and first estimation results the reader is referred to Molle (1983).

2. THEORY AND MODEL

The model originated from two basic ideas. First, the existence was assumed of a threshold value of the growth rate, with respect to which regions can be classified into two groups: nongrowing or very slowly growing regions on the one hand, and the more rapidly growing ones on the other hand (regional growth is to be understood, for example, as growth in value added or growth in employment). From the outset the model was meant to be sector-specific, the growth of each individual sector being treated independently. More specifically, the possibility was assumed of relating the regional growth of individual sectors to a general synthetic indicator of regional characteristics. Of such an indicator—if it exists and can be constructed—the threshold value referred to above represents a critical value. If for a particular sector a region scores below this value, being poorly endowed with relevant growth-inducing factors, then this region will be counted as a non-grower or slow grower as far as the sector in question is concerned; a region that scores above the critical value will be classified among the growing regions. The threshold value is more than a dividing line between regions. It also serves as a thermometer by which the pathological state of a non-growing or slow-growing region can be diagnosed, and identifying the specific actions that could bring the region closer to, or even beyond, the threshold value, to get the growth process going.

These concepts are schematically represented in Figure 1, in which the S-curve suggests an empirical description of the relation observed between sectoral regional growth and the sector-specific indicator of regional attractiveness. The crosses in Figure 1 correspond to possible positions of, respectively, non- or slow-growing regions and faster-growing regions in the diagram. Figure 1 further suggests a potential saturation level for regional growth, which varies with the individual sectors (and possibly with the state of the economic systems).

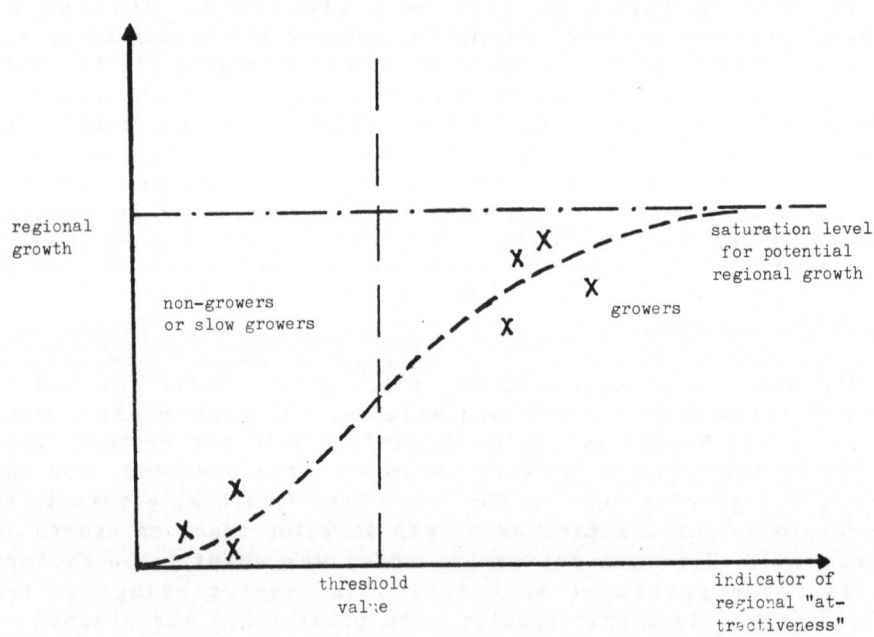

Figure 1. Growth curve identifying the threshold between growing and non-growing regions.

The second general idea from which the model was developed is that of the 'attractiveness' of a region, already introduced above. The assumption is that new firms will tend to settle in a certain region, and existing firms will expand their activities there, if this region is sufficiently endowed with factors conducive to such settlement and expansion. These factors have been called 'factors of location,' the set of all potential factors of location making up the regional profile. Regional economists have described and substantiated the concept of regional profiles. The following sets of elements, or sub-profiles, could be distinguished: a so-called attraction profile, consisting of all the intermediary and final supply and demand elements; a 'classical' profile made up of the various factors of production (availability of capital, characteristics of the labor market, the factor 'land,' access to sources of energy); an environmental profile; and, a political profile, including the intensity with which instruments of regional policy are operating in the region. Regions vary in the density of these sets of elements, and these sets can be relevant in varying degrees to the location or expansion of different sectoral activities.

Economic theory offers various hypotheses regarding the relations between some of these location factors and the growth of industrial sectors, but so far relatively little has been done to specify and quantify these relations empirically, especially on a regional scale. There is every reason, therefore, to develop a method to deal with the level of disaggregated sectors. The method should identify what elements of a regional profile are relevant for the growth of a specific industrial sector, and arrange them in a quantitative hierarchy. Only then will it be possible to design consistent and efficient regional policies for reshaping the regional profile of a backward region, raising it up to or above the threshold value and initiating a growth process.

Besides the two concepts of a threshold value of regional 'attractiveness,' and a regional profile, a third one of paramount importance is the interdependence and interwovenness between activities in different regions. Regions cannot be regarded as closed systems, nor is sectoral regional growth an isolated phenomenon; spatial interaction is not an intra-regional mechanism. Indeed, both input and output markets of most modern industrial sectors are spatially dispersed far beyond the frontiers of the region where the industries are settled, and industrial technologies are also interdependent beyond regional borders. Industries may settle in a given region not because that region itself is so attractive, but because it is next to an important market area for their products that is too congested to admit new firms. Or a region may be chosen as a spatial compromise between various contiguous regions of which one offers an output market, the second offers access to primary inputs, the third offers ancillary services, and so on. Obviously, such interregional effects are very important elements to include, at least implicitly, in the model.

The first task to be done, then, is to formulate a classification rule assigning regions to either of two classes, comprising, respectively, growers and non-growers in terms of the sectoral evolution of, say, the employment variable. A natural approach to tackle this classification problem is discriminant analysis. The discriminant model, trying to minimize the expected cost of misclassification, will be summarized only briefly here. Given a classification rule, errors of classification can be of two kinds: a vector of measurements from population π_1 can be classified as coming from population π_2, and vice versa. Either error gives rise to a certain cost, $C(2|1)$ and $C(1|2)$, respectively, and, if the vector of measurements x is a random vector, the probabilities of these errors, $P(2|1,R)$ and $P(1|2,R)$, can be defined, where R represents the classification rule defining the classes R_1 and R_2. Let the a priori probabilities of the two populations be denoted by q_1 and q_2. Thus the problem is to minimize the expected cost of misclassification:

$$C(2|1)P(2|1,R)q_1 \quad + \quad C(1|2)P(1|2,R)q_2 \quad . \quad (2.1)$$

If the densities of vector x are $p_1(x)$ in population π_1 and $p_2(x)$ in population π_2, equation (2.1) can be rewritten as

$$C(2|1)q_1 \int_{R_2} p_1(x) \ dx \quad + \quad C(1|2)q_2 \int_{R_1} p_2(x) \ dx \quad . \quad (2.2)$$

Further, if the a priori probabilities q_1 and q_2 are known, the Bayes procedure minimizing equations (2.1) and (2.2) leads to the choice of R_1 and R_2 according to the criteria

$$R_1: \quad p_1(x)/p_2(x) \geq C(1|2)q_2/[C(2|1)q_1] \quad , \text{ and } \quad (2.3a)$$

$$R_2: \quad p_1(x)/p_2(x) < C(1|2)q_2/[C(2|1)q_1] \quad . \quad (2.3b)$$

If the a priori probabilities q_1 and q_2 are not known, a unique classification procedure can still be found by applying the minimax principle to the expected cost of misclassification, $C(2|1)P(2|1,R)$ if the observation is from π_1, or $C(1|2)P(1|2,R)$ if it is from π_2. The outcome of the model further depends upon how the distribution of the random vector of measurements x is specified, and on the parameters of the distributions. If these parameters are unknown, as will often be the case, they have to be estimated.

The assumption here will be that the elements of the vector x follow a multivariate normal distribution, with mean vector μ_1 in the population π_1, and mean vector μ_2 in the population π_2, but with the same variance-covariance matrix Σ. None of these parameters are known a priori. Therefore, one can write the ratio of densities as

$$p_1(x)/p_2(x) = \exp[-(x-\mu_1)^T\Sigma^{-1}(x-\mu_1)/2]/\{\exp[-(x-\mu_2)^T\Sigma^{-1}(x-\mu_2)/2]\} \quad (2.4a)$$

$$= \exp\{-[(x-\mu_1)^T\Sigma^{-1}(x-\mu_1)-(x-\mu_2)^T\Sigma^{-1}(x-\mu_2)]/2\} \quad . \quad (2.4b)$$

According to equation (2.3a), class R_1, for example, is defined by the inequality

$$-[(x-\mu_1)^T\Sigma^{-1}(x-\mu_1) - (x-\mu_2)^T\Sigma^{-1}(x-\mu_2)]/2 \geq$$

$$\ln\{q_2 C(1|2)/[q_1 C(2|1)]\} \quad , \quad (2.5)$$

or, after rearrangement of the terms,

$$u = x^T\Sigma^{-1}(\mu_1-\mu_2) - (\mu_1+\mu_2)^T\Sigma^{-1}(\mu_1-\mu_2)/2$$

$$= x^T a_1 + b \geq \ln\{q_2 C(1|2)/[q_1 C(2|1)]\} \quad . \quad (2.6)$$

The first term on the left-hand side of equation (2.6) is the so-called discriminant function, which is linear in the elements of the vector x from the assumption of a common variance-covariance matrix of the two distributions. If the assumption of common variance-covariance is dropped, the discriminant function can be shown to become a quadratic function of the elements of the x vector (McGillivray, 1970). This result is apparent from equation (2.5), where the bilinear terms do not drop out. Generalization along these lines is a topic for further investigation.

The problem being dealt with here differs from classical discriminant analysis in that a sample of regions is to be divided into two classes, namely growers and non-growers, without prior information about the two regimes. The above framework can be used, however, to set up a model by which one can estimate the value of the threshold parameter separating the two regimes as well as explain functionally the growth process under study. Such a model can be developed in the following fashion. Starting from

$$P_1(x) = q_1 p_1(x)/[q_1 p_1(x) + q_2 p_2(x)] \quad , \quad (2.7)$$

the posterior probability that a measurement vector x comes from population π_1 is given by

$$P_1(x)/[1 - P_1(x)] = [p_1(x)/p_2(x)](q_1/q_2) \quad . \quad (2.8)$$

Substituting into equation (2.4), equation (2.8) becomes

$$P_1(x)/[1 - P_1(x)] = [\exp(x^T a_1 + b)](q_1/q_2) \quad . \quad (2.9)$$

234

Hence,

$$P_1(x)/[1 - P_1(x)] = \exp(x^T a_1 + a_0) = \exp[g(x)] , \quad (2.10)$$

where

$$a_0 = b + \ln(q_1/q_2) , \quad (2.11)$$

and so

$$P_1(x) = \exp[g(x)]/\{1 + \exp[g(x)]\} . \quad (2.12)$$

In the absence of prior information one has to set $q_1 = q_2$ in equation (2.9), so that $a_0 = b$ in equation (2.11). From equation (2.7) this clearly means that in this case the posterior probabilities rest on sample information alone. Since in the context of the present application no prior information is assumed to be available, only this case will be treated below.

Suppose further that a function of regional growth, y, can be found that serves as an adequate 'proxy' variable for $P_1(x)$. After some further rearrangements the following estimable linear model is obtained in terms of the elements of the vector of measurements:

$$u = -\ln(y^{-1} - 1) = g(x) = x^T a_1 + b. \quad (2.13)$$

Given a sample of observations of y and x, the parameters of equation (2.13) then can be estimated by standard estimation techniques.

The distribution of u follows from the multinormality of vector x as defined above [the effect of the possible presence of a random disturbance term in equation (2.13) being ignored]. If x is from population π_1, $u \sim N(-\alpha/2, \alpha)$, where

$$\alpha = (\mu_1 - \mu_2)^T \Sigma^{-1} (\mu_1 - \mu_2) = a_1^T \Sigma a_1 , \quad (2.14)$$

the probability of misclassification, if the measurements correspond to a region from population π_1, is then

$$P(2|1) = {}_{-\infty}\int^c (2\pi\alpha)^{-1/2} \exp[-(z-\alpha/2)^2/(2\alpha)] \, dz$$

$$= {}_{-\infty}\int^k (2\pi)^{-1/2} \exp(-w^2/2) \, dw , \quad (2.15a)$$

where $w = (z - \alpha/2)/\alpha^{1/2}$ and $k = (c - \alpha/2)/\alpha^{1/2}$. Similarly, the probability of a classification error of a vector of measurements from π_2 is

$$P(1|2) = {}_k\int^\infty (2\pi)^{-1/2} \exp(-w^2/2) \, dw . \quad (2.15b)$$

For the minimax solution, which is relevant if no prior

information is available, one must choose the threshold parameter c
such that the ratio between the probabilities of misclassification
equals the reciprocal of the ratio of the corresponding costs; that
is,

$$P(2|1)/P(1|2) \; = \; C(1|2)/C(2|1) \qquad . \qquad (2.16)$$

Clearly, from equations (2.15) and (2.16), or from equation
(2.6), if the costs of misclassification are equal, c = 0. If the
costs of misclassification are different, and given estimates of
the parameters a_1 and Σ, the combination of the relations (2.14),
(2.15) and (2.16) determines a unique value for the threshold
parameter c. The classification problem being assumed solved, the
maximum likelihood estimator of Σ would be

$$\hat{\Sigma} = [\sum_{i=1}^{i=n} (x_{1i}-\bar{x}_1)(x_{1i}-\bar{x}_1)^T + \sum_{i=1}^{i=n} (x_{2i}-\bar{x}_2)(x_{2i}-\bar{x}_2)^T]/(n_1+n_2-2) \; . \qquad (2.17)$$

A first approximation to equation (2.17) is provided by

$$\hat{\hat{\Sigma}} = [\sum_{i=1}^{i=n} (x_i-x)(x_i-x)^T]/(n-2) \qquad , \qquad (2.18)$$

which is the pooled sample variance–covariance matrix. The initial
classification then proceeds as follows: given

$$\hat{u} \; = \; x^T\hat{a} + \hat{b} \qquad , \qquad (2.19)$$

for a particular vector of measurements x, choose R_1 and R_2
according to

$$R_1: \; \hat{u} \geq c \qquad , \text{ and} \qquad (2.20a)$$

$$R_2: \; \hat{u} < c \qquad . \qquad (2.20b)$$

This process then could be iterated. This first classification
rule would be used as a starting point to compute Σ according to
equation (2.17), followed by the calculation of a new value for c
by equations (2.14), (2.15), and (2.16). These two steps would be
repeated until a stable classification is obtained.

To summarize, by combining a discriminant–analytic approach
with a modal–split type model, the above model not only leads to
determination of a threshold value to distinguish the two regimes,
but also provides a functional explanation of the growth phenomenon
in the form of the sigmoid specification (2.13). The fundamental
equation of the model, namely (2.13), can be developed further to

yield a specification, giving rise to an interesting dynamic version of the model. In equation (2.13), the variable y is to be understood as the theoretical equilibrium value of the 'proxy' variable for $P_1(x)$, given a specific realization of the valuation function $g(x)$. Suppose the actual observation of this variable is denoted by $\tilde{y}$, and the difference between y and $\tilde{y}$ does not correspond to the usual random disturbance term. Similarly, let $\tilde{u}$ be the logarithmic transformation of $\tilde{y}$ according to equation (2.13); given this notation, it is in terms of u and $\tilde{u}$ that the dynamics are to be defined.

3. TIME AND SPATIAL INTERACTION

3.1. Dynamics

The adjustment specification is as follows:

$$\tilde{u}_t - \tilde{u}_{t-1} = \alpha(u_{t-1} - \tilde{u}_{t-1}) + \beta(\Delta u_t - \Delta\tilde{u}_t) , \quad (3.1)$$

where $\tilde{u} = \log(\tilde{y})$. Equation (3.1) states that the change in observed values between time period t−1 and time period t can be decomposed into two parts: a first component in terms of the discrepancy between observed and equilibrium values in the original period, and a second in terms of the discrepancy between the corresponding changes. One expects, a priori, that the level effect will be dominant, which implies $\beta < \alpha$. Rewriting equation (3.1) leads to

$$\tilde{u}_t = m\tilde{u}_{t-1} + nu_t + pu_{t-1} , \quad (3.2)$$

with
$$m = (1 + \beta)^{-1}(1 - \alpha + \beta) , \quad (3.3)$$
$$n = (1 + \beta)^{-1}\beta , \text{ and} \quad (3.4)$$
$$p = (1 + \beta)^{-1}(\alpha - \beta) . \quad (3.5)$$

In fact, one additional propensity parameter for each explanatory variable will have to be estimated. This means that one degree of freedom is left with respect to equations (3.3) thru (3.5). The hypothesis is made that all variables develop linearly through time. This hypothesis leads for some arbitrary variable to

$$\alpha[du + (n-1)/n \, du + \ldots + 1/n \, du) \quad (3.6a)$$
$$\cong (\alpha\Delta u/n^2)(1 + \ldots + n) = \alpha n(n+1)/2n^2 \, \Delta u \quad (3.6b)$$
$$\cong \alpha/2 \, \Delta u \quad \text{for large n,} \quad (3.6c)$$

implying that $\beta = \alpha/2$.

3.2. Spatial Discounting

A 'good' specification of a spatial—interaction or 'friction' function should satisfy the following requirements:

(i) $0 \leq f_{rs}(d_{rs}, \gamma) \leq 1$,

(ii) $\max f_{rs}(d_{rs}, \gamma) = 1$,

(iii) $\lim_{d_{rs} \to \infty} f_{rs}(d_{rs}, \gamma) = 0$, except for special values of γ, as will be shown below,

(iv) f_{rs} is a function of only one parameter (friction parameter),

(v)   the function should be independent of the units of measurement of the interregional distances, and

(vi)   the estimation of the friction parameter should be relatively easy.

In these preceding conditions, f_{rs} is the friction function and d_{rs} is the distance separating regions r and s. The first three conditions mean that the degree of interregional dependence should be non—negative and should permit the measure to be equal to one (normalization) for maximum dependence, and to tend towards zero as the distance between two regions increases indefinitely. As indicated, the combination of conditions (i) and (ii) defines the normalization of the function. The function could be normalized in other ways, such as by imposing the condition that the integral of the function over all possible distances be equal to one, in which case the friction function would operate as a frequency function (i.e., a normalized weighted combination). The friction function should have a functional form permitting an easy and unambiguous interpretation [conditions (iv) and (v)], and keeping the specification of the model from becoming too complex or too highly non—linear, and the estimation of its parameters unduly complicated.

A number of candidate functions have been studied, including a (modified) Tanner function (Ancot, 1979) and the Poisson function. Neither or these last two functions satisfied completely conditions (i) thru (vi). The friction function finally adopted resulted from the following considerations. As a function of the distance variable d_{rs}, the (modified) Tanner function is formally a linear function divided by an exponential function, and the Poisson—like function is an exponential function divided by a factorial. The function proposed here is a logarithmic function divided by a linear function, corresponding to the following specification:

$$f_{rs}(d_{rs}, \gamma) \overset{\Delta}{=} \exp(1-\gamma^*)[\ln(1+\gamma d_{rs})+\gamma^*](\gamma d_{rs}+1)^{-1} , \quad (3.7)$$

where $\overset{\Delta}{=}$ denotes the definitional equality symbol, and

$$\gamma^* = \gamma/(1 + \gamma) , \; \gamma \geq 0 . \quad (3.8)$$

This function reaches a maximum value equal to 1 at the point

$$d_{rs}^{max} = [\exp(1 - \gamma^*) - 1]/\gamma \quad . \quad (3.9)$$

The shape of the curve and some properties of the function are illustrated in Table 1, for certain selected values of the friction parameter γ and the distance variable d_{rs}. From Table 1, for small values of γ the maximum of the function is seen to occur at large distances from the reference region, which implies high spatial interaction. Conversely, for large values of γ the maximum of the function occurs at a very short distance. In this second case the function decreases very steeply with increasing distances, a pattern typical of the absence of spatial interaction.

Now, if $_p x_r^{(k)}$ represents the value in region r of the k-th potentialized variable as defined, that expression now can be specialized by introducing the explicit friction function (3.7):

$$_p x_r^{(k)} = \sum_{s=1}^{s=R} \{\exp(1-\gamma_k^*)[\ln(1+\gamma_k d_{rs}) + \gamma_k^*](1 + \gamma_k d_{rs})^{-1} x_s^{(k)}\} , \quad (3.10)$$

where R denotes the total number of regions. Then econometric

TABLE 1

SELECTED NUMERICAL VALUES ILLUSTRATING
THE SHAPE OF THE FRICTION FUNCTION

Friction of Distance Parameter γ	Selected Distances (in an arbitrary primary unit) Between Regions r and s						maximum
	0	.25	.67	1.50	4.00	∞	
0.00	0.000	0.000	0.000	0.000	0.000	1	∞
0.05	0.123	0.154	0.202	0.289	0.497	0	31.8400
0.10	0.226	0.280	0.362	0.498	0.758	0	14.8200
0.40	0.584	0.741	0.842	0.965	0.975	0	2.6100
1.00	0.824	0.974	0.984	0.934	0.696	0	0.6500
7.00	0.992	0.777	0.522	0.327	0.166	0	0.0190
20.00	0.999	0.480	0.265	0.148	0.069	0	0.0024
55.00	1.000	0.254	0.125	0.066	0.029	0	0.0003
150.00	1.000	0.097	0.056	0.029	0.012	0	4.4×10^{-5}
400.00	1.000	0.044	0.025	0.012	0.005	0	6.2×10^{-6}
∞	1.000	0.000	0.000	0.000	0.000	0	0.0000

estimation can proceed by an interative procedure based upon the transformation

$$d^*_{rs} \stackrel{\Delta}{=} d_{rs} + \gamma_k^{-1} \quad , \quad (3.11)$$

so that

$$1 + \gamma_k d_{rs} \stackrel{=}{=} \gamma d^*_{rs} \quad , \quad (3.12)$$

yielding

$$f_{rs}(d^*_{rs}, \gamma) = \exp(1-\gamma^*_k)[\ln(\gamma_k d^*_{rs})+\gamma^*_k](\gamma_k d^*_{rs})^{-1} \quad . \quad (3.13)$$

Using equation (3.13), one then can write the expression for each potentialized variable as the sum of two terms:

$$_p x_r^{(k)} = \sum_{s=1}^{s=R} [\exp(1-\gamma^*_k) \ln(d^*_{rs})(\gamma d^*_{rs})^{-1} s_s^{(k)}] +$$

$$\sum_{s=1}^{s=R} \{\exp(1-\gamma^*_k)[\ln(\gamma_k)+\gamma^*_k](\gamma d^*_{rs})^{-1} x_s^{(k)}\} \quad , \quad (3.14)$$

which is a linear function of

$$\exp(1-\gamma^*_k)\gamma_k^{-1} \quad \text{and} \quad \exp(1-\gamma^*_k)\gamma_k^{-1}[\ln(\gamma_k)+\gamma^*_k].$$

4. ESTIMATION

Provisionally ordinary least squares has been used and no correction for spatial autocorrelation of the random terms has been applied. This second aspect probably has negligible impact on the estimation, as the first-moment specification explicitly includes spatial interdependence. As for the first aspect, since static interdependence is present, the authors envisage the use later of dynamic simultaneous least squares (Ancot, Kuiper and Paelinck, 1982a, applied in Ancot, Kuiper and Paelinck, 1982b).

4.1. Non-negativity

On the assumption that the equations are of type (3.2), ordinary linear methods could be used to estimate the parameters. However, the equations differ from those in a usual regression model in that they represent discriminant functions rather than causal behavioral equations. Some of the reaction coefficients are to be interpreted as fundamentally non-negative parameters in the sense that variables apparently relevant for the discrimination process are identified as location factors and thus can only be favorable to the growth process. In other words, their coefficients must be positive. In this sense these coefficients are referred to as propensities, and the propensity parameter space is limited to the non-negative orthant.

A possible procedure for estimating the parameters (apart from the developments in Ancot and Paelinck, 1981) is to resort to quadratic programming (i.e., QP). The quadratic program corresponding to the above problem can be expressed as follows:

$$\text{MIN:} \quad \sum_{r=1}^{r=R} e_r^2 \quad , \qquad (4.1)$$

$$\text{where } e_r = u_r - \sum_{k=1}^{k=K} \eta_k \left(\sum_{s=1}^{s=R} f_{rs} x_s^{(k)} \right) - \sum_{1=1}^{1=N-K} \omega_1 x_r^{(1)} - b \quad , \quad (4.2)$$

$$\text{S.T.:} \quad (1) \quad \eta \geq 0 \quad , \text{ and} \qquad (4.3a)$$

$$(2) \quad \omega \geq 0 \quad . \qquad (4.3b)$$

Inclusions and exclusions of explanatory variables in the discriminant function then are governed by the binding character of constraints (4.3a) and (4.3b) in the optimum solution. Here an interesting contribution has been forthcoming. Consider the following QP:

$$\text{MAX:}_{x} \quad a^T x + x^T H x/2 \qquad (4.4)$$

$$\text{S.T.:} \quad x \geq 0 \quad , \qquad (4.5)$$

with matrix H being negative definite. The Kuhn-Tucker conditions for problem (4.4) are

$$a + Hx \leq 0 \quad , \qquad (4.6)$$

$$\hat{x}(a + Hx) = 0 \quad , \text{ and} \qquad (4.7)$$

$$x \geq 0 \quad . \qquad (4.8)$$

Proposition. The linear complementarity problem (4.6) through (4.8) can be solved by the following linear program (LP):

$$\text{MIN:}_{x} \quad b^T x \qquad (4.9)$$

$$\text{S.T.:} \quad (1) \quad a + Hx \leq 0 \quad , \qquad (4.10)$$

$$(2) \quad x \geq 0 \quad , \tag{4.11}$$

$$(3) \quad \text{with arbitrary } b > 0 \quad , \text{and} \tag{4.12}$$

$$(4) \quad \text{equation (4.9) bounded from below.}$$

Proof. From equation (4.10) extract the i-th inequality

$$h_{ii}x_i \leq - a_i - \sum_{j \neq i} h_{ij}x_j \overset{\Delta}{=} r_i \quad . \tag{4.13}$$

Now (whatever the x_js, $j \neq i$), if $r_i \geq 0$, the optimum solution in x_i is $x_i^o = 0$. If $r_i < 0$, $x_i^o = r_i/h_{ii} > 0$, given $h_{ii} < 0$ from the negative definiteness of matrix H. The Kuhn-Tucker product conditions (4.7) are satisfied, and so are conditions (4.6) and (4.8). One should note that the solution is unique (cf., Mangasarian, 1981; we are indebted to Professor van der Hoek for this reference and discussion of our results), that condition (4.12) is sufficient and that the boundedness of problem (4.9) is mentioned simply as a general requirement in linear programming.

Also noteworthy is that redefining inequality (4.6) as

$$w \overset{\Delta}{=} - a - Hx \geq 0 \tag{4.14}$$

and adding the conditions $x^T w = 0$ and $x \geq -a$ implies that for $a \leq 0$, $x = 0$ together with $w = -a$ constitute a condition (we owe this remark to Professor van der Hoek, too).

Extension. Consider now the QP with an objective function the same as (4.4). Now let the constraints be

$$\text{S.T.:} \quad (1) \quad Bx \leq r \quad , \tag{4.15}$$

$$(2) \quad x \geq r \quad , \text{ and} \tag{4.16}$$

$$(3) \quad \text{with matrix } H \text{ being negative definite.}$$

The Kuhn-Tucker conditions for the linear program (4.14) thru (4.16) are

$$a + Hx - B^T\lambda \leq 0 \quad , \tag{4.17}$$

$$\hat{x} (a + Hx - B^T\lambda) \overset{o}{=} 0 \quad , \tag{4.18}$$
$$\overset{o}{=} \text{ denoting equality at the optimum,}$$

$$Bx \leq r \quad , \tag{4.19}$$

$$\hat{\lambda}(Bx - r) \overset{o}{=} 0 \quad , \text{ and} \tag{4.20}$$

$$x, \lambda \geq 0 \quad . \tag{4.21}$$

Consider further the LP

$$\underset{x}{\text{MIN:}} \quad \alpha^T x + \beta^T \lambda \tag{4.22}$$

$$\text{S.T.:} \quad (1) \quad \alpha + Hx - B^T \lambda \leq 0 \quad , \tag{4.23}$$

$$(2) \quad bx - \varepsilon B \leq r \quad , \tag{4.24}$$

$$(3) \quad x, \lambda \geq 0 \quad , \text{ and} \tag{4.25}$$

$$(4) \quad \alpha, \beta, \varepsilon > 0 \quad , \tag{4.26}$$

where ε is small, and condition (4.26) in α and β again being sufficient.

By the same argument as in the proof above, the x- and B-vectors satisfy the Kuhn-Tucker product conditions (4.18) and (4.20), and represent ε-solutions to equations (4.14) thru (4.16). These solutions could diverge from the exact (4.14) thru (4.16) solution in pathological cases [e.g., simultaneously intersecting hyperplanes, in the case of linearly dependent constraint normals (see Wymenga and van der Hoek, 1981)], but in the limit they should converge to that solution. Use can be made of Weierstrasz's theorem, which states that every bounded infinite subset of R^k has a limit point in R^k (this suggestion was made to us by Th. ten Raa). The limit point can be obtained by taking the sequence of sets x and λ corresponding to values of $\varepsilon = 1/n$, where n is an integer that is approaching infinity.

4.2. Multicollinearity

The presence of the lagged variable $\tilde{u}_{t-1}$ in equation (3.2) gives rise to the following estimation. In preliminary tests the estimation of equation (3.2) in the absence of lagged terms gave rise to a coefficient of determination of the order of 0.94, and statistically significant coefficients. However, when the lagged endogenous variable is introduced, the coefficient of determination rose to 0.99, and the significance of the other terms dropped considerably. This change is due to the intercorrelation between the lagged endogenous and the exogenous terms in the second case, due in turn to the high correlation between the current and lagged endogenous variables.

To avoid this difficulty, an instrumental variable has been substituted for the lagged endogenous term. This variable is obtained by the following two-stage procedure:

(i)   first, regress the lagged endogenous variable on the

exogenous variables (vector x_t), and

(ii) second, compute the residuals of the regression and use them as an instrumental variable for $\tilde{u}_{t-1}$ [it is indeed this part of $\tilde{u}_{t-1}$ that is uncorrelated with the x_t-vector (Johnston, 1972, p. 131)].

5. CONCLUSIONS

The proof of the pudding is in the eating; this eating is the projection made with the help of the model. Projections first have been made along the following lines. Define y_{rc} as the (observed) European employment share of region r belonging to country c. No sectoral index is used to simplify the notation and exposition. By construction

$$y_{rs} = \sigma_c / \{1 + \exp[-g(\mathbf{f}_{rc})]\} \quad , \quad (5.1)$$

where σ_c is a specialization index for country c, and $\mathbf{f}_{rc}$ is a vector of relevant locational factors observed in region r belonging to country c. This vector corresponds to vector x in equation (2.12).

The following equation can be derived from equation (5.1):

$$dy_{rc} = \sigma_c y_{rc}(1 - c^{-1} y_{rc})dg_{rc} + \sigma^{-1} y_{rs} d\sigma_c \quad . \quad (5.2)$$

In equation (5.2) dg is the variation in the explanatory variables between two periods, such as they appear in the econometric exercise. To compute $d\sigma_c$, recall that

$$\underset{r \varepsilon c}{\Sigma} dy_{rc} = dy_c^* \text{ , with } dy^* \text{ given ,} \quad (5.3)$$

where dy* is known from the national sectoral projections, so that

$$d\sigma_c = \sigma_c [dy_c^* - \sigma_c \underset{r \varepsilon c}{\Sigma} y_{rc}(1 - y_{rc})dg_{rc}]/y_{rc} \quad . \quad (5.4)$$

These projections, first applied as a test of the model, have not given plausible results, so the projection model had to be refined. This refinement had to do with iterative analysis. Indeed, contemporary y_{rc}-variables appear in explanatory variables (market tension and supply tension). And so, since the model is non-linear, a starting value, call it y_0, should be introduced to prime an iterative process. Moreover, let

$$y_{irc} = \sigma_c / \{1 + \exp[-g(y_0) + (a/\sigma_a + b/\sigma_b)(y_{irc} - y_0)]\} \quad (5.5)$$

with $g = g_{-1} + \Delta g$, and $\quad (5.6)$

$$\Delta g = \ldots + (a/\sigma_a)(m-y_0-\mu) + (b/\sigma_b)(s-y_0-\mu) + \ldots , \quad (5.7)$$

with, for equation (5.7), $\mu \equiv 0$, and $\qquad\qquad\qquad\qquad\qquad$ (5.8)

$$a,b > 0. \qquad\qquad (5.9)$$

Now let

$$x \stackrel{\Delta}{=} (a/\sigma_a + b/\sigma_b)(y_{irc} - y_0) \quad , \qquad (5.10)$$

so that equation (5.5) can be rewritten as

$$y_{irc} = \sigma_c/[1 + \exp(-g+x)] \quad , \qquad (5.11)$$

which after using a MacLaurin expansion, around $x = 0$, gives:

$$y_{irc} = y_{irc}^1[1-(1-\sigma_c^{-1}y_{irc}^1)(a/\sigma_a + b/\sigma_b)(y_{irc}-y_0)] . \quad (5.12)$$

In equation (5.12) y_{irc}^1 is the 'first-round' computed value. Let

$$s \stackrel{\Delta}{=} 1 - \sigma_c^{-1}y_{irc}^1 \quad , \text{ and} \qquad (5.13)$$

$$\alpha \stackrel{\Delta}{=} a/\sigma_a + b/\sigma_b \quad . \qquad (5.14)$$

Then equation (5.12) becomes

$$y_{irc} = y_{irc}^1(1 + \alpha sy_0)/(1 + \alpha sy_{irc}^1) \quad . \qquad (5.15)$$

Now the following properties hold:

$$y_0 > y_{irc}^1 \nrightarrow y_{irc} > y_{irc}^1 \quad , \text{ and} \qquad (5.16)$$

$$y_0 < y_{irc}^1 \nrightarrow y_{irc} < y_{irc}^1 \quad , \qquad (5.17)$$

under the sufficient condition that $\alpha > 0$ ($s > 0$ is always satisfied). The elegance of equation (5.15) is that both y_0 and y_{irc}^1 appear. Equation (5.15) further has the property that the correction is region-specific.

The last problem to be tackled is that of projecting FLEUR-figures over shorter periods than its typical ten-year span. Recall that the dynamics of FLEUR is given by equation (3.1), with for a 'regular' series of u and $\tilde{u}$,

$$\beta = \alpha/2 \quad . \qquad (5.18)$$

Now suppose that

$$u_{t-1} - \tilde{u}_{t-1} \stackrel{\Delta}{=} x \quad , \text{ and} \qquad (5.19)$$

$$\tilde{u}_t - \tilde{u}_{t-1} \overset{\Delta}{=} \Delta\tilde{u} \quad , \tag{5.20}$$

permitting the first part of equation (3.1) to be written as

$$\Delta\tilde{u} = \alpha x \quad . \tag{5.21}$$

Hence

$$d\tilde{u}_i = \alpha*[(n - i)/n]x \quad , i=1,2,\dots,n \quad , \text{ and} \tag{5.22}$$

$$\int_T d\tilde{u} = \Delta\tilde{u} = [\alpha*/2n](n-1)nx = \alpha*(n-1)x/2 \quad , \tag{5.23}$$

from which is obtained

$$\alpha* = 2n^{-1}\alpha \quad \text{for large n.} \tag{5.24}$$

Now summing over periods 1 thru n/2 gives:

$$\Delta\tilde{u}(1/2) = \alpha*(1 + 2 + \dots + n/2)n^{-1}x \tag{5.25a}$$

$$=(\alpha*/2)(n/2)(n/2+1)n^{-1}x \tag{5.25b}$$

$$= \alpha*(n/8 + 1/4)x \tag{5.25c}$$

$$= (\alpha/4)x \quad \text{for large n.} \tag{5.25d}$$

Finally,

$$\Delta\tilde{u}(1/2) = (\alpha/4)(u_{t-1} - \tilde{u}_{t-1}) + (\alpha/8)(\Delta u - \Delta\tilde{u}) , \tag{5.26}$$

which, as a linear difference equation, leads to

$$\tilde{u}_{t-1/2} = [(1 - \alpha/8)/(1 + \alpha/8)]\tilde{u}_{t-1} +$$

$$[(\alpha/8)/(1 + \alpha/8)]u_{t-1} + u_{t-1/2}) . \tag{5.27}$$

For periods 1 to $\rho^{-1}n$, equation (5.27) has an obvious generalization, leading for large ρ to right-hand-side coefficients approaching, respectively, 1 and 0, as $\alpha/8$ is generalized to $\alpha/(2\rho^2)$. It can be shown that the coefficient for a continuous model relating the time derivative du/dt to $(u - \tilde{u})$ is $\alpha/(1 + \alpha/2)$.

6. REFERENCES

Ancot, J. P. 1979, Une Approache par Analyse Discriminante à des Problèmes de Seuils Régionaux et d'Analyse de Localisation, Recherches Economiques de Louvain, 45: 281-297.

_____ and J. Paelinck, 1981, Recent Research in Spatial

Econometrics, in Dynamic Spatial Models, edited by D. Griffith and R. MacKinnon, Alphen aan den Rijn: Sijthoff and Noordhoff, pp. 344–364.

_____, J. Kuiper and J. Paelinck, 1982a, Réflexions sur la Simulation de Modèles Dynamiques, in Actes du VIIIe Colloque d'Econométrie Appliquée, Lille, forthcoming.

_____, J. Kuiper and J. Paelinck, 1982b, Urban Development in Developing Countries: Some Further Econometric Results, Environment and Planning A, forthcoming.

Johnston, J., 1972, Econometric Methods (2nd ed.), New York: McGraw–Hill.

McGillivray, R., 1970, Demand and Choice Models of Modal Split, Journal of Transport Economics and Policy, 4: 192–207.

Mangasarian, O., 1979, Simplified Characterization of Linear Complementarity Problems Solvable as Linear Programs, Mathematics of Operations Research, 4: 268–273.

Molle, W., 1983, Industrial Location and Regional Development in the European Community, Farnborough: Gower.

Netherlands Economic Institute, 1981, Stage V of the FLEUR Study, The Model: Specification, Estimation, Testing Aspects and First Results, Report No. 3, Rotterdam.

Rudin, W., 1964, Principles of Mathematical Analysis, New York: McGraw–Hill.

Wymenga, R. and van der Hoek, 1981, Linearly Dependent Active Constraint Normals in Nonlinear Programming, Report 8125, Econometric Institute, Erasmus University, Rotterdam.

A SIMULATION MODEL OF REGIONAL DEMOECONOMIC DEVELOPMENT IN NORTH RHINE-WESTPHALIA: THEORY, STRUCTURE, AND CALIBRATION

Claus Schoenebeck

University of Dortmund
Federal Republic of Germany

1. INTRODUCTION

This paper reports on the results of a project that investigated the way in which economic, technical and social change relates to the development of regional settlement systems, within a framework of migration and investment decisions. For this purpose a multilevel dynamic simulation model was developed, consisting of (1) a macro-analytic model of demoeconomic development in 34 labor market regions in North Rhine-Westphalia, (2) a micro-analytic model of intraregional location and migration decisions in 30 zones of the urban region of Dortmund, and (3) a micro-analytic model of land use development in one or more districts of Dortmund. For the general structure of this three-level model system see Wegener (1980) and Wegener et al. (1983). In this paper only the top level of the multilevel hierarchy, the model of regional demoeconomic development in North Rhine-Westphalia, will be discussed. First, various theories about regional development are evaluated, in the light of empirical data, to establish a conceptual framework for regional model building. Second, the major submodels of the simulation model as well as the links between them are discussed. This discussion focuses on submodels submodels evaluating regional labor markets, market potentials, and housing, and on two submodels simulating the aging and migration of population and the locational behavior of industries. Finally, a method of

I would like to thank Michael Wegener, Giorgio Leonardi, Folke Snickars and Peer Just for many beneficial and stimulating discussions about the work reported on in this paper.

calibrating the model is explained, and some results are presented and evaluated.

2.   INTERREGIONAL COMPETITION: THEORY AND FACT

2.1. State of the Art

One of the problems regional and local planners in Europe are faced with is the question of agglomeration. Will agglomeration continue, slow down, or even be followed by a phase of deglomeration? Surprisingly enough, recent empirical research leads to contradictory conclusions.

The Netherlands Economic Institute has been engaged in a series of studies that analyze the urbanization process in various countries in Europe (see van den Berg and Klaassen, 1978, Molle and Klaassen, 1978; Klaassen et al., 1979). In these studies the following three succeeding stages of urban development have been identified:

(1)   a phase of urbanization with high growth rates in the agglomeration core,

(2)   a phase of suburbanization as a process of relocating from the core to its surroundings, and

(3)   a phase of deurbanization characterized by high losses of both the core area and its suburbanized periphery.

After classifying the population trajectories of 115 European metropolitan areas, the authors concluded that, although suburbanization is still dominant today, there is a noticeable transition to deurbanization. Unless a powerful reurbanization policy is implemented, deurbanization may cause 'garbage cities' or metropolitan ghost towns. However, since the theoretical foundations of these studies seem to be weak, some of their conclusions should be considered highly speculative. The authors have assumed that

(1)   for the phase of deurbanization in large settlement systems, the development process is dominated by residential preferences of households, in particular high income households (Klaassen and Scimeni, 1979),

(2)   footloose establishments, mostly of the tertiary or quaternary sector, tend to follow the migration decisions of households (Klaassen and Scimeni, 1979),

(3)   spatial mobility of population and employment tends to

benefit medium-sized towns because these settlements are supposed to have the most favorable living conditions (van den Berg et al., 1979), and

(4) the shift of both population and employment from larger to medium-sized towns is a mutually reinforcing process (van den Berg et al., 1979).

In contrast, investigations comparing the locational quality of metropolitan areas with other types of regions conclude that the attraction of metropolitan areas will continue to grow. It is assumed that favorable labor market conditions attract migrants from less developed regions, especially when overall unemployment is high. Accordingly, the disparities between metropolitan and rural regions tend to increase.

What are the main reasons for the attractiveness of highly urbanized areas for industry and services? Traditional arguments emphasize the quality and diversity of the labor force, good access to intermediate or final consumption markets, opportunities for information exchange, personal contacts, and the substantial fixed infrastructure. Current research has stressed the quality aspects of the infrastructure. Direct access to an international airport or to a political or decision-making center substantially adds to the attractiveness of a region, because such access offers opportunities in the competitive advantages of technical, organizational, and financial decision-making (Ewers et al., 1979). Such high-level infrastructure seems to be most beneficial for international corporations with extensive investment in research and development (RD). And, according to growth pole theory, industries with large RD expenditures are the key ones underpinning regional growth.

Furthermore, the spatial concentration of such corporations in the central business districts (CBDs) of large metropolitan regions is regarded as a comparative and thus self-reinforcing advantage. Especially for 'headquarter' industries with a high demand for information, location in such areas provides the necessary direct and undisturbed access needed for decision-relevant information flows (Buttler et al., 1977). The theory of innovation diffusion stresses the point that spatial diffusion processes proceed stepwise in time, and each step is accompanied by information biases and losses. This reinforces the comparative advantage of metropolitan regions.

If all these arguments are true, the situation for all non-urbanized, peripheral regions is in fact hopeless. These regions would have the potential to compete with urban areas only if they succeeded in providing a high-level metropolitan infrastructure and attracting headquarter industries. But then they would no longer be rural.

2.2. Demoeconomic Trends in North Rhine-Westphalia

Before presenting a conclusion for the preceding discussion, it may be helpful to present some empirical evidence of recent demoeconomic trends in North Rhine-Westphalia. Theoretical and empirical considerations then will be evaluated jointly, in order to establish a conceptual framework for the model approach presented here.

In North Rhine-Westphalia there are two large polycentric agglomeration areas comprising more than 50 percent of the total population of the area. One is the 'Rhine corridor,' extending along the Rhine River from Duesseldorf to Cologne and Bonn, and the other is the 'Ruhr region,' a conglomeration of industrial towns dominated by mining, iron, and steel industries that extends from Duisburg in the west to Dortmund/Hamm in the east. A statistical comparison of the economic development of this Rhine corridor and this Ruhr region, over a period of 20 years, reveals that losses of workplaces in the Ruhr region have been accompanied by gains of employment in the Rhine corridor. And as in the past, the main characteristics of the Ruhr economy today are its relatively low level of sectoral diversification, below average expenditures for research and development, a presistent dominance of standardized mass production, and a quantitatively and qualitatively underdeveloped service sector.

The development of population follows the same path (see Figure 1). The regions with high negative net migration are those with a large proportion of coal mining and steel manufacturing industries, which are characterized by high unemployment, unattractive and highly polluted environments, and poor accessibility to recreational areas in the countryside. Additionally, as in most urban areas, birth rates are low. Regions with high positive net migration consist of those of the Rhine corridor, mainly because of attractive labor market conditions and a large supply or high-quality infrastructure, and all rural areas having good accessibility to more than one urban area (e.g., the Coesfeld region, which is close to the Ruhr region and to Muenster).

Consequently neither the deurbanization hypothesis (i.e., Rhine and Ruhr should diminish) nor the urbanization hypothesis (i.e., Rhine and Ruhr should grow) fit reality. Rather, current demoeconomic trends are composed of many different counter-current as well as supplementary processes.

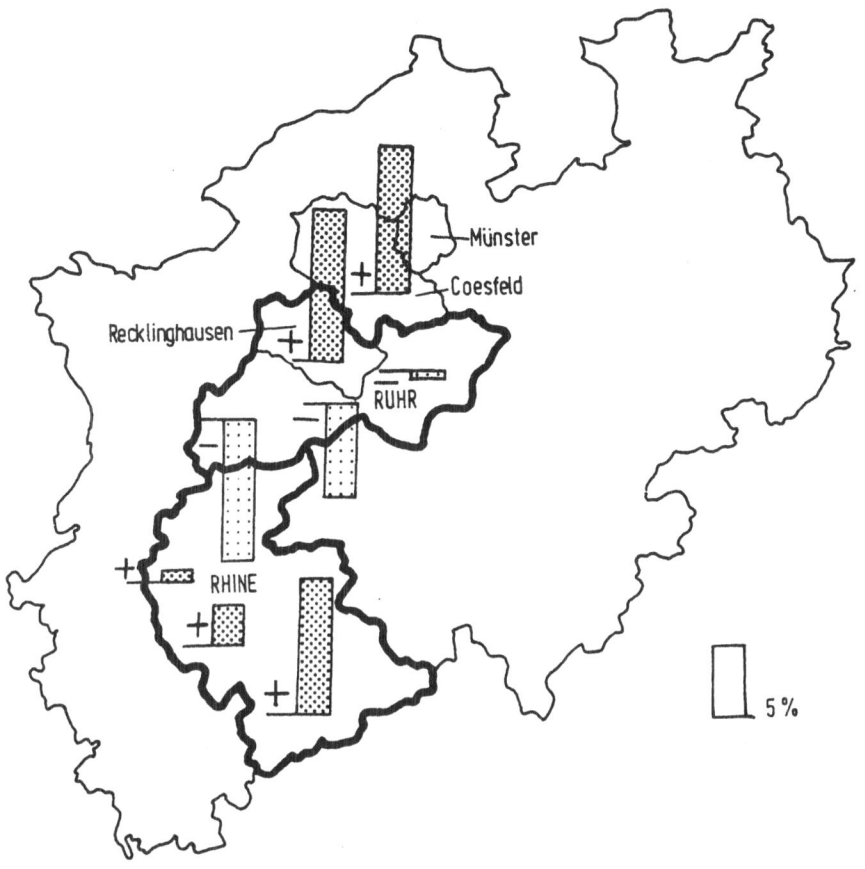

Figure 1.   Percentage population change 1970–1979 in and around the
Rhine/Ruhr agglomerations

2.3. Basic Hypotheses of the Demoeconomic Model

Since demoeconomic trends are so diverse, a holistic
understanding of the processes involved is important. A model of
demoeconomic processes must have the requisite level of complexity
to be able to grasp the variety of causes and effects inherent in
reality. Moreover, such simple assumptions as 'employment follows
labor' or 'labor follows employment' are not social laws, but have
to be modeled in a way that allows them to respond to changes in
the surrounding environment. With this in mind, the following set
of basic hypotheses has been formulated for the demoeconomic model.
First, the development of a regional economy depends greatly upon
its present industrial mix. An above average share of growing

industries is a positive asset with at least medium-term effects. Conversely, a large share of declining or stagnating industries tends to bode ill for the future development of a region. Second, regional economic development also is related to regional attractiveness differentials in terms of accessibility, the supply of business-oriented infrastructure, access to markets, financial aids and taxes, the labor market situation, and wage levels. Favorable scores on these conditions, ceteris paribus, tend to attract more investment-creating job opportunities. In contrast, unfavorable scores tend to result in a slowdown of economic change and eventually in a loss of jobs. Third, regional populations change through aging and migration. While aging is a well-defined, steady process, migration flows are highly selective and variable over time. Migration can be seen as a kind of voting in favor of the place of destination to the detriment of a place of origin. The criteria by which populations evaluate the attractiveness of a region as a place of residence include accessibility, the supply of household-oriented infrastructure, the housing supply, the environmental quality, the labor market situation, and the wage levels. Fourth, migration decisions of population and location decisions of enterprises are interrelated, although from different points of view. In part, both decisions are based upon the evaluation of regional labor markets. At times of high overall unemployment, job considerations become of primary importance for the migration decisions of workers, while at the same time the importance of labor market factors for location decisions of industry decreases. Finally, regional labor markets are highly segmented. Conditions for a specific market segment, say the labor market for computer engineers, may be quite different from overall conditions. Computer engineers may be in great demand, while overall unemployment is severe. Such segmentation and imbalances may have an important influence on both migration and location decisions.

Next a macro-analytic demoeconomic model of regional development, based upon the above hypotheses, is presented. This model simulates the demographic and economic development of 34 labor market regions of the state of North Rhine-Westphalia between 1970 and 1990.

3.   MODEL STRUCTURE

It is helpful to distinguish three dimensions of the model's structure: spatial organization, the processing of time, and the model content. The discussion below will focus on the model content; therefore, only the first two dimensions are summarized here (for details see Schoenebeck and Wegener, 1977; Wegener, 1980). North Rhine-Westphalia (with a population of about 17 million) has been subdivided into 34 labor market (i.e., functional) regions, following the regionalization of Klemmer and Kraemer (1975). The populations of the regions range from 140,000 (Hoexter) to 1,700,000 (Cologne/Leverkusen plus hinterland).

The model is of the recursive-dynamic type, operating with a two-year simulation period. The base year is the year of the 1970 census, the last published census for the Federal Republic of Germany. As in all recursive models, a distinction can be made between status description parts, referring to a point in time, and process simulation parts, referring to a time interval. The subsequent discussion will make use of this distinction.

3.1. Status Description

The status description part of the model is subdivided into description and evaluation submodels. At first, different aspects of the regions' situations are analyzed to determine the relative position of each region in the system. In a second step, these regional characteristics are evaluated from the point of view of particular groups of actors. The resulting group-specific regional attractiveness indicators are used later in the process simulation parts of the model to underpin the decision-making behavior of migrants and locating industries.

The concept of regional attractiveness will be discussed first. Its formal properties coincide with those of the additive model of multi-attribute utility theory, namely

$$U_{ni} = \sum_m w_{mn} v_{mn} a_{mi} / \sum_m w_{mn} \quad , \qquad (3.1)$$

where U_{ni} is the attractiveness of evaluation object i (which here is a region) as seen by actor type n, a_{mi} is the m-th attribute of that evaluation object (e.g., the regional labor market), and w_{mn} and v_{mn} are group-specific importance weights and value or utility functions, respectively.

If, for simplicity, we define $w'_{mn} = w_{mn} / \sum_m w_{mn}$ as the normalized weights, then equation (3.1) is written more succinctly as

$$U_{ni} = \sum_m w'_{mn} v_{mn} a_{mi} \quad . \qquad (3.2)$$

The attributes a_{mi} are either indicators of amenities supplied in region i, namely

$$a_{mi} = f_m(b_{ki}) \quad ,$$

or accessibility measures

$$a_{mi} = \{\sum_j c_{ij}[f_m(b_{kj}) \exp(-\beta c_{ij})]\} / [\sum_j f_m(b_{kj}) \exp(-\beta c_{ij})] \quad , \quad (3.3)$$

where $f_m(b_{ki})$ is a generating function specifying how a_{mi} is related to the k-th variable b of region i, and c_{ij} is an indicator for travel time or cost between regions i and j (Wegener, 1980).

The following discussion will concentrate on the three most important attractiveness indicators on the regional scale: the indicators evaluating regional labor markets, market potentials, and housing. Calibration of the parameters of these attractiveness indicators is discussed in Section 4.

Attention first will be turned to the evaluation of labor markets. Job opportunities are a major factor driving interregional, or long-distance, migration. At the same time, the availability of qualified workers can be of utmost importance for firms seeking new locations. Therefore, information on supply of and demand for the regional labor market is a prerequisite for modeling spatial decision-making behavior of population and industry. The analysis of regional labor market conditions involves three problems: the identification of homogeneous labor market segments, the separate forecasting of regional unemployment for each market segment, and the evaluation of labor market conditions for different population groups and industrial sectors.

Forecasts of labor demand in the whole of North Rhine-Westphalia for both male and female workers were based on research by MAGS (1977), Battelle Institute (Bluem and Frenzel, 1977), and Brune et al. (1978). On the labor supply side, regional labor force participation rates by age, sex, and nationality are updated on the basis of such rates for the whole of North Rhine-Westphalia using research by Kuehlewind and Thon (1977). At present, labor demand and supply are not disaggregated by skill, although this clearly would be desirable.

Next, regional unemployment can be estimated. Regional unemployment rates are the prime indicators used in the model to evaluate the labor market. Estimating them seems to be a trivial problem, at first, if both regional labor demand and supply are known. This contention is true when the region is very large (e.g., a state), and commuting across its boundaries is negligible. In this case, the unemployment rate $\bar{u}$ may be defined as

$$\bar{u} = (L^* - E^*)/L^* \quad , \qquad (3.4)$$

where L^* is the total labor force for the state, and E^* is the total number of jobs presently filled for the state. However, at a smaller spatial scale, interregional commuting tends to be neither balanced nor negligible. In general, interregional commuting has a smoothing effect on regional labor market inequalities. Consequently, the regional unemployment rate may be defined as

$$u_i = (L_i - \sum_j T_{ij})/L_i \quad , \tag{3.5}$$

where L_i is the labor force resident in i, and T_{ij} are home-to-work trips originating in region i. Thus T_{ii} describes the number of workers living and working in region i, and $\sum_j T_{ij}$ is the total number of workers living in region i and working in region i, or elsewhere. Unfortunately, the work trip matrix, **T**, is not known.

Consider E_j and L_i as the two marginal totals for estimating the trip matrix **T**. Estimation in this case must be constrained such that

(1) the number of work trips originating in region i is less than or equal to the labor force in region i,

$$\sum_j T_{ij} \leq L_i \qquad \text{for all i, i = 1, 2, ..., I, and} \tag{3.6}$$

(2) the number of work trips ending in region j equals the number of jobs in region j,

$$\sum_j T_{ij} = E_j \qquad \text{for all j, j = 1, 2, ..., I.} \tag{3.7}$$

Of course, no negative values are allowed for T_{ij} (i.e., $T_{ij} \geq 0$). Among all those matrices that satisfy these constraints, only one agrees with the predefined spatial discounting assumption. This matrix **T** will be considered to be the solution to the problem.

There are basically two procedures for calculating the work trip matrix **T**. First, a dummy region may be introduced with

$$\sum_i L_i - \sum_j E_j$$

jobs, no labor force, and at a location at an infinite distance from all other regions. Then a standard doubly-constrained interaction model solution can be determined. The second method employs a constrained entropy maximizing approach in the form of either a pseudo-production-constrained model or a pseudo-attraction-constrained model (cf., Leonardi, 1981). A discussion of these two algorithms appears in the Appendix. Once the work trip matrix **T** is estimated, the regional unemployment rates are determined by means of equation (3.5). This calculation is done separately for the male and the female labor market segments.

The criteria used in the model to evaluate regional labor markets depend upon the preference system of the specific model actor considered. Both the labor force and the employers are concerned about the level of regional unemployment and regional wage differentials, but for quite different reasons. For the employers, the number of workers commuting to adjacent regions also is relevant. These commuters are considered as a potential that might be attracted by job opportunities in their home region.

The group-specific weights w_n attached to regional unemployment and the regional commuting level, respectively, depend upon the level of overall unemployment and are such that $w_n^{min} \leq w_n \leq w_n^{max}$, where w_n^{min} and w_n^{max} are group-specific minimum and maximum values that are specified exogenously.

Attention now will be turned to a discussion of the evaluation of input and market potentials. Access to inputs and to markets are important locational factors for industry. The markets of different industries are as different as their production processes. Because there is a continuous change in production caused by new technologies and changing demand patterns, markets change, too. Consequently, it is necessary to define the sector-specificity and time-dependence of markets. For this purpose two potentials are calculated for each sector s in each region i. $R_{si}(t)$ is the regional market potential with respect to inputs (purchases), and $Z_{si}(t)$ is the regional market potential with respect to outputs (sales). These two potentials are defined as follows:

$$R_{si}(t) = \sum_{s'} a_{s's} \sum_{j} [E_{s'j}(t) \exp(-\beta_s^r c_{ij})] \quad , \text{ and } \quad (3.8)$$

$$Z_{si}(t) = \sum_{s'} a_{ss'} \sum_{j} [E_{s'j}(t) \exp(-\beta_s^z c_{ij})] \quad , \quad (3.9)$$

where $E_{s'j}$ is the number of jobs in sector s' in region j, $a_{ss'}$ are input-output coefficients expressing trade relationships between sectors s and s', and the $\exp(-\beta_s c_{ij})$ are sector-specific deterrence functions of interregional transportation costs, c_{ij}. In these equations, all regional economic structures are weighted with a sector-specific set of input-output coefficients. Spatial discounting is introduced by the sector-specific spatial deterrence functions. Two different sets of β parameters, β^r and β^z, are needed for input and output markets. In the absence of regional input-output matrices for estimating the coefficients, a national input-output matrix must be used, although regional peculiarities will not be captured in such a matrix.

These two market potentials are interpreted in the same way. High values of $R_{si}(t)$ and $Z_{si}(t)$ indicate good access to input or output markets, respectively, whereas low values indicate that the region is remote with regard to its relevant markets.

The final attractiveness indicator for each region relates to housing. The attractiveness of a region as a place to live is evaluated as a function of its location, its supply of household-oriented infrastructure (e.g., transport, education, recreation, and health care), its housing supply, and its environmental quality. The location of a region is measured in terms of its relative accessibility to the other regions of North Rhine-Westphalia, and to the remaining parts of the Federal Republic of Germany. The supply of household-oriented infrastructure is represented by the number of transport facilities (e.g., autobahn access, express train departures, airports), educational facilities (e.g., universities, and graduate and professional schools), recreational facilities (e.g., parks and woodlands, lakes and waterways), and health care facilities (e.g., hospital beds per capita). The housing supply of the region (e.g., in terms of the average unit size, and the proportion of units equipped with bath and toilet) is evaluated in relation to regional demand for housing. The environmental quality of a region is evaluated in terms of its share of polluting industries.

These indicators, each illustrating a specific aspect of the regional quality of life, are evaluated separately by each actor group depending upon its specific preference system (for calibration of the utility model, see Section 4).

3.2. Process Simulation: The Demographic Submodel

In the process simulation part of the model, the changes of model variables between two points in time, say t and t+1, are projected. Starting from the state of the model system at time t, the process simulation produces a new status at time t+1. For instance, starting in 1970 and simulating in two-year intervals, 1980 is reached after five runs of the simulation model. In this section the demographic submodel is described. The economic submodel will be described in the next section. The demographic submodel combines a model describing the aging process of regional population stocks with a model explaining and predicting migration. The distinction is made between aging and migration because of the different character of the two processes. Aging is a continuous process, primarily determined by the initial population distribution. Basically migration is a piecemeal process caused by changing regional attractiveness differentials. The two processes, however, are closely interrelated. Aging causes a change of needs and preferences that in turn affect migration patterns. In contrast, migration alters the age structure of regional population stocks. The aging submodel, the migration submodel, and a justification for their sequential treatment will be described now.

The aging submodel serves to project by one simulation period, the population of each region, classified by age, sex, and nationality, and adjusts for births and deaths. The aging submodel uses cohort-survival techniques adapted to five-year age groups, and is based upon exogenously updated life tables, and dynamic, age-specific, and regionalized fertility estimates. Some extensions to this conceptualization are planned, such as the inclusion of transitions of foreign born persons to citizens. The aging submodel distinguishes between the following three kinds of transitions:

(1) changing cohort (i.e., transitions from cohort a to cohort a+1),

(2) births (i.e., transitions into cohort 1), and

(3) deaths (i.e., transitions out of any cohort a but not into another cohort).

Modeling the rates of transition from age group a to age group a+1 is straightforward if the length of the simulation period agrees with the size of the cohorts (i.e., after one period all survivors of cohort a have changed to the following cohort a+1). However, where the length of the simulation period is less than the number of years defining the cohorts, there is no correct solution for calculating these transition rates.

At first glance, it seems to be a good approximation to estimate the annual number of transitions $\tilde{C}_a^{sni}(t,t+1)$ from five-year cohorts a to a+1 as one-fifth of the initial population stock $P_a^{sni}(t)$ of age group a. Subscripts s, n and i indicate sex and nationality, and region, respectively. For a simulation period of Δt years, this yields

$$\tilde{C}_a^{sni}(t,t+1) = (\Delta t/5) P_a^{sni}(t) \qquad , a = 1, 2, \ldots, 19. \quad (3.10)$$

This approximation assumes a flat distribution of one-year cohorts within the five-year cohorts. Unfortunately, the shape of population pyramids in reality is not flat-sided enough to make this assumption acceptable. Empirical tests revealed that, in general, a much better approximation is obtained when the average of the origin and the destination cohort sizes is used (Wegener, 1982), so that

$$\tilde{C}_a^{sni}(t,t+1) = (\Delta t/5)[P_a^{sni}(t) + P_{a+1}^{sni}(t)]/2 \qquad , \quad (3.11)$$
$$a = 1, 2, \ldots, 19.$$

Equation (3.10) is preferred only where $P_a^{sni}(t)$ is found to be a peak or a dip, compared with its adjacent age groups $P_{a-1}^{sni}(t)$ and $P_{a+1}^{sni}(t)$.

The number of babies born who survive to the end of the first simulation period is estimated on the following basis:

(1) periodically updated, age-specific fertility estimates $f_a^n(t,t+1)$ for each female age group between the ages of 16 and 50, for the whole of North Rhine-Westphalia,

(2) a constant h describing the proportion of boys among the newborn babies, and

(3) periodically updated, sex-specific survival rates for age group 1, $q_1^s(t,t+1)$.

The updated fertility estimates $f_a^n(t,t+1)$ are regionalized by a multiple regression model leading to modified fertility rates $g_{ai}^{ni}(t,t+1)$ for each region i. Then

$$\overleftarrow{C}^{1ni}(t,t+1) \;=\; \sum_{a=4}^{a=10} [g_a^{ni}(t,t+1) \; h[q_1^s(t,t+1)]^\lambda \; P_a^{2ni}(t)] \quad , \text{ and} \tag{3.12}$$

$$\overleftarrow{C}^{2ni}(t,t+1) \;=\; \sum_{a=4}^{a=10} [g_a^{ni}(t,t+1)(1-h)[q_1^s(t,t+1)]^\lambda \; P_a^{2ni}(t)] \quad , \tag{3.13}$$

where $\lambda = [1/(2)^{1/2}]\Delta t$. $\overleftarrow{C}^{sni}$ are newborn male, for s = 1, and female, for s = 2, babies, respectively, of nationality n in region i having survived the first simulation period. The multiplication of the exponent Δt of the survival rate by $(2)^{1/2}$ takes into account the fact that births are distributed evenly over the period Δt [i.e., the number of newborn babies increases cumulatively (cf., Wegener, 1982)]. To compute the number of deaths between time periods t and t+1, periodically updated age- and sex-specific survival rates $q_a^s(t,t+1)$ are used (cf., Willekens and Rogers, 1978). These numbers are computed on a yearly basis so that

$$\overrightarrow{C}_a{}^{sni}(t,t+1) \;=\; \{1 - [q_a^s(t,t+1)]^{\Delta t}\} \; P_a^{sni}(t) \quad , \tag{3.14}$$
$$a = 1, 2, \ldots, 20.$$

Finally, based on transitions, the regional population stock for all cohorts during time period t is initially updated using the following three equations:

$$P_1^{sni}(t+1) \;=\; P_1^{sni}(t) \;+\; \overleftarrow{C}_1^{sni}(t,t+1) \;-$$
$$\widetilde{C}_1^{sni}(t,t+1) \;-\; \overrightarrow{C}_1^{sni}(t,t+1) \quad , \tag{3.15}$$

$$P_a^{sni}(t+1) \;=\; P_a^{sni}(t) \;+\; \widetilde{C}_{a-1}^{sni}(t,t+1) \;-$$
$$\widetilde{C}_a^{sni}(t,t+1) \;-\; \overrightarrow{C}_a^{sni}(t,t+1), \tag{3.16}$$
$$a = 2, 3, \ldots, 19, \text{ and}$$

$$P_{20}^{sni}(t+1) = P_{20}^{sni}(t) + \widetilde{C}_{19}^{sni}(t,t+1) - \overrightarrow{C}_{20}^{sni}(t,t+1) \quad . \tag{3.17}$$

Attention now will be turned to a discussion of the migration submodel. In each time period the migration submodel results in 24 matrices describing migration flows between the 34 labor market

regions, plus in-migration to and out-migration from North Rhine-
Westphalia for six age groups g (i.e., 1-15, 16-20, 21-35, 36-50,
51-65, 65+ years) by sex and nationality (cf., Gatzweiler, 1975).
Children belonging to age group 1-15 are assumed to migrate with
their parents. Forecasting group-specific migration flows for the
20 adult age groups proceeds through four general steps that now
will be described. First the general propensity to migrate (i.e.,
the total interregional migration volume and the total number of
people migrating into and out of North Rhine-Westphalia) are
determined. It is assumed that temporal variations in the
propensity to migrate depend upon two complementary factors, namely
job security and regional diversity. With respect to job security,
it is assumed that high unemployment rates $\bar{u}$ correspond to a low
level of spatial mobility (Bartels and Liaw, 1981). Strong
empirical evidence in favor of this assumption can be found in the
data of North Rhine-Westphalia for the period 1970 thru 1979 (see
Figure 2, in which the index of spatial mobility denotes the number
of migrants per capita as a percentage of the total number of
migrants in 1970). With respect to regional diversity, it is

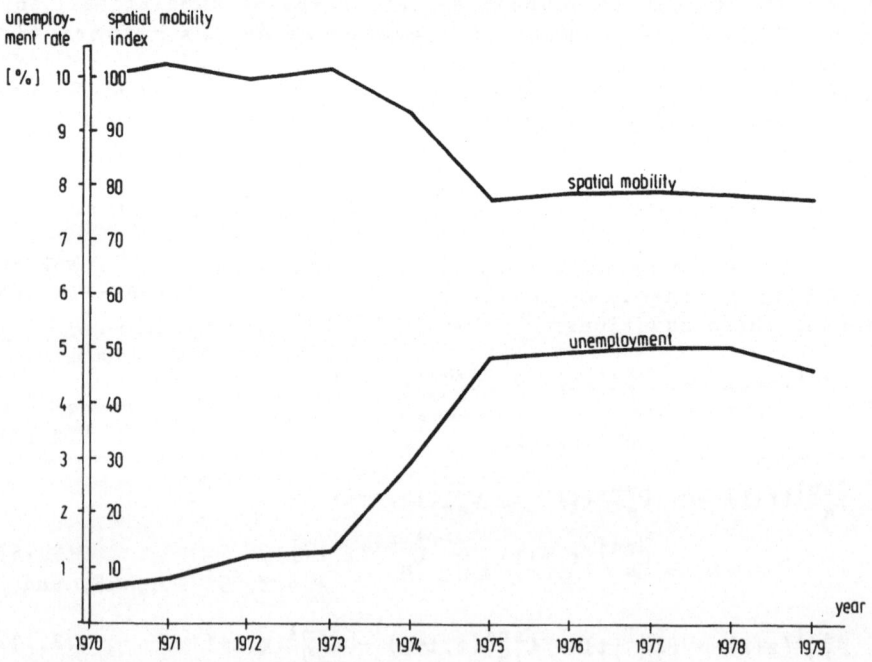

Figure 2. Spatial mobility as a function of unemployment in North
Rhine-Westphalia.

assumed that spatial mobility decreases with increased similarity in the living conditions in the regions. The chance of improving one's living conditions by changing place of residence, and consequently, the stimulus to migrate, is reduced by increased interregional homogeneity.

The variance of the regional attractiveness indicators $U_{gsni}(t)$ defined for each combination of adult age group g, sex s and nationality n, as in equation (3.1), is taken to be an appropriate measure of regional diversity. So mobility rates $m^*_{gsn}(t,t+1)$ of g, s and n should be determined as a function of the general unemployment rate $\bar{u}(t)$ and the variance $v_{gsn}(t)$ of the attractiveness indicators, such that

$$m^*_{gsn}(t,t+1) = f_{gsn}[\bar{u}(t), v_{gsn}(t)] \quad , \quad (3.18)$$

where the mobility rates $m^*_{gsn}(t,t+1)$ are expressed as multiples of the mobility of the observed migration volume $M^*_{gsn}(1,2)$ of the first simulation period. Thus

$$M^*_{gsn}(t,t+1) = m^*_{gsn}(t,t+1) \; M^*_{gsn}(1,2) \quad (3.19)$$

is the total interregional migration volume for each of the adult age groups gsn occurring between times t and t+1. Total in-migration and out-migration from North Rhine-Westphalia are presently exogenously specified.

In the second step the number of people migrating out of each region is estimated as a proportion of total interregional migration $M^*_{gsn}(t,t+1)$. It is assumed that the regional differences in the propensity to migrate depend upon regional attractiveness and home ownership. It is assumed that less attractive regions have relatively greater out-migration than do attractive ones. This is the traditional, yet not undisputed, 'push hypothesis' (cf., Dejon, 1982). It is further assumed that home owners are less inclined to migrate than tenants (cf., Deutschmann, 1972). The attractiveness of a region for migration consists of two subsets of attributes, with the first one expressing the housing situation (e.g., housing supply, accessibility, public services, retail and private services, recreational facilities, and environmental quality), and the second one expressing the labor market situation (e.g., labor supply and demand, and wage levels). The relative weighting of these two subsets is done dynamically in response to the statewide labor market situation in order to take into account the fact that at times of high unemployment migration decisions basically are determined by job considerations.

Based on these hypotheses, the equation for regional out-migration is

$$M^0_{gsni}(t,t+1) =$$

$$\{[1-U_{gsni}(t)]^{\gamma}P_{gsni}(t)\ M^*_{gsn}(t,t+1)\}/\{\textstyle\sum_i [1-U_{gsni}(t)]^{\gamma}P_{gsni}(t)\}\ , \tag{3.20}$$

where $M^0_{gsni}(t,t+1)$ is the number of migrants of age group g, sex s, and nationality n leaving region i between times t and t+1, $P_{gsni}(t)$ denotes the population group gsn living in region i at time t, and γ is a meaasure of sensitivity of the population to regional attractiveness differentials. The expression $1 - U_{gsni}(t)$ ensures the inverse relationship between the regional attractiveness $U_{gsni}(t)$ and the level of regional out-migration.

In the third step the total out-migration by origin established in the second step [i.e., the $M^0_{gsni}(t,t+1)$] are distributed by the following production-constrained spatial interaction model, reinterpreted within the framework of random utility theory:

$$p_{gsnij}(t) =$$

$$E_j(t)[U_{gsnj}(t)]^{\alpha}\ \exp(-\beta c_{ij})/\{\textstyle\sum_j E_j(t)[U_{gsnj}(t)]^{\alpha}\ \exp(-\beta c_{ij})\}\ , \tag{3.21}$$

where $p_{gsnij}(t)$ denotes the probability that a migrant originating in region i ends up in region j, $E_j(t)$ are jobs in region j, and α is a measure of the migrants' sensitivity with regard to regional attractiveness differentials. Accordingly the migration flows are calculated as

$$M_{gsnij}(t,t+1) = p_{gsnij}(t)\ M^0_{gsni}(t,t+1) \ , \tag{3.22}$$

where $M_{gsnij}(t,t+1)$ are flows of population group gsn from region i to region j between times t and t+1. One should note that the attractiveness indicators are used in equation (3.21) as pull or attraction variables and in equation (3.20) as push or deterrence variables.

After having estimated the interregional migration of the adult age groups, the number of children migrating with their parents is estimated for each of the five parent age groups. These children form the sixth migration age group. In addition, interstate migrations into and out of North Rhine-Westphalia are distributed to each region in proportion to its share of interregional migration.

Finally, in the fourth step all changes of the regional population age structure caused by migration are executed. For this purpose, the interregional migration flows have to be split up to correspond to the five-year age structure of the population, taking account of the different migration propensities of each age group a. To achieve this, two matrices are defined, whose respective elements are $M^{in}_{asnj}(t,t+1)$ for the number of in-migrants to region j for each five-year age group a, and $M^{out}_{asni}(t,t+1)$ for the number of out-migrants from region i for each five-year age group a. These two matrices are computed for each of the six migration age groups g, separately. For each region i, sex s, and nationality n, the share d_{asni} of each five-year age group a belonging to the larger migration age group g is calculated and weighted with its specific migration rate $b_{a'sn}$ (cf., Castro and Rogers, 1979; Rogers and Castro, 1981), so that

$$d_{asni} = b_{a'sn} P_{a'sni} / [\sum_{a' \varepsilon G} b_{a'sn} P_{a'sni}] \qquad , \qquad (3.23)$$

where {G} is the set of five-year age groups a' belonging to migration age group g. The d_{asni} serve to split the migration flows as follows:

$$M^{in}_{asnj}(t,t+1) = \sum_i d_{asni} M_{gsnij}(t,t+1) \quad , \text{for all } j, \text{and} \quad (3.24)$$

$$M^{out}_{asni}(t,t+1) = \sum_j d_{asni} M_{gsnij}(t,t+1) \quad , \text{ for all } i. \qquad (3.25)$$

One should note that in both equations (3.24) and (3.25) the split factors d_{asni} of the origin region are used to maintain consistency of the population age structures. With the two matrices determined, the population age structures are updated using the following equation:

$$P_{asni}(t+1) = P_{asni}(t) + M^{in}_{asni}(t,t+1) - M^{out}_{asni}(t,t+1). \quad (3.26)$$

The discussion now will turn to the linkages between the aging and migration submodels. The aging and migration submodels as well as all transitions within the aging submodel are processed sequentially. When compared with the multistate projection technique (Rogers, 1975; Willekens and Rogers, 1978; Rogers, 1981), this sequential type of model needs some justification. The following four kinds of transitions occur in the aging and migration submodels:

(1) aging (i.e., transitions from age group a to age group a+1),

(2) births (i.e., entry into age group 1),

(3) deaths (i.e., exit from any age group a without entering another age group), and

(4) migrations (i.e., transitions from region i to region j).

Since each kind of transition occurs in a continuous stream over the whole projection interval, all transitions should be treated simultaneously in a projection model. This is how the multistate projection model works.

Processing the four kinds of transitions sequentially in a projection model has many advantages in terms of model organization, computing time, and computer storage space. However, the sequential model at the same time creates some problems. Depending upon the order in which they are processed, the transitions are applied to different populations at risk, and this will of course affect the results. To minimize such distortions, the sequential model used here divides the simulation period into four equal subperiods. Starting with a population age structure status at t, the following sequence of steps is performed:

(1)   aging the population from t to t+0.25 and accounting for deaths,

(2)   calculating births from t to t+0.5 and accounting for deaths,,

(3)   aging the population and accounting for deaths from t+0.25 to t+0.5,

(4)   updating the populations at t+0.5 by births from t to t+0.5,

(5)   calculating migration from t to t+1,

(6)   updating the population at t+0.5 by migration from t to t+1,

(7)   aging the population from t+0.5 to t+0.75 and accounting for deaths,

(8)   calculating births from t+0.5 to t+1 and accounting for deaths,

(9)   aging the population from t+0.75 to t+1 and accounting for deaths, and

(10)   updating the population at t+1 by births from t+0.5 to t+1.

Applications to different data sets showed that the multistate and sequential models used in this fashion produce results that are equivalent for all practical purposes.

3.3. Process Simulation: The Economic Forecasting Submodel

The economic submodel predicts employment by 40 sectors

including nonservice, service and retail employment for each of the
34 labor market regions in North Rhine-Westphalia, making use of
sectoral forecasts for the whole state by Battelle (1980). The
Battelle study produced forecasts of sectoral employment until 1995
based upon two scenarios of sectoral change. However, only the
growth rates of the pessimistic scenario were used. In this
pessimistic scenario, the authors assumed that the slowing down of
growth in the seventies was no accident, but rather the start of a
continuous phase of depression. However, viewed from 1982 even
this scenario seems to be far too optimistic. Adapted to the
classification of the model, the sectoral employment projections of
this scenario (where 1970 is set to 100) are listed in Figure 3.

The development of regional employment is assumed to be
determined basically by changes in the economic structure (i.e.,

no.	industry	employment (1970 = 100) [2]			
		1975	1980	1985	1990
1	agriculture, forestry	77	58	47	38
2	energy	135	121	137	146
3	mining	77	75	73	74
4	chemical industry				
5	mineral oil processing	109	103	110	114
6	synthetics processing				
7	rubber, asbestos				
8	stones and minerals				
9	ceramics	75	70	58	51
10	glass				
11	iron and steel				
12	nonferrous metals	106	85	83	80
13	foundries				
14	cold rolling				
15	steel and metal construction				
16	machine manufacturing	80	94	94	96
17	EDP/office equipment				
18	vehicle manufacturing				
19	electrical products				
20	precision mechanics	77	73	66	61
21	sheet metal products				
22	musical instruments, toys				
23	lumber, sawmilling				
24	wood processing				
25	pulp, paper	100	85	84	81
26	paper processing				
27	printing, copying				
28	leather, clothing	68	55	43	36
29	textile industry	68	55	42	36
30	food and drugs	87	77	71	66
31	small industry and crafts [1]	89	83	79	77
32	building industry	82	82	75	72
33	wholesale	91	82	77	75
34	trading, brokering				
35	retail	97	98	99	101
36	transport, telecommunications	103	97	99	100
37	banks, insurances	116	122	133	141
38	restaurants, hotels	99	103	103	106
39	other services	106	124	129	141
40	nonprofit org., public service	114	125	135	148
	Total	95	93	93	94

1) under 10 jobs per firm

2) Source: Battelle (Blüm et al. 1980)

Figure 3.   Employment forecast for North Rhine-Westphalia 1970-90.

the sectoral composition of the economy) and changes in the relative attractiveness of competing regions. Using this distinction between structural or capacity effects and locational influences, the level of employment in sector s at time t+1 in region i is calculated as follows:

$$E_{si}(t+1) = \frac{[E_s^*(t+1)/E_s^*(t)]E_{si}(t) \ f[U_{si}(t)] \ E_s^*(t+1)}{\sum_i [E_s^*(t+1)/E_s^*(t)]E_{si}(t) \ f[U_{si}(t)]} , \quad (3.27)$$

where $E_s^*(t)$ is the total employment for sector s at time t, and $U_{si}(t)$ and $f[U_{si}(t)]$ are sector-specific attractiveness indicators and locational factors for region i, respectively. The locational factors are weights ensuring that sectoral regional growth $E_{si}(t+1)/E_{si}(t)$ is above average, in the case of favorable regional conditions, and below average in the case of below average regional attractiveness values. One should note that no distinction has been made between investment and shut-down decisions (i.e., it has been assumed that regions of high attractiveness attract an above average share of total investments, while at the same time the shut-down rate is comparatively low).

In a forecasting context, the updating of the attractiveness indicators $U_{si}(t)$ and thus also of the locational factors $f[U_{si}(t)]$ is crucial. Good operational explanatory models of these location quality indicators are not really available, and for this reason a priori assumptions have to be made. In many regional models attractiveness differentials are omitted entirely because of the lack of theory and data. In the shift-share model, for instance, this omission is equivalent to setting all locational factors to unity. As such equation (3.27) simplifies to

$$E_{si}(t+1) = [E_s^*(t+1)/E_s^*(t)] \ E_{si}(t) \quad . \quad (3.28)$$

The implication, of course, is that only the capacity effects are considered. This is a very crude and unrealistic assumption. As an alternative, proxies for the sector-specific regional locational factors can be based on regional employment growth rates of the past, yielding $E_{si}(t-1)/E_{si}(t-2)$. Replacing $f[U_{si}(t)]$ in equation (3.27) by this growth rate proxy and rearranging terms leads to

$$E_{si}(t+1) =$$

$$\{[E_{si}(t) \ E_{si}(t)/E_{si}(t-1)]/\sum_i [E_{si}(t) \ E_{si}(t)/E_{si}(t-1)]\} \ E_s^*(t+1) \quad . \quad (3.29)$$

This model would be appropriate if the relative locational attractiveness of each region remained the same as in the previous period. But this will likely only be true over short periods. Over a longer number of time periods, the locational factor $f[U_{si}(t)]$ has to be reviewed periodically in the light of changing regional conditions and changing locational preferences of industry.

In the economic forecasting submodel presented here, the attractiveness of a region as a location for enterprises of a certain industry is represented by attributes such as location, business-serving infrastructure, availability of financial aids, access to markets, labor supply and labor demand, and wage levels. In accordance with the above conclusion, the sector-specific regional attractiveness indicators $U_{si}(t)$ are calculated anew for each period. As was done with the attractiveness for migration, the labor market criteria are weighted dynamically in response to the general labor market situation in the whole state in order to take account of the fact that the relative importance of labor market considerations for industrial location decisions depends so strongly upon the scarcity of labor.

4. CALIBRATION OF THE ATTRACTIVENESS INDICATORS

The attractiveness indicators are the key explanatory variables for the decision behavior of the model actors [i.e., 40 employer groups and 20 adult population groups by age (5 groups), sex (2 groups) and nationality (2 groups)]. Since quantitative information on actor-type specific preferences generally is not available, calibration of the attractiveness indicators deserves special attention. In this section, an approach for overcoming the problem of lack of data will be presented. To recapitulate first, though, the concept of regional attractiveness was described as an additive model of multiattribute utilities such that

$$U_{ni}(t) = \sum_m w_{mn} U_{mni}(t) \quad , \quad (4.1)$$

$$m = 1, 2, \ldots, M \quad \text{and} \quad n = 1, 2, \ldots, N \quad ,$$

where $U_{mni}(t)$ is a partial utility describing a specific aspect, m, of the situation in region i as seen by actor-type n, and w_{mn} is an actor-type specific importance weight associated with this utility. This approach begins by decomposing the M partial utilities into two subsets, those that are eventually time invariant and those that are not. In the first group all partial utilities $U_{kni}(t)$, k = 1, 2, ..., K and K ≤ M, are assembled that express basically time-invariant regional attractiveness differentials. In a twenty-year time perspective, utilities associated with (1) the region's location in the state, (2) recreational facilities, (3) the environmental quality, (4) part of the transport facilities, and (5) utilities associated with image factors (e.g., a scenic landscape, a picturesque town, a certain cultural tradition) may be considered to belong to this group. With the associated weights w_{kn} assumed to be (almost) constant over time, a quasi-static part of the N actor-specific models is defined as

$$\hat{U}_{ni}(t) = \sum_k w_{kn} U_{kni}(t) \qquad , \qquad (4.2)$$

$$k = 1, 2, \ldots, K \qquad \text{and} \qquad n = 1, 2, \ldots, N \qquad ,$$

where consequently $\hat{U}_{ni}(t)$ will be the same for each t. The second group is defined as a subset of partial utilities $U_{hni}(t)$, h = 1, 2, ..., H and H = M − K, that are likely to change over time. These utilities basically are associated with labor market conditions, housing supply, and market access, as seen by service and nonservice industries. These defined the dynamic part of the N actor–type specific utility models. Hence, equation (4.1) can be rewritten as

$$U_{ni}(t) = [\sum_h w_{hn} U_{hni}(t)] + (1 - \sum_h w_{hn}) \hat{U}_{ni} \qquad , \quad (4.3)$$

$$h = 1, 2, \ldots, H \qquad \text{and} \qquad n = 1, 2, \ldots, N \qquad .$$

One should note that each of the dynamic utility components, $U_{hni}(t)$, can be identified. In contrast, the static part is represented by only one single constant $\bar{U}_{ni}$.

Calibration of the parameter of equation (4.3) proceeds stepwise. Census data for two points in time, t and t+1, are needed. At first, all H utility components of the dynamic part are estimated. In the absence of complete data, heuristic methods must be applied. The residual static utilities $\hat{U}_{ni}$ then can be derived iteratively with

$$\hat{U}_{ni}^{(\tau+1)} = \hat{U}_{ni}^{(\tau)} +$$

$$f[g_{ni}^o(t,t+1) - g_{ni}^\tau(t,t+1)\ \bar{g}_n^o(t,t+1)/\bar{g}_n^{(\tau)}(t,t+1)] \qquad , \quad (4.4)$$

where τ is an iteration counter, $g_{ni}^o(t,t+1)$ and $g_{ni}^{(\tau)}(t,t+1)$ are observed and predicted population or employment growth rates of actor–type n for region i, for time period t to t+1, respectively, and $\bar{g}_n^o(t,t+1)$ and $\bar{g}_n^{(\tau)}(t,t+1)$ are the corresponding growth rates for the whole of North Rhine–Westphalia. The predicted growth rates represent results of the following iterative model runs:

(1) at first, the demoeconomic model is executed to determine regional growth rates of population and employment between points in time t and t+1 using arbitrary starting values for the $\hat{U}_{ni}^{(\tau)}$,

(2) as $g_{ni}^o(t,t+1)$ and $g_n^o(t,t+1)$ are known, the $\hat{U}_{ni}^{(\tau+1)}$ can be calculated by equation (4.4), and

(3) the old $\hat{U}_{ni}^{(\tau)}$ are replaced by the new $\hat{U}_{ni}^{(\tau+1)}$, and the model is run again, with $\tau = \tau+1$, so that a new set of

regional growth rates can be obtained, and, as before, all utility values are corrected.

This adjustment process tends to converge after a few iterations [i.e., the differences between predicted and observed growth rates (for all n and i) tend to zero]. At the end of the process the demoeconomic model very closely reproduces the observed regional population and employment levels at time t+1.

One should note that time t+1 may be any multiple of the simulation model's forecasting period after the base year 1970, depending upon available data (i.e., 1972, 1974, 1976, 1978, or 1980). As the matrix $\bar{U}_{ni}$ is assumed to express time-invariant regional attractiveness differentials, the definition of t+1 should not effect the results very much. It follows that calculating $\hat{U}_{ni}$ for successive periods 1970-1972, 1972-1974, ..., 1978-1980 will be a good check to test the temporal stability of the static utilities $\hat{U}_{ni}$.

Thus far calibration results have been based on two points in time, namely 1970 and 1980. The results of these calibrations appear in Figure 4, and illustrate the high degree of agreement between the observed and predicted population and employment levels in the year 1980 (expressed as a percentage of the 1970 numerical values). However, as long as the stability of the $\hat{U}_{ni}$ over different time intervals has not been tested, these results must be considered preliminary.

5. CONCLUSION

In this paper the demoeconomic development of a region has been assumed to be a highly dynamic and selective process. From this it is concluded that multiregional demoeconomic models must have a requisite level of complexity in order to grasp the variety of causes and effects inherent in real-world processes. But, however desirable, the implemention of a comprehensive modeling approach must overcome a number of serious difficulties. One of these difficulties lies in the limited availability of disaggregated regional data. Another difficulty is more fundamental and is related to the lack of good theory or even hypotheses about the development of regional settlement structures. Seen in this framework, the simulation model presented in this paper is an ambitious initial undertaking towards a holistic and integrated approach.

The model is fully operational as described here. With the exception of some calibration data, the data base is complete and most of the parameters have been estimated. Work presently is continuing on both the analysis of demoeconomic processes of the

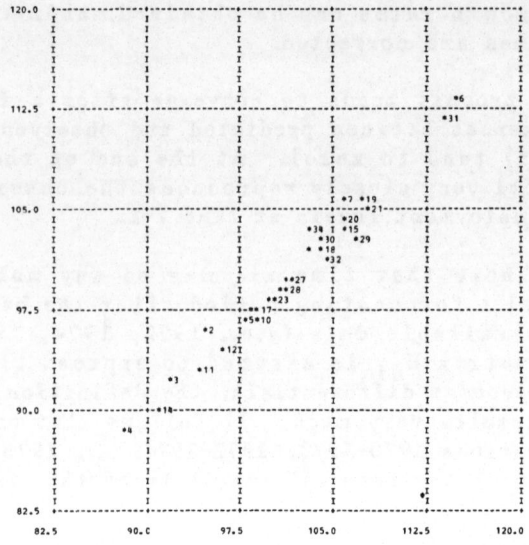

observed mean	above:	102.0	below:	94.2
predicted mean	above:	100.4	below:	94.2
r—squared	above:	0.9748	below:	0.9726
mean absolute percent error	above:	1.7	below:	1.0
number of observations	above:	34	below:	34

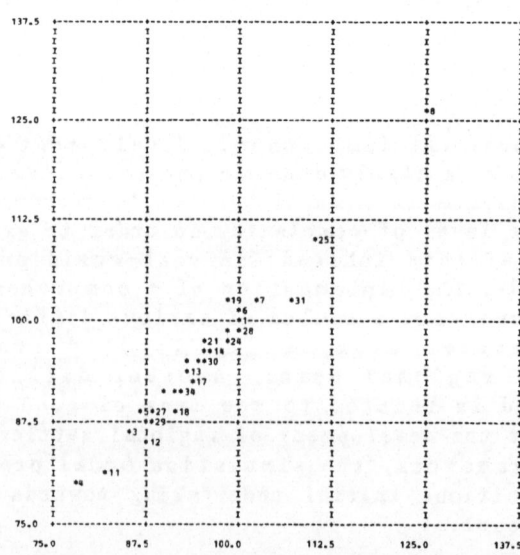

Figure 4. Observed versus predicted variables: population (above) and employment (below), 1980 figures as a percent of 1970 figures, observed (x—axis) and predicted (y—axis).

past decade, and the finishing of the calibration task.
Forecasting applications are planned. The validity and consistency
of a number of hypotheses about long-term regional development will
be tested by using different sets of data. The eventual aim is to
use the model not only as a research tool, but also as a planning
tool for forecasting intermediate and long-term effects of
alternative regional policies.

6. REFERENCES

Bartels, C. and K. Liaw, 1981, The Dynamics of Spatial Labor
    Mobility in the Netherlands, Working Paper WP-81-87,
    Laxenburg, Austria: International Institute for Applied
    Systems Analysis.

Bluem, A. and U. Frenzel, 1977, Quantitative und Qualitative
    Vorausschau auf den Arbeitsmarkt der Bundesrepublik
    Deutschland, Nuernberg: Institut fuer Arbeitsmarkt- und
    Berufsforschung, BeitrAb, 8.1, 8.2.

_____, _____, S. Huebner, E. Kaiser and L. Lichtwer, 1980, Der
    Arbeitsmarkt in Nordrhein-Westfalen und im Ruhrgebiet,
    Research Report BF-R-64.269-3, Frankfurt am Main: Battelle-
    Institute.

Brune, R., H. Hennies-Rautenberg and K. Loebbe, 1978,
    Wirtschaftsstrukturelle Bestandsaufnahme fuer das Ruhrgebiet –
    1. Fortschreibung, Essen: Rheinisch-Westfaelisches Institut
    fuer Wirtschaftsfoerderung.

Buttler, F., K. Gerlach and P. Liepmann, 1977, Grundlagen der
    Regionaloekonomie, Hamburg: Rowohlt.

Castro, L. and A. Rogers, 1979, Migration Age Patterns:
    Measurement and Analysis, Working Paper WP-79-16, Laxenburg,
    Austria: International Institute for Applied Systems
    Analysis.

Dejon, B., 1983, Attraction-regulated Dynamic Equilibrium Models of
    Migration of the Multiplicative Type, in Evolving Geographical
    Structures, edited by D. Griffith and A. Lea, The Hague:
    Martinus Nijhoff, pp. 12-23.

Deutschmann, H., 1972, The Residential Location Decision: Study of
    Residential Mobility, Socio-Economic Planning Science, 6:
    349-364.

Ewers, H., H. Seidenfus, J. Mueller and N. Sparding, 1979, Regionale Entwicklungsunterschiede des Sektoralen Strukturwandels in der Bundesrepublik Deutschland, Working Paper 19, Muenster: Sonderforschungsbereich 26.

Gatzweiler, H., 1975, Zur Selektivitaet Interregionaler Wanderungen, Bonn: Bundesforschungsanstalt fuer Landeskunde und Raumordnung.

Klaassen, L., W. Molle and J. Paelinck (eds.), 1979, Dynamics of Urban Development, Rotterdam: Netherlands Economic Institute.

Klemmer, P. and D. Kraemer, 1975, Regionale Arbeitsmaerkte: Ein Abgrenzungsvorschlag fuer die Bundesrepublik Deutschland, Bochum: Brockmeyer.

Kuehlewind, G. and M. Thon, 1976, Projektion des Deutschen Erwerbspersonenpotentials fuer den Zeitraum 1975 bis 1990, in Mittel- und Laengerfristige Arbeitsmarktprojektionen des IAB, edited by W. Klauder, et al., Nuernberg: Institut fuer Arbeitsmarkt- und Berufsforschung, BeitrAB 16.

Leonardi, G., 1981, A Unifying Framework for Public Facility Location Problems, Environment and Planning A, 13: 1001-1028, 1085-1108.

MAGS (Ministerium fuer Arbeit, Gesundheit und Soziales des Landes NW), 1977, Perspektiven der Arbeitsmarktpolitik in Nordrhein-Westfalen, Arbeit und Beruf 17, Karlsruhe.

Molle, W. and L. Klaassen, 1978, Trends of Regional and Urban Development in North-Western Europe, paper presented at the Lustrum-Seminar of the Erasmus University, Rotterdam, November 8-10.

Rogers, A., 1975, Introduction to Multiregional Mathematical Demography, New York: Wiley.

_____, (ed.), 1981, Advances in Multiregional Demography, Research Report RR-81-6, Laxenburg, Austria: International Institute for Applied Sytems Analysis.

_____ and L. Castro, 1981, Model Schedules in Multistate Demographic Analysis: The Case of Migration, Working Paper WP-81-82, Laxenburg, Austria: International Institute for Applied Systems Analysis.

Schoenebeck, C., 1982, Sectoral Change and Interregional Mobility: A Simulation Model of Demoeconomic Development in North Rhine-Westphalia, Collaborative Paper CP-82-10, Laxenburg, Austria: International Institute for Applied Systems Analysis.

_____, and M. Wegener, 1977, Kleinraeumige Standortwahl und Intraregionale Mobilitaet: Die Raum-Zeit-Struktur des Modells, Working Paper 7, Muenster: Sonderforschungsbereich 26.

van den Berg, L. and L. Klaassen, 1978, The Process of Urban Decline, Working Paper 1978/6, Rotterdam: Netherlands Economic Institute.

_____, -----, W. Molle and J. Paelinck, 1979, Synthesis and Conclusions, in Dynamics of Urban Development, edited by L. Klaassen, Rotterdam: Netherlands Economic Institute.

Wegener, M., 1980, A Multilevel Economic-demographic Model for the Dortmund Region, paper presented at the workshop on Urban Systems Modeling, Moscow, USSR, September 30-October 3 (forthcoming in Sistemi Urbani).

_____, 1982, Aspects of Urban Decline: Experiments With a Multilevel Economic-Demographic Model for the Dortmund Region, Working Paper WP-82-17, Laxenburg, Austria: International Institute for Applied Systems Analysis.

_____, C. Schoenebeck, F. Gnad, M. Vannahme and H. Tillmann, 1983, Siedlungsstruktur und Strukturwandel der Wirtschaft. Abschluss- und Ergebnisbericht, Sonderforschungsbereich 26 Muenster, Dortmund: Universitaet Dortmund (forthcoming).

Willekens, F. and A. Rogers, 1978, Spatial Population Analysis: Methods and Computer Programs, Research Report RR-78-18, Laxenburg, Austria: International Institute for Applied Systems Analysis.

7. APPENDIX: TWO ALTERNATIVE METHODS FOR ESTIMATING MATRIX T

Two procedures for estimating matrix T are presented here. In the first procedure, a fictitious dummy region $I+1$ with

$$\sum_i L_i - \sum_j E_j$$

jobs and $L_{I+1} = 0$, is defined, thus ensuring that

$$\sum_j E_j = \sum_i L_i \quad .$$

All distances between region I+1 and all other I regions have to be set to nearly infinitely high values. Taking L_i and E_j (with $i,j = 1, 2, ..., I+1$) as the marginal vectors of a doubly-constrained spatial interaction model, the work trip matrix T is estimated by a multiproportional method, such as the RAS algorithm. All workers 'migrating' to this dummy region I+1 are assumed to be unemployed. Thus,

$$u_i = T_{i,I+1}/L_i \quad . \tag{A.1}$$

In the second procedure, the problem is defined as the following constrained entropy-maximizing one (Leonardi, 1981):

$$\text{MAX:} \quad - \sum_{i,j} T_{ij} \log[T_{ij}/\exp(-\beta c_{ij})] -$$
$$- \sum_i (L_i - \sum_j T_{ij}) \log(L_i - \sum_j T_{ij}) \quad , \tag{A.2}$$

subject to the inequality constraint given by equation (3.6) and the equality constraint given by equation (3.7), where c_{ij} is a measure of separation between regions i and j, and β is the distance decay parameter. The following are two alternative ways for solving this maximization problem iteratively (Leonardi, 1981):

$$\omega_j^{(\tau+1)} =$$
$$E_j/\{[(\sum_i L_i \exp(-\beta\, c_{ij})]/[1 + \sum_j \omega_j^{(\tau)} \exp(-\beta c_{ij})]\} \quad , \tag{A.3}$$

or

$$\sigma_i^{(\tau+1)} =$$
$$L_i/\{1 + [\sum_j E_j \exp(-\beta\, c_{ij})]/[\sum_i \sigma_j^{(\tau)} \exp(-\beta c_{ij})]\} \quad , \tag{A.4}$$

where τ is an iteration counter. The initial values $\omega^{(0)}$ and $\sigma^{(0)}$ must be positive or zero. The iterative procedure stops if either or both of the following conditions hold:

$$\xi^\omega > \omega_j^{(\tau+1)} - \omega_j^{(\tau)} \quad , \text{ for all j, and} \tag{A.5}$$

$$\xi^\sigma > \sigma_i^{(\tau+1)} - \sigma_i^{(\tau)} \quad , \text{ for all i.} \tag{A.6}$$

ξ^ω and ξ^σ are predefined error limits.

In the case of calculating the vector ω, the work trip matrix T is estimated by means of the pseudo-production-constrained model

$$T_{ij} = [L_i\, \omega_j \exp(-\beta c_{ij})]/[1 + \sum_j \omega_j \exp(-\beta c_{ij})] \quad . \tag{A.7}$$

Calculating matrix σ, the work trip matrix is estimated by means of the pseudo-attraction-constrained model

$$T_{ij} = [E_j \ \sigma_i \ \exp(-\beta c_{ij})]/[\sum_i \sigma_i \ \exp(-\beta c_{ij})] \qquad . \qquad (A.8)$$

A good test of the functional form of the model is to calculate the limiting case where all $\exp(-\beta c_{ij})$ are set to unity. Under these circumstances, work trip origins should distribute over space in proportion to labor supply, resulting in equal unemployment rates u_i in all regions [see equation (3.5)]. All three models fulfill this condition.

SECTION 4

SPATIAL PRICING: EFFICIENCY AND DYNAMIC ASPECTS

Economic geographers and spatial economists have long been interested in various types of spatial pricing, and in particular with efficiency properties of this type of pricing. Although the static problem has been thoroughly researched, very little work has been done on the dynamic problem. As is so often the case, some new and very important issues are uncovered when dynamics are explored. All three papers in this section deal with pricing problems in space and time, and each one is interested in questions of efficiency. Yet, as they treat different specific problems using different models, and ask a variety of different questions, they could be considered exemplary of the wide range of ways in which spatio-temporal pricing can be handled.

In the first paper Curry deals with a very general and fundamental spatial pricing problem using a conceptual framework akin to the dissipative structure paradigm of Prigogine and his colleagues. He makes use of potential theory and notions of areal entropy and income entropy to show that, in general, spatial pricing can be expected to lead to inefficient trade patterns between regions. While pricing systems that are operating perfectly tend to lead to stable equilibria in trade relations, imperfections can lead to unstable relations or dissipative structures. If prices are disturbed radically, so that they are far from equilibrium, the non-linearities in the system dynamics become important, and it becomes very unlikely that the former stable equilibrium would be converged upon. To illustrate this theory Curry developes an evolutionary model relating to the shock to the world economy caused by the oil price increase in 1976. He argues that more than likely this shock has given rise to a significant bifurcation in the space-economy, leading to an inefficient steady state. Hence, without a similarly large shock of an opposite nature, the economy is very unlikely to reach a more efficient steady state. Curry's paper indicates that very profound insights into the dynamics and evolution of at least the spatial pricing problem can be obtained using theory and models akin to the evolutionary paradigm of dissipative structures. Some of Curry's thoughts on how to construct dynamic spatial models have a parallel in the works of Haag and Weidlich, and Dendrinos, which appear earlier in this volume.

Sheppard, in the second paper, develops a multiple commodity urban system growth model, following the tradition of Marx, but making use of some of the powerful matrix algebraic techniques developed by modern quantitative classical theorists. Employing a model that could be characterized as a generalized Sraffarian input-output model, Sheppard is able to show that the dynamic equilibrium for this problem is an unstable one. The system can be easily shocked into a state in which supplies and demands do not match, and further investment will not occur at prevailing rates of profit. He also developes a neo-Keynesian model in which all surplus in one period is reinvested in the next period, and shows that this investment strategy leads to another crisis. In general, a contradiction exists between balanced growth and the realization of an appropriate rate of return on investment. Sheppard then examines the following four strategies, which attempt to change the structure of production relations in order to avoid this contradiction : increasing interactions with outside regions, changes in non-labor inputs, changes in the use of labor, and changes in interurban trading patterns. The last strategy, essentially one of altering transport inputs, is treated most fully, using a discrete dynamic price model. In the last part of the paper Sheppard utilizes a simulation model to explore the dynamic nature and sensitivities of trade patterns, coupled with two pricing systems, namely an economic Marxian pricing mechanism and a geographic pricing scheme based on potentials. This latter pricing system is similar to Curry's model, which has been mentioned above. A large number of interesting results are generated. Perhaps the most remarkable is the similarity in the time paths of the two pricing systems. Although this paper gives some interesting insights, it leads to an even larger number of important questions.

In the third paper Haining is interested in the relationship between spatial structure and spatio-temporal processes, and uses prices of gasoline over a set of gas stations as an example problem. After summarizing the key features of binary interaction and Markov models, Haining reviews research into the analysis of patterns of diffusion, employing the model based methodology of map description and the possibility of drawing inferences about processes from patterns. This analysis stresses models capable of distinguishing between different processes or effects, such as intra-site effects and neighborhood effects. Next an autologistic model is fitted to gasoline prices of 85 stations to test whether or not spatial price competition was evident during each of four points in time. These results are compared in a comparative static manner. Haining concludes with an abstract discussion of flows on a two-dimensional surface and how they relate to structural attributes of such surfaces. Griffith, in the preceding section, also is concerned about the role played by spatial configuration.

INEFFICIENCY AND INSTABILITY OF TRADE PATTERNS

Leslie Curry

University of Toronto
Canada

1.   INTRODUCTION

The motivation of this paper is a probing of the range of validity of competitive pricing. In the presence of externalities pricing systems are inefficient allocators. Sheppard and Curry (1982) have discussed conditions under which spatial price equilibrium will not distribute a commodity from suppliers to markets efficiently. However, the aim of this paper is to go further, to show that the world trade system is inefficient because of an inherent feature of market forces. In particular, the inefficiencies of large scale inflation and unemployment have existed for a number of years, a regime that seems to have been inaugurated with the shock of the oil prices increase.

Instead of modelling this situation directly, the work of Prigogine and his colleagues and expositors will serve as a foundation. It was fortunate that the use of a number of the concepts originating in chemistry and physics, such as entropy and potentials, have been well developed in economic geography. Curry (1982a) recently modified some of the definitions of thermodynamics to encompass the geographical usages, and thus the language of equilibrium thermodynamics can be employed to discuss economic problems. It was therefore a relatively short step to borrow Prigogine's ideas. Basically Prigogine says that large fluctuations can drive systems to new steady states, and that these states can provide different spatial orderings (Prigogine and Nicolis, 1973), which in itself is hardly remarkable. Unfortunately the degree of abstraction and the level of technical discourse needed to justify this statement is formidable. This paper will attempt to employ the recently formulated grammar of

macro-economic geography in Prigogine's arguments, and thus put
them in a much wider context.

2. SPATIAL PRICE EQUILIBRIUM AND REALITY

The spatial price equilibrium model (SPE) may be regarded as
the epitome of an efficient geographical system. Given separate
supply and demand schedules for a single commodity in each of a
number of regions together with the transport costs between each
pair of regions, either by the action of competitive traders or by
programming algorithms an equilibrium set of prices is obtained.
Each region's production, consumption, imports and exports have
adjusted to reach an overall optimum with minimal prices. Places
not on the trading network but which have been evaluated in terms
of it also have their prices in equilibrium; hence, there is a
topological choice of network involved. A large number of
'inefficiencies' can make the real world depart from this
condition. The regional tatonnement process is unwieldy,
especially if estimating the equilibrium prices is difficult.
Ignorance of prices and of opportunities will occur. the
inherently monopolistic aspect of local space more than likely is
to be encountered.

SPE can be represented in terms of potential theory (Sheppard
and Curry, 1982), which may be summarized by Ωr, where Ω is the
minimum average potential (equilibrium price gradient) and r is its
corresponding flow up the gradient used to form the mean. The real
world prices map can be described by its areal entropy, S. To
obtain S, pass the map through a series of successively larger
scale filters (i.e., band passes) so that it is broken up into
scale components. Absolute values at each scale are summed, and
the grand total is obtained and used to normalize the absolute
values so that they become probabilities. Then Shannon's entropy
measure, $-p_i \log(p_i)$, is applied to obtain the areal entropy of the
map, namely

$$S = - \sum_{n=1}^{n=N} \sum_{r=1}^{r=R} p_{nr} \log(p_{nr}) \quad , \qquad (2.1)$$

where the summation is over scales from r = 1 to r = R and over
areal units at the same scale from n = 1 to n = N (Curry, 1972).
Therefore

$$\Omega r = - S T \quad , \qquad (2.2)$$

where T is the coefficient of inefficiency that makes the two sides
equal $(0 \leq T \leq 1)$. The efficient map Ω also is described in terms
of areal entropy. Differentiating equation (2.1) leads to a

discussion of the difference between efficient and inefficient conditions, and yields

$$\Omega \ \partial r = T \ \partial S - \partial U \qquad , \ r \ \text{and} \ S \ \text{are fixed,} \qquad (2.3)$$

$$r \ \partial \Omega = \partial E - T \ \partial S \qquad , \ \Omega \ \text{and} \ S \ \text{are fixed,} \qquad (2.4)$$

$$- S \ \partial T = \Omega \ \partial r + \partial A \qquad , \ r \ \text{and} \ T \ \text{are fixed, and} \ (2.5)$$

$$- S \ \partial T = \partial G - r \ \partial \Omega \qquad , \ \Omega \ \text{and} \ T \ \text{are fixed,} \qquad (2.6)$$

where ∂E is the loss of flows due to opportunities not being exploited, ∂U is the increase in price gradients due to a lack of exploitation of opportunities, ∂A represents inefficient, or 'wrong,' price gradients, and ∂G represents the inefficient flows resulting from the wrong gradients. Equation (2.1) was analyzed, and two Lagrange multipliers developed, one for the general efficiency control T, and the other for allocating effort between A and U (i.e., between chasing opportunities for known prices and getting the prices correct, thus reducing transaction costs) [Curry, 1982a].

The notion of entropy now will be extended from being a purely descriptive value for map patterns. Various writers have suggested that it be used to describe freedom of choice. In a SPE situation there is clearly no freedom of choice whatsoever, but as the aim of optimization is relaxed, more possibilities emerge. There is only one possible price map for competitive SPE, but σ, the difference between its entropy and that of the real world map, represents the freedom of action that traders have had in forming the latter map. Now freedom of choice is only given up in order to gain more income. But income represents another sort of freedom of choice, and thus can be given an entropy measure. Unfortunately the SPE areal entropy does not capture this effect, since it only represents the map pattern of prices.

Equations (2.1) and (2.2) are tied to the SPE system, and it is difficult to see how income can fit into them. Let income move out of the trading system into another, namely the household system, so that income entropy leaves the system, although it is generated there. Income entropy can be added to σ to obtain total entropy ε developed in the SPE system. Entropy stays in the system to the extent that there is inefficiency, and as efficiency improves more entropy leaves the system as income.

3. TRADE AND FLUCTUATING PAYMENTS

It is necessary to generalize the SPE model, in an informal way since this generalization will be the topic of a later paper,

which developes a model of trade. In order to do so trade
relations between regions need to be discussed using potential
notions and employing a distance argument. Assume that each of a
set of commodities is established in SPE independently. Add up all
the potentials weighted by the quantities of flows. Incomes are
fixed by the goods sold and together with preferences, these fix
the various amounts of goods demanded. Since incomes and payments
must be in balance, adjustments will occur. The goods imported,
exported and used internally will affect one another. Price of
agents and consumption levels will adjust, and technologies will
change. Those industries that have the greatest advantage relative
to other industries internally, and relative to other regions
externally, will persist. The same filtering process will apply to
demand. By heroic assumption this has all been done and an
equilibrium established. At least it is possible to conceive of
such a model with equilibrium relative prices represented by
potentials, and an aggregate minimum potential being achieved. The
friction of distance will be an aggregate for all the goods moving
between pairs of regions. If these regions are conceived of as
having separate currencies, then the aggregate relative potential
differnces represent rates of exchange. Let this equilibrium be
for a time average around which there will be fluctuations.

People and areas sell where they can and buy from where they
would. The sense of exchange of goods is only seen a posteriori.
Individuals will balance their books only over a period, and
regions with their multilateral means of balancing their accounts
wil follow suit. This entails a constantly fluctuating supply and
demand for credit as payments and receipts lag or lead.
Intensification of these fluctuations occurs when either labor
alone or entire plants become(s) idle or worked overtime.

The control of fluctuations operates via prices. For the
normal range of fluctuation of the difference between payments
coming in and going out, an inventory of credit can be established
and its corresponding price will match the seriousness of the
exchange situation and bring it back into line. The rate of
exchange and the price of credit are intimately related. In static
deterministic theory, the rate of exchange is fixed by the ratio of
export revenues and import payments. The price of credit, under
the assumption made here, is fixed by the size of fluctuations in
this ratio. A fluctuating exchange rate is capable of
stabilization in the same way as credit by means of a managed pool.
The price of credit is the price of borrowing, from the bank say,
and this is related to the price of lending to the bank. A
relatively high interest rate increases a flow of funds from other
areas that is similar to increased exports, and causes the currency
to be dearer and reduces exports or increases imports. Similarly,
a debit balance on a foreign account can be corrected either by
reducing the exchange rate on the home currency or by increasing

interest rates to attract outside funds. When imperfections in the
operation of the economy are allowed, the control of fluctuations,
while operating through the same channels, is less exact, and the
stability of the system is called into question. A criterion for
stability is needed and now will be established.

4. STABILITY IN LINEAR PROCESSES

Consider the situation in which no economic action will be
undertaken unless it maintains or improves freedom of choice (i.e.,
total entropy ε does not decrease). Thus, $d\varepsilon \geq 0$. For a given T
this implies that $d\sigma \geq 0$. As long as the general level of
efficiency does not change, so that the proportion of total entropy
found in the inefficiencies of the commodity distribution is
constant, then $d\sigma/dt \geq 0$ under average conditions. Purposeful
economic activity must involve doing the same things over again, or
aiming at increasing income so that $d\varepsilon \geq 0$. This is not an
instantaneous condition. For example, it will not always be true
when fluctuations occur. But it is difficult to conceive of
economic activity designed to reduce income and increase
inefficiency within the present context. Since $d\sigma/dt \geq 0$, an
irreversible process is defined.

Isolated systems tend to an equilibrium where macroscopic
processes cease, whereas open systems approach a stationary state
where processes continue at a constant rate with a balance between
inflows and outflows, and hence no aggregate changes occur. For an
equilibrium position of an isolated system undergoing irreversible
change, equation (2.3) implies that

$$T \, \partial S - \Omega \, \partial r - \partial U \geq 0 \quad .$$

To be stable any small fluctuation of the variables satisfies the
inequality

$$- T \, \partial S + \Omega \, \partial r + \partial U \geq 0 \quad ,$$

so that $dS(\text{or } d\sigma) \leq 0$ is the stability criterion (i.e., $dS/dt \geq 0$
implies stability). In isolated systems equilibrium corresponds to
maximum entropy, and hence to a zero entropy increase. Here this
has to be so by definition. It also can be shown to be so using
the inferential version of entropy (Curry, 1982a). In an open
system, as income and its concomitant areal entropy increase, there
must be exchanges with other systems (households or resource
sectors, for example) to produce a balance. Freedom of choice can
only be gained (i.e., constraints weakened) by increasing
constraints elsewhere. In equilibrium, internal entropy
production, income change ∂S_d, is exactly compensated by an entropy
flux to the environment, other economic sectors ∂S_e. For example,

an increase in efficiency, causing a reduction in the areal entropy of the price map, leads to an increase in income, that moves out of the system. None of these changes occur in equilibrium. Consequently,

$$\partial S/\partial t = \partial S_e/\partial t + \partial S_d/\partial t = 0 \quad .$$

In open systems steady states can occur that are not equilibria, and the corresponding stability criterion becomes

$$\partial S_d/\partial t = \partial S/\partial t - \partial S_e/\partial t \geq 0 \quad .$$

Onsager (Zotin, 1972) states that the rate of entropy production is constant and minimal and external parameters do not change.

Since $d\sigma$ is the difference between actual and efficient price structures, so that it is equal to TdS, it can be thought of as the result of the dissipation of possible income. Its rate of production, P, is the product of actual flows and their conjugate price gradients, such that

$$P = d\sigma/dt = \sum_i \Omega_i r_i \quad . \qquad (4.1)$$

The two terms on the right-hand side of equation (4.1) are related by the coefficients of conductivity L_{ik} between regions i and k as

$$r_i = \sum_k L_{ik} \Omega_k \quad . \qquad (4.2)$$

Equation (4.2) gives

$$d\sigma/dt = \sum_k L_{ik} \Omega_i \Omega_k \geq 0 \quad ,$$

which defines the dissipation function

$$\gamma(\Omega,\Omega) = (0.5) \sum_k L_{ik} \Omega_i \Omega_k \quad . \qquad (4.3)$$

The first and second derivatives of equation (4.3) are, respectively,

$$\partial\gamma/\partial\Omega_i = \sum_k L_{ik} \Omega_k = r_i \quad , \text{ and}$$

$$\partial^2\gamma/\partial\Omega_i\Omega_k = \partial r_i/\partial\Omega_k = L_{ik} \quad .$$

Writing r_i as a function of the forces $\Omega_i = r_i(\Omega_1,\Omega_2,....)$, expanding as a Taylor series in powers of Ω_1 and Ω_2, and then truncating after the first term of the Taylor series yields the linear relation $r_i = \sum_k L_{ik}\Omega_k$, which was given above as equation (4.2). This equation will give good results only for small deviations of forces and fluxes. Since the dissipation function is quadratic, the external variational principle must involve a maximum, so that

$$\delta[P(r,\Omega) - \gamma(\Omega,\Omega)] = 0 \quad,$$

and hence

$$P(r,\Omega) - \gamma(\Omega,\Omega) = \max.$$

Because $\gamma(\Omega,\Omega) \geq 0$ it also is clear that $P(r,\Omega) \geq 0$.

If the values of some forces Ω_1, Ω_2, ..., Ω_v are fixed so that the system cannot attain an equilibrium state, then the unrestricted forces beyond v (i.e., v+1) are defined by

$$\partial(d\sigma/dt)/\partial\Omega_{v+1} = \sum_k (L_{v+1,k} + L_{k,v+1}) \Omega_k$$

$$= 2 \sum_k L_{v+1,k} \Omega_k = 2r_{v+1} = 0 \quad,$$

where $L_{ik} = L_{ki}$. This system will be in a stationary state relative to the unrestricted variables, and therefore in a state of minimal entropy production that also is stable (Schakenberg, 1977). These results are simply an extension of the equilibrium state into the non-equilibrium domain (Turner, 1980). Consequently, qualitatively new states cannot appear near the equilibrium point of linear irreversible processes. But a steady state of low S can be associated with P_{min} in the same way as the equilibrium can be. This principle assumes importance for non-linear processes.

5.   STABILITY IN NON-LINEAR PROCESSES

As was said earlier, near equilibrium the first power expansion of some mathematical function may be used. To handle non-linearities that will occur further away from equilibrium it is necessary to be involved with at least the second power, or in the present case, $\delta^2\sigma$. This term represents excess entropy only because it is an extra contribution over and above the linear term, and has no substantive interpretation on its own. The term σ denotes the amount of possible income dissipated by spatial inefficiences, and $\delta\sigma$ denotes the change in this dissipation. Hence, $\delta^2\sigma$ has the same meaning as $\delta\sigma$.

According to Glansdorff and Prigogine (1971) $\delta^2\sigma$ is a local stability criterion for non-equilibrium states since it is a Lyapunov function--a quadratic negative definite expression. Thus, when $\delta^2\sigma \geq 0$ in the linear range close to equilibrium, a steady state would be stable and fluctuations would die out. the same authors postulate an evolutionary criterion from the same factor (Turner, 1980). Writing

$$\partial P = \partial P_r + \partial P_\Omega \quad ,$$

the natural evolution of a non-equilibrium system is prescribed by

$$\partial P_\Omega / \partial t \leq 0 \quad . \tag{5.1}$$

In the limit of linearity this equation reduces to minimum entropy production. In this case this reduction is as follows:

$$\partial P_\Omega = \partial P_r = 0.5 \, \partial P \leq 0 \quad .$$

Away from equilibrium a potential function will not exist and, even when the equality in equation (5.1) holds, stability is not ensured and more than one steady state is possible. As was mentioned earlier, trading is founded on estimates of future prices from the equilibrium price, but when prices are severely disturbed no basis for calculating risk can exist. Potentials do not exist. Glansdorff and Prigogine developed the notion of the local potential for circumstances when the equilibrium potential no longer has meaning. This notion allows the use of variational techniques for non-linear, irreversible processes. It involves each unknown function twice, once as a mean value and once as a fluctuating value. It takes on a minimum value when the average quantity coincides with the most probable quantity.

With movement away from equilibrium and into non-linearities the criterion defined by equation (5.1) no longer applies, and so the following excess entropy production has to be considered (Prigogine and Lefever, 1973, 1975):

$$\partial (\delta^2 \sigma) / \delta t = \delta P = \sum_i \delta r_i \, \delta \Omega_i \quad ,$$

where $\delta r_i \delta \Omega_i$ are the excess flows and forces due to the fluctuation from the mean regime. When $\delta P > 0$ the system is stable, but for larger fluctuations it passes through zero and reaches instability. Actually this criterion can be used near equilibrium, also, but making this distinction allows the two situations to be distinguished. Thus, excess entropy must increase in time for stability, and consequently it is consistent with the idea that economic actions are aimed at increasing freedom of choice. Some types of non-linear processes can provide negative inputs and thus a destabilizing influence, though.

After a large fluctuation a new price structure must occur, giving a new potential surface; but, it will not be that of the competitive (or mathematical programming) solution, and therefore will incorporate inefficiencies. Two features of stabilization under SPE are of particular importance. First it optimizes over any size of space designated so that smaller sections of this space do not necessarily benefit from being included. SPE emphasizes the whole interest rather than the regional interest. To the extent that local interests are taken into account, the distribution pattern will deviate from the overall one, and consequently the areal entropy will be larger. If SPE is subject to a large fluctuation, so that spatial prices are far from equilibrium, its previous stability will be different from that of the 'inefficient' system that is likely to be more local in character. Thus, when income dissipation in 'inefficiencies' is sufficiently dominant, large fluctuations will be damped down. There is a local equilibrium in which entropy (income) is the same function of the local macroscopic variables as in macroscopic equilibrium.

The second feature is that, in achieving an equilibrium, a regional tàtonnement process must go on, equivalent to the powering of an interactions matrix. Thus, the individual trader could go through all the changes in prices, gaining and losing money, while the efficient structure is learned. This clearly implies that changes are small and infrequent. Or, the trader could make a best guess at what the equilibrium price will be, guided by experience. These are really the same situations, and imply small infrequent changes. In the single commodity case it is even possible that there could be adaptation to large changes, provided these changes were rare. When considering trade as a set of interacting commodities, though, this is highly unlikely. Price is a potential in the sense that all prices have adjusted, and the matrix has been powered until a steady state is reached. This value is what the economic actors base their decisions on. If prices are disturbed radically so that they are far from equilibrium, no potential exists that would characterize the state of the system.

6. OIL PRICE EFFECTS

It was noted in the introduction that the coincidence between the oil price rise and the development of a persistently ailing world economy stimulated this paper. The general principles underpinning this relationship have been set out above. However, a fairly relevant, specific model will be adapted from Glansdorff and Prigogine (1971) to deal directly with this issue. Only processes at a single point in space are used here. There are two classes of country, denoted by incomes X and Y, which are related in the following way:

$$dX/dt = A + X^2Y - BX - X$$
$$dY/dt = -X^2Y + BX$$

$$(6.1)$$

where X represents oil exporters and Y represents, say, Western Europe. The $(+X^2Y)$ term is the export income of oil with its strong self-multiplying property due to growing confidence in monopoly power based upon the European market, while the $(-X^2Y)$ term is the loss of income due to European imports of oil. Hence, Europe is a purely equilibrium economy that gains according to the size of its exports to the oil countries $(+BX)$, which in turn loose this income by their imports $(-BX)$. These are obviously the major items that will affect the stability of the system. Finally, let the oil countries' income grow by a constant amount, A, in normal circumstances, on the one hand, and let this income be reduced in proportion to its own size by general imports (i.e., $-X$) on the other.

In the steady state

$$\left. \begin{array}{l} X(t) \;=\; X_0 \;+\; x \exp(wt) \\ Y(t) \;=\; Y_0 \;+\; y \exp(wt) \end{array} \right\} \qquad , \qquad (6.2)$$

where $X_0 = A$ and $Y_0 = B/A$, and w refers to the selected spectral frequency domain. From the linearized perturbation equations around this steady state the dispersion relation is

$$w^2 \;+\; (X_0^2 + B + 1 - 2X_0Y_0)w \;+\; X_0^2 \;=\; 0 \qquad . \qquad (6.3)$$

The value of B for which the coefficient of w vanishes corresponds to a transition point. The solution is unstable when

$$B \;>\; B_c \;=\; 1 \;+\; A^2 \qquad . \qquad (6.4)$$

It may be shown that $\delta_m P$ is positive definite when $B \leq 1$, and thus stability occurs due to the negative contribution of the X^2 term. As B is increased up to B_c, excess entropy production vanishes. Thus, the argument is not that a large shock such as the rise in oil prices will change the structure of prices, and thus lead to a reorientation of trade. There would be nothing novel in that. Rather, a new steady state is established, which is a dissipative system, so that there is a considerable increase in (from the present argument, spatial) inefficiencies. These inefficiencies would include the areal results of inflation and unemployment, for example, with both leading to a lack of control of the allocation of resources. Inflation operating on prices and encouraging 'wrong' pricing can be said to result in the rise of transaction costs, δA, while unemployment implies a rise in opportunity costs, δE. The fact that they occur together does imply an underlying 'cause' for both, and probably will be associated with a rise in regional autonomy and specialization in one or the other type of inefficiency. This model, then, copes with important aspects of present reality, which are not included in most other (economic) models.

If this assessment is valid, then how can a more efficient

steady state be retrieved? An event such as a sizeable war, such
as World War II, probably would be a sufficient fluctuation, but
this is not to be recommended. The trouble is that the price
system is working as it 'should' in both the efficient and the
inefficient steady states, and it is unlikely that using the price
system is going to improve things. Perhaps a suspension of the
working of the price system may be necessary. This could be the
large fluctuation required to reach a more efficient steady state.
A sufficiently detailed planning model using projected consumption
levels and with no inflation and full employment as its aims might
be devised and allowed to set prices. After the new steady state
is reached price controls would be dropped, but the planning model
would be maintained in case of future shocks.

7. GEOGRAPHICAL CHANGE WITH LARGE FLUCTUATIONS

A trade system undergoing fluctuations is extremely complex,
with individual areas experiencing their own changes in supply and
demand, and these changes moving out in space to influence other
areas. In a continuous economy flows of funds can be usefully
represented as a diffusion process (Curry, 1976, 1978); the long
jumps over space involved in some trade links can be replaced by
the set of short links of a diffusion process. However, the
evolution of a system of independent impulses and diffusion
requires a set of coupled non-linear partial differential
equations. For those systems in which solutions can be obtained,
there are usually multiple solutions.

Lefever will continue to be followed in analyzing this effect,
but the oil export-import interpretation will be dropped for a
general spatial interrelation. So far only a single point in space
has been dealt with, or rather, it can now be said, a collection of
identical points in uniform space. As was seen, each point can
generate an instability so that in a system of a very large number
of small regions, the result of a process without spatial
dependence is very local areal differentiation that is equivalent
to homogeneity. A diffusion process is introduced now, which is
the only way to increase the spatial scale of the differentiation.
The conditions under which there will be a balance between local
differentiation as the result of single point processes, and
spatial homogeneity as a result of the mixing by diffusion are
sought. Presumably there are conditions in which stable, non-
uniform distributions occur (i.e., strong regionalization
develops). Equation (6.1) now may be written to include the one-
dimensional diffusion terms as

$$\partial X/\partial t = A + X^2Y - BX - X + D_x(\partial^2 X/\partial r^2)$$
$$\left. \right\} . \quad (7.1)$$
$$\partial Y/\partial t = -X^2Y + BX + D_y(\partial^2 Y/\partial r^2)$$

To study stability in terms of space dependent perturbations, instead of the previous analysis in time characterized by equation (6.2), this equation must be altered to read

$$X = X_0 + x \exp[wt + (ir)/\lambda]$$
$$Y = Y_0 + y \exp[wt + (ir)/\lambda] \quad \} \qquad . \qquad (7.2)$$

Here r is the geometric coordinate and λ is the wavelength of the inhomogeneity, and as before

$$|x/X_0| \ll 1 \qquad \text{and} \qquad |y/Y_0| \ll 1 \qquad .$$

Substituting equation (7.2) into the systems of equations (7.1) yields the dispersion equation

$$w^2 + (A^2 + 1 - B + a + b)w + A^2(1 + a) + (1 - B)b + ab = 0 , \quad (7.3)$$

where $a \equiv D_x/\lambda^2$ and $b \equiv D_y/\lambda^2$.

With $\lambda \to \infty$ so that perturbations are extended to the point of homogeneity, equation (7.3) reduces to equation (6.2). There are two instabilities that can occur now. First, an instability results when the coefficient for w vanishes, as before, plus the effect of the additional expression when it has a positive value. Thus,

$$B_c' < B \qquad , \text{ and}$$

$$B_c'(\lambda) = 1 + A^2 + a + b \qquad .$$

The minimum value of $B_c'(\lambda)$ is still given by equation (6.3) with $\lambda \to \infty$. Second, an instability emerges when the additional expression vanishes, and the instability conditions become

$$B < B_c'' \qquad , \text{ and}$$

$$B_c''(\lambda) = (A^2 + b)(1 + a)/b \qquad .$$

The wavelength giving a minimum of $B_c''(\lambda)$ is

$$\lambda_c^2 = (D_x D_y)^{0.5}/A \qquad , \text{ and}$$

$$B_c = [1 + A(D_x/D_y)^{0.5}]^2 \qquad .$$

The instability that develops will depend upon the value of B and the ratio of the diffusion coefficients, D_x/D_y. In general, the rate of diffusion has to be comparable to the rate of the non-spatial process for spatial inhomogeneities to arise. If it is too rapid, the mixing leads to increasing wavelengths, and so to homogenization. Unequal diffusion coefficients (i.e., small D_x/D_y) result in regionalization. For equal diffusion coefficients instability usually results in a limit cycle in time. This will not be discussed here, since it does not appear relevant in the present context.

Instability occurs when $P(\partial S) > 0$, and hence for the trading sysem of equations (7.1) it can be shown that

$$P(\partial S) = [(1-B)/A]x^2 + (A^3/B)y^2 + (D_x/\lambda^2 A)x^2 + (D_y/\lambda^2 B)y^2 .$$

The $(-B/A)x^2$ term is the only negative term. It involves the industrial self-multiplying term and profoundly affects stabilty. Diffusion aids stability. At the marginal critical state, $P(\partial S)$, the excess entropy vanishes.

8. CONCLUSIONS

The development of regional differentiation is interesting from several points of view. In the first place, it was remarked that one type of divergence from an efficient system is the development of regional autonomy, so that this divergence represents a dissipative system. In this sense the world system is basically a dissipative system. In other words, it is not organized on the basis of a single SPE system. It probably cannot be organized in this manner because such a system would be insufficiently stable, and thus would develop into the sort of world economy that presently exists under the impact of large fluctuations. One could say the same thing about a large subsystem of the world; it too would develop regional polarizations. In fact, is this not just what currently exists? Half the world is rich and half is poor. At each scale one examines there is areal differentiation, so that a spectrum of spatial scales is needed to describe geographical structure. As Prigogine and Lefever (1975) note,

slight and probably random positive or negative fluctuations because of the multiplying effect, can determine which region will eventually be positive or negative in a stable regionalisation. Once regionalisation occurs, small fluctuations can no longer reverse the configurations; they must be of the same size as the areal differences themselves to do this.

Of course regional differentiation occurs via historical processes, but if the focus is regional differentiation then dynamics are not of concern.

Another remarkable feature is that if labor markets are cast in terms of the Ising model from physics, they too break up into polarized regions under certain efficiency conditions (Curry, 1982b). There is a fundamental difference between these two systems, nevertheless. Ising instability occurs directly from equilibrium conditions--at least at the macro-level given certain boundary conditions--whereas in the present case steady states are separate from equilibrium and incorporate inefficiencies. The former always develop, the latter may, depending upon special circumstances. Regional polarization also occurs in the Lotka-Volterra ecology of vacancy chains with diffusion terms (Curry, 1981), and in the geographical specialization of intensity of land use under trade (Curry, 1970). Excess entropy production was not considered in the Ising and Lotka-Volterra models, though.

It is evident that the Prigogine investigations usually will not be relevant to direct modelling. Fortunately, the Prigogine stability criterion can be applied here immediately because of the remarkable analogy between concepts useful for analyzing abstract geographical structures and those of thermodynamics. With more specific or specialized spatial systems models it would not be possible to say what excess entropy production was doing. However, the Prigogine viewpoint does provide a philosophical stance--a sort of existence theorem if you like--of the possibility of spatial structures developing as dissipative systems from previous steady states, subject to large fluctuations.

9. REFERENCES

Curry, L., 1970, Geographical Specialisation and Trade, Studies in Regional Science, 2: 85-95.

_____, 1972, A Spatial Analysis of Gravity Flows, Regional Studies,. 6: 131-147.

_____, 1976, Fluctuation in the Random Spatial Economy and the Control of Inflation, Geographical Analysis, 8: 339-353.

_____, 1978, Demand in the Spatial Economy: II. Homo Stochasticus, Geographical Analysis, 10: 309-344.

_____, 1981, Macro-Ecology of Vacancy Chains, in Dynamic Spatial Models, edited by D. Griffith and R. MacKinnon, Alphen aan den Rijn: Sijthoff and Noordhoff, pp. 159-185.

_____, 1982a, Inefficiency of Spatial Prices, unpublished paper, Department of Geography, University of Toronto.

_____, 1982b, Inefficiencies in the Geographical Operation of Labor Markets, unpublished paper, Department of Geography,

University of Toronto.

Glansdorff, P. and I. Prigogine, 1971, Thermodynamic Theory of Structure, Stability and Fluctuations, New York: Wiley.

Prigogine, I. and R. Lefever, 1973, Theory of Dissipative Structures, in Synergetics Proceedings of the 1972 Symposium on Synergetics, edited by H. Haken, Stuttgart: B. G. Teubner, pp. 126-135.

_____ and _____, 1975, Stability and Thermodynamic Properties of Dissipative Structures in Biological Systems, in Stability and Origin of Biological Information, edited by I. Miller, New York: Wiley, pp. 26-57.

_____ and G. Nicolis, 1973, Fluctuations and the Mechanism of Instabilities, in From Theoretical Physics to Biology, edited by M. Marois, Paris: S. Karger, pp. 89-109.

Sheppard, E. and L. Curry, 1982, Spatial price equilibria, Geographcal Analysis, 14: 279-304.

Schnakenberg, J., 1977, Thermodynamic Network Analysis of Biological Systems, New York: Springer-Verlag.

Turner, J., 1980, Non-equilibrium Thermodynamics Dissipative Structures and Self-organization: Some Implications for Biomedical Research, in Dissipative Structures and Spatio-temporal Organization Studies in Biomedical Research, edited by G. Scott and J. McMillin, Ames, Iowa: Iowa State University Press, pp. 13-52.

Zotin, A., 1972, Thermodynamic Aspects of Developmental Biology, Basel: S. Karger.

PASINETTI, MARX AND URBAN ACCUMULATION DYNAMICS

Eric Sheppard

International Institute for Applied Systems Analysis
Austria
(on leave from the University of Minnesota, United States)

1. INTRODUCTION

Modeling change and accumulation in an urban system is a complex task that traditionally has been simplified by using a neo-classical model where a homogeneous good is produced from inputs of capital, labor, and technology (cf., Henderson, 1977). As ever, there is a price to be paid in this simplification, a price that is perhaps too high.

The aggregate production function used to represent economic activity in each city leads to some results that may seem pleasing. The costs of capital and labor, based upon marginal productivities, are inversely related to their scarcity. It is then possible to conclude that the rewards to entrepreneurs and workers are merely a technical issue, independent of the social relations of society. Technical choice is also a simple affair; the higher the cost of labor (i.e., the scarcer the labor) in a city, the less of it is used. Capital and labor then move between cities in response to the different local prices that can be earned, until all pockets of scarcity and surplus are ironed out, and prices are the same everywhere.[1] In the end, each city, now endowed with identical factor supplies, will grow at the same rate. This is, in short, a model with very strong equilibrating forces, describing an urban system that moves steadily into a state of balanced growth. This result provides a pleasant picture, namely that of a harmonious society allocating resources on a purely technical basis in a way that achieves stable growth, on the average, in all cities and all industries. The effect, however, of these assumptions is not just to simplify reality, it leads to conclusions that do not hold once the assumptions are relaxed.

In an urban system many commodities are produced with very different characteristics and uses. Further, the capital that firms use to produce commodities is not some homogeneous good, but rather is an attempt to capture machines, raw material and finance and lump them into one phenomenon. If this is not done, and instead one goes to the effort of analyzing an urban system with many commodities produced in many locations, then a very different world emerges (Harcourt, 1972; Garegnani, 1970; Sraffa, 1960). Wages and profits no longer are technical indices void of any emotive content. Both cannot be priced without some conception of the relations between, and power held by, the owners of capital and labor. Regions with high labor costs now may find it more profitable to invest in labor intensive activities, destroying the equilibrating forces of the neo-classical model. Perhaps most important, society no longer is a harmonious collection of individuals whose interests balance out to give collective welfare. It is made up of two broad groups with conflicting interests that struggle over the division of the economic surplus. Equilibrium in the growing urban system still can be discussed, but it is a classical equilibrium in the tradition of Ricardo and Marx rather than Walras. It also is an unstable equilibrium that consequently is of no empirical interest. These surely are dramatic reversals in one's view of the urban system, when it is considered that they emerge from such an apparently harmless generalization of moving from a single commodity to a multiple commodity world.

A multiple commodity approach is far more complex to model, but it at least does not require one's imagination about reality to be stretched quite so severely when modeling its economic dimensions. Social and political issues can move to the center, instead of being relegated to market imperfections. One no longer has to believe that the unstable history of urban growth and change is in fact a stable system that keeps being knocked about by external forces. This paper attempts to outline the dynamics of such a multiple commodity urban growth model. It follows in the tradition of Marx, and of those economists who attempted to distill out some of Marx's and Ricardo's logic by use of matrix algebra (Sraffa, 1960; Pasinetti, 1977; Morishima, 1973). The achievements to date are limited, particularly when held up to the claims implied in these opening paragraphs. However, steady progress is being made, there exist strong links with geographical theory, and finally this approach provides an explicit alternative to the much criticized aggregate neo-classical approach.

2. DEFINITIONS

The following notation will be used below:

a_{ij}^{mn} = the amount of good m produced in city i that is used per unit of good n produced in city j,

a^{mn} = the total amount of good m used in the production of a unit of good n,

b_{ij}^{n} = the amount of consumer good n produced in city i that is consumed by a worker and his dependents per day in city j,

b^{n} = the total amount of good n consumed by a worker and his dependents per day,

l_{m} = the number of hours of direct labor necessary to produce a unit of good m,

p_{j}^{n} = the production price of good n in city j,

x_{j}^{n} = the number of units of good n produced in city j,

λ_{j}^{n} = the labor value of a unit of good n produced in city j,

r = the rate of profit,

T = the length of the working day (not to be confused with the superscript T denoting matrix transpose),

b, l = vectors of dimension M-by-1 containing the elements l_{n} and b_{n}, respectively,

p, x, Ω = vectors of dimension MJ-by-1 containing the elements p_{j}^{n}, x_{j}^{n}, and λ_{j}^{n}, respectively (vectors of prices, quantities, and labor values),

A = the inter-urban input-output matrix,

τ_{ij}^{k} = the number of units of transportation needed to ship all non-wage goods from city i to city j,

τ_{ij}^{c} = the number of units of transportration needed to ship all wage goods from city i to city j, and

τ_{ij}^{cn}, τ_{ij}^{kn} = special instances of τ_{ij}^{c} and τ_{ij}^{k}, respectively, defined per unit of output of good n in city j.

Subscripts to indicate time will be introduced as necessary. There are M commodities and J cities.

It is assumed that there is a uniform real wage and identical technologies used for production of a particular good in each city, such that

$$\sum_i b^n_{ij} = b^n_j \quad , \quad \text{for all j, and}$$

$$\sum_i a^{mn}_{ij} = a^{mn} \quad , \quad \text{for all j.}$$

3. BASIC RELATIONS

The basic relationships reflect a competitive capitalist economy in dynamic equilibrium with no significant degree of corporate ownership. Three circuits can be described representing the circulation of money, goods, and labor inputs. ·The money circuit is driven by the assumption that a flexible capital market and a lack of direct investment lead to the profit rate being equalized in all industries. Then, for industry n in city j,

$$p^n_j = (1 + r)[\sum_i \sum_m a^{mn}_{ij} p^m_i + \sum_i \tau^k_{ij} p^f_i + w_j 1_n] \quad , \quad (3.1)$$

where w_j is the wage paid per hour of work in city j and p^f_i is the price of transport hired in city i. Since the wage is assumed to be given by a fixed bundle of consumption goods, then

$$w_j = \omega [\sum_i \sum_m b^m_{ij} p^m_i + \sum_i \tau^c_{ij} p^f_i] \quad , \quad (3.2)$$

where $\omega = 1/T$. Substitution of equation (3.2) into equation (3.1) leads to a model of price determination paralleling Marx's transformation problem, but with inputs also valued by their price (Marx, 1967). The price of good n at the factory gate in city j is given by the cost of inputs plus a rate of profit on capital invested in those inputs.

Consider the following inter—urban coefficient definitions:

(a) if m is a capital good, a^{mn}_{ij} is the amount of good m bought from city i used to produce a unit of good n in city j,

'b) if m is a wage good, $a^{mn}_{ij} = \omega b^m_{ij} 1_n$, the amount of good m from city i consumed per unit of production of good n in city j, and

(c) if m is transportation, $a^{mn}_{ij} = (\tau^{cn}_{ij} + \tau^{kn}_{ij})$, the total units of transportation necessary to assemble wage and capital goods from city i used per unit of production of good n in city j. If c^m_{ij} is the number of units of transportation needed to ship a unit of good m from city i to city j, then

$$a^{fn}_{ij} = \tau^{cn}_{ij} + \tau^{kn}_{ij} = \sum_{m \neq f} a^{mn}_{ij} c^m_{ij} \quad . \quad (3.3)$$

Substituting the above definitions into equation (3.1), the price circulation scheme in matrix form becomes

$$p^T = (1 + r)p^T A \quad . \tag{3.4}$$

Equation (3.4) calculates relative production prices at the factory gate for each good in each city under the assumption of a common rate of profit. The solution is given by the principal eigenvalue of matrix A [which equals $1/(1+r)$] and its associated left-hand eigenvector, which are guaranteed to be non-negative by the theorems of Frobenius and Perron (Pasinetti, 1977). Indeed if matrix A is indecomposable, then both this scalar and vector will be positive.

The circuit of labor inputs is obtained by defining the labor value of a good as the sum of the hours of direct labor involved in production and the labor value of non-labor inputs (Pasinetti, 1977)[2], giving

$$\lambda_j^n = 1_n + \sum_i \sum_m a_{ij}^{mn} \lambda_i^m \quad , \tag{3.5}$$

where the second summation only includes capital goods and the transport necessary to move them. In matrix form this may be expressed as

$$\Omega^T = L^T + \Omega^T A* \quad , \tag{3.6}$$

or

$$\Omega^T = L^T(I - A*)^{-1} \quad , \tag{3.7}$$

where matrix $A*$ is an input-output matrix of non-wage inputs, obtained by setting all values b_{ij}^n in matrix A equal to zero. It can be shown that positive profits only occur if the labor value of the hourly wage is less than the labor value of an hour of work, or that positive profits only occur if and only if the rate of exploitation of labor is positive (Morishima, 1973). Further, the rate of exploitation depends upon location. The rate of profit is always less than the highest urban rate of exploitation observed in the system (Sheppard, 1981).

The rate of exploitation, and hence the rate of profit, will only be positive if the economy is productive in the sense that some physical surplus of outputs over inputs results in each production cycle. But this quantity circuit cannot be described without making some assumption about the dynamics of production (i.e., about how the surplus is used). Thus, the entire model is inherently dynamic. A natural starting point is to seek some dynamic equilibrium. Static equilibrium is only feasible if all excess product is wastefully consumed, which is not a characteristic of competitive capitalism. A dynamic equilibrium may be defined that satisfies two conditions: (1) a common rate of growth in all industries and all cities, and (2) the condition that all outputs find a use as inputs to production in the next time period. These two conditions respectively are expressed mathematically as

$$x_{t+1} = (1 + g)x_t \quad , \text{ and} \tag{3.8}$$

$$x_t = Ax_{t+1} \quad , \tag{3.9}$$

where the scalar g represents the rate of growth. Combining equations (3.8) and (3.9) gives

$$x_{t+1} = (1 + g)Ax_{t+1} \quad . \tag{3.10}$$

The solution to equation (3.10) again is derived from the largest eigenvalue of matrix A and the associated right-hand eigenvector. One can conclude that if the rate of growth is equal to the rate of profit, and if the relative quantities produced in all industries and cities match the principal right-hand eigenvector of matrix A, then smooth crisis-free capital accumulation and growth can occur in an inter-urban capitalist system of production with constant input coefficients.

It should be noted that it has been implicitly assumed that all profits are saved for reinvestment (i.e., the propensity of capitalists to save from profits, s_c, equals one), and that workers save nothing (i.e., $s_w = 0$). If these two assumptions are not true, and

$$w_j = (1 + s_w)[\omega \sum_i \sum_m b_{ij}^m p_i^m + \sum_i \tau_{ij}^c p_i^f] \quad ,$$

then in dynamic equilibrium

$$g = s_c r + s_w w_k \quad , \tag{3.11}$$

where w_k is the ratio of wages to total costs (see the Appendix of this paper). However, for simplicity the traditional Marxian saving hypothesis (s_c equals one and s_w equals zero) will be retained throughout. Also assumed here is that there is no joint production[3].

4.   UNSTABLE EQUILIBRIA AND ACCUMULATION CRISES

The previous section has established the existence of an equilibrium growth path maintainable without forcing any changes in inter-industry relations. However equilibria are of substantial interest only if they are stable. In this instance it will be demonstrated that this equilibrium is unstable. As a preliminary to the analysis, though, the dynamics of production will be rewritten in a mathematical state space with orthogonal axes. As long as matrix **A** does not change, this is possible, and it has the advantage of creating mathematical entities that are independent of one another. One of the difficulties of analyzing industries directly is the interdependence between producers. The dynamics of one producer depends upon the dynamics of another. This can be avoided by transformation (Goodwin, 1976).

Assume that matrix **A** is indecomposable and primitive (i.e., it has no two eigenvalues that are identical). Define **H** to be a matrix containing the left-hand eigenvectors of matrix **A** as rows. Matrix $\mathbf{H}^{-1}$ then contains the right-hand eigenvectors of matrix **A** as columns, assuming that the eigenvectors have been normalized to sum to one. Then

$$\mathbf{H}^{-1}\mathbf{A}\mathbf{H} = \pi \qquad , \qquad (4.1)$$

where π is a diagonal matrix containing the eigenvalues of matrix **A**. Premultiplying each side of equation (3.9) by matrix **H**, and applying equation (4.1), yields

$$x_t^e = \pi x_{t+1}^e \qquad , \qquad (4.2)$$

where $x_t^e = \mathbf{H}x_t$. Each element of vector x_t^e is a linear combination of elements of vector x_t. Each such element shall be called a production configuration, since it represents a particular geographical combination of production sectors each weighted by the appropriate eigenvector loadings. Each configuration is independent of all other configurations, which is illustrated by the fact that vector x_t^e is changed into vector x_{t+1}^e via a diagonal matrix. Indeed for each element of equation (4.2),

$$x_{i,t+1}^e = \mu_i^{-1} x_{it}^e \qquad , \qquad (4.3)$$

where μ_i is the i-th diagonal entry of matrix π.

4.1. Demand-side Disequilibrium

One necessary condition for continued crisis-free accumulation in this sytem is that producers can sell all their output in their own or other cities. It is this requirement that is expressed in equations (3.9) and (4.2). Analysis of the dynamics of this situation will give insight into whether or not this requirement is sufficient to generate a stable growth path. If so, then a perfectly informed multi-commodity market would be sufficient for crisis-free growth. But this is not the case.

Matrix π contains the eigenvalues of matrix **A**, which all must be less than one in absolute value if the economy is productive with a positive rate of profit. Therefore, all the diagonal

300

elements of matrix π^{-1} are greater than one in absolute value. This condition is sufficient for one to deduce that the dynamic system described by equation (4.2), and rewritten as

$$x^e_{t+1} = \pi^{-1} x^e_t \quad , \quad (4.4)$$

is unstable. Each production configuration x^e_{it}, representing a particular geographical combination ofproduction sectors, is either growing exponentially or declining exponentially. This occurs in such a manner that any perturbation of the system away from equilibrium will be self-reinforcing. The equilibrium that would maintain smooth accumulation, then, is of no substantive interest. As the system moves away from equilibrium, production in certain sectors is forced down to zero (see Figure 1). Thus the result of

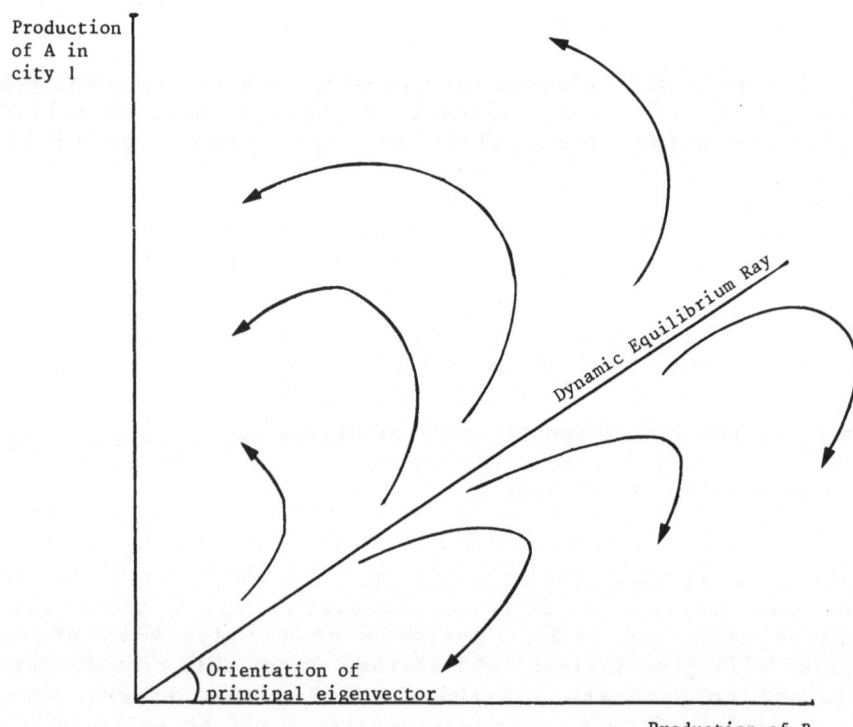

Figure 1. Conceptualization of demand driven dynamics: two interdependent producers in two cities.

attempts to avoid a disproportionality crisis, where supplies and demands do not match, is an evolution of the economy towards a realization crisis, where investment cannot occur at the given rates of profit and growth, due to a lack of demand for some products (see Harris, 1978; Roemer, 1981). The aggregate inter-urban fluctuations that result are simply the sum of the twelve production configurations (one for each sector in each city) weighted by their respective rates of change.

As a descriptive account, this demand-driven dynamic is clearly inadequate. First, it requires perfect information in order to coordinate past outputs with future demands. Second, the extreme instabilities seem unrealistic. However it does illustrate an inevitable conflict between market clearing and profit realization in an economy where inter-industry relations are unchanging.

4.2. Supply-constrained Accumulation

The logical alternative to assuming that supplies meet demands is to build a model of accumulation based upon some conceptualization of how investment decisions are made. This is reminiscent of a neo-Keynesian approach. Of course, a large number of alternatives are possible. Analysis in this section shall be restricted to alternatives that do not include alterations in the inter-industry and capital-labor relations captured in matrix **A**. This constraint has the disturbing implication that no substitution is possible between two suppliers of the same good from different locations. However, this restrictive assumption will be relaxed later in this paper. Note that if matrix **A** is constant, then prices and the rate of profit do not change over time.

A simple investment hypothesis is that in order to maximize accumulation, and as a result of competition to maintain market share, all producers in all cities reinvest all revenues in the next production period. Since rates of profit are equal everywhere, there is no incentive initially to invest in other than one's own firm. Define R^n_{it} as the revenue in industry n of city i at time t. By assumption, this revenue equals investment, I, in the next time period:

$$R^n_{it} = I^n_{i,t+1} \quad , \tag{4.5}$$

$$R^n_{it} = p^n_i x^n_{it} \quad , \text{ and} \tag{4.6}$$

$$I^n_{i,t+1} = x^n_{i,t+1} \left[\sum_m \sum_j a^{mn}_{ji} p^m_j \right] \quad . \tag{4.7}$$

Substituting equations (4.6) and (4.7) into (4.5), and rearranging terms yields

$$x^n_{i,t+1} = [p^n_i/(\sum_m \sum_j a^{mn}_{ij} p^m_j)] \, x^n_{it} \qquad (4.8)$$

$$= (1 + r) \, x^n_{it} \quad , \qquad (4.9)$$

since $p^n_i = (1 + r)(\sum_j \sum_m a^{mn}_{ij} p^m_j)$. Thus, not surprisingly, all industries will grow at the rate of growth g, equal to r, no matter what the supply of goods was at time t. This investment behavior would attempt to force the economy onto the optimal growth rate without adjusting quantities so that supplies will match demand. The obvious result would be the obverse of that described in the previous section; attempts to avoid a realization crisis lead rapidly to a disproportionality crisis as mismatches between supply and demand evolve.

This result can be seen by transforming equation (4.9) onto orthogonal axes using the H matrix introduced above, so that

$$Hx_{t+1} = (1 + r)Hx_t \qquad , \text{ and} \qquad (4.10)$$

$$x^e_{t+1} = (1 + r)x^e_t \quad . \qquad (4.11)$$

The realization crisis is avoided if each production configuration grows at the same rate. Comparing equations (4.11) and (4.4), since $(1+r)^{-1}$ is the largest eigenvalue of matrix A, it must be (in absolute value) the smallest eigenvalue of matrix π^{-1}. Thus one eigenvalue in matrix π^{-1} is equal to $(1+r)$. Call this the k-th entry of matrix π, associated with the (real, positive) k-th geographical production configuration, x^e_k. This component of the aggregate geography of production in the urban system is unique, in that it grows in such a way that both realization and disproportionality crises are avoided. Such crises are endemic to the remaining configurations.

In many of these other configurations the rate of growth that avoids a disproportionality crisis is positive and greater than the rate of growth feasible if all profits are reinvested. In these configurations, then, there is under-accumulation. Production cannot grow fast enough to allow reproduction and growth. In the remaining configurations the rate of growth necessary to avoid a disproportionality crisis is negative, and thus less than the rate of growth necessary if profits are to be reinvested. These configurations then will show over-accumulation, or a building up of stocks that indicates a realization crisis.

It is essential to recall that the configurations discussed here exist only in a mathematical sense. However it is perhaps illuminating to know that the geography of accumulation may be described as the sum of a series of geographical configurations (of varying importance), some of which have a tendency toward under-accumulation and others showing pervasive over-accumulation. The aggregate pattern is given by the sum of these series, but the counter-acting tendencies in these two groups do not cancel one another out. Rather, the geography of change, even when only slightly perturbed from equilibrium, is crisis-ridden. The urban system seems caught between crises of disproportionality and realization. Consequently, even if it were possible to relate the state space of eigenvectors to real industries, and to discuss the possibliity of inducing capitalist entrepreneurs to shift investment away from configurations of over-accumulation towards configurations of under-accumulation, the net effect [replacing the dynamics of equation (4.4) by the dynamics of equation (4.11)] would be to substitute a disproportionality crisis for a realization crisis.

5. STRATEGIES FOR CONFLICT AVOIDANCE

In the context of these assumptions then, with matrix **A** unchanged over time, a contradiction exists between balanced growth and realization of investments due to the unstable nature of the dynamic equilibrium growth ray. The capital theorist Hahn (1966) has refers to this problem as 'the golden nail in the coffin of capitalism,' since the invisible hand is conspicuous by its absence (quoted by Shell, 1973, p. 208). The natural way to resolve this theoretical problem is identical to the practical solution practiced by capitalist entrepreneurs, namely to institute changes in the structure of production represented by matrix **A**. The result is some sort of non-linear model of change, with feedback mechanisms to influence inter-industry and inter-urban interactions.

Four broad types of alterations to matrix **A** can be conceived as strategies for avoiding accumulation crises. These are: increasing interactions with areas outside the urban system, changes in the quantities of non-labor inputs to production, changes in the use of labor, and changes in inter-urban trading patterns. Any changes in the elements of matrix **A** will alter the rate of profit and the prices. Consideration will be restricted here to those changes that are profitable for individual capitalists. In the case of changing non-labor inputs, commonly known as technical change, this question has been investigated at length (see the summary in Roemer, 1981). A change that is profitable for any individual capitalist is one that reduces costs and raises his individual rate of profit. Such a change also will raise the collective rate of profit after adjustment to competitive equilibrium, as long as all other factors (such as the real wage) are unchanged. Comparative static analysis implies that a falling rate of profit is inconsistent with individual profit maximizing actions by capitalists. With joint production the answer is less clear, since if capitalists are driven to choose the lowest cost-price method in the short run, regardless of long-run tendencies, then the collective rate of profit may fall (Shaikh, 1980).[3] But in the absence of joint production such contradictions do not occur. Changes in matrix **A** of interest to individual capitalists will lead to reduced costs of inputs and adjusted prices such that the collective rate of profit increases. In the quantity sphere the key issue is whether the resulting actions will help alleviate crises of accumulation.

This issue will be briefly discussed for three of the four strategies listed above, with the exclusion of technical change. The first two, namely expansion outside the urban system and changes in the use of labor, are only outlined. The strategy of changing trade patterns is covered in more detail.

5.1. Expansion Beyond the Urban System

An obvious response to a disproportionality crisis is to seek external markets where surplus products may be sold and shortages may be bought. If disproportionality can be alleviated in this way, then realization crises also can be avoided. Technically this solution would amount to expanding the dimensions of matrix **A** to include the rest of the world, with a reduction in flows within the urban system. The effect on the rate of profit is indeterminate since matrices of different dimensions are being compared. But if this strategy were feasible it could delay crises in the accumulation sphere. Indeed this finding parallels the classical Marxian model of imperialism as an expansion into foreign markets to resolve over-accumulation or under-consumption problems (Barratt-Brown, 1974).

5.2. Capital-labor Conflicts Over the Work Process

Comparative static analysis of this topic is illuminated by results from the theorems of Frobenius and Perron. If matrix **A** is indecomposable, then a reduction of any element in matrix **A** reduces the size of the principal eigenvalue of **A**. Since r and g are inversely related to this eigenvalue, a reduction of any element in matrix **A** will icrease the rate of profit and the equilibrium rate of growth, ceteris paribus. Thus, such reductions are in the interest of individual capitalists, and one arena for such changes is in the workplace.

These alterations may be realized as: reductions in the real wage (b), increases in the work week or reductions in time off (reducing ω), and rationalization (such as reducing non-wage inputs by eliminating wastage). Regardless of the form taken, such actions represent an attempt by capitalist entrepreneurs to expand their share of the total income in the urban system relative to that share commanded by labor. This competition over division of the surplus implies that the interests of these two groups are diametrically opposed, creating a necessary and sufficient foundation for conflict between capital and labor. This result is summarized in Figure 2.

The shape of the curves in Figure 2 has no meaning except to indicate that this trade-off, or 'factor-price frontier,' is not linear apart from the case of a neo-classical, aggregate, single commodity world (Garegnani, 1970). It is for this reason that the following two central neo-classical aggregate hypotheses are false: (1) the wage and profit rates equal marginal productivities and are thus purely technological parameters rather than social indices, and (2) economies with higher capital costs choose capital intensive technologies (Harcourt, 1972; Harris, 1978).

This conflict and its geographical dimensions are a crucial

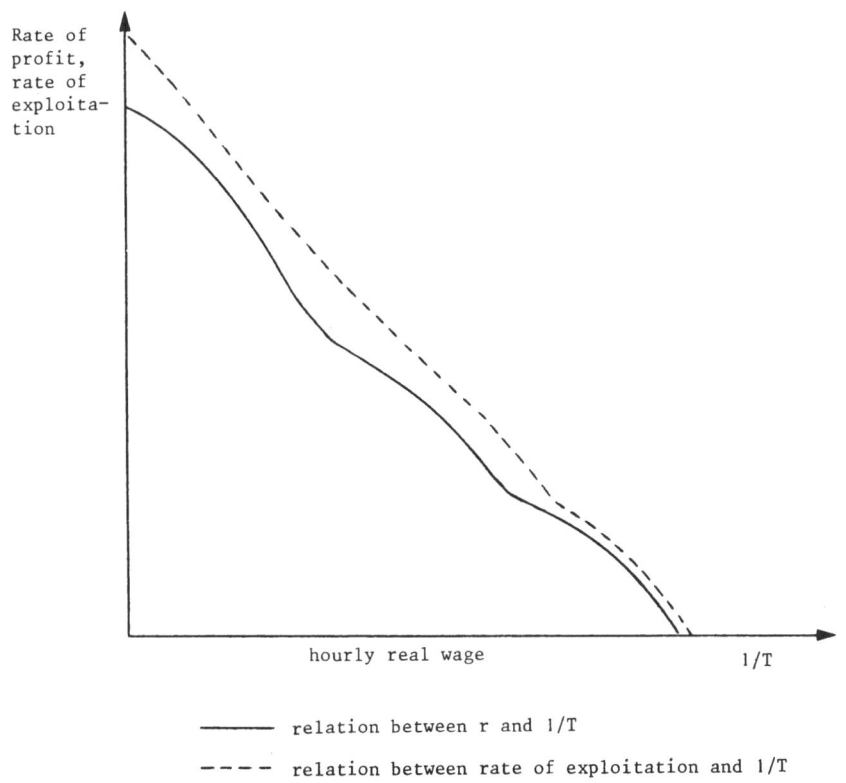

Figure 2.   Conflict between capital and labor.

element in any theory of accumulation in an urban system. For example, there is ample evidence that high labor costs, both direct and indirect, have played a significant role in decisions to relocate production facilities outside the traditional manufacturing belt of the United States. However, to capture this complex phenomenon it is necessary to link the model of production with a demographic model. Population growth and migration, and labor market considerations are vital elements in attempts to describe how wages and unemployment vary locally, and how the struggles between labor and capital are realized in different cities. This represents a major task for future research.

5.3. Trading Responses

Of all the means of changing the structural matrix **A**, trading is the easiest to implement. It is typically easier to change the location from which supplies are purchased than to introduce new technologies, negotiate new working conditions, or enter new markets. Consideration will be restricted here to those changes in trading patterns that are effected without influencing these three other arrangements. This special situation is also of theoretical interest because it introduces the possibility of producers responding in their demands to the relative prices of inputs. This form of limited substitution may have the potential of reducing the likelihood of under- or over-accumulation.

A decision to change trading patterns is essentially a decision to change transport inputs. Producers will choose from a set, possibly a continuous set, of trading options that option which most reduces the costs of inputs, or (equivalently in the absence of joint production) maximizes the short-run profit gains. Mathematically, a change in trading patterns will be instituted if, in industry n of city j,

$$\sum_{i,m} a_{ij}^{mn} p_i^m < \sum_{i,m} a_{ij}^{mn*} p_i^m + k \quad , \qquad (5.1)$$

where k is the cost of change and $a_{ij}^{mn}*$ refers to the new trading coefficients. By assumption, other aspects of production methods are unchanged. Therefore,

$$\sum_i a_{ij}^{mn} = \sum_i a_{ij}^{mn*} = a^{mn} \quad , \qquad (5.2)$$

for all goods m except transportation. Of course, producers in this model will choose their suppliers on the basis of a comparison of delivered prices of good m delivered in city j from a producer in city i, where

$$q_{ij}^n = p_i^n + c_{ij}^n p_i^f \quad . \qquad (5.3)$$

These prices at any time T will influence purchase decisions of inputs to be used at time T+1. Schematically,

$$a_{ij}^{mn}(T+1) = f(\mathbf{p}_T) \quad , \text{ and} \qquad (5.4)$$

$$\mathbf{p}_{T+1} = h(\mathbf{A}_{T+1}) \quad , \qquad (5.5)$$

where f and h are functions to be determined. Equation (5.4) represents trading responses to previous prices, whereas equation (5.5) is the equation of price formation at the next time period. Given the complexity of the price formation model [see equation (3.4)], this result represents a dynamic formulation that apparently is too complex to solve analytically. Its properties will be examined by performing some elementary deterministic simulations.

6. A DYNAMIC SIMULATION OF INTER-URBAN TRADING AND PRICING

Equation (5.4) will not be specified as a deterministic profit maximizing choice, whereby all purchases would come from the cheapest producer. It is presumed that each industrial sector in each city is represented by a relatively large number of firms. Firms have different levels of information, have contracts with suppliers that may or may not be easy to break, and cannot perfectly predict future prices. As a result, different firms will behave differently. and individual firms will not want to rely on only one supplier. A relatively simple framework to capture these characteristics is given by

$$a_{ij}^{mn}(T+1) = a^{mn} \{\exp[-\beta q_{ij}^m(T)]\}/\{\sum_i \exp[-\beta q_{ij}^m(t)]\} \quad , \quad (6.1)$$

for all $m \neq f$.

In random utility theory β may be regarded as a parameter reflecting the responsiveness of purchase orders to price differentials (Williams, 1977). Without wishing to invoke this choice theory with its array of assumptions, it can at least be conjectured that as β tends to infinity a greater and greater proportion of purchases will be from the cheapest supplier, and thus the trading pattern will be more efficient (cf., Curry, 1983).

Having a simulation model available will enable one to test two seemingly different theories of price formation [representing equation (5.5)], the one economic and the other geographic in origin. The economic formulation is the Marxian pricing mechanism that forms a foundation of the paradigm employed in this paper, and is given by

$$\mathbf{p}_T^T = (1 + r)\mathbf{p}_T^T \mathbf{A}_T \quad . \qquad (6.2)$$

The geographic formulation stems from recent research by Sheppard and Curry (1982). In an examination of spatial price equilibrium, an equivalence was found between the prices obtained as duals of an optimal solution to single commodity spatial price equilibria, and potentials generated in analog model solutions to the identical problem. This was used to conjecture that for any type of interaction model describing trade, the potential field associated with that interaction model would describe the prices of goods. Since these results are not fully proven, and since the link to economic pricing is not clear, a comparison of this finding to Marxian pricing may be illuminating.

The matrix **A** has all eigenvalues less than one in absolute value. Thus

$$(I - A)^{-1} = \sum_{k=0}^{k=\infty} A^k \quad . \tag{6.3}$$

A potential matrix, **U**, can be associated with this trading process (Kemeny, Snell and Knapp, 1966; Sheppard, 1979) in the following way

$$U = (I - A)^{-1} - I \quad , \tag{6.4}$$

were an element u_{ij}^{mn} of matrix **U** may be interpreted as the mean aggregate number of times that a unit of good m from city i passes through the production process for good n in city j, given that good m is sold from city i. Hence,

$$u_{ij}^{mn} = \sum_{k=1}^{k=\infty} (a_{ij}^{mn})^k \quad . \tag{6.5}$$

Here $(a_{ij}^{mn})^k$ is the appropriate element from matrix A^k, representing the probability that in the k-th iteration of the process of production of commodities by means of commodities a unit of good m starting initially from city i enters into production of good n in city j. This is also the k-th order multiplier effect.

The column sums of matrix **U** represent the relative centrality of each industry/city in the spatial economy of the urban system. These sums are the origin potentials, interpreted here as prices $\hat{p}_j^n$, and given by

$$\hat{p}_T = i^T U_T \tag{6.6}$$

$$i^T [(I - A_T)^{-1} - I] \quad , \tag{6.7}$$

where i is a vector of ones, and $\hat{p}_T$ is a vector of 'potential prices' at time T.

6.1. Simulation Results

In order to examine the dynamic interrelations between price formation and trading patterns a simulation model was constructed for three hypothetical cities and four hypothetical industries (two capital goods [1 and 2], a transport good [t], and a wage good [c]). Inputs included a 4-by-4 matrix of technological coefficients, with both identical and different matrices for each city being tried, a matrix of initial trading patterns between cities, a matrix of inter-urban transportation coefficients, a consumption level of the wage good, parameters ω and β, and a variable parameter τ, representing the degree of distance friction, by which all transport costs were multiplied. Thus, for example,

$$\tau_{ij}^{kn} = \tau \sum_m a_{ij}^{mn} t_{ij}^m \quad , \tag{6.8}$$

where t_{ij}^m is the amount of transport needed to move a unit of good m from city i to city j (taken from the matrix of transportation coefficients), a_{ij}^{mn} is the technological coefficient for inputs of good m from city i into good n in city j, and the summation is taken over all capital goods. The simulation experiment that was conducted consisted of 75 runs.

So far seven salient results have been obtained. First, in every case the dynamic model evolves rapidly (in less than 15 time periods), and without oscillations, to an equilibrium pattern of trading and prices. The dynamic model of trading and prices thus seems to be stable. Recall, however, that the dynamic model of production levels and accumulation is unstable when vector **p** and matrix **A** are constant. The substantive implication to be drawn from this, then, is that even if trading adjustments in the short run are able to alleviate accumulation crises, in the medium run a stable trading pattern results that is just as vulnerable to such crises as before. Thus, adjustments of the trading patterns in and of themselves will not resolve the crises of accumulation desribed above.

Second, the time taken before spatial price equilibrium is reached is significantly longer when the initial trading propensities are very low, than when the cities are initially strongly connected by trade. This supports the conjecture of Sheppard and Curry (1982) that the likelihood of achieving a spatial price equilibrium is inversely related to the connectivity of the trading pattern.

Third, in comparing Marxian pricing to potential pricing, simulations have revealed a remarkable correspondence between the two price vectors. After one time period there is a difference in

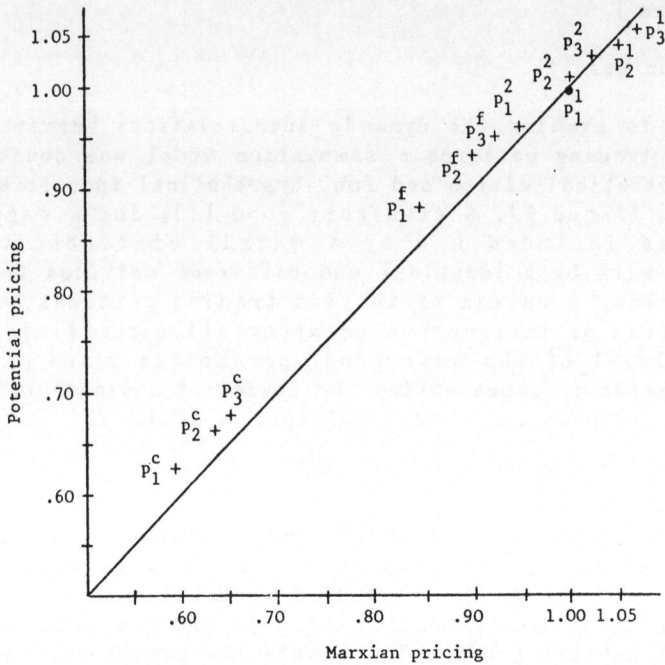

Figure 3-a. Comparison of Marxian and potential prices[4]: after two time periods.

relative prices of between 1% and 5%. At equilibrium the difference always was less than 10% in comparisons of some 20 simulation runs with different initial conditions (see Figure 3). This result suggests the conjecture that the two rules might under certain conditions be equivalent, which would lead to an economic interpretation of the geographic approach and vice versa. It was, however, observed that under potential pricing an equilibrium was achieved significantly more quickly (taking no more than 5 time periods).

The remaining four findings stem from a particular set of starting conditions, given in Table 1, where technological coefficients differed for each city and transportation costs to and from City 3 were greater than those between Cities 1 and 2. First, there is a systematic relation between the equilibrium rate of profit in the spatial economic system, and three other fundamental parameters. Figure 4 illustrates that as the efficiency of trading increases [the exponent on the choice model of equaiton (6.1)], so does the rate of profit. Thus the more efficient the trading

(a) Intra-urban Technological Coefficients

For	From commodity	To commodity:			
		1	2	c	f
CITY 1	1	0.300	0.200	0.100	0.200
	2	0.300	0.300	0.100	0.100
	c	0.800	1.600	2.400	3.200
CITY 2	1	0.250	0.175	0.225	0.200
	2	0.200	0.400	0.050	0.145
	c	1.200	1.000	3.050	2.750
CITY 3	1	0.148	0.300	0.150	0.268
	2	0.400	0.178	0.122	0.098
	c	0.500	2.230	3.090	2.690

(b) Transportation Input Requirements

From	For commodity:		
	1	2	c
City 1 to city 1	1.00	2.00	4.00
City 1 to city 2	2.00	3.00	5.00
City 1 to city 3	2.50	3.40	8.00
City 2 to city 1	2.10	4.20	6.80
City 2 to city 2	1.50	2.00	4.40
City 2 to city 3	2.80	4.10	8.44
City 3 to city 1	3.00	4.00	5.80
City 3 to city 2	3.20	5.40	8.70
City 3 to city 3	0.90	1.50	2.40

(c) Other Coefficients (unless otherwise stated)

$\Omega = 10$ ($\omega = 0.1$)	working day in hours
$\tau = 0.095$	distance friction
$\beta = 4.5$	trading efficiency index

TABLE 1: INPUT DATA FOR SIMULATION RUNS

312

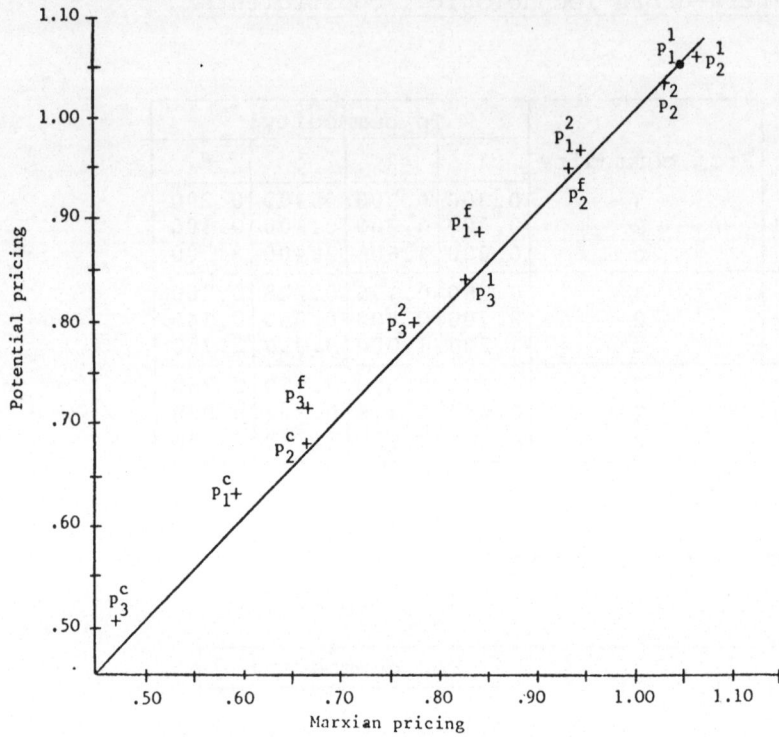

Figure 3-b. Comparison of Marxian and potential prices: spatial equilibrium prices.

pattern, the higher the profits to be made in equilibrium. Furthermore, a maximum rate of profit is quite rapidly achieved. In the simulation, a choice parameter of 13 was sufficient to ensure a maximal rate of profit. The inference to be drawn from this is that inefficiencies in the geography of trade need not bring about drastic reductions in the profits made. Figure 5 shows that as the difficulty of inter-urban transportation increases, the rate of profit steadily decreases. From these results it appears that reduced transport costs are more effective in increasing profit rates than is the introduction of a completely efficient trading pattern. Thus the achievable rate of profit in Figure 4 is less than the maximum profit rate in Figure 5. Finally, Figure 6 shows that a longer workday leads to higher profits. It can be shown (Morishima, 1973; Sheppard, 1981) that the length of the workday is proportional to the rate of exploitation of labor by capital, as represented by Marx (1967). Hence this result can be interpreted as stating that increased exploitation leads to increased profits. Apparently, comparing Figures 4 and 6, this effect is very much like that of more efficient trading. It seems

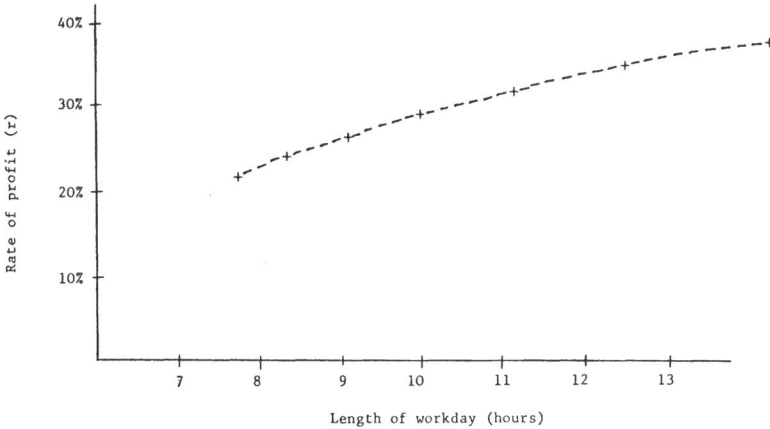

Figure 4. Rate of profit in equilibrium, by the efficiency of the trading pattern.

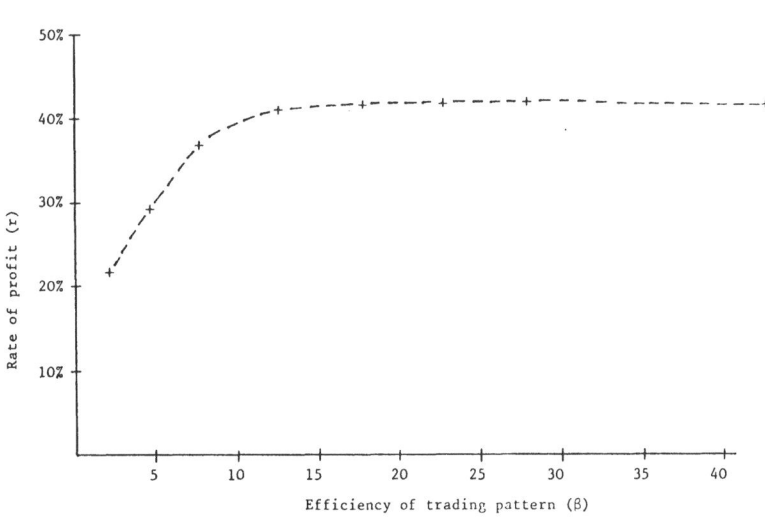

Figure 5. Rate of profit in equilibrium, by the difficulty of transportation.

314

that the same maximal profit rate is achievable either by increasing the efficiency of trading, or by increasing exploitation. From the results presented here, the most effective (and presumably socially most acceptable) means of increasing profits and growth is to reduce spatial friction. This suggests that better transportation technology indeed can play a fundamental role in stimulating capitalist expansion in an urban system.

Second, the relative prices in equilibrium are systematically influenced by the efficiency of the trading system. This is shown in Figure 7. The influence seems to be location specific, in that the prices for all goods in a particular city change in an

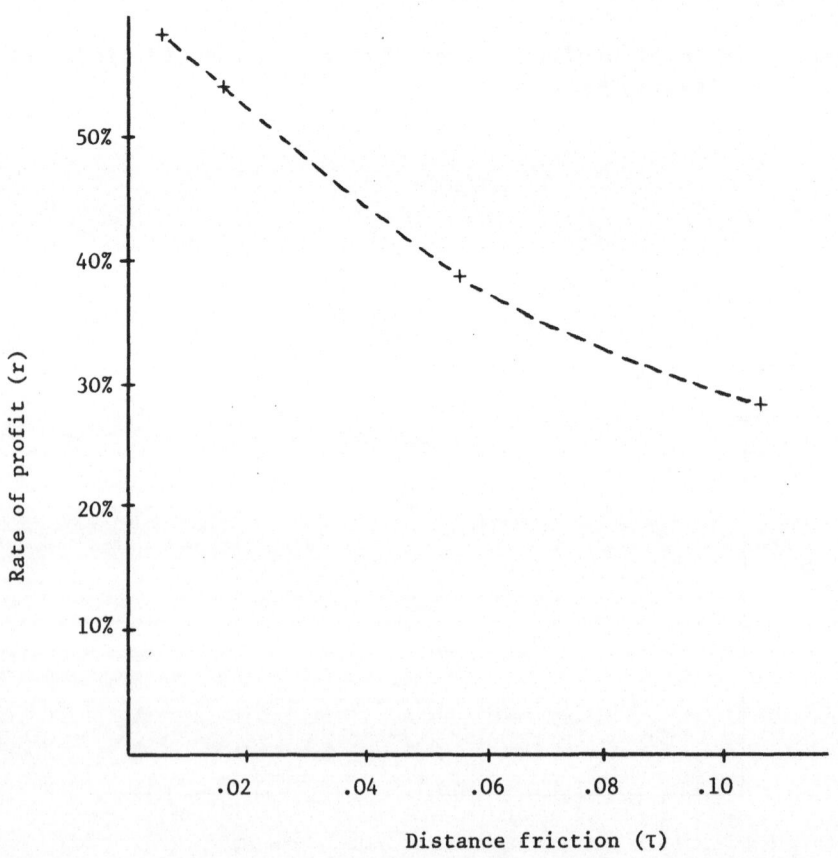

Figure 6. Rate of profit in equilibrium, by difficulty of transportation.

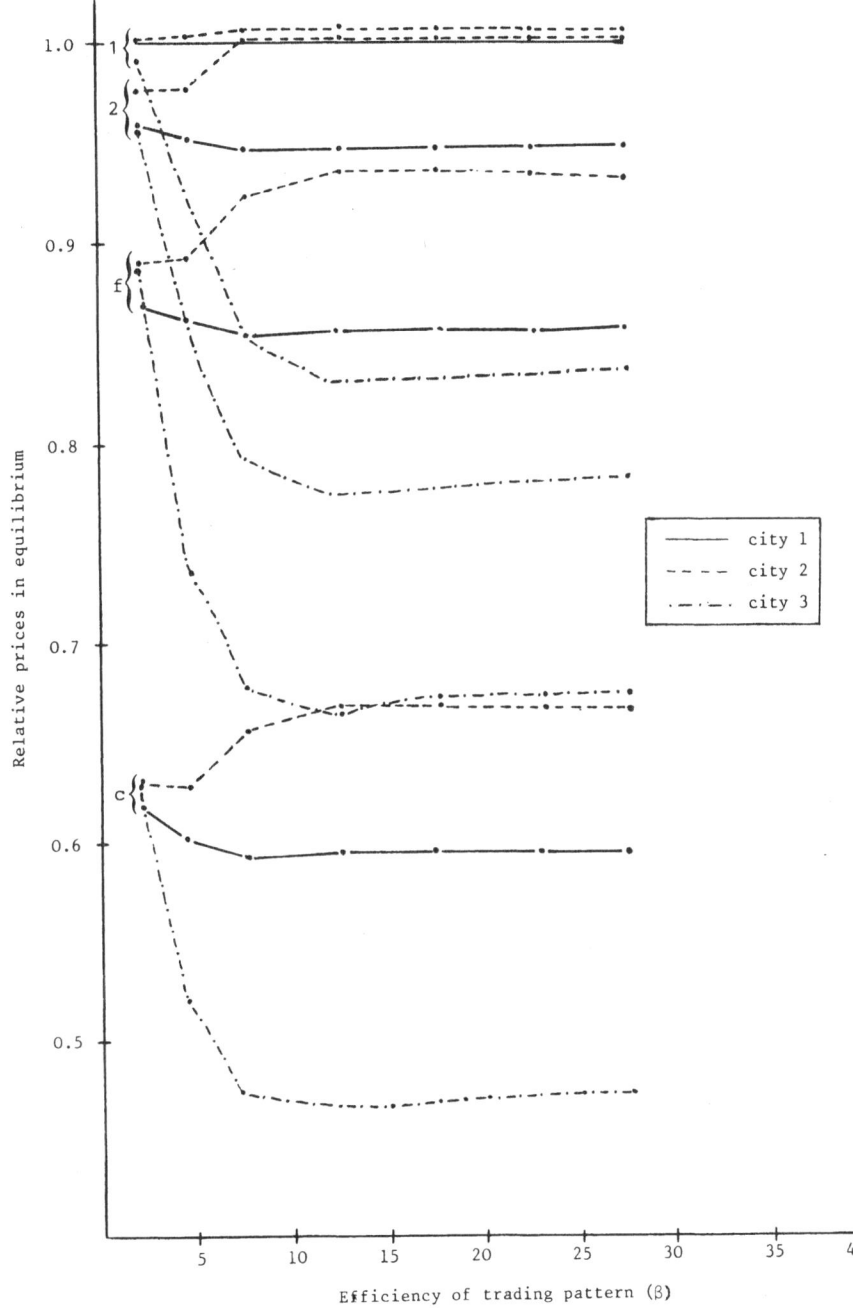

Figure 7. Spatial equilibrium prices at various levels of efficiency of trading.[5]

identical direction as the efficiency of trading is altered. When
trading is highly inefficient, the prices for a particular good are
approximately identical in all cities. This is a situation where
trading patterns are more or less random. No city enjoys a
particular price advantage due to relative location, since prices
do not strongly affect trading patterns. This approximates the
situation that would be found in a spaceless economy, and thus
relative prices reflect inter-industrial demands and flows and not
inter-urban demands. As the trading pattern becomes more
efficient, transportation costs influence trading patterns through
their effect on delivered prices. With maximal efficiency, a
pricing pattern would be expected where the prices for a particular
good differ at various locations by an amount equal to the cost of
transportation. Just as increasing efficiency of trade rapidly
allows profits to be maximized, so spatial price differences
rapidly evolve to a pattern determined by transport costs, and by
the spatial pattern of supply and demand. These trends are
comparable to those discussed by Curry (1983).

Of particular interest here are the prices charged in the
third city. These prices fall to such a degree that the
differential from prices in other cities cannot be explained solely
by transport cost differentials. What in fact is happening is that
this third city, in order to maintain the same rate of profit as
elsewhere, is becoming more and more self-sufficient. This is the
most remote city (see Table 1) in the system, and in order to avoid
being faced by higher costs than capitalists elsewhere as a result
of engaging in trade, capitalists in this city prefer to buy (and
thus sell) locally. This result is only possible, however, when no
great spatial differences in technological advantage exist.

Third, Figure 8 shows the effect of variations in distance
friction on spatial price equilibrium. As in Figure 7, when
transport costs are close to zero, space is effectively absent from
the economy, and prices simply reflect inter-sectoral flows.
Prices are differentiated by sector, but not by location. As the
effort necessary to overcome distance friction increases, spatial
differences in prices increase. Once again, however, the trends
are the same at each location. Thus the prices in the first city
become higher, whereas those in the third city become lower than is
the case for the second city. In this case, the self-sufficiency
introduced in City 3 becomes more evident. As the cost of
transport increases, City 3 is placed at a greater and greater
disadvantage relative to other cities. The high spatial costs of
importing supplies would push prices up to a point where no other
cities would be willing to buy from City 3. Since imports
represent a competitive disadvantage for subsequent exports, the
city chooses to become more and more autarchic. It seems that the
resulting low prices, due to low transport costs, can only be
maintained as long as the city does not trade.

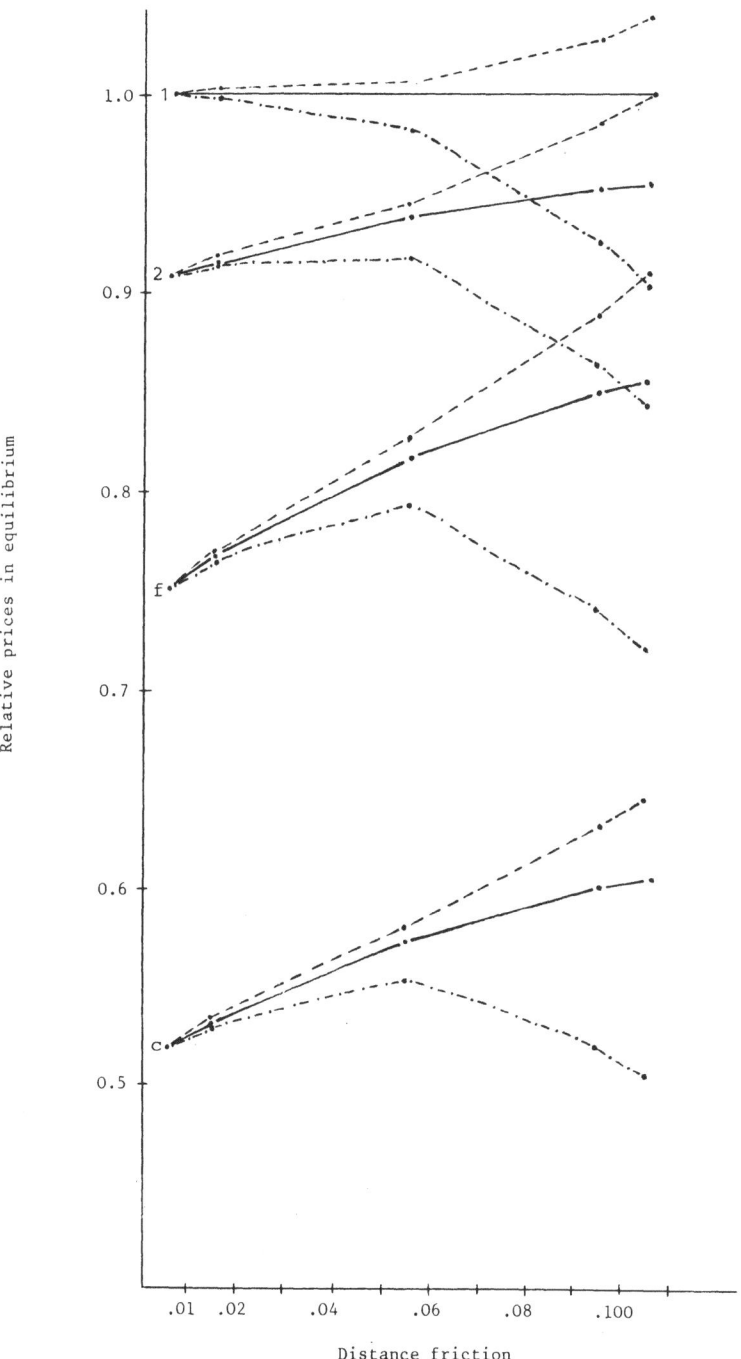

Figure 8.   Spatial equilibrium prices for various levels of the
difficulty of transportation.[5]

In Figure 8, the prices of different goods in any one city seem, if anything, to converge as distance friction increases. This is a result of common trading strategies adopted by all producers in a given city. In this case, Cities 1 and 2 trade relatively more with one another (than with City 3), with City 1 as the principal supplier and City 2 as the customer.

Finally, in Figure 9 the effects of changing the working day on spatial price equilibria can be seen. At first it seems puzzling that there is a link between the workday, or rate of exploitation, and spatial pricing patterns. To trace out what this means, it is necessary to take into account the intermediate effect of changes in the rate of exploitation on relative prices. The consumer and transportation goods have significantly higher direct labor inputs than do the other two goods. This can be seen in Table 1, because in a one-consumer good model the input of the consumer good is proportional to direct labor. The effect of increasing the rate of exploitation is to reduce the prices of those goods with the highest direct labor input, because it is here that labor costs are most important. Thus, in Figure 9, the price of the consumer good and of transportation fall relative to other goods as the rate of exploitation increases. But this is not the only effect. It can be seen that, in addition to this, the prices of each good are converging toward more uniform spatial pricing. This can be explained by reference to Figure 8. Because transport costs fall relatively as the rate of exploitation increases, this implies that in economic terms the costs of trade are falling. This will have the same effect as reducing distance friction, space becomes a less important determinant of prices. In Figure 8, low distance friction leads to a spatial uniformity in pricing. In Figure 9 the falling transport price due to increased exploitation has the same effect. Once again, when trading is relatively costly (a low exploitation rate) prices in City 3 are lowest due to a more self-sufficient trading pattern. This suggests the general conclusion that remote locations must be more self-sufficient when trade is difficult, in order to maintain rates of profit equal to those elsewhere. It then can be imagined that if in fact remote cities become locked into a situation where they are highly dependent on trade, then rates of profit actually made there would be less than those prevailing in more accessible parts of the economy.

7. CONCLUSION

The purpose of this paper has been to present a neo-Marxian alternative to the usual neo-classical models used to describe economic change in an urban system. This neo-Marxian perspective has allowed the research to concentrate on how macro-social structures act to influence economic change in an urban system,

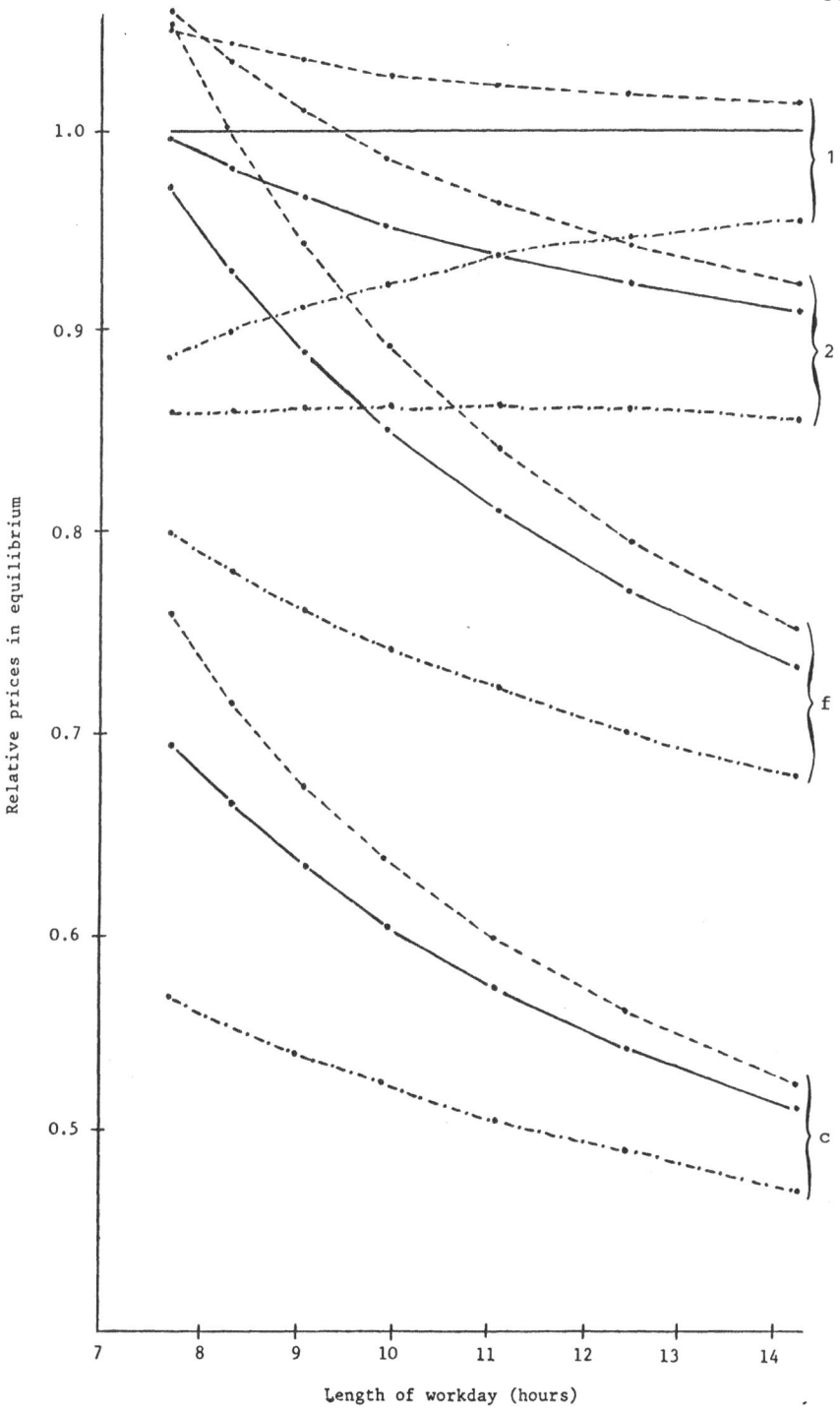

Figure 9. Spatial equilibrium prices at various lengths of the
working day.[5]

under the constraint that expanded reproduction of commodities must continue. It was shown that any simple dynamic equilibrium is unstable, and some analytical tools enabled classification of the resulting instabilities into types of accumulation crises.

Given this instability, of necessity attention must be focused on change in disequilibrium. This paper has only given one or two halting steps in this direction. Many key issues remain, such as investment and location behavior of entrepreneurs, the role of corporate power, and the evolution of conflict between labor and capital. The utility of this approach will depend upon whether or not such important questions can be incorporated into the analysis in order that a realistic statement quantifying growth in an urban system can be assembled.

8.   REFERENCES

Barratt-Brown, M., 1974, The Eonomics of Imperialism, London: Penguin.

Curry, L., 1983, Inefficiency and Instability of Trade Patterns, in Evolving Geographical Structures, edited by D. Griffith and A. Lea, The Hague: Martinus Nijhoff, pp. 278-292.

Garegnani, P., 1970, Heterogeneous Capital, The Production Function and the Theory of Distribution, Review of Eonomic Studies, 37: 407-438.

Goodwin, R., 1976, Use of Normalized General Coordinates in Linear Value and Distribution Theory, in Advances in Input-Output Analysis, edited by K. Polenski and J. Skolka, Cambridge, Mass.: Ballinger, pp. 581-602.

Hahn, F., 1966, Equilibrium Dynamics With Heterogenous Capital Goods, Quarterly Journal of Economics, 80: 633-646.

Harcourt, G., 1972, Some Cambridge Controversies in the Theory of Capital, Cambridge: Cambridge University Press.

Harris, D., 1978, Capital Accumulation and Income Distribution, Stanford: Stanford University Press.

Henderson, J., Economic Theory and the Cities, New York: Academic Press.

Kemeny, J., J. Snell and A. Knapp, 1966, Denumerable Markov Chains, Pinceton, N. J.: Van Nostrand.

Marx, K., 1967, Capital: A Critique of Political Economy, III. The Process of Capitalist Production as a Whole, Moscow: International Publishers.

Morishima, M., 1973, Marx's Economics, Cambridge: Cambridge University Press.

Pasinetti, L., 1977, Lectures on the Theory of Production, London: MacMillan.

Roemer, J., 1981, Analytical Foundations of Marxian Economic Theory, Cambridge: Cambridge University Press.

Shaikh, A., 1980, Marxian Competition versus Perfect Competition: Further Comments on the So-called Choice of Technique, Cambridge Journal of Economics, 4: 75-83.

Shell, K., 1973, Discussion of the Paper by F. H. Hahn, in Models of Economic Growth, edited by J. Mirlees and N. Stern, London: Methuen, pp. 193-206.

Sheppard, E., 1979, Geographical Potentials, Annals, Association of American Geographers, 69: 438-447.

_____, 1981, Spatial Economic Development in Capitalist Economies, paper presented at the 28th North American meeting of the Regional Science Association, Montreal.

_____ and L. Curry, 1982, Spatial Price Equilibria, Geographical Analysis, 14: 279-304.

Sraffa, P., 1960, The Production of Commodities by Means of Commodities, Cambridge: Cambridge University Press.

Williams, H., 1977, On the Formation of Travel Demand Models and Economic Evaluation Measures of User Benefit, Environment and Planning A, 9: 285-344.

9. FOOTNOTES

[1]

    If it is recognized that spatial factor mobility involves a cost, then the argument is modified so that spatial factor prices differ by the cost of movement. The effects of this on the conclusions drawn about the stability of regional growth trends, however, is minimal.

[3]

    A homogeneous labor force is assumed. The identical real wage is paid for an hour of work, and likewise the labor value of an hour of work is identically equal to one. The latter also may be regarded as the labor value of one hour of average productivity.

3

    Joint production is defined as the production of more than one distinctively different product simultaneously by a single production process.

4

    For these comparisons, the model summarized in Table 1 was used, with $\beta = 17.5$, $\tau = 0.095$ and $\omega = 0.1$. The relative prices are scaled so that $p_1 = \hat{p}_1 = 1.0$.

5

    In this Figure, the symbols 1, 2, c and f refer to the four commodities produced: capital good one, capital good two, the consumer good and transportation. The lines that are grouped by each symbol refer to the spatial equilibrium prices for this good in the three cities.

10. APPENDIX: RELATION BETWEEN THE RATE OF PROFIT AND THE RATE OF GROWTH WHEN WORKERS SAVE

In dynamic equilibrium total investment (i.e., the value of increased production) should equal total savings, such that

$$g \, p^T \, Ax = s_c \, r \, p^T ax + s_w \, W \qquad , \qquad (A.1)$$

where W is the value of the total real wage. Accordingly,

$$W = p^T \, B \, L \, x \qquad , \qquad (A.2)$$

where B is a matrix with entries b_{ij}^n, and L is a diagonal matrix with entries l_n in each row representing industry n. Therefore,

$$g = s_c \, r + s_w (p^T \, B \, L \, x)/(p^T \, A \, x) \qquad (A.3)$$

$$= s_c \, r + s_w \, w_k \qquad , \qquad (A.4)$$

where w_k is the ratio of the wage bill to total costs.

SPATIAL STRUCTURE AND SPATIAL-TEMPORAL PROCESSES

Robert Haining

University of Sheffield
England

1. INTRODUCTION

There are many ways of describing pattern and testing for structure in spatial data. The best known statistics have been discussed, together with a summary of earlier contributions and important extensions, in two books by Cliff and Ord (1973, 1981). The first of these was concerned largely with various indices for testing for significant structure in the organization, or 'spatial autocorrelation,' of spatial data. The later book shifted the emphasis to a more explicit recognition of the underlying models that generated spatial structure.

However, the emphasis in both of these books is on static properties of spatial systems, and there is little attempt to develop links between static measures and dynamic properties of spatial systems. Part of the reason for this may lie in the early emphasis on developing statistical indices of map structure coupled with the fact that the key concept of their work, namely 'spatial autocorrelation,' was given an empirical (or data base) definition. The view taken here is that the shift of emphasis towards a 'model-based' definition of spatial variation offers a twofold advantage: (1) better map description (though perhaps over a more limited range of spatial structures), and (2) an opportunity to develop links with spatial-temporal process modelling. Used in this context, 'model based' refers to a definition of spatial structure in terms of specific closed-form models, such as those found in Whittle (1954), Besag (1974) and Haining (1978). Since specific organizational properties are amongst the properties associated with these types of models, a model may be selected that is compatible with a set of desired organizational properties. Hence, to specify these models is to specify specific spatial structures. In turn, these models may be used to devise model-specific tests of spatial structure, which frequently are more powerful than the generalized tests investigated by Cliff and Ord (Haining, 1977).

The construction of better methods of describing and distinguishing between map structures is of undoubted importance. It provides important impetus to this area of research. The advantage of model based tests for describing spatial structure have been discussed elsewhere (Haining, 1982a, 1982b). In the third section of this paper there is a summary review of research into the analysis of patterns of diffusion using a model based methodology of map description. This research also highlights the second theme of this paper, which is the development of links with spatial-temporal system properties, by using model descriptions to construct 'forecasting' maps of the further progress of diffusion.

In the fourth and fifth sections other aspects of the structure-process relationship are developed. The index approach to describing map structure usually has been allied with an aspatial modelling strategy. That is, models of geographic events did not explicitly incorporate spatial interactions, but rather left those to be tested for (usually in the residuals) after the model had been fitted. The alternative, discussed here, is to incorporate spatial interactions into the specification of the model, so that model validation includes direct tests for the spatial interactions. In the fourth section this approach is exemplified by a study of intra-urban gasoline pricing.

In the fifth section the interrelationships between structure and process are discussed from a different perspective. If interactions (represented by spatial flows) mold spatial structure, then it is surely the case that spatial structure may mold spatial flows. A formalism is presented for generating flows from spatial structure, and vice versa. Some results of a preliminary investigation of the influence of spatial structure on movement patterns are presented.

The organization of this paper is presented in Figure 1. The class of spatial models that dominate the discussion here is the class of binary interaction models. In the next section a brief resume of this class of models is presented.

2. BINARY INTERACTION MODELS

In this section a class of binary interaction models is introduced, in particular those that satisfy a two-dimensional generalization of the Markov property. These models provide a probability structure for a set of spatially distributed random

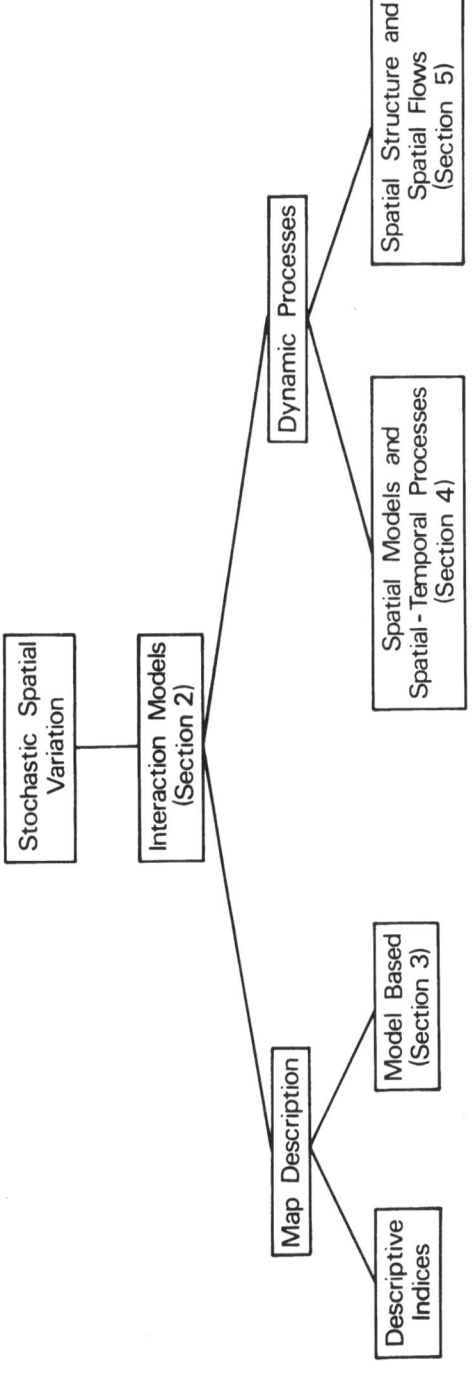

Figure 1. Organization of the sections of this paper.

variables, and hence can be thought of as providing a description of a spatial configuration. The following discussion should be supplemented with Haining (1982a, 1982b) and references therein.

Assume a region R consisting of n sites, say V, that are fixed and labelled. Let X_i be the random variable associated with the i-th site, and denote its observed value as x_i, which may take on the value of either 0 or 1. Any particular configuration of site values can be represented by the vector $x^T = (x_1, x_2, ..., x_n)$. The set of all possible configurations is denoted by Ω.

To each $x \varepsilon \Omega$ assign a real number $I(x)$, which is called the weight of x. $I(x)$ must be positive. Define

$$Z = \sum_{x \varepsilon \Omega} I(x)$$

so that $I(x)/Z$ can be interpreted as the probability of the configuration x.

The interesting problem is to use this structure to describe spatial configurations where site values reflect inter-site dependence. With this in mind, an interaction function, say T, is introduced which assigns a real number to each subset of V based on the observed site values within the subsets. Let

$$I(x) = \prod_B T(B) \quad ,$$

where the product is over all subsets B of V [by definition $T(\emptyset) = 1$].

Consider the system of sites presented in Figure 2. Then

$$I(x) = T(x_1) \ldots T(x_8) \, T(x_1,x_2) \ldots T(x_1,x_8) \ldots T(x_7,x_8)$$
$$\ldots T(x_1,x_2,x_3) \ldots T(x_6,x_7,x_8) \ldots T(x_1,\ldots,x_8) \quad .$$

Non-spatial models are obtained by setting all T functions involving two or more site values to unity. Spatial models can be developed by relaxing this assumption. An important sub-class of spatial models is the group of spatial Markov models. In order to explain what these models are, it is necessary to define the terms 'neighbor' and 'clique,' which now will be done.

Consider an edge system imposed on the set of sites of Figure 2. A particular case is illustrated in Figure 3. This edge system is arbitrary here, but in empirical analyses its specification will be of considerable importance, as the later sections of this paper will make clear. In some cases it is the purpose of statistical analysis to idenfity the edge system for the set of sites, in other cases an edge system is assumed prior to statistical analysis, which then is concerned only with parameter estimation and

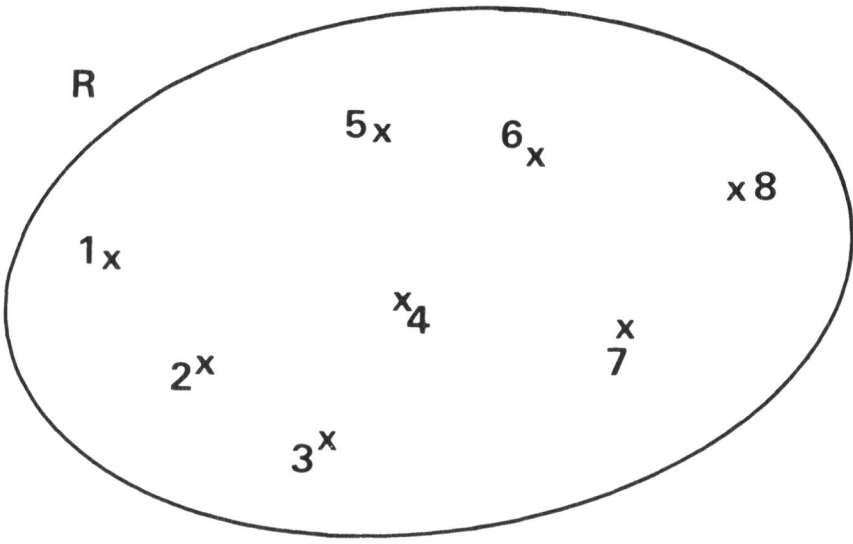

Figure 2. An illustrative system of sites.

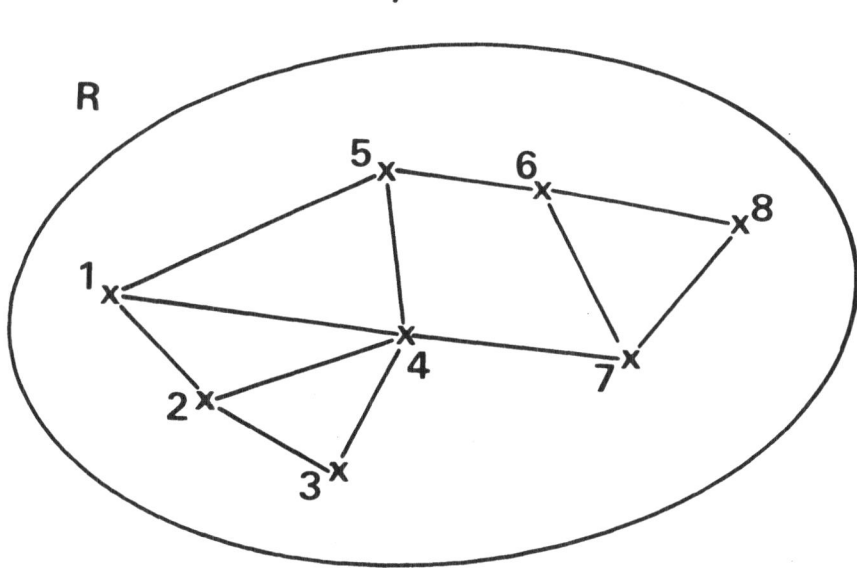

Figure 3. An illustrative system of edges for the system of sites appearing in Figure 2.

associated hypothesis and goodness of fit tests.

Two sites are said to be neighbors (or nearest neighbors) if they are joined by an edge. So in Figure 3, sites 1, 4 and 6 are neighbors of site 5, and vice versa. A clique is a set of sites such that every pair of sites in the set are neighbors. So in Figure 3 there are 8 cliques of size one (conventionally all single points are defined as cliques of size one), 12 cliques of size two, and 4 cliques of size three. It turns out that the general form of $I(x)$ for a Markov model on the sites of R in Figure 3 then is given by (Besag, 1974):

$$I(x) = T(x_1) \ldots T(x_8) \, T(x_1,x_2) \, T(x_1,x_4) \, T(x_1,x_5) \, T(x_2,x_3)$$
$$\ldots T(x_7,x_8) \, T(x_1,x_2,x_4) \ldots T(x_6,x_7,x_8).$$

Thus the weight function, $I(x)$, consists of products of interaction functions T that refer only to subsets of sites that form cliques.

A partial form of this model is obtained by setting all T functions associated with cliques of size three or larger equal to unity (Besag, 1974). This procedure specifies the class of 'auto' Markov models, and the associated $I(x)$ function for Figure 3 becomes

$$I(x) = T(x_1) \ldots T(x_8) \, T(x_1,X_2) \, T(x_1,x_4) \, T(x_1,x_5) \ldots T(x_7,x_8) \ .$$

With the additional assumption that all the T functions are positive and have the form

$$T(B) \ = \ \exp[R(B)] \qquad ,$$

and, since $I(x) = \prod_B T(B)$, it follows that

$$I(x) \ = \ \exp[\sum_B R(B)\} \qquad .$$

In the case of binary variables and allowing only cliques of size one or two, it can be shown that (Besag, 1974)

$$T(x_i) \ = \ \exp(\alpha_i \, x_i) \qquad ,$$

and

$$T(x_i \, x_j) \ = \ \begin{cases} \exp(\beta_{ij} \, x_i \, x_j) & \text{if sites i and j} \\ & \text{are neighbors,} \\ 1 & \text{otherwise,} \end{cases}$$

where $\beta_{ij} = \beta_{ji}$. Therefore the model has the form

$$I(x) \ = \ \exp(\sum_{i=1}^{i=n} \alpha_i x_i + \sum_{1 \le i < j \le n} \beta_{ij} x_i x_j) \qquad . \qquad (2.1)$$

Additional details are provided in Haining (1982a) and references therein.

The first term on the right-hand side of equation (2.1) is often referred to as an intra-site effects, or external effects, term, while the second term captures inter-site effects or internal interaction effects. Equation (2.1) is called the autologistic model because of its close similarities with the classical logistic model. Besag (1974, 1975) has discussed methods of parameter estimation for the sets of $\{\alpha_i\}$ and $\{\beta_{ij}\}$ terms, as well as other aspects of model fitting for equation (2.1).

3.   INTERACTION MODELS FOR PATTERN DESCRIPTION

Suppose there is a set of binary observations, $x_1,\ldots,x_n$, recorded at the n sites (V) of a region R. These n values are represented by the vector x. The spatial configuration of these values may be summarized in terms of the earlier model (2.1). Such a model may be particularly appropriate if it is suspected that the distribution of values displays spatial pattern.

Describing map patterns in terms of a model such as (2.1) is, operationally, straightforward. It is first necessary to specify the edge system for the n sites. This edge system may be specified a priori, particularly if the nature of site interaction is known or in those cases where a particular type of spatial organization is being tested for in the data. Alternatively, a number of different edge systems may be imposed, with the one chosen being that which gives the model the best overall fit to the data. In this first case, the edge system is specified a priori, whereas in this second case organization is assumed in the data and the objective is to identify the linkage system that results in the best description of the data. Particularly in the second case it is usual to insist that the linkage system chosen be 'reasonable,' within the context of the questions being asked, in terms of the likely structure of site interaction.

In either of these two cases it is necessary to estimate the $\{\alpha_i\}$ and $\{\beta_{ij}\}$ parameters of equation (2.1) The significance of $\{\beta_{ij}\}$ values indicates the presence of spatial organization in the data. Positive $\{\beta_{ij}\}$ values indicate contagion, or the clustering of similar values across the map. Negative $\{\beta_{ij}\}$ values indicate a propensity for similar values to repel one another so that the map is characterized by alternatively high and low values. If the objective of map analysis is to test for the presence of clustering or alternating values across the map, it has not yet been established whether fitting a model such as (2.1) has any statistical advantages to offer over, say, using one of the well established spatial autocorrelation tests (Cliff and Ord, 1981).

It would be interesting, for example, to contrast the power of existing autocorrelation tests with likelihood ratio tests derived from model (2.1). However, there may be a number of interpretative advantages, which now will be briefly considered.

A structure such as equation (2.1) offers the opportunity to provide a succinct description of the entire data set, providing in addition a summary measure of the organizational properties of the spatial values. Where the structure of the model can be related to attributes of the geographical system, parameters may be given a substantive interpretation. In such cases it may even be possible to study and explain shifts in parameter values through space and time. Parameters that can be interpreted in this way also provide a more stable and worthwhile basis for problems of estimation and forecasting. Parameter changes also may indicate properties of the underlying process, generating hypotheses for further study (Haining, 1982b).

The following is an example which suggests the possible role of interaction modelling in the study of diffusion processes. Suppose at some point in time an innovation is introduced into a population, or conversely an epidemic breaks out. The following problem presents itself. Given the spatial distribution of adopters/infectives at time t, what is the likely areal pattern at time (t+1)? Which sites will receive the strongest stimuli to adopt (in the case of an innovation), or which sites are most at risk (in the case of an epidemic)?

The reason for pursuing this example is because these processes have mechanisms that have parallels with the formal structure of equation (2.1). Hudson (1972) classified the stimuli to adopt innovations into two groups, namely aspatial media influences (i.e., the impact of external sources of information upon a susceptible population with different levels of susceptibility), and inter-personal communication influences (i.e., the passing on of information involving interaction between groups of individuals). In the case of epidemics, certain infectious diseases diffuse through a combination of environmental stimuli and face-to-face contact. The former relate to intra-site 'treatments,' while the latter relate to personal interactions. Hence both processes are driven by a combination of internal and external interactions.

These distinctions have parallels with the structure of the autologistic model, where the weight function contains terms associated with intra- and inter-site effects (respectively, cliques of size one and cliques of size two or more). It is not being suggested here that the $\{\alpha_i\}$ and $\{\beta_{ij}\}$ parameters of equation (2.1) can be related directly to the processes of adoption or infection, but only that their estimation may give some insight into the importance of neighborhood effects. Therefore, consider the application of the autologistic model, with a weighting function given by equation (2.1), for describing a pattern of innovation adoption. The following discussion identifies map pattern properties that result from such a model description of the data.

In equation, (2.1) set X_i equal to unity if the i-th site has adopted, and zero otherwise. Assume $\alpha_i = \alpha$ for all i. The maximum likelihood estimator for α, when the interaction parameters are set to 0, is the natural logarithm of the ratio of adopters to non-adopters. If the full model (2.1) is estimated so that all the α_i and β_{ij} parameters are estimated simultaneously, then changes in the value of the α_i parameters and the magnitude of the β_{ij} parameters provide measures of the relative importance of neighborhood properties in the map. If the definition of neighborhood (i.e., the edge structure imposed on the set of sites) is chosen so that it is consistent with known properties of regional inter-personal contact, a firmer ground for making forecasts is likely.

The following example is based upon data from Hagerstrand (1967) on the adoption of TB controls in a small area of Northeast Sweden thru 1940. There were 386 farmsteads in the area, of which 139 had adopted by 1940.

In all the autologistic models intra-site effects were assumed to be uniform, so that $\alpha_i = \alpha$ for all i. Three different edge structures were tried. First,

$$\beta_{ij} = \beta \exp(- \gamma d_{ij}), \qquad (3.1)$$

where d_{ij} is the distance separating the i-th and j-th farmsteads, and γ is a distance decay parameter ($\gamma = 0.35, 0.80, 1.20$ and 1.80). This definition was based upon empirical information about inter-personal communication. Second,

$$\beta_{ij} = \begin{cases} \beta & \text{if i and j are neighbors,} \\ 0 & \text{otherwise.} \end{cases} \qquad (3.2)$$

Farmstead j was taken as a neighbor of farmstead i if it was one of the six nearest neighbors of i. Since the model requires that $\beta_{ij} = \beta_{ji}$, and since site i might not be one of the six nearest neighbors of site j, additional links were permitted to satisfy this symmetry requirement. Third,

$$\beta_{ij} = \begin{cases} \beta & \text{if i and j share a boundary in a} \\ & \text{Dirichlet partitioning of the surface,} \\ 0 & \text{otherwise.} \end{cases} \qquad (3.3)$$

Full details of this analysis are given in Haining (1982c).

The first edge structure gave increasingly better fits as γ increased, suggesting that the pattern could be adequately described in terms of local patterns. The third edge structure gave a better fit than any of the four negative exponential schemes of the first edge structure. However, the best fit was provided by the nearest neighbor edge structure. The results of fitting specifications (3.1) and (3.3) suggest that proximity is important in the structure of the adoption map. Directionality, however, seems considerably less important, as is evidenced by the poorer fit of specification (3.3) as compared to (3.2). These results lend some support to the argument that farmers do react to the decisions of their nearest neighbors.

The next step was to obtain maps of the probability of adoption at each site given the existing (i.e., 1940) distribution of adoption. Such a map can be obtained from evaluating the conditional probability:

$$P(X_i = 1 | \text{the values at all other sites}, \{\hat{\alpha}_i\}, \{\hat{\beta}_{ij}\})$$

$$\equiv P(X_i = 1 | \cdot) \quad .$$

It can be shown (see Haining, 1982a) that

$$P(X_i = 1 | \cdot) =$$

$$[\exp(\hat{\alpha}_i + \sum_{\substack{j=1 \\ j \neq i}}^{j=n} \hat{\beta}_{ij} x_j)] / [1 + \exp(\hat{\alpha}_i + \sum_{\substack{j=1 \\ j \neq i}}^{j=n} \hat{\beta}_{ij} x_j)] \quad , \quad (3.4)$$

where $\{\hat{\alpha}_i\}$ and $\{\hat{\beta}_{ij}\}$ denote the estimated parameter values. Figures 4 and 5 are based on the evaluation of equation (3.4) for the negative exponential weighting scheme ($\gamma = 1.2$) and the nearest neighbor weighting scheme, respectively.

4. INTERACTION MODELS AND SPATIAL PROCESSES: INTRA-URBAN GASOLINE PRICING

In the preceding section interaction models were used to describe spatial configurations. In this section interaction models will be treated as equilibrium state models for spatial-temporal processes. The relationships between these models and certain spatial-temporal processes have been discussed in Preston (1974) and Bartlett (1975), and reviewed in Haining (1982a). The following general points are worth emphasizing, though.

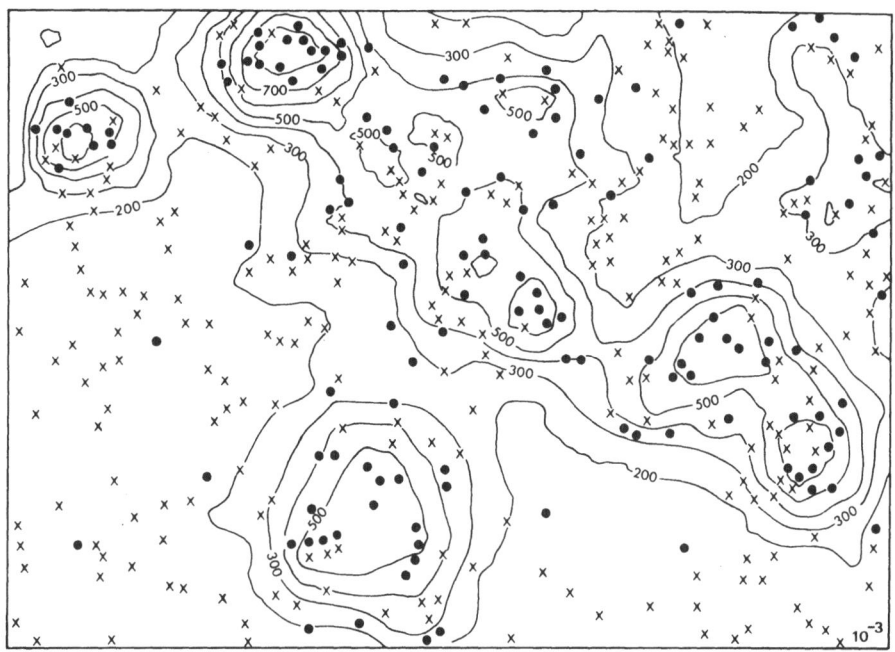

Figure 4. Observed pattern of adopters of TB controls for an area
of Northeast Sweden, together with the probability-
of-adoption surface based on the negative exponential
weighting scheme, where γ=1.2 (. denotes adopters and x
denotes non-adopters).

The autologistic model may be derived from a process in which
individual sites (V) in region R transfer back and forth between
the two states (0 and 1) in response to both external influences
and the states of neighboring individuals. The process starts with
an initial sprinkling of 'ones,' and then individuals transfer
back and forth with no time lags limiting the rate at which
transfers take place. This process also assumes time
reversibility. Models of this type have been proposed as models
for spatial competition, particularly between individual plants
(Mead, 1967). In this section the use of these models for studying
intra-urban price competition will be explored. The underlying
theory suggests that it is the equilibrium state of the process
that is of most interest. The theory and results presented here
represent excerpts from a research project still in progress, the
results of which will be more fully reported on at a later date.

According to the Cournot-Enke problem, large scale price
variation for a single homogeneous good results from the existence

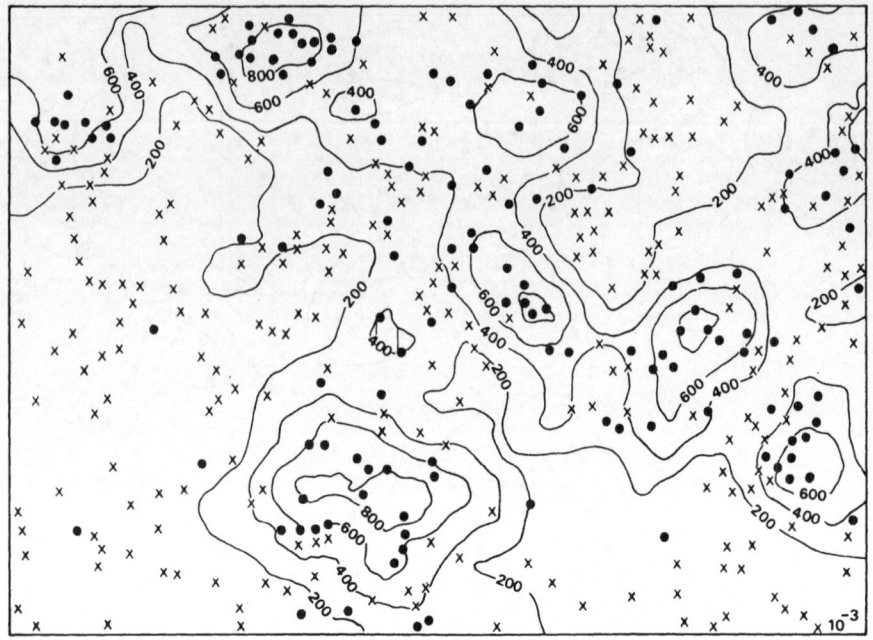

Figure 4.   Observed pattern of adopters of TB controls for an area
            of Northeast Sweden, together with the probability-
            of-adoption surface based on the nearest neighbor
            weighting scheme, (. denotes adopters and x denotes
            non-adopters).

of n distinct local markets, each with its own local supply and
demand curves. The existence of inter-market price differences
sets up trading relationships of exports and imports between these
markets. The spatially competitive equilibrium price level results
from these different local market prices, as well as transport cost
differentials between the markets [see Samuelson (1952) for a
discussion of this problem].

    At the intra-urban level, say for a set of n gasoline retail
outlets, spatial price variation cannot be explained in terms of
either the existence of spatially discrete market areas or transfer
costs between markets. Market interdependence exists not only by
virtue of the movement behavior of buyers, but also by the spatial
price awareness of sellers vis-a-vis their competitors.
Fluctuations in local area commodity price levels at least may be
imagined as resulting from competitive or co-operative interaction
among traders in response to the neighborhood distribution of
prices, and the volatile demand characteristics of their market

catchment areas (Haining, 1982a).

Consider n retail outlets for gasoline within a single urban area. Given the supply and demand characteristics for gasoline, the following pair of equations may be specified:

$$\left.\begin{array}{c} \mathbf{D}_t = \mathbf{A}\,\mathbf{p}_t + \mathbf{c} \\ \mathbf{S}_t = \mathbf{B}\,\mathbf{p}_{t-1} + \mathbf{e} \end{array}\right\} \qquad , \qquad (4.1)$$

where $\mathbf{D}_t$ is the demand vector (of length n) at time t, $\mathbf{S}_t$ is the supply vector at time t, and $\mathbf{p}_t$ and $\mathbf{p}_{t-1}$ are the n-dimensional price vectors at time t and t-1, respectively. The vectors $\mathbf{c}$ and $\mathbf{e}$ are also of length n, and are vectors of constants. $\mathbf{A}$ and $\mathbf{B}$ are n-by-n matrices. Equation (4.1) is the familiar interacting markets model with a lagged supply function. Assuming that clearance takes place in each market means

$$\mathbf{D}_t - \mathbf{S}_t = \mathbf{0} \qquad . \qquad (4.2)$$

From equations (4.1) and (4.2) it follows that

$$\mathbf{p}_t = \mathbf{A}^{-1}\,\mathbf{B}\,\mathbf{p}_{t-1} + \mathbf{A}^{-1}\,(\mathbf{e} - \mathbf{c}) \qquad . \qquad (4.3)$$

Next let $\mathbf{p}_e$ denote the equilibrium price vector. It can be shown that

$$\mathbf{p}_e = (\mathbf{A} - \mathbf{B})^{-1}\,(\mathbf{e} - \mathbf{c}) \qquad . \qquad (4.4)$$

This equilibrium solution is stable providing that the eigenvalues of $(\mathbf{A}^{-1}\mathbf{B})$ are less than 1 in absolute value.

Rearranging the terms of equation (4.4) yields

$$\mathbf{p}_e = \mathbf{B}^{-1}\,\mathbf{A}\,\mathbf{p}_e + \mathbf{B}^{-1}\,(\mathbf{c} - \mathbf{e}) \qquad , \qquad (4.5)$$

which expresseas the equilibrium price vector in terms of itself. Now writing equation (4.5) as

$$\mathbf{p}_e = [\overline{\mathrm{Diag}}(\mathbf{B}^{-1}\,\mathbf{A}) + \mathrm{Diag}(\mathbf{B}^{-1}\,\mathbf{A})]\mathbf{p}_e + \mathbf{B}^{-1}\,(\mathbf{c} - \mathbf{e})$$

$$\equiv [\mathbf{Y} + \mathbf{W}]\mathbf{p}_e + \mathbf{B}^{-1}\,(\mathbf{c} - \mathbf{e}) \qquad , \qquad (4.6)$$

where matrix $\mathbf{Y}$ is a diagonal matrix whose entries are the diagonal elements of matrix $(\mathbf{B}^{-1}\mathbf{A})$, and matrix $\mathbf{W}$ is the off-diagonal or non-diagonal entries in $(\mathbf{B}^{-1}\mathbf{A})$, which is denoted as $\overline{\mathrm{Diag}}$.

Let the subscript i refer to the i-th element/row of the appropriate vectors/matrices. Then equation (4.6) implies that

$$p_{i,e} = (1 - y_{ii})^{-1} \mathbf{W}_i p_e + (1 - y_{ii})^{-1} (\mathbf{B}^{-1})_i (c - e)$$
$$i=1,2,\ldots,n \qquad (4.7)$$

which expresses the equilibrium price at the i-th outlet as a function of equilibrium prices elsewhere in the system.

Returning to equation (4.1), suppose that the diagonal entries of matrix $\mathbf{A}$ are negative and that the off-diagonal entries are positive. Assume also that only the diagonal entries of matrix $\mathbf{B}$ are non-zero, and that the diagonal entries are positive. In other words, assume that there is no market interaction in terms of supply. This places the burden on the individual retailer to adjust prices to ensure market clearance, since the major oil companies follow a schedule of delivery times. With these assumptions equation (4.7) now can be rewritten as

$$p_{i,e} = \sum_{\substack{j=1 \\ j \neq i}}^{j=n} a_{ij}/(b_{ii} - a_{ii})\, p_{j,e} + (c_i - e_i)/(b_{ii} - a_{ii})\ , \quad (4.8)$$

where $\{a_{ij}\}$ and $\{b_{ii}\}$ denote the appropriate entries in matrices $\mathbf{A}$ and $\mathbf{B}$, respectively. It follows immediately from equation (4.8) that

$$a_{ij}/(b_{ii} - a_{ii})\ > \ 0 \qquad ,$$

which implies a system of co-operative spatial pricing.

Given the movement behavior of buyers, an additional assumption that could be made is that demand is a stochastic variable. Thus, suppose the demand equation (4.1) is rewritten as

$$\mathbf{D}_t = \mathbf{A}\, p_t + c + u \qquad ,$$

where u is distributed as $IN(0, \sigma^2 I)$. It can then be shown that equation (4.8) becomes

$$p_{i,e} = \sum_{\substack{j=1 \\ j \neq i}}^{j=n} a_{ij}/(b_{ii} - a_{ii})\, p_{j,e} + (c_i - e_i)/(b_{ii} - a_{ii}) + u_i' \qquad ,$$

where u_i' is $IN[0, \sigma^2/(b_{ii} - a_{ii})^2]$, or that

$$p_{i,e} \propto \sum_{\substack{j=1 \\ j \neq i}}^{j=n} a_{ij}\, p_{j,e} + c_i - e_i + u_i \qquad . \quad (4.9)$$

This last equation suggests fitting an autoregressive model with a site varying mean, after appropriate simplification of the $\{a_{ij}\}$ parameters (Whittle, 1954). This is an interaction model, too, but with variables that are real valued. Results will be reported here, however, for a set of simpler tests that are suggested by both equations (4.8) and (4.9).

Data were collected at monthly invervals, for the period of January thru March of 1982, for a set of 85 gasoline retail outlets in Southwest Sheffield. This was supposedly a period of intensifying price competition in response to excess world supply of oil, coupled with dwindling consumer demand in the face of the economic recession. An edge system was superimposed on this set of 85 sites. The pattern of linkages reflected both the principal radial routeways of Sheffield, and local interaction that was assumed to exist between retail outlets that were in close proximity to one another, though perhaps not on the same principal routeway. The distribution of prices of January, 1982, is shown in Figure 6. These prices are in pence per imperial gallon of four star gasoline. It can be seen, for example, that just over 50% of the outlets sold gasoline at between 153.9 and 156.3 pence. Spatial price competition between retail outlets might be expected at the low end of the price range. However, competition also might be expected at several levels in the price distribution reflecting local patterns of competition. For instance, the clustering of outlets charging between 156.7 and 157.1 pence seemed worth analyzing.

In each analysis outlets whose price was in the given range were coded 1, while the rest were coded 0. The autologistic model was fitted to test whether or not spatial price competition was evident within the stated price ranges. Table 1 presents the results for various price levels using a pseudo-log likelihood estimation procedure for the parameters α and β. In all cases, $\alpha_i = \alpha$ for all i, and the β_{ij} were defined as in equation (3.2).

Exact significance levels are difficult to establish here. Let $L_n(\tilde{\alpha}, \tilde{\beta})$ denote the maximized log–likelihood function for a two parameter model based on n independent observations. Then

$$L_n(\tilde{\alpha}, \tilde{\beta}) \; - \; L_n(\tilde{\alpha}, \beta=0)$$

is, for large n, $(\frac{1}{2}) \, \chi^2$ distributed with 1 degree of freedom. In the present case the observations are not independent so that the critical region is not exact. However, the reduction in the log–likelihood function is given in Table 1, and the critical value is approximately 3.31 for a 90% test (i.e., 10% level of significance). All the interaction parameters were positive (except one, which was not significant), suggesting a spatial clustering of prices at those price levels. This finding is

338

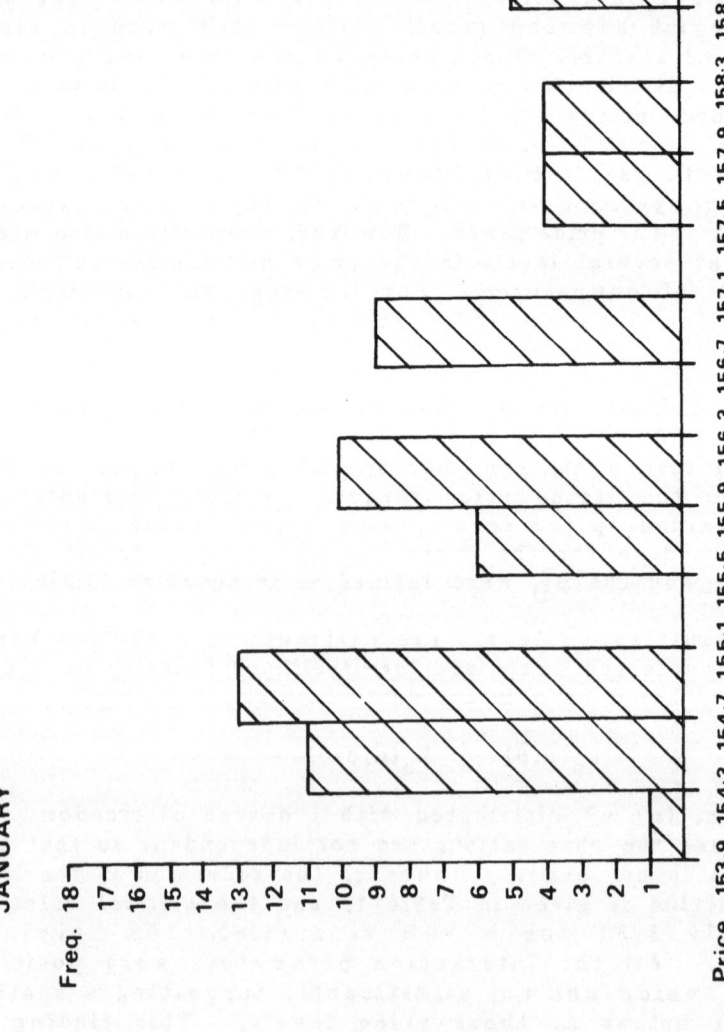

Figure 6. The distribution of gasoline prices for retail outlets in Southwest Sheffield, January, 1982.

TABLE 1

RESULTS FROM THE PSEUDO-LOG LIKELIHOOD ESTIMATION PROCEDURE
FOR THE AUTOLOGISTIC MODEL

Price Range	Parameter Estimates		Maximum of Log-Likelihood Function	χ^2 Variate $L_n(\tilde{\alpha},\tilde{\beta}) - L_n(\tilde{\alpha},\tilde{\beta}=0)$
	$\tilde{\alpha}$	$\tilde{\beta}$		
153.9, 156.3	-0.50	0.80	-42.52	4.06*
155.0, 156.0	-0.87	0.40	-45.39	1.19
156.9, 157.0	-1.49	-0.35	-28.08	0.47
157.8, 158.0	-2.35	0.97	-23.64	1.11
158.7, 160.0	-2.15	1.52	-23.65	4.90*

* Denotes a significant difference from zero at the 0.05 level.

consistent with the implications of equations (4.8) and (4.9). Spatial price competition is in evidence at the low end of the price level, and, perhaps surprisingly, also at the higher end of the price levels (the very highest prices have not been included on the frequency chart). Further tests for interaction effects within the ranges given in Table 1, particularly within the very competitive range of (153.9,156.3), did not yield significant results.

The preliminary results obtained here confirm the existence of spatial price competition in terms of the principal routeways into and out of Sheffield, but with this competitiveness seeming to be quite localized along these routeways. Competitiveness tends to exist across at least a range of 1 or 2 pence, rather than at finer levels.

5. INTERACTION MODELS AND SPATIAL PROCESSES: FLOWS ON A TWO-DIMENSIONAL SURFACE

In this section relationships will be considered between the structural attributes of two-dimensional surfaces, on the one hand, and flows that may exist across them, on the other hand. As a concrete example of the sort of questions that this section addresses, consider the gasoline price study of the previous section. The following questions might be of interest. First, given the distribution of retail outlets together with their prices, what implications are there for the frequency distribution of trip lengths that involve gasoline purchasing? Second, given an established trip frequency distribution, what implications are there for the spatial pattern of gasoline prices? It is suggested here that the autocorrelation function that summarizes average map structure may provide an approach to questions of this sort. This analysis starts by introducing a number of graph theoretic concepts that help to organize the material of this section in the case of lattice systems (Biggs, 1977). Later in this section the case of continuous spatial systems will be considered.

Consider a system of sites (V) and edges (E) in a region R. Suppose that an orientation is imposed on each edge $e \varepsilon E$. Now define the matrix **D** such that it has as many rows as there are sites (V) in region R, and as many columns as edges (E). The sites and edges are labelled and coincide with the labelling of the rows and columns in matrix **D**. Then

$$d_{ve} = \begin{cases} 1 \text{ if site v is the positive end of the edge e,} \\ -1 \text{ if site v is the negative end of the edge e, and} \\ 0 \text{ if site v is not incident with edge e.} \end{cases}$$

Since the edges are oriented, the two ends of the edge may be distinguished, and by convention an arrowhead points in the direction of the positive end.

Let $\emptyset(e)$ denote the flow on the edge $e \varepsilon E$. Let $w(v)$ denote the state of the site $v \varepsilon V$. Consider two additional functions, namely s, which assigns a state to each site on the basis of the flows defined for each edge $e \varepsilon E$, and f, which assigns a flow to each edge on the basis of the site values defined for each $v \varepsilon V$. For example, let

$$s(v) = \sum_{e \varepsilon V} d_{ve} \, \emptyset(e) \qquad , \text{ and} \qquad (5.1)$$

$$f(e) = \sum_{v \varepsilon V} d_{ve} \, w(v) \qquad .$$

Clearly the value $s(v)$ for any $v \varepsilon V$ represents the net accumulation of flow at site v, while $f(e)$ for any $e \varepsilon E$ represents the difference in site values at the two ends of edge e. Hence

$$f(e) = w(x) - w(y) \qquad ,$$

if $x, y \varepsilon V$ and x is the positive end and y the negative end of edge e.

The following questions now can be posed for fixed sites V and edges E in region R:

(1) For any given $\{\emptyset(e)\}_e$ and matrix **D**, what is the structure of $\{s(v)\}_v$?

(2) For any given $\{w(v)\}_v$ and matrix **D**, what is the structure of $\{f(e)\}_e$?

In both questions the aim is to relate configurational properties of maps to other, dynamic, attributes. It also will be of interest to assess to what extent configurational shifts result in changes in other, dynamic, attributes.

Consider a system of sites and suppose for convenience that they are arranged as a rectangular lattice. Suppose the site values follow a first order autoregressive model, that is,

$$X_{i,j} = \rho(X_{i-1,j} + X_{i+1,j} + X_{i,j-1} + X_{i,j-1}) + e_{i,j} \quad (5.3)$$

$$\text{for all } i,j \quad ,$$

where $e_{i,j}$ is $IN(0,\sigma^2)$ and ρ is a parameter. Letting $\rho \to 0$, equation (5.3) becomes

$$X_{i,j} = e_{i,j} \quad ,$$

so that site values represent independent drawings from a normal curve. In equation (5.2) let e denote an edge linking any site (i.j) with one of its four nearest neighbors specified by equation (5.3). Then, for example,

$$f(e) = (x_{i,j} - x_{i-1,j}) \quad \text{if } x_{i,j} > x_{i-1,j} \quad ,$$

where the lower case elements refer to observed values. Thus the orientation of the graph is specified by the distribution of $\{x_{i,j}\}$, and the flows are specified by the difference in values at the two ends. Consequently flows follow the (price) gradient from high values to low values. The frequency distribution of trip totals at 'lag one' is thererfore a half normal curve truncated at 0.

As ρ changes, map structure changes. As ρ increases, map structure will show evidence of contagion or clustering of similar sized values, so that the difference in neighboring site values will tend to decrease. The frequency distribution of trip totals will be pushed further to the left. As ρ decreases from 0, map structure will show evidence of negative autocorrelation, or a propensity for similar sized values to be separated. Now the frequency distribution of trip totals will be pushed to the right. These tendencies will be most marked when ρ reaches its upper limit for stationarity. On an unbounded lattice equation (5.3) is stationary for the case in which $|\rho| < 0.25$. For $\rho = 0.25$, for example, the model reduces to a linear trend surface and flows occur only down the gradient and in proportion to the slope of the trend.

Suppose that additional edges are imposed on the lattice, linking sites that are separated by greater distances (i.e., greater 'lags' if connectivity is defined by the path mapped out by only following each site's four nearest neighbors). Since the autocorrelation function for equation (5.3) is monotonically decreasing as distance increases (Whittle, 1954), site values tend towards independence. Thus, for all values of ρ at sufficiently long distances, the trip frequency distribution function will be a truncated normal distribution.

In the above, flows have been assigned in accordance with equation (5.2) on the basis of the gradient between two sites. Two further observations can be made. First, flows need to be weighted by the cost of overcoming the intervening distance. This would eliminate the undersirable property encountered in the analysis above of trip functions converging to an identical form for all distances. Second, average trip frequency distributions can be obtained by using an average measure of surface gradient, which would be more suitable for continuous surfaces. The autocorrelation function is a measure of average surface gradient at all distances, since large values of the autocorrelation function denote similar values (low gradient) and low values of the autocorrelation function imply dissimilar values (steeper gradient).

Consider a stationary isotropic spatial surface $\{Z(h)\}_h$; that is, suppose Z is a random variable and h is a specified location in continuous two-dimensional space. Let f(x) denote the average flow between any two points separated by distance x on this surface. Let $f(x) = E[Z(h) - Z(h + x)]^2/c(x)$, where c(x) is a strictly increasing function, with c(0) = 0. It follows then that

$$f(x) = 2\sigma^2(Z) [1 - \rho(x)]/c(x) \quad ,$$

where $\sigma^2(Z)$ is the variance of the surface and $\rho(x)$ is the autocorrelation of the surface at lag x. Note that as

$$x \to \sigma^\infty \quad , \quad f(x) \to {}^2(Z)/c(x) \to 0 \quad ,$$

while $\lim_{x \to} f(x) = 0$, if $\rho(x) \to 1$ faster than $c(x) \to 0$. Now f(x) may be re-expressed so that

$$f(x) = r(x)/c(x) \quad .$$

What is the behavior of f(x) for different types of r(x), that is different types of autocorrelation functions? An appropriate function for exploring this kind of relationship is the logistic curve. Let

$$r(x) = A \exp(\beta x)/[1 + A \exp(\beta x)] \quad , \quad (5.4)$$
$$\beta > 0, \ 0 < A < 1 \quad .$$

For A close to 0, equation (5.4) might represent the average gradient for a surface with an autocorrelation function that is cusp-shaped near the origin, and then declines monotonically to zero (see, for example, Whittle, 1954, p. 447-8). For an A that is larger, equation (5.4) might represent a surface with a 'spiked' autocorrelation function (Whittle, 1954, p. 445).

Let c(x) represent a cost function, such that the cost of travel increases with distance. This cost function also might be related to the probability of encountering an intervening gradient on the surface that would deter the trip. For present purposes let

$$c(x) = \exp(\alpha \ x) \quad .$$

Then

$$f(x) = A \exp[(\beta - \alpha)x]/[1 + A \exp(\beta \ x)] \quad . \quad (5.5)$$

A variety of frequency functions have been generated for equation (5.5), with their results presented in Figure 7. The case for β = 2.0, $(\beta-\alpha)$ = 1.0 and A = 1.0 can be used as a reference curve. Trip frequency decreases monotonically with increasing lag. As A decreases, or as $(\beta-\alpha)/\beta$ increases, a trip frequency peak is observed that is not at the origin. The situation where the ratio $(\beta-\alpha)/\beta$ is increasing may result from a fall in the value of α, implying a fall in the cost of travel. The situation where A is decreasing suggests a smoothly varying surface with near neighbor variate dependencies 'cusped' rather than 'spiked.' Not surprisingly, the total number of trips undertaken decreases as A decreases.

6. CONCLUSIONS

This paper has suggested research areas in spatial modelling that are linked with research problems in the study of dynamical systems. The first of these involved map description as a basis for developing spatial-temporal forecasts. The second of these considered map structure as the equilibrium state of a spatial-temporal process relating to intra-urban gasoline pricing. The analysis of static map structure became an essential element in model testing, even though the analysis involved a space-time model. The third problem discussed is concerned with the influence

344

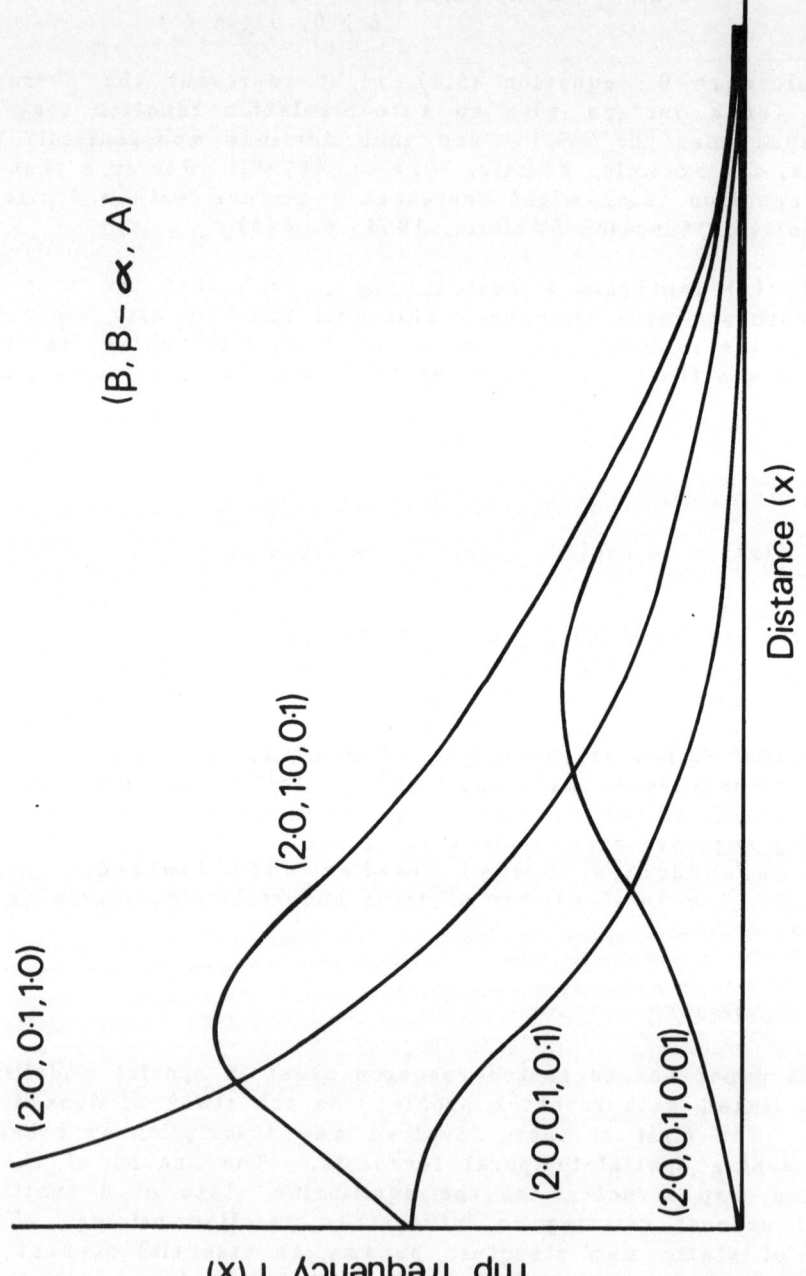

Figure 7. The shape of equation (5.5) for different values of α, β and A.

of spatial structure on spatial flows, and it was suggested that the autocorrelation function, as a measure of average map gradient, might be used to explore changes in trip frequency functions. The last problem treated seems to raise particularly interesting questions with regards to the impact of structural change in the spatial surface on structural changes on the associated flow patterns, and vice versa. The structure of spatial flows may change rapidly in time, whereas spatial structure changes more slowly. To what extent does a given flow structure re-inforce or modify an existing spatial structure? At the end of the last section it was shown that spatial surfaces with spiked autocorrelation functions generated monotonically decreasing trip frequencies, while surfaces with smooth autocorrelation functions generated frequency curves that peaked away from the origin. If interaction is a determinant of spatial structure, then it would be interesting to ascertain whether or not these flow patterns are unstable. Given a change in the underlying spatial structure brought about by a change in the parameters of the surface, what effect will this have on the structure of flows defined on the surface? The study of relationships between fast (flow) and slow (pattern) moving spatial structures may be an important area of research at the interface between static and dynamic geographical problems.

7. REFERENCES

Bartlett, M., 1975, The Statistical Analysis of Spatial Pattern, London: Chapman-Hall.

Besag, J., 1974, Spatial Interaction and the Statistical Analysis of Lattice Systems, Journal of the Royal Statistical Society, 36, Series B: 192-236.

_____, 1975, The Statistical Analysis of Non-lattice Data, The Statistician, 24: 179-195.

Biggs, 1977, Interaction Models, Cambridge: Cambridge University Press.

Cliff, A. and J. Ord, 1973, Spatial Autocorrelation, London: Pion.

_____, 1981, Spatial Processes: Models and Applications, London: Pion.

Hagerstrand, T., 1967, Innovation Diffusion as a Spatial Process, Chicago: University of Chicago Press.

Haining, R., 1977, Model Specification in Stationary Random Fields, Geographical Analysis, 9: 107–129.

_____, R., 1978, The Moving Average Model for Spatial Interaction, Transactions, Institute of British Geographers, 3 (new series): 202–225.

_____, R., 1982a, Interaction Models and Spatial Diffusion Processes, Geographical Analysis, 14: 95–108.

_____, R., 1982b, Map Analysis: Description or Modelling, Environment and Planning A, 14: 1097–1106.

_____, 1982c, Spatial and Spatial-temporal Interaction Models and the Analysis of Patterns of Diffusion, Transactions, Institute of British Geographers, forthcoming.

Hudson, J., 1972, Geographical Diffusion Theory, Evanston, Ill.: Northwestern University, Studies in Geography No. 19.

Mead, R., 1967, A Mathematical Model for the Estimation of Interplant Competition, Biometrics, 23: 189–205.

Preston, D., 1974, Gibbs States on Countable Sets, Cambridge: Cambridge University Press.

Samuelson, P., 1952, Spatial Price Equilibrium and Linear Programming, American Economic Review, 42: 283–303.

Whittle, P., 1954, On Stationary Processes in the Plane, Biometrika, 41: 434–449.

SECTION 5

TOWARDS DYNAMIC MODELS OF SPATIAL SEARCH

A key ingredient in the evolution of self-organizing systems is the interaction of individuals or elements making up the system. This interaction helps bifurcations to occur, thus generating a family of trajectories that can be followed by the system through time and over space. Perhaps the most commonly talked about problem area in geography, spatial economics and allied fields of study that treat this type of interaction has to do with migration and residential search. It has been appreciated for a long time now that an understanding of these particular interactions is indispensible to an understanding of the spatial organization of the system within which they occur. At first this understanding was sought through the formulation of static models. But, recently it has been recognized increasingly that static models of residential movement are insufficient, and hence an impressive array of explicitly dynamic models has been developed. The first paper in this section, by Rogerson, discusses how the theory of search relates to models of job market competition. In doing so, individual decisions and competition levels are treated simultaneously. It is here that principal interaction effects are captured. The interplay of job vacancies, job changes, hiring rates, voluntary quits and unemployment is focused on. Clearly this interplay is strongly related to wage levels and rates of inflation. Rogerson concludes by examining interregional mobility and regional labor markets. He demonstrates that modelling the system is analytically tractable if the distributions of wages and offers both conform to a Pareto distribution. His results are especially interesting in that they are consistent with the Phillips curve of economics. The structure of Rogerson's model has something in common with that of Leonardi in the first section of this volume, since both deal with the problem associated with externalities due to scarce capacity.

The second paper of this section, by Clark, presents a very useful review of residential mobility and housing search models, most of which to date are relatively static. In contradistinction to the nature of these models, this review stresses the dynamic processes underlying individual household decisions. Most of the theoretical literature employs concepts of stress and inconvenience thresholds, which furnish a basis for the decision concerning whether or not to search, how long to search, and whether or not to relocate. In concluding, Clark implies that by now sufficient work has been completed on the static problem, so that substantial future rewards could be gained from theoretical and empirical work that is explicitly dynamic.

348

The two papers of this section shed light on the role of interaction in formulating an evolutionary spatial model. They also indicate the extensive gap between conventional wisdom and the evolutionary model as a final goal.

THE EFFECTS OF JOB SEARCH AND COMPETITION ON
UNEMPLOYMENT AND VACANCIES IN REGIONAL LABOR MARKETS

Peter Rogerson

Northwestern University
United States

1.   INTRODUCTION

Regional labor market conditions depend in a complex manner on
the decisions of both individuals on the supply side and firms on
the demand side of the market.   Individuals must decide whether to
enter or leave the labor force, whether to quit or retain their
present jobs, and how to search for new employment.   Firms must set
desired employment levels and determine the levels of vacancies and
their associated wages.   Figure 1 illustrates how these decisions
contribute to the resultant stocks and flows in a single region
labor market.

In order to better understand the operation of labor markets,
one must be able to adequately explain the nature of the decisions
made by individuals and firms.   The theory of search (Lippman and
McCall, 1976) is concerned with the optimal behavior of individuals
who are concerned with maximizing discounted future income.   This
theory has been the object of much attention, and various
modifications have been made to it in order to examine such
phenomena as quit behavior, on-the-job search, and individual risk
aversion. Optimal policies for firms concerned with maximizing
profits have been given less attention from a labor market
perspective.   Mortensen (1970) and Eaton and Watts (1977) are
exceptions to this.

This paper has several objectives.   The second section of the
paper combines the theory of search with models of job market
competition.   Intuitively, search decisions and expectations are a
function of market competition.   Likewise, competition levels are a
result of individual decisions.   It therefore is appropriate that

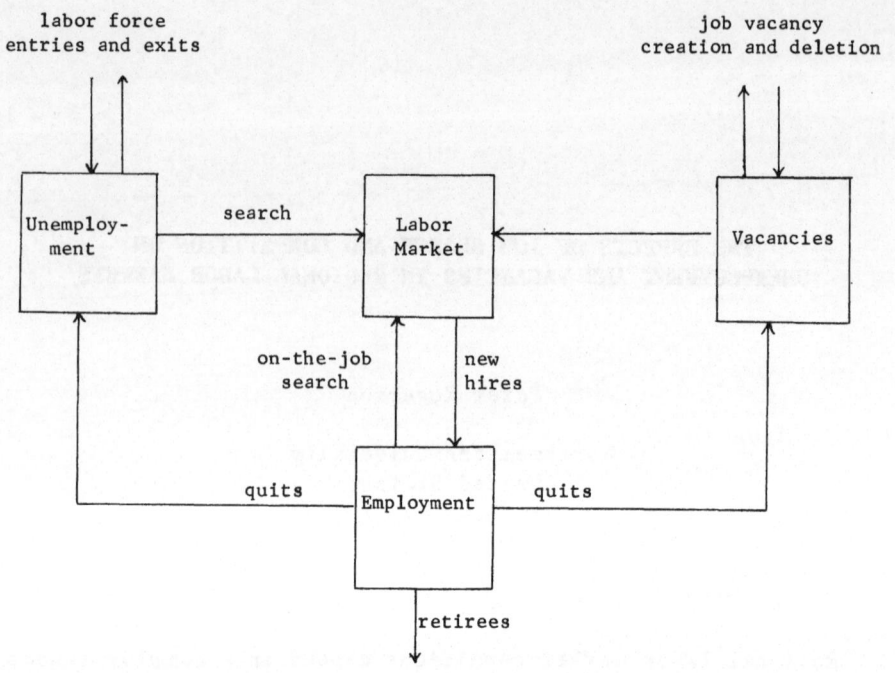

Figure 1.   Stocks and flows in the labor market.

individual decisions and competition levels be determined
simultaneously. Rogerson (1982a) takes an initial step in this
direction for the case of an exogenously specified number of job
vacancies. In contradistinction, vacancies are determined within
the model developed in this paper as a function of initial
conditions and subsequent job changes.

The third section is concerned with the relationship between
unemployment and vacancies. General conditons yielding a negative
relation between the two are derived in terms of separation and
hiring rates. Previous specifications of separation and hiring
functions are shown to obey these conditions only under certain
circumstances. It is shown that the search and competition model
can lead to a positive relation between unemployment and vacancies,
where the unemployment stock is composed of the queue of voluntarty
quits.

For a labor force of a given size, increasing the number of
vacancies may not necessarily lead to an increase in the number of
new hires. More vacancies are contacted by potential employees,

but heightened expectations lead to a greater probability of offer rejection that may more than offset the increased number of contacts. An increase in employment may be achieved through an appropriate combination of vacancy creation and wage increases. The fourth section examines the effects of wage offer inflation on the equilibrium, and demonstrates the existence of a Phillips curve.

The fifth section of the paper examines the interaction between interregional mobility and regional labor market conditions. In the sixth section, it is demonstrated that the system is analytically tractable if it is assumed that wages and offers are distributed according to a Pareto distribution. Some useful properties of the distribution are used to derive the specific forms of the relationships discussed in previous sections.

2. JOB SEARCH AND COMPETITION

The theory of optimal search derives a stopping rule for an individual engaged in sequential search. The theory assumes that individuals receive one offer per time period for a fixed cost, c, and that job seekers have knowledge of the parameters of the stationary wage distributiion, $f(x)$. It is then optimal for risk neutral job seekers maximizing expected income over an infinite time horizon to calculate a reservation wage, ξ, and to take the first job that has a wage greater than ξ. ξ is defined by

$$\xi = -c + \xi F(\xi) + \int_{\xi}^{\infty} x \, dF(x) \qquad , \qquad (2.1)$$

where $F(x)$ represents the cumulative distribution function associated with the wage distribution, $f(x)$, which represents the probability density function. Given equation (2.1), ξ has the property that the marginal returns from additional search are equal to the marginal cost of search, or

$$c = \int_{\xi}^{\infty} (x - \xi) \, dF(x) \qquad .$$

Equation (2.1) may be modified to account for the uncertainty associated with generating a job offer, yielding

$$\xi = -c/p + \xi F(\xi) + \int_{\xi}^{\infty} x \, dF(x) \qquad , \qquad (2.2)$$

where p is the parameter of a geometric random variable denoting the probability of generating an offer during the period, $1/p$ is the number of periods an individual can expect to wait before receiving an offer, and

$$1/[p \int_{\xi}^{\infty} dF(x)]$$

is the expected term of unemployment.

For individuals with relatively low wages it will be optimal to quit and search for a job with higher wages. Specifically, individuals are assumed to look for another job if their current wage w is less than their expected return from search by an amount greater than the cost (K) associated with quitting:

$$\xi - w > K \quad .$$

With a constant labor force size (L) composed of employed individuals (E) and unemployed individuals (U), the number of quits during a period is given by

$$S = sE \int_0^{\xi-K} e(x) \, dx = s(L - U) \int_0^{\xi-K} e(x) \, dx \quad , \quad (2.3)$$

where s is the frequency at which employed individuals consider quitting, and e(x) is the distribution of employee wages. On-the-job search behavior has been examined (Lippman and McCall, 1976) by assuming that on-the-job search must be less intensive, and therefore more costly per unit of time than full-time search. Using Markovian decision theory, Lippman and McCall show that below a given wage ξ' it is optimal to quit, and between ξ' and a given wage ξ^* it is optimal to engage in on-the-job search.

Only quits to search full-time will alter the stocks of vacancies and unemployment. Since the focus of this paper is on the effects of search and competition on market conditions, full-time search will be focused on, and, although recognizing the importance of on-the-job search, this latter phenomenon will not be discussed in any depth.

The hiring process adopted here may be envisioned as follows. Each of U unemployed job seekers randomly contacts one of V vacancies during a period. The employer then offers the job randomly to one of the applicants, who in turn accepts the offer if it exceeds his reservation wage. Here initially all job seekers are assumed to have the same reservation wage. As a result of the random contacting, the number of vacancies visited during a period is $V[1 - \exp(-\lambda)]$, where $\lambda = U/V$ (Rogerson and MacKinnon, 1981). The number of new hires now may be given by

$$H = V[1 - \exp(-\lambda)] \int_\xi^\infty f(x) \, dx \quad . \quad (2.4)$$

The probability that an individual obtains a job offer may be derive, from explicit assumptions regarding the search process, to be

$$p = [1 - \exp(-\lambda)]/\lambda \quad . \quad (2.5)$$

The probability that an individual obtains a job during the period in question may be derive, from explicit assumptions regarding the search process, to be

$$p \; {}_{\xi}\!\int^{\infty} f(x) \; d(x) \quad .$$

Thus only vacancies with wages greater than ξ are filled.

Much note has been taken of the fact that under the assumptions of the standard search model, there is nothing to prevent the wage offer distribution from collapsing to a single point. If all reservation wages are identical, firms have no incentive to offer wages greater than ξ, and if they offered a wage less than ξ they would lose all of their employees. Butters (1977), Eaton and Watts (1977), Kormendi (1979), and many others have shown that if reservation wages vary across individuals, or if information is imperfect, dispersion in the offer distribution will be sustained in equilibrium. Given a distribution of reservation wages, say $h(\xi)$, equations (2.3) and (2.4) may be rewritten as

$$S \; = \; {}_{0}\!\int^{\infty} s \; E \; {}_{0}\!\int^{\xi-K} e(x) \; h(\xi) \; dx \; d\xi \quad , \text{ and}$$

$$H \; = \; {}_{0}\!\int^{\infty} V[1 - \exp(-\lambda)] \; {}_{\xi}\!\int^{\infty} f(x) \; h(\xi) \; dx \; d\xi \quad .$$

The reservation wage used in equations (2.3) and (2.4) may be taken as the mean of this distribution such that

$$\bar{\xi} \; = \; {}_{0}\!\int^{\infty} \xi \; h(\xi \; d\xi) \quad .$$

In adopting equations (2.3) and (2.4), it therefore implicitly is assumed that

$${}_{0}\!\int^{\infty} sE \; {}_{0}\!\int^{\xi-K} e(x) \; h(\xi) \; dx \; \xi = sE \; {}_{0}\!\int^{\bar{\xi}-K} e(x) \; dx \quad , \text{ and } \quad (2.6)$$

$${}_{0}\!\int^{\infty} V[1 - \exp(-\lambda)] \; {}_{\xi}\!\int^{\infty} f(x) \; h(\xi) \; dx \; d\xi \; =$$

$$V[1 - \exp(-\lambda)] \; {}_{\bar{\xi}}\!\int^{\infty} f(x) \; dx \quad . \quad\quad\quad (2.7)$$

If the cumulative distribution functions of $e(x)$ and $f(x)$ are linear in ξ, equations (2.6) and (2.7) will hold by Jensen's inequality. Also suppose that the distribution of reservation wages is symmetric. Then equations (2.6) and (2.7) will hold if ${}_{0}\!\int^{\bar{\xi}-K} e(x) \; dx$ and ${}_{0}\!\int^{\infty} f(x) \; dx$ are symmetric with respect to $\bar{\xi}$. Although there are no a priori reasons to expect either of these conditions to hold, equations (2.6) and (2.7) should provide reasonable approximations.

In an economy with a fixed labor force (L) and a fixed number of jobs (J),

$$E = J - V = L - U \quad , \text{ and}$$

$$L - J = U - V = k \quad .$$

So k is the necessary level of unemployment when $U > V$, and is the necessary level of unfilled vacancies when $U < V$. Assuming that a quit generates a vacancy and that a new hire eliminates one, for a closed economy

$$\dot{V} = S - H \quad , \text{ and} \tag{2.8}$$

$$\dot{U} = S - H \quad . \tag{2.9}$$

Equations (2.2) thru (2.5), (2.8) and (2.9) now may be used with initial conditions V^0, U^0 ($U^0 - V^0 = k$) to specify the trajectories of vacancies and unemployment. In equilibrium, $\dot{V} = \dot{U} = 0$ implies that quits are equal to new hires. Hence,

$$s(L - U) \, {}_0\!\int^{\xi - K} e(x) \, dx = V[1 - \exp(-\lambda)] \, {}_\xi\!\int^{\infty} f(x) \, dx \quad .$$

Figure 2 illustrates the behavior of the system in the V–U plane. Starting from an initial V,U pair (e.g., A_1, A_2, A_3 or A_4), the system moves along the line $U - V = k$. Point B illustrates a stable equilibrium point, and point C an unstable one. The line BC represents the locus of potential equilibrium points. Determination of the slope of the equilibrium line is discussed in the next section. It is shown in Appendix A1 that an equilibrium solution is stable under the condition that new hires respond faster than separations to changes in unemployment and vacancies. In this situation, the equilibrium point (e.g., point B in Figure 2) will be restored following any disturbance characterized by $dV = dU$. If there is an exogenous change in either final demand or in the size of the labor force, a new equilibrium will be established (e.g., point D in Figure 2).

For the system of equations (2.3) and (2.4),

$$\partial H / \partial U =$$

$$\exp(-\lambda) \, {}_\xi\!\int^{\infty} f(x) d(x) + V[1 - \exp(-\lambda)][d/dU \, {}_\xi\!\int^{\infty} f(x) d(x)] > 0 ,$$

$$\partial S / \partial U = -s \, {}_0\!\int^{\xi} e(x) dx + s(L - U) \, d/dU \, {}_0\!\int^{\xi} e(x) dx < 0 ,$$

$$\partial H / \partial V =$$

$$[1 - \exp(-\lambda) - \lambda \exp(-\lambda)] \, {}_\xi\!\int^{\infty} f(x) dx +$$

$$V[1 - \exp(-\lambda)] \, d/dV \, {}_\xi\!\int^{\infty} f(x) dx \lesseqgtr 0 \quad , \text{ and}$$

$$\partial S / \partial V = s(L - u) \, d/dV \, {}_0\!\int^{\xi} e(x) dx > 0 \quad .$$

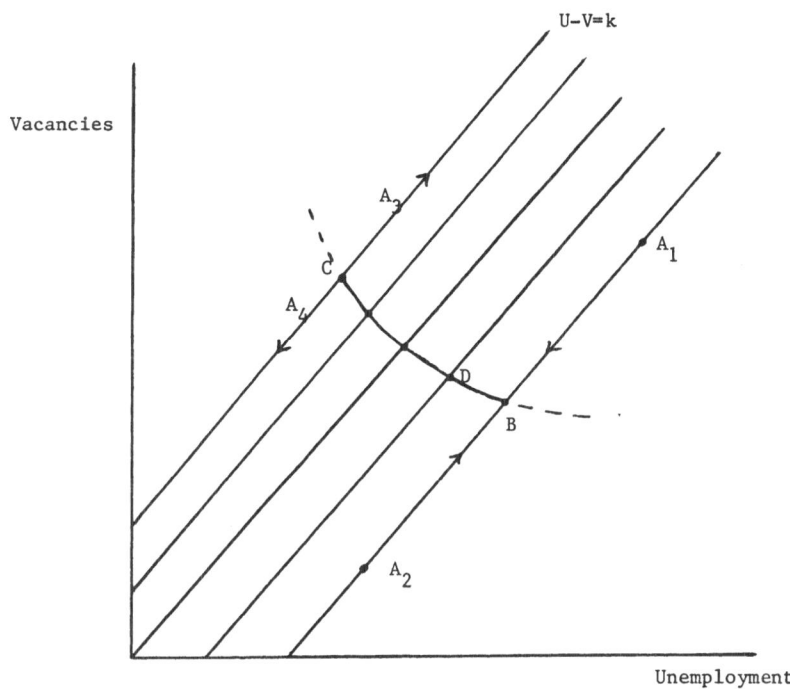

Figure 2. Equilibrium locus of unemployment and vacancies.

As unemployment increases, the number of vacancies contacted increases and job seekers' expectations decline, leading to an increase in the number of new hires. Two factors will lead to a decrease in separations as unemployment increases. First, employment decreases imply that even at the same separation rate there will be fewer total separations. Second, the increase in unemployment leads to higher competition, thereby lowering reservation wages and voluntary quits. As vacancies increase, the number of contacts increases, but expectations rise due to a decline in the level of competition. Increased reservation wages lead to a higher probability of rejecting offers, and therefore the net effect on new hires is indeterminate. If expectation increases are low relative to the increase in firms contacted, new hires will increase. An increase in vacancies leads to more voluntary quitting through the positive effect of vacancies on reservation wages.

Since $\partial H/\partial U - \partial S/\partial U > 0$, equation (A1.1) of the appendix implies that a sufficient condition for stability is that $\partial H/\partial V > \partial S/\partial V$. In other words, following an exogenous disturbance of unemployment and vacancies, the change in new hires is greater than the change in separations. An exogenous increase in V and U, for example, would lead to a greater increase in hires than in separations. By equations (2.8) and (2.9), this change would lead to downward pressure on U and V until the original equilibrium was restored.

3. THE VACANCY-UNEMPLOYMENT RELATIONSHIP

Intuitively, it may be argued that a relatively large number of job vacancies is an indication of an excess demand for labor. Hence, unemployment is likely to be relatively low. Conversely, when vacancies are relatively low, unemployment is likely to be relatively high. Considerable empirical evidence has been gathered to support a negative relationship between unemployment and vacancies, but there have been few attempts to derive such a relationship from more basic principles.

Holt (1978) argues that the V-U relationship may be derived by initially assuming that job separations are negatively related to job competition, such that

$$S = s E \lambda^{-q} \qquad , \qquad (3.1)$$

where λ is the ratio of unemployment to vacancies (U/V) and q is the elasticity of separation. This relationship reflects the fact that quits are relatively more important than layoffs in determining how separation rates change with changes in job competition. Holt takes new hires as positively related to both the number of vacancies and the number of unemployed, so that

$$H = h V^v U^{(1-v)} \qquad , \qquad (3.2)$$

where v is the elasticity of new hires with respect to the competition ratio, λ. For a given growth rate,

$$G = (H - S)/E \qquad .$$

It then may be shown that a negative relationship between U and V will result:

$$(V/E)^{v-q} (U/E)^{1-v+q} = (s + G)/h \qquad . \qquad (3.3)$$

Moreover, since the exponents sum to unity, then as the exponent for U increases, the one for V necessarily must decrease. Clearly, the derivation of this result relies on the assumption that $s \gg G$

and $q \ll 1$.

It is possible to derive more general conditions under which the negative relation between U and V will occur. Following Holt, let the separation and new hire functions be given by

$$S = S(V,U) \qquad and \qquad H = H(V,U) \qquad ,$$

with the following derivatives representing first-order conditions:

$$\partial S/\partial V > 0 \quad , \quad \partial S/\partial U < 0 \quad , \quad \partial H/\partial V > 0 \quad , and \quad \partial H/\partial U < 0 . \quad (3.4)$$

Assume that the rate of growth (G) associated with a fixed labor force (L) is equal to 0. In moving from one V–U pair to another, the change in new hires must equal the change in separations, so that

$$dH = dS \qquad . \qquad (3.5)$$

Furthermore, one may write

$$dS = \partial S/\partial V \, dV + \partial S/\partial U \, dU \qquad , and \qquad (3.6)$$

$$dH = \partial H/\partial V \, dV + \partial H/\partial U \, dU \qquad . \qquad (3.7)$$

Substituting equations (3.6) and (3.7) into equation (3.5) yields

$$dV/dU = [\partial S/\partial U - \partial H/\partial U]/[\partial H/\partial V - \partial S/\partial V] \qquad . \qquad (3.8)$$

Since the numerator on the right-hand side of equation (3.8) is always negative, for dV/dU to be negative it is necessary that $\partial H/\partial V > \partial S/\partial V$. As vacancies expand for a given level of unemployment, new hires will increase faster than separations. To maintain constant employment there must be a compensating decrease in unemployment, thereby cutting the growth of new hires and increasing separations.

In Figure 3, S_i and H_i represent lines of equal separations and new hires, respectively, so that

$$H_1 = S_1 < H_2 = S_2 < H_3 = S_3 \qquad .$$

Let point A represent the current state of the economy (V_1,U_1). In seeking a new state characterized by higher vacancies (V_2), for the given unemployment level U_1, the number of new hires will increase faster than the number of separations $(H_3 - H_1 > S_2 - S_1)$. To re-establish equilibrium (characterized by a constant employment level), there must be a decrease in unemployment, $U_2 U_1$, corresponding to the increase in vacancies, $V_2 V_1$. For the equations (3.3) and (3.4) given by Holt, the necessary conditon

358

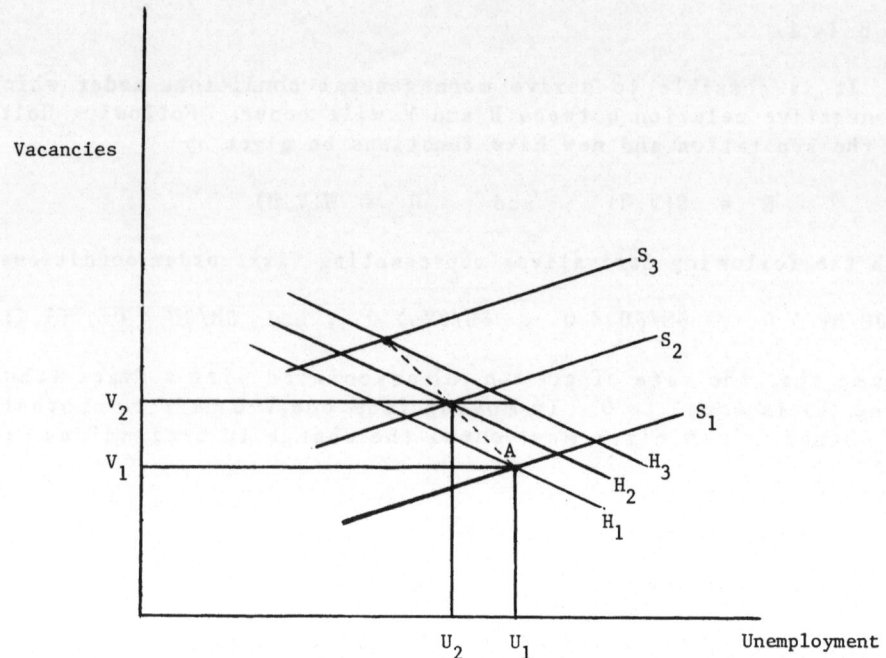

Figure 3. The vacancy-unemployment relationship.

$\partial H / \partial V > \partial S / \partial V$ reduces to

$$q \; < \; h \, v \, U \, \lambda^{q-v}/(s \, E) \; = \; h \, v \, U \, \lambda^{q-v}/[s(1 - u)] \qquad ,$$

where u is the unemployment rate. At low rates of unemployment, the elasticity of separation, q, must be relatively small. For example, if $h = 0.2$, $v = 0.5$, $s = 0.01$, $\lambda = 1$ and $U/E = 0.05$, q must be less than 0.5.

The functional forms of the new hire and separation relationships given by Holt are largely ad hoc in the sense that they are not governed by any specified theory of individual behavior. As an alternative, consider the following market equilibrium given by Mortensen (1973) and by Rogerson and MacKinnon (1981):

$$\delta(L - U) \; = \; V[1 - \exp(-\lambda)] \qquad , \qquad (3.9)$$

where δ is the exogenous separation rate and $V[1 - \exp(-\lambda)]$ is the number of new hires resulting from random matching of jobs with searchers. The first-order conditions (3.4) hold, with the exception that separations are independent of vacancies ($\partial S/\partial V = 0$). Again $\partial H/\partial V > \partial S/\partial V$ implies a negative relationship between vacancies and unemployment. One should note that in both this and Holt's formulations, the negative V–U curve consists of stable equilibrium points. For the search and competition model, the slope of the V–U line will depend upon the sign and magnitude of $\partial H/\partial V$. For $\partial H/\partial V$ positive and greater than $\partial S/\partial V$, a negative relationship between unemployment and vacancies will result. If $\partial H/\partial V$ is less than $\partial S/\partial V$, the relationship will be positive.

4. WAGE INFLATION

An increase in vacancies may lead to an increase in frictional unemployment of voluntary job seekers with no wage inflation. The decline in competition entices others into the market and firms will find that job seekers are more selective in accepting offers. There actually may be fewer new hires, with the decreased satisfaction from search accompanying the increase in expectations more than compensating for the increase in jobs. To increase the flow of new hires to firms, jobs must be offered at higher wages. Following an approach similar to that of Mortensen (1970, 1973), let new hires be given by

$$ H = V[1 - \exp(-\lambda)] \int_{\xi\exp(hg')/\exp(hg)}^{\infty} f(x)\, dx \quad , $$

where g' is the expected inflation in wage offers whereas g is the actual inflation of offers, and h denotes the length of the time period. The reservation wage after accounting for expected inflation is $\xi\exp(hg')$. Since wages actually increase by $\exp(hg)$ over the period, if $g > g'$ the number of new hires will be greater than the number of new hires that would result under conditions of perfectly anticipated inflation.

Likewise, if employee wages also increase at a rate, say g, the number of quits may be given by

$$ S = s\, E_0 \int_0^{\xi\exp(hg')/\exp(hg)} e(x)\, dx $$

$$ = s\, E_0 \int_0^{\xi\exp[h(g'-g)]} e(x)\, dx \quad . $$

When the actual inflation rate exceeds the anticipated rate, quits will be lower than they would be under conditions of correctly anticipated inflation.

Exogenous changes in product demand lead firms to adjust their employment by simultaneously setting wage offers and vacancy levels. This dual adjustment mechanism is emphasized by Holt

360

(1978). The changes in new hires and separations for such exogenous shocks are, respectively, as follows:

$$dH = \partial H/\partial g \ dg + \partial H/\partial V \ dV + \partial H/\partial U \ dU \quad \text{, and}$$

$$dS = \partial S/\partial g \ dg + \partial S/\partial V \ dV + \partial S/\partial U \ dU \quad .$$

Figure 4 shows that the equilibrium locus of unemployment and vacancies will shift from AB to A'B' as the rate of inflation increases. This figure presents an equilibrium plane in the 3-space of inflation, unemployment and vacancies. For a given level of vacancies, the equilibrium relationship between inflation and unemployment is given by

$$dU/dg = [\partial H/\partial g - \partial S/\partial g]/[\partial S/\partial U - \partial H/\partial U] < 0 \quad . \quad (4.1)$$

A Phillips curve therefore is obtained along the line CD in Figure 4. More generally, allowing vacancies to vary, equation (4.1) may

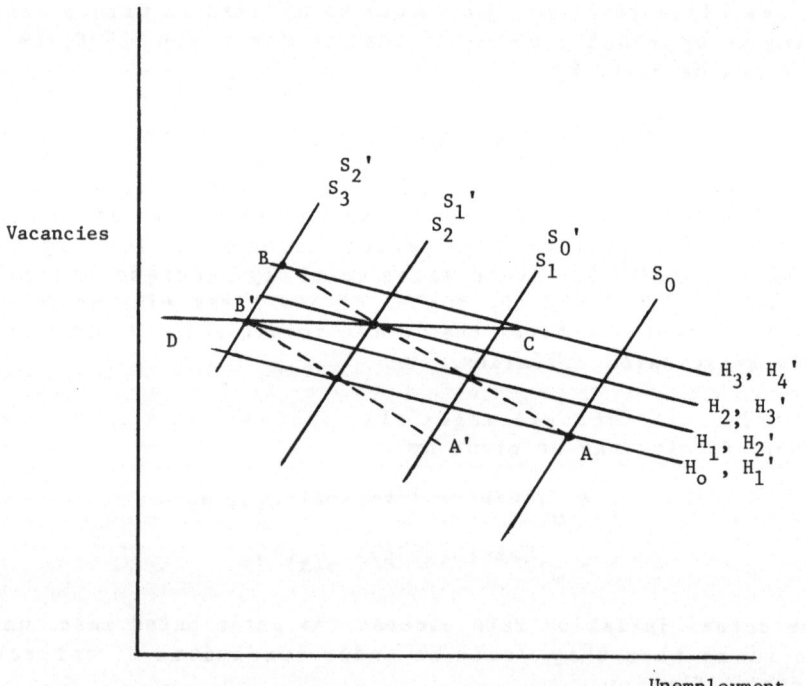

Figure 4. Effects of wage inflation on hires, quits, vacancies, and unemployment.

be expressed as

$$dU/dg = [\partial H/\partial g - \partial S/\partial g]/[\partial S/\partial U - \partial H/\partial U] +$$

$$[\partial S/\partial V - \partial H/\partial V]/[\partial H/\partial U - \partial S/\partial U](dV/dg) \quad .$$

Figure 5 shows that with an increase in demand, given an initial equilibrium A, a new, higher equilibrium employment level may be achieved through either wage increases (point B), by increasing the difference between the number of jobs and the size of the labor force (point C), or some combination of both (along the line BC).

5. SEARCH BEHAVIOR IN A MULTIREGIONAL ENVIRONMENT

Little has been done to examine search models in a spatial setting. Exceptions include the recent work of Weibull (1978), Smith et al. (1979), and Rogerson (1982b). Holt (1970) has shown

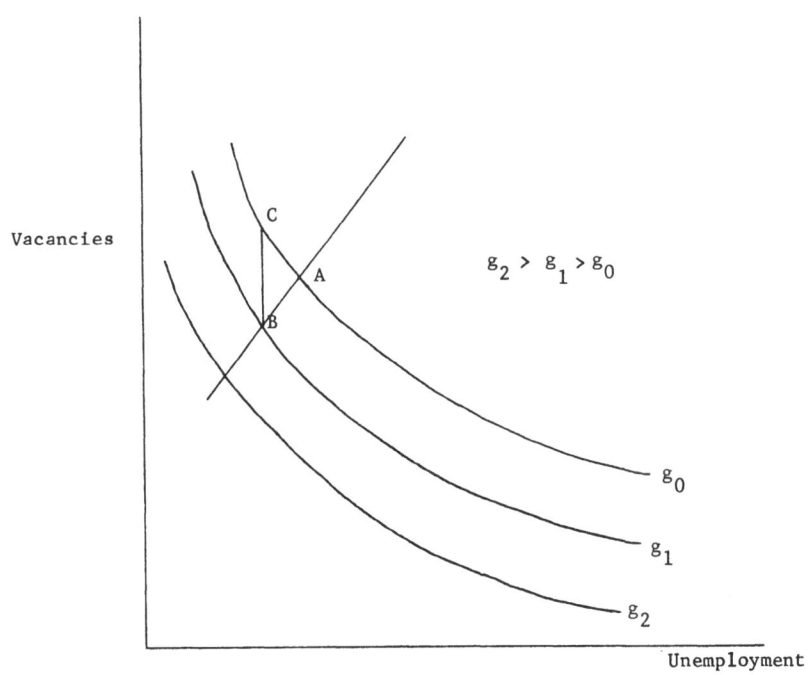

Figure 5. Equilibrium adjustment through wage and job changes.

that when markets are segmented and do not interact, aggregate levels of unemployment and vacancies will be higher than when they do interact. His analysis is based upon the hiring function $H = hUV$. He finds that segmenting the market into N regions with equal numbers of unemployment and vacancies cuts the total number of new hires linearly, by a factor of N. Total unemployment and vacancies would be increased by a factor of $N^{1/2}$. His result is dependent upon the fact that halving the number of vacancies and the number of unemployed decreases the number of new hires by a factor of four. This analysis is incomplete, since it does not consider job separations and the resulting equilibrium U-V point. To illustrate, take the relationships given by equation (3.9) and assume a two-region system with

$$U_1 = 1000 \quad , \quad U_2 = 1405 \quad , \quad V_1 = 1500 \quad , \quad V_2 = 905 \quad ,$$

$$L_1 = 20000 \quad , \quad L_2 = 20000 \quad , \quad S_1 = 0.0384 \quad , \text{ and}$$

$$S_2 = 0.0384 \quad .$$

Both regions are in equilibrium, with $H_1 = S_1 = 730$, and $H_2 = S_2 = 710$. The unemployment rates of regions 1 and 2 are $U_1/L_1 = 0.05$ and $U_2/L_2 = 0.0702$, respectively. The aggregate unemployment rate for the system is $(U_1 + U_2)/(L_1 + L_2) = 0.0601$. If mobility, information and interaction between the two regions were perfect, unemployment and vacancies each initially would equal 2405, but new hires would exceed separations ($H = 1520$, $S = 1444$). The aggregated one-region system would settle at the equilibrium point where $U = V = 2291$ and $H = S = 1448$. The new equilibrium unemployment level is $2291/40000 = 0.0583$, which is less than that in the segmented market case.

To illustrate in a more general manner the effect of segmentation on U and V, it is convenient to solve equation (3.9) for V explicitly, using the fact that $U - V = k$. Using the first three terms in the expansion of $\exp(-\lambda)$, V may be solved for approximately by using the quadratic formula

$$V =$$

$$\frac{\delta(L-k) - k/e \pm \{[k/e - \delta(L-k)]^2 + 2(1 + \delta + 1/e)k^2/e\}^{1/2}}{2(1 + \delta + 1/e)} \quad . \quad (5.1)$$

If equation (5.1) is used for each of n non-interacting regions, aggregate vacancies are given by

$$\sum_{i=n}^{i=n} V_i = [\delta(L - k) - k/e]/[2(1 + \delta + 1/e)] \pm$$

$$\sum_{i=1}^{i=n} \{[k_i/e - \delta(L_i-k_i)]^2 + 2(1 + \delta + 1/e)k_i^2/e\}^{1/2}/[2(1 + \delta + 1/e),$$
(5.2)

where $L' = \sum_{i=1}^{i=n} L_i$ and $k = \sum_{i=1}^{i=n} k_i$. Note that since only one root

$$2(1 + \delta + 1/e)k^2/e > 0 \quad ,$$

of equations (5.1) and (5.2) will be positive. The difference in aggregate vacancies between the non-interacting and interacting cases, then, is

$$\sum_{i=1}^{i=n} V_i \; - \; V \; =$$

$$\sum_{i=1}^{i=n} \{[k_i/e - \delta(L_i-k_i)]^2 + 2(1 + \delta + 1/e)k_i^2/e\}^{1/2}/[2(1 + \delta + 1/e)$$

$$- \{[k/e - \delta(L-k)]^2 + 2(1 + \delta + 1/e)k^2/e\}^{1/2}/[2(1 + \delta + 1/e)].$$
(5.3)

This function is minimized when the k_i are equal for all n regions. Figure 6 shows the results of simulating equation (5.3) under a variety of conditions. As regional variation in economic conditions increases, the detrimental effect on aggregate unemployment and vacancies of non-interacting markets becomes more pronounced. For a given amount of regional variation, the effect of market segmentation or low mobility becomes more pronounced as the system-wide excess of unemployment over vacancies increases. One should note that when $k_i = k/N$ for all regions, enhanced mobility does not reduce the aggregate unemployment and vacancy rate.

Search theory may be generalized to include regional variations in wage distributions (Rogerson, 1982b). It is assumed that wage distributions, say $f_i(x)$, are independent and time invariant, and that the cost to an individual of generating an offer from j, say c_{ij}, is known. Let p_{ij} be the probability that an individual in location i applies for a job in location j. The probability of generating an offer, conditional upon applying in j, is p_j, where from equation (2.5)

$$p_j = [1 - \exp(-\lambda_j)]/\lambda_j \quad , \tag{5.4}$$

where $\lambda_j = \sum_{i=1}^{i=n} U_i \, p_{ij}/V_j$. The optimal individual strategy is to

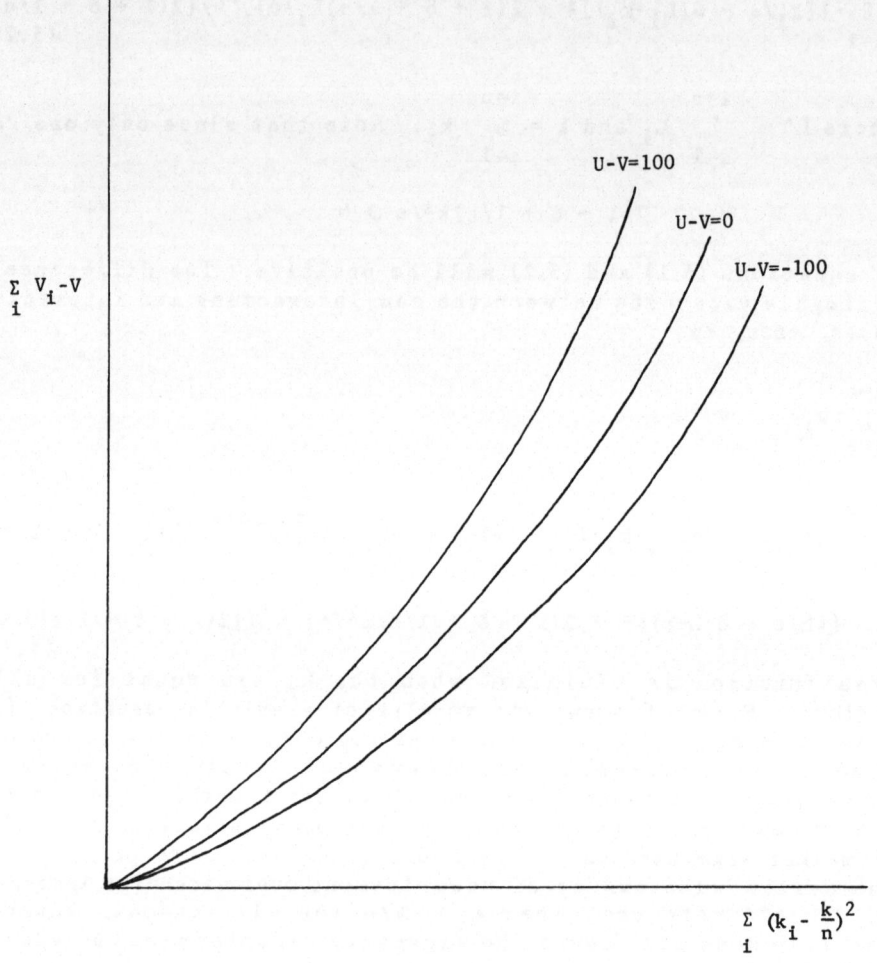

Figure 6. Effects of market segmentation on aggregate vacancies.

calculate a reservation wage of

$$\xi_i = \sum_{i=1}^{i=n} -c_{ij} \, p_{ij} / (\sum_{i=1}^{i=n} p_{ij} \, p_j) \ +$$

$$\sum_{i=1}^{i=n} \{[p_{ij} \ p_j/(\sum_{i=1}^{i=n} p_{ij} \ p_j][\xi_i \ F_j(\xi_i) + {}_{\xi_i}\!\!\int^{\infty} x \ dF_j(x)]\} \ . \qquad (5.6)$$

The first term on the right-hand side of equation (5.6) gives the expected cost of search. The second term is a weighted average of the expected return from each region, where the weights are the conditional probabilities of receiving offers from particular destinations. Hiring and separation functions may be given as follows:

$$H_j = \{V_j[1 - \exp(-\lambda_j)]/[\sum_{i=1}^{i=n} U_i \ p_{ij}]\} \sum_{i=n}^{i=n} u_i \ p_{ij} \ {}_{\xi_i}\!\!\int^{\infty} dF_j(x) \ , \text{ and} \qquad (5.7)$$

$$S_j = s \ (L_j - U_j) \ {}_0\!\!\int^{\xi_j} dF_j(x) \ . \qquad (5.8)$$

The change in unemployment then is

$$\dot{U}_j = S_j - \sum_{i=1}^{i=n} M_{ji} \ , \qquad (5.9)$$

where M_{ji} is the number of unemployed individuals in location j that have accepted employment in location i, namely

$$M_{ij} = U_i \ p_{ij} \ p_j \ {}_{\xi_i}\!\!\int^{\infty} dF_j(x)$$

$$= [U_i \ p_{ij}/(\sum_{i=1}^{i=n} U_i \ p_{ij}) \ V_j \ [1 - \exp(-\lambda_j)] \ {}_{\xi_i}\!\!\int^{\infty} dF_j(x) \ .$$

Finally, the change in vacancies is

$$\dot{V}_j = S_j - H_j \ . \qquad (5.10)$$

The system of equations (5.5) thru (5.10) may be interpreted as the multiregional version of the search and competition model specified in the second section of this paper. Equilibrium is characterized by $S_j = H_j = \sum_{i=1}^{i=n} M_{ji}$. Detailed examination of the dynamic properties of this system is beyond the scope of the present paper and is the focus of current work. A comparative static analysis of this spatial search model is summarized in Rogerson (1982b).

6.   PARETO–DISTRIBUTED WAGES

The Pareto distribution often is used to describe income distributions (Cowell, 1977) and has the desirable feature that the reservation wage may be solved for directly (Rogerson, 1982a). The distribution is given by

$$f(x) = \alpha \, \beta^{\alpha} \, x^{-(\alpha+1)} \quad , \quad \alpha > 0 \text{ and } \beta < x < \infty \, ,$$

where α and β are parameters to be estimated. The corresponding cumulative distribution function is

$$F(x) = 1 - (\beta/x)^{\alpha} \quad .$$

The assumption that wage offers are Pareto–distributed allows the following explicit solution to be obtained for equation (2.2) for the reservation wage:

$$\xi = [c \, (\alpha - 1)/(p \, \beta^{\alpha})]^{1/(1-\alpha)} \quad . \tag{6.1}$$

Substituting equation (2.5) into equation (6.1) results in the reservation wage of

$$\xi = \{[c \, (\alpha - 1) \, \lambda]/[\beta^{\alpha} \, (1 - \exp(-\lambda))]\}^{1/(1-\alpha)} \quad .$$

In Appendix A2 derivatives of new hires and separations with respect to unemployment and vacancies are given. It is shown that $\partial H/\partial V < 0$. Stability of the equilibrium requires that

$$\alpha \, s \, E \, \{(\lambda-1)[1 - \exp(-\lambda)] - (\lambda^2-\lambda)\exp(-\lambda)\}/\{V[1 - \exp(-\lambda)]^2 - s$$

$$< \{(\alpha \, V - \lambda)[1 - \exp(-\lambda)] - (\lambda \, V - \lambda^2)\exp(-\lambda\} \quad . \tag{6.2}$$

When $\lambda = 1$, it may be shown that equation (6.2) holds. For $\lambda \neq 1$, it is more difficult to interpret equation (6.2) directly, but simulation experiments have shown equilibrium points to be stable for a wide range of parameters.

The negative relationship between new hires and vacancies has the interesting implication of a positive relationship between vacancies and unemployment when offers and wages are Pareto–distributed (see Figure 7). This finding does not contradict previous theory and empirical observation, as it may initially appear to. It should be kept in mind that in the search and competition model, equilibrium unemployment is a result of voluntary quits. Regions with many vacancies will be characterized by relatively high quit rates, and therefore will have a relatively long queue of voluntary unemployment. One should note from Figure 7 that regions with low vacancies and unemployment are characterized by low labor turnover, and that regions with high

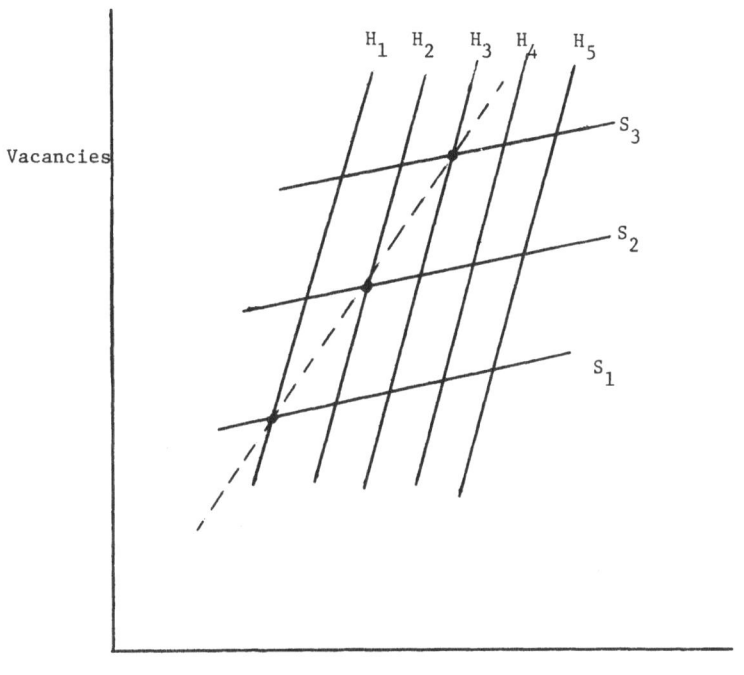

Figure 7. Vacancies and unemployment in the search and competition model with Pareto-distributed wages.

vacancies and unemployment have a higher level of turnover.

7. SUMMARY

A major objective of this paper has been to suggest how several elements of search theory could be synthesized to examine the macro characteristics of regional labor markets. In particular, search models may be combined with market competition models to determine simultaneously the equilibrium levels of vacancies and unemployment. Under the condition that new hires respond less (more) quickly than separations to exogenous changes in vacancies, a positive (negative) relationship between vacancies and the queue of involuntary unemployment exists. The level of turnover in a region is positively related to its equilibrium level of vacancies and unemployment. To specify these relationships more directly, the Pareto distribution may be used; it is both analytically tractable and empirically supported.

When the positive effect of wage offer inflation on new hires and the negative effect of wage inflation on separations are accounted for, an equilibrium plane of triplets (U*,V*,g*) may be derived. Under quite general conditions, this plane is characterized by a Phillips curve, which demonstrates that the micro-level theory being developed here is consistent with an aggregate Phillips relation. Lack of regional mobility will lead to higher aggregate levels of unemployment and persistence of regional differences in labor surpluses and deficiencies. How regional labor market conditions evolve as a consequence of labor force migration may be examined by combining the single region search and competition model with spatial models of job search.

There is a strong need to incorporate the decision as to where to search within the spatial model. In the current formulation, people move based upon their success in finding jobs in regions with varying economic conditions, and they base their decision of when to search upon market expectations. Where they search is the result of an exogenously specified stochastic process. More realistically, the location of search is dependent upon both the information transmitted by recent migrants, friends and relatives and current economic conditions.

8. REFERENCES

Butters, G., 1977, Equilibrium Distributions of Sales and Advertising Prices, Review of Economic Studies, 44: 465-491.

Cowell, F., 1977, Measuring Inequality: Techniques for the Social Sciences, New York: Wiley.

Eaton, B. and M. Watts, 1977, Wage Dispersion, Job Vacancies, and Job Search in Equilibrium, Economica, 44: 23-35.

Holt, C., 1970, How Can the Phillips Curve be Moved to Reduce Both Inflation and Unemployment, in Microeconomic Foundations of Employment and Inflation Theory, edited by E. Phelps, New York: Norton and Co., pp. 224-256.

_____, 1978, Wages and Job Availability in Segmented Labor Markets, paper presented at a Conference held by the International Economic Association on 'Unemployment in Western Countries Today,' Strasbourg, France.

Kormendi, R., 1979, Dispersed Transactions Prices in a Model of Decentralized Pure Exchange, in Studies in the Economics of Search, edited by S. Lippman and J. McCall, New York: North-Holland, pp. 53-81.

Lippman, S. and J. McCall, 1976, The Economics of Job Search: A Survey, *Economic Inquiry*, 34: 155–189.

Mortensen, D., 1970, A Theory of Wage and Employment Dynamics, in *Microeconomic Foundations of Employment and Inflation Theory*, edited by E. Phelps, New York: Norton and Co., pp. 167–211.

Mortensen, D., 1973, Job Matching Under Imperfect Information, in *Evaluating the Labor–Market Effects of Social Programs*, edited by O. Ashenfelter and J. Blum, Princeton: Princeton University Press, pp. 194–231.

Rogerson, P., 1982a, Job Search and Competition, *Proceedings* of the Thirteenth Annual Conference on Modeling and Simulation, Pittsburgh, 13: 1039–1043.

_____, 1982b, Spatial Models of Search, *Geographical Analysis*, 14: 217–228.

_____ and R. MacKinnon, 1981, A Geographical Model of Job Search, Migration, and Unemployment, *Papers* of the Regional Science Association, 48: 89–102.

Smith, T., W. Clark, J. Huff and P. Shapiro, 1979, A Decision-making and Search Model for Intraurban Migration, *Geographical Analysis*, 11: 1–22.

Weibull, J., 1978, A Search Model for Microeconomic Analysis—With Spatial Applications, in *Spatial Interaction Theory and Planning Models*, edited by A. Karlqvist *et al.*, New York: North-Holland, pp. 47–73.

9. APPENDIX A1: STABILITY CONDITIONS

Stability conditions for the equilibrium point may be derived by examining the eigenvalues of the matrix of the first partial derivatives of the system:

$$F(U,V) = \dot{U} = S(U,V) - H(U,V)$$
$$G(U,V) = \dot{V} = S(U,V) - H(U,V)$$

The characteristic equation is

$$\det \begin{pmatrix} \partial F/\partial U - \lambda & \partial F/\partial V \\ \partial G/\partial U & \partial G/\partial V - \lambda \end{pmatrix} = 0 \quad ,$$

or $\lambda(\lambda - \partial F/\partial U - \partial G/\partial V) = 0.$

The eigenvalues of this characteristic equation are.

$$\lambda_1 = 0 \quad , \text{ and}$$

$$\lambda_2 = (\partial S/\partial U - \partial H/\partial U) + (\partial S/\partial V - \partial H/\partial V) \quad .$$

Stability requires that λ_2 be negative. Therefore it is required that

$$\partial S/\partial V - \partial H/\partial V < \partial H/\partial U - \partial S/\partial U \qquad (A1.1)$$

at the equilibrium. Equation (A1.1) may be written as

$$\partial S/\partial U + \partial S/\partial V < \partial H/\partial U + \partial H/\partial V \quad . \qquad (A1.2)$$

Since any disturbance from the equilibrium is governed by the constraint $dU = dV$, equations (A1.1) and (A1.2) imply that the change in separations must be less than the changes in new hires, such that

$$dS < dH \quad .$$

10. APPENDIX A2: COMPARATIVE STATIC ANALYSIS FOR PARETO-DISTRIBUTED WAGES

When separations and new hires are given by equations (2.3) and (2.4), respectively, and wages and offers are Pareto-distributed, the first-order conditons given by equation (3.4) are satisfied, with the exception that $\partial H/\partial V < 0$. Hence,

$$\partial S/\partial V =$$

$$-s \, E \, \beta^\alpha \, [c(\alpha-1)/\beta^\alpha]^{\alpha/(\alpha-1)} \, (\partial/\partial V) \, \{\lambda/[1 - \exp(-\lambda)]\}^{\alpha/(\alpha-1)} =$$

$$\Omega_s \, [\alpha\lambda/(\alpha-1)]\{\lambda/[1 - \exp(-\lambda)]\}^{1/(\alpha-1)}\{\lambda\exp(-\lambda) -$$

$$[1 - \exp(-\lambda)]\}/\{V[1 - \exp(-\lambda)]^2\} \quad ,$$

where the following conditions hold:

$$\Omega_s = -s \, E \, \beta^\alpha \, [c(\alpha-1)/\beta^\alpha]^{\alpha/(\alpha-1)} \quad ,$$

$$\partial S/\partial V > 0, \text{ since } 1 - \exp(-\lambda) > \lambda\exp(-2) \quad ,$$

$$\partial S/\partial U = \Omega_s \, \partial/\partial U \, \{\lambda/[1 - \exp(-\lambda)]\}^{\alpha/(\alpha-1)} -$$

$$s\{1 - \beta^\alpha[c\lambda(\alpha-1)/(\beta^\alpha(1 - \exp(-\lambda)))]\}^{\alpha/(\alpha-1)}$$

$$= \Omega_s[\alpha/(\alpha-1)]\{\lambda/[1-\exp(-\lambda)]\}^{1/(\alpha-1)}$$

$$\{V[1-\exp(-\lambda)]-U\exp(-\lambda)\}/\{V^2[1-\exp(-\lambda)]^2\} \quad -$$

$$s\{1 - \beta^\alpha[c\lambda(\alpha-1)/(\beta^\alpha(1 - \exp(-\lambda)))]^{\alpha/(\alpha-1)}\} \quad ,$$

and $\partial S/\partial U < 0$, since $V[1 - \exp(-\lambda)] > U \exp(-\lambda)$. Now

$$\partial H/\partial U \; =$$

$$\beta^\alpha [c(\alpha-1)/\beta^\alpha]^{\alpha/(\alpha-1)} \; \partial/\partial U\{V[1-\exp(-\lambda)]\{\lambda/[1 - \exp(-\lambda)]\}^{\alpha/(\alpha-1)}\}$$

$$= \; V\Omega_H \; [\lambda^{\alpha/(\alpha-1)}/(1-\alpha)][1 - \exp(-\lambda)]^{\alpha/(1-\alpha)} \; \exp(-\lambda) \; +$$

$$[\alpha/(\alpha-1)][1 - \exp(-\lambda)]^{1/(1-\alpha)} \; \lambda^{1/(\alpha-1)} \quad ,$$

where the following conditons hold

$$\Omega_H \; = \quad \beta^\alpha \; [c(\alpha-1)/\beta^\alpha]^{\alpha/(\alpha-1)} \quad ,$$

$$\partial H/\partial U = V\Omega_H\lambda^{1/(\alpha-1)}[1-\exp(-\lambda)]^{\alpha/(1-\alpha)}\{\alpha[1-\exp(-\lambda)]-\lambda\exp(-\lambda)\} \; ,$$

$$\partial H/\partial U > 0, \text{ since } [1 - \exp(-\lambda)] > \lambda \exp(-\lambda) \quad \text{and} \quad \alpha > 1.$$

$$\partial H/\partial V = \Omega_H \; \partial/\partial V \; \{V[1 - \exp(-\lambda)]^{1/(1-\alpha)} \; \lambda^{\alpha/(\alpha-1)}\}$$

$$= \; \Omega_H \; \{V[1-\exp(-\lambda)]^{1/(1-\alpha)}[\alpha/(\alpha-1)]\lambda^{1/(\alpha-1)}(-UV^{-2}) \; +$$

$$\lambda^{1/(\alpha-1)}\{[V/(1-\alpha)][1-\exp(-\lambda)]^{\alpha/(1-\alpha)}[-UV^{-2}\exp(-\lambda)$$

$$+ \; [1-\exp(-\lambda)]^{1/(1-\alpha)}\}$$

$$= \; \Omega_H \; \lambda^{\alpha/(\alpha-1)}[1-\exp(-\lambda)]^{\alpha/(1-\alpha)}\{[1-\exp(-\lambda)] \; -$$

$$\alpha[1-\exp(-\lambda)]/(\alpha-1) \; - \; \exp(-\lambda)/(1-\alpha)\}$$

$$= \; \Omega_H \; \{\lambda^{\alpha/(\alpha-1)}[1-\exp(-\lambda)]^{\alpha(1-\alpha)}/(\alpha-1)\}\{\lambda \; \exp(-\lambda) \; -$$

$$[1-\exp(-\lambda)]\} \; < \; 0 \quad .$$

STRUCTURES FOR RESEARCH ON THE DYNAMICS OF RESIDENTIAL MOBILITY

W. A. V. Clark

University of California, Los Angeles
United States

1. INTRODUCTION

The exponential growth rate of new published and unpublished research on residential mobility has created a situation in which synthesis of existing models is at least as important as the creation of new models. The proliferation of different models and different research strategies calls for an organization and evaluation of the literature, a clarification of the research accomplished to date, and the identification of problems for research in the future. The focus on an examination of evolving geographical structures by this NATO Advanced Study Institute serves as an appropriate context for such an organizational approach. Two recent papers (Clark, 1981, 1982) have reviewed current research on migration and mobility. In these reviews, the major concern was with the way in which research could be classified rather than with the organization of a modelling strategy. That is, there was more of a concern with arranging research topics under general classifications than considering the elements of a research paradigm. The approach of this paper is with the research paradigm itself, with the mobility, housing search, and relocation outcomes of individual households.

This paper is organized in the following manner. First, a synthesis of the work that has been accomplished in the past five years by the research group at UCLA and UC/Santa Barbara will be presented, set within a framework of what appear to be the important questions being posed in research on the dynamics of mobility. The second part of the paper will evaluate other research contributions to the mobility paradigm. Finally, a comparison will be made between the structural forms used to model mobility, and then the limits of dynamic models will be reexamined, especially with reference to the debates about the behavioral and structural approaches to mobility.

2. A SYNTHESIS OF RESEARCH STRATEGIES

For the past half dozen years, the research group at UCLA and

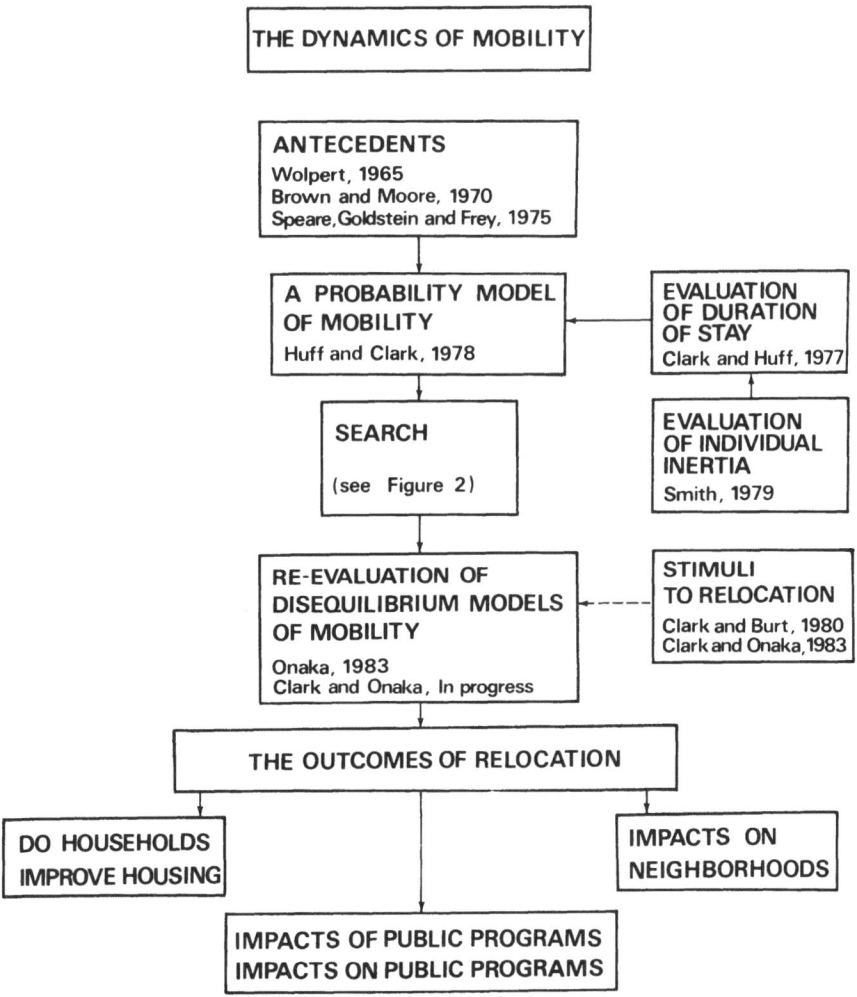

Figure 1. A guide to the literature on the dynamics of mobility.

Santa Barbara has undertaken a series of studies designed to develop a coherent approach to residential relocation behavior. These studies have been designed to answer specific questions about the dynamics of residential mobility, which include the probability or relocation, the process of relocation or search, and the actual move decision. Several discussions of mobility have suggested that there is a pre—move stage (the probability of a move), an active search and evaluation stage, the decision to move and the outcomes of the move. For convenience, the diagrammatic presentation is divided into two figures. Figure 1 includes the probability of a move and the decision to move, and Figure 2 (which in the paradigm is embedded in Figure 1) covers the search process. There is increasing agreement in the literature that a simple division between the decision to move and the search process is a useful simplifying schema for structuring mobility research. As will be noted later, some economic models evaluate the probability of search and movement, where a disequilibrium generates search, and following search—a move, although search might be rejected because of the high costs associated with search.

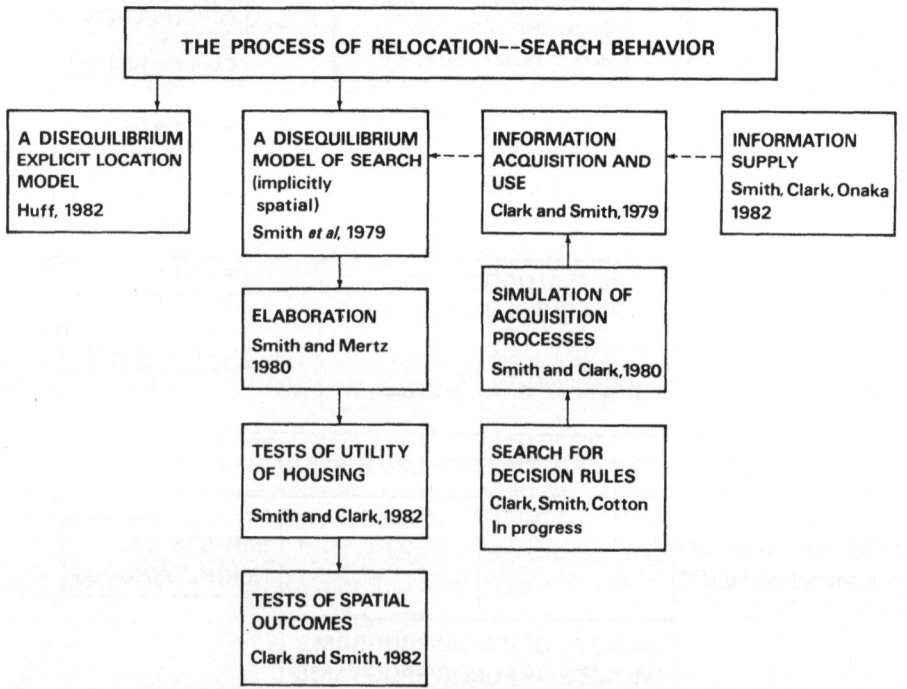

Figure 2.   An overview of the literature on search behavior and the process of relocation.

2.1. Models of the Decision to Move

Most of the research on the dynamics of mobility can be traced from attempts to relate the notions of dissatisfaction to the probability of moving. Initial approaches to understanding residential mobility by Wolpert (1965), Brown and Moore (1970), and Speare, Goldstein and Frey (1975) emphasized the stimulus of dissatisfaction (generated by both family circumstances and the housing environment) in relocation behavior.

While Wolpert, and Brown and Moore outlined structures for models of the relocation process, Speare, Goldstein and Frey attempted a complete symbolic specification. Central to their model is the notion of a threshold. A person or household will consider moving, in this model formulation, only if the amount of satisfaction falls below a threshold. Their modelling strategy includes two components. First is the following logit formulation of the decision to consider moving:

$$P_c = \{1 + \exp[k(s-t)]\}^{-1} \quad , \qquad (2.1)$$

where P_c = the probability of considering moving,

s = the residential satisfaction of a current location,

t = the threshold for dissatisfaction, and

k = a constant for the nature of transition.

Second is the following linear formulation of the decision to move or stay, which involves a comparison of residential satisfaction at the place of origin with the expected level of satisfaction at the alternative destination selected by the search process:

$$P_{m.c} = k \, (sd - so) - c \quad , \qquad (2.2)$$

where $P_{m.c}$ = the probability of moving for those who consider moving,

sd = the expected level of satisfaction at a destination,

so = the residential satisfaction at an origin,

c = the cost of moving, and

k = a constant to convert satisfaction into units comparable to the cost of moving.

In their final specification, summarized here in the following equation, these researchers noted that it is difficult, if not impossible, to combine the formulations for the stages of mobility decision making. Hence, they have suggested the following simplified model, in which the probability of moving depends upon both the relative residential satisfaction and the costs of moving:

$$P_m = f(s' - hc) \quad , \tag{2.3}$$

where P_m = the probability of moving,

   s' = the relative residential satisfaction,

   c = the cost of moving, and

   h = a constant.

The antecendents, especially the Speare formulation, influenced the development of a model in which there was a trade-off between dissatisfaction, on the one hand, and inertia, from increasing lengths of residence at one location, on the other (Huff and Clark, 1978). The level of dissatisfaction is assumed to increase at a decreasing rate over time as the household falls out of adjustment with its environment, changes its expectations, or has changes in its family composition. At the same time, there are factors that deter households from moving. This inertia also is assumed to increase over time, at a decreasing rate. The balance of dissatisfaction and inertia generates a propensity to move as a function of length of stay (Huff and Clark, 1978). Thus,

$$p(t) = \begin{cases} k[S(t) - R(t)] & , \text{ if } S(t) > R(t) \\ 0 & , \text{ if } S(t) \leq R(t) \end{cases} \quad , \tag{2.4}$$

where $S(t)$ = the pressure to move at time t, and

   $R(t)$ = the resistance to move at time t.

While this model is quite similar to that of Speare, Goldstein and Frey (1975), it requires specifications for both $S(t)$ and $R(t)$. Two exponential functions are utilized to provide estimates of the following probability of moving over time:

$$p(t) = k\{(\tilde{S} - \tilde{R}) - [\tilde{S} - S(0)] \exp(-\sigma t) + [\tilde{R} - R(0)] \exp(-\rho t)\} \quad , \tag{2.5}$$

where $\tilde{R}$ = the maximum resistance, and

   $\tilde{S}$ = the maximum stress.

The model as developed is a black box model. Inertia can be

specified reasonably well. In fact, it seems clear from other research that even though there are few, if any, duration-of-stay effects on an aggregate population (Clark and Huff, 1977; Pickles, et. al., 1982), the evaluation of individual inertia and risk-taking does emphasize that inertia on an individual basis (seen as risk) is an important variable in the individual mobility decision-making process (Smith, 1980). Specifications for stress are more speculative.

Economists have been impatient with the general dissatisfaction models, because of the rather vague concepts of dissatisfaction and stress contained in these models(Weinberg, Friedman and Mayo, 1981). However, economists have utilized what is a very similar conceptual approach, but couched in the specific terms of housing dissatisfaction. In fact, the following housing expenditure disequilibrium model represented by Hanushek and Quigley (1978a, 1978b), and developed by a number of other economists, is focused specifically on both the housing market and the difference between the actual and optimal amounts of housing consumed:

$$P_{t,t+1} = f[(H_{t+1}^d - H_t)/H_t^d] \quad , \quad (2.6)$$

where H_t = the actual housing consumption,

H_t^d = the equilibrium housing consumption at time t, and

H_{t+1}^d = the equilibrium housing consumption at time t+1.

Mobility is seen as a response to the change in the demands for housing and services. In this sense the models are derivative from Rossi's (1955) initial suggestions. Even so, the above equation is concerned only with the stress component of the more general concept suggested in the cumulative inertia–cumulative stress model.

Recently we have attempted to extend the disequilibrium model (Onaka, 1983; Onaka and Clark, in progress). The models presented in these papers are designed to provide a better empirical foundation for the housing disequilibrium research that utilizes the insights of both the cumulative inertia–cumulative stress models and the housing disequilibrium models. The research in Onaka (1983) and proposed in Onaka and Clark is based upon a model that extends the concept of housing consumption disequilibrium to include dissatisfaction with individual attributes of housing, hence going beyond a single index of housing services. In particular, this model gets around the assumption of Hanushek and Quigley (1978a, 1978b) that the socioeconomic characteristics of the household are reflected only in the equilibrium housing expenditure. Presumably there are interaction effects between these characteristics and the estimated disequilibrium.

One of the central issues in any mobility model is the role of stimuli to movement, or triggers to movement. Clark and Onaka (1983) have argued that a clear distinction has to be made between the effects of stages in the life cycle and changes in the life cycle on the mobility decision. The former clearly has an effect on aggregate mobility rates and on the likelihood of mobility for an individual household. Younger renters certainly are more mobile than older homeowners. (For a review, see Quigley and Weinberg, 1977.) At the same time, the life cycle stage of a household is likely to be a key determinant of changes in the life cycle, with these changes affecting the type and frequency of stimuli to movement. Thus, household formation and dissolution, often the critical stimuli in the relocation process, vary considerably by life cycle stage. In general, moves result from a complex interrelationship between present status of the household and stimuli or shocks to the household. It is insufficient to study housing dissatisfaction alone, as in the Hanushek and Quigley approaches. In addition, a stimulus for moves also is created by exogenous events that cannot be controlled by the household. Such forced moves in turn raise the issue of behavioral (i.e., individual choice) versus structural (i.e., societal) explanations of mobility so important in discussions of mobility in the British housing market. This issue will be examined further later in this paper. While family formation and dissolution may be the most obvious stimulus to movement, the relationship of residence to workplace is an additional potential stimulus to relocation. Most research, however, emphasizes the constraining influence of work-residence relationships rather than the triggering effect of job changes (Clark and Burt, 1980).

In sum, the set of research articles noted in Figure 1 contains a coherent approach to the probability of relocation and the decision to move. The decision to move, of course, is dependent upon the identification of an alternative location. The probability of moving is associated with a decision to search, or in Maclennan and Woods' (1982) terms a decision to enter the market. Accordingly, the box labelled SEARCH in Figure 1 now can be filled on the basis of a discussion of the process of relocation or the process of search.

2.2. The Process of Relocation

Most of the models that have attempted to develop an overall conceptual structure for the dynamics of mobility have viewed search as an outcome of the decision to move. For some time housing search has been viewed as a constraint on the moving process. Search simply was embedded as an independent variable in the model of mobility and search, as is illustrated by Weinberg,

Friedman and Mayo (1981). Recently, though, there has been more interest in the process of search itself.

Our research on housing search has utilized an optimal stopping rule approach, while incorporating measures of disequilibrium (or stress) as a stimulus to search (Smith, Clark, Huff and Shapiro, 1979). This model may be described as an implicitly spatial approach, one that identifies areas in which individuals search (see Figure 2). The model has been elaborated by Smith and Mertz (1980), who were especially concerned about specifying the way in which a searcher's beliefs change as housing market vacancies are viewed, and in turn about evaluating the impact of this change in beliefs on the existence of a critical frontier separating acceptable from non-acceptable vacancies. These authors then extended their research to show that the way in which a given set of vacancies are ordered by an agent may significantly affect the search time. A development of this basic model has implications both for individual behavior and for intervention (by agents) in the housing market. A full discussion of the mathematical form of the model is contained in Clark (1981).

This disequilibrium model of implicit spatial search recently has been tested with housing search data for the Los Angeles housing market (Smith and Clark, 1982; Clark and Smith, 1982). The first paper deals with utility functions for housing for each individual. Utility functions are derived from an experimental setting in which house price, floor space, construction quality, and neighborhood quality are varied. The expected utilities were derived from an approximation that took the following form:

$$V(X) \simeq \alpha_0 + \sum_{i=1}^{i=4} \alpha_i (X_i - \bar{X}_i) + \sum_{i,j=1}^{i,j=4} \beta_{ij} (X_i - \bar{X}_i)(X_j - \bar{X}_j), \quad (2.7)$$

where X_i, $i = 1, 2, 3, 4$, are the components of the housing characteristics vector X, $\bar{X}_i$ are the mean values of the components over the experiment, and α_i and β_{ij} are the coefficients of the Taylor series expansion. The first component of the vector X is taken to be the price of housing. It was possible to establish that the estimated utility fucntions are essentially linear, that they can be used to predict the ratings of real houses, and that the values for the expected value of future search are consistent with respondent stated expectations of search.

The second paper reports on a direct test of the stress function calculated from individual data on households, both during the process of search using a longitudinal sample, and for households who already bought a house using a retrospective sample. The tests showed that households begin search when stress is positive (they think they can do better than their present location) and cease search when stress is negative. In general, the expected utility approach provides a reasonable approximation to some aspects of housing market judgement and decison processes (Clark and Smith, 1982, p. 736).

A parallel approach, which is more explicitly concerned with the spatial element of the search, focuses on the sequential process of individual search within a confined area. In fact, the search model suggested by Huff (1982) is a two-step process, which consists first of an area where search is to be concentrated, and second of the actual selection of vacancies within a targeted area. In Huff's model, the decision to search in an area is assumed to be a function of the household's expected utility from searching in this area. However, his model is not dependent upon a knowledge of household beliefs. It is a normative model in that households looking for the same kinds of residences generate the same kinds of search patterns.

Much of the recent focus on the process of relocation and housing search has emphasized the role of information, and in particular, the role that information plays in both the evaluation and choice decisions in the process of residential search. An examination of information acquisition and information use has been carried out in two papers, both of which explore an elementary theory of market information that focuses on the nature of various channels used for information transmission (Clark and Smith, 1979, Smith and Clark, 1980). A general market information flow system was used to set up a structure of information channels and to investigate how individuals use these channels. The theory is laid out in the initial paper and some elementary tests based on simulations are presented in both papers. Results reported in these papers imply that the temporal efficiency of search is to a large extent dependent upon the costs of general information, the costs of collecting information about agents, and the costs of contacting agents. The spatial efficiency of search, that is, the number of areas that are searched, and the ratio of acceptable areas to unacceptable areas that are searched, seems to depend only on the costs of general information. Naturally, the nature of the costs of obtaining information is importnat. Unfortunately it was possible to explore only a limited set of costs in the simulation tests. A number of repetitive structures of the sequences of search behavior were developed from these simulations, though. For example, one type of structure, which emerged quite clearly from the simulations, involved a situation in which general information is collected, followed by the collection of messages about agents, and then either additional general information is collected, or the message about the agent might be used to contact an agent. If the real estate agent is a poor one, another agent message is collected, as well as more general information. If a good agent is obtained, more general information is collected together with repeated searches with the agent. This type of structure arose in response to relatively low costs of general information, low costs of information about agents, low costs of initial search with agents, but relatively high costs of additional searches with an agent.

One of the concerns of current reserch in progress is to relate these theoretical perspectives to an empirical base. Presently results are being analyzed from a study of 60 individuals who recently had bought houses and who participated in an experimental game. The game was conformable with the model structured in the earlier papers (Clark and Smith, 1979; Smith and Clark, 1980). The investigation is focusing on decision rules of the 'if ... then' form. Although there are a large number of possible rules, the use of some simple production system algorithms promises to be a fruitful approach to the problem (Clark, Smith and Cotton, in progress).

An additional goal of the research scheme is to balance the interest in the consumption of information with an analysis of the impacts of the way in which information is supplied. Most of the focus in recent research has been on information acquisition, and the use of information by consumers. This approach can be traced to the descriptive studies of information sources by Rossi (1955). But the way in which information is presented is equally important. A limited study of the nature of newspaper real estate advertisements emphasizes the importance of the supply side (Smith, Clark and Onaka, 1982). The results of that exploratory analysis of advertisements suggests the following inference. There is a clear difference between the advertisements of realtors who advertise one vacancy from a portfolio of vacancies and owners who advertise their individual houses (vacancies). Moreover, there appear to be marked differences in responses to market conditions. In tight markets it appears that realty agents may treat price and location information as partial substitutes to optimize the degree of ambiguity that will in turn tempt readers to respond to advertisements (to enter the information gathering stage). More price information is given when the housing market is relatively stable, but much less information is given when there is likely to be uncertainty about the price distribution. Owners, in general, seem to provide more specific and detailed information about vacancies.

2.3. Outcomes of Relocation

Until very recently, mobility research did not proceed beyond this point of information analysis. However, there are compelling arguments in favor of the position that it is just as important to examine the outcomes of relocation as it is to examine the mobility process itself (Clark and Moore, 1982; Moore and Clark, 1980).

Until now, most studies of the outcomes of relocation have been
aggregate studies, and our own resarch has been limited to general
issues of whether or not households improve their housing and the
broad impacts of relocation on neighborhoods that result.   In
general, this focus raises the issue of the links between mobility
research and policy.   But, it is not a topic to be pursued in this
discussion of modelling strategies.   It is sufficient to note that
the outcomes of mobility can be divided into impacts on households,
impacts on neighborhoods, and impacts of programs.   Clearly,
mobility is related to public programs, and in the long run the
connection between models of mobility and public programs must be
taken into account.   Many of the programs depend upon mobility for
producing some desired effects.   Examples include the investments
in the Experimental Housing Allowance Program, and the
rehabilitation component of the Community Development Block Grant
Program, both of which have sought to affect household relocation
and consumption decisions directly, especially for low and moderate
income households.   Households do not respond to just the
characterisics and costs of dwellings', they also respond to their
neighborhoods.   The nature of the response is yet to be spelled
out, but there are a number of programs that induce shifts in
services and amenities,   thus changing a neighborhood's
attractiveness.   Rent control and school desegregation programs are
examples of programs that have changed the externalities for
households, and generated redistributional effects that may or may
not be consistent with the program goals.   Examples of studies that
have attempted to relate mobility to contextual effects include the
recent set of papers by Frey (1979, 1980).

The second part of this paper takes the resarch work
contemporaneous to the work at UCLA and Santa Barbara, and fits it
into the same research structure.   A similar structure is employed,
and only those papers that have developed models or empirical tests
of models are discussed.

3.   ALTERNATIVE APPROACHES TO MOBILITY AND SEARCH MODELS

The research discussed to this point is not independent of a
large number of other streams of thought on analytical models of
residential mobility.   Indeed, the research approaches summarized
in Figures 1 and 2 have been especially influenced by the work of
economists who have investigated the housing allowance demand and
supply experiments.

3.1. Review of Existing Models

It is possible to identify a stream of research that has the
same antecedents as the models discussed in Figures 1 and 2.
Models that are closest to those suggested by Speare, Goldstein and

Frey (1975) and Huff and Clark (1978) have been formulated by Hanushek and Quigley (1978a, 1978b); these latter models have been further developed by Brummell (1979, 1981) [see Figure 3]. As Brummell notes, the Quigley and Weinberg (1977) and Hanushek and Quigley (1978a, 1978b) approaches have used economic consumer theory to analyze mobility, while ignoring concepts that were developed in the earlier behavioral generalizations by Speare, Goldstein and Frey (1975). Hanushek and Quigley focused only on the nature of the housing market. Brummell's purpose was to integrate both the consumer behavior approach, which is based on the notion that consumers attempt to obtain a combination of quantities of goods that maximizes their satisfaction or utility, subject to certain constraints, and concepts from the antecedent studies, such as aspirations, needs and stress. The model is one in which the decision to move is based upon the difference beween experienced and aspiration 'place utilities.' But as Brummell notes, the optimzing approach need not imply a perfect economic rationality; the decision maker is only intendedly rational. The heart of the model, like that of Huff and Clark (1978) and Speare, Goldstein and Frey (1975), is the formulation of residential stress as the difference between aspiration and experienced place utilities. Stress is defined as

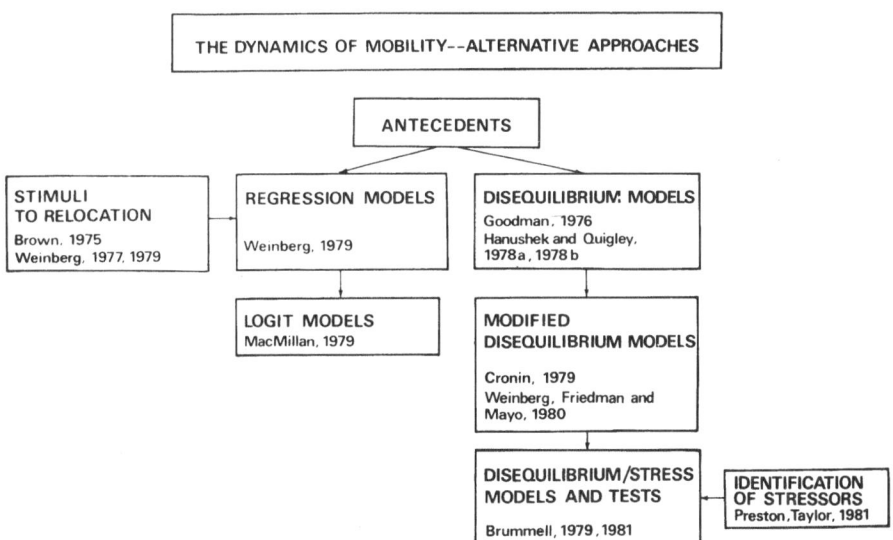

Figure 3. A taxonomy of the literature on alternative perspectives on the dynamics of mobility.

$$S = U_t^* - U_t^o \quad , \tag{3.1}$$

where U_t^* = aspiration place utility at time t, and

U_t^o = experienced place utility at time t.

In turn, U_t depends upon a vector of attributes of the residential environment and a vector of attributes of other goods. Both experienced and aspiration utilities can be defined. As Brummell notes, how the household responds to stress varies with the levels and the sources of stress (Brummell, 1979, p. 343). At low levels of stress it is unlikely that the household would consider moving, but at higher levels there is a clear possibility of relocation. Brummell suggests that it is whether or not the value of S exceeds a household's stress threshold level that determines the probability of seeking a new residence.

Although there is also a trade-off in the Hanushek and Quigley formulations--the disequilibrium between present housing and the optimal consumption of housing--it is worth reiterating that Hanushek and Quigley direct their model only to issues of housing and housing costs. Expanding equation (2.6), their basic model, permits an incorporation of the distinction between the effects of change in equilibrium housing demand between times t and t+1, and the current disequilibrium in consumption. Hence,

$$P_{t,t+1} = f'[(H_{t+1}^d - H_t^d)/H_t^d + (H_t^d - H_t)/H_t^d] \quad , \tag{3.2}$$

where the variables are defined as in equation (2.6). In the Weinberg, Friedman and Mayo (1981) variation of the disequilibrium model, search is included (as an independent variable) and mobility is discussed specifically.

Even though the economic disequilibrium models have yielded relatively poor explanatory results, the latest work by Weinberg, Friedman and Mayo (1981) recognizes the model might profitably be expanded by: (1) decomposing the household bundle into components rather than treating it simply as housing expenditure, (2) better specification of the cost of search and moving, and (3) the use of demographically disaggregated demand functions to better identify the relevant benefits of moving. Some of these suggestions are at the heart of the work by Onaka and Clark, which is in progress.

A quite different strain of modelling, again with the same general antecedents, is a separate study by Weinberg (1979) of the determinants of intraurban household mobility, which uses the conclusions of previous literature to construct a general regression model of the probability of moving. In the conceptual background to the paper, the same disequilibrium approach suggested earlier by Weinberg and Quigley (1977) and Hanushek and Quigley

(1978) is outlined, but the research emphasis is on the shocks that change a household's equilibrium. Weinberg (1979) suggests that these disequilibrating influences can include changes in prices or in income (i.e., changes in the slope of the budget line), and changes in the slope of the indifference curves (i.e., changes in preferences resulting from changes in life cycles). From this emphasis on shocks, he postulates an empirical model in which the change in a household's location is dependent on a variety of independent variables, and then he examines this contention through an ordinary least squares solution. The ordinary least squares approach is justified by arguing that weighted least squares and Zellner-Lee joint estimation provides roughly equivalent results. With similar objectives, McMillan (1979) uses a logit formulation to examine mobility as a result of a variety of independent variables.

The ongoing work on search processes has focused more on theoretical development than on empirical testing (see Figure 4). The work on search behavior and search processes can be divided into those developments related to stopping rule models, including the work of Flowerdew (1976), Weibull (1978), Phipps (1978) and Meyer (1980), the work that is more specifically related to housing markets (McFadden, 1978; van Lierop, 1981; Maclennan and Wood, 1982; Wood and Maclennan, 1982), and regression models of search processes.

Flowerdew (1976) was one of the first to outline the use of stopping rule models for housing search. Although he did not carry out any empirical analysis, his study was a stimulus to further work on stopping rule models. In an important paper, Weibull (1978) suggested one of the first models that incorporated space explicitly in a model of search, again without empirical tests. Similarly, Phipps (1978) adopted a stopping rule model for an experiment on student apartment selection. In this case, however, there was no specifically spatial component in the model. Rather, students examined up to 50 multi-attribute descriptions of apartments, one by one, and had to try to select the best apartment they could. Phipps concluded that only a few subjects followed the optimal strategy, perhaps because the costs of the search were not included. In contrast to these two papers, Meyer (1980) has outlined a much less formalized model of decision making under uncertainty, basing his approach on an argument made by Slovic, Fischhoff and Lichtenstein (1977) that normative mdoels are difficult to test because of calibration problems. In a preliminary discussion of the model, again applied to hypothetical apartment searches, Meyer concluded that preferences change as the distribution of the utility of opportunities is learned, and that stopping is based on a process of making inferences about the distribution, the time available for search, and the quality of the alternative (the apartment unit that is being viewed).

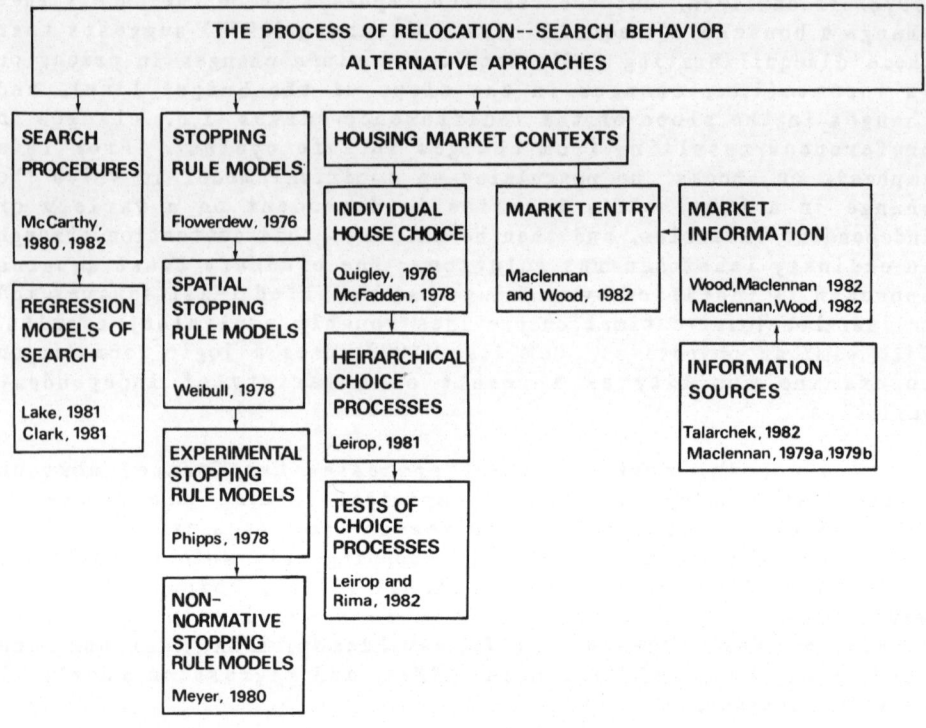

Figure 4.   A review of the literature on alternative ways of conceptualizing search behavior and the process of relocation.

In contradistinction with these derivative models, the work which is more specifically related to housing markets can be viewed in two ways. One way, which mostly is concerned with individual house choice but which utilizes a logit formulation, derives from the work of Quigley (1976) and MacFadden (1978). The other way focuses on a more complete housing market context, but again highlights individual behavior (Maclennan and Wood, 1982; Wood and Maclennan, 1982).

Quigley (1976) and MacFadden (1978) use the technical structure of the multinomial logit model to evaluate housing choice. The models are not explicitly concerned with mobility or search, but rather focus on the choice of housing type. Even so, these models were important in stimulating the hierarchical choice models of van Lierop (1981). Both van Lierop (1981) and Maclennan and Wood (1982) postulate relatively complex schematic representations of housing search theory. Both of these approaches are important in that they identify a large number of the elements involved in housing market search. They also are very complicated and will be difficult to model given the relatively simple tools available at the present time.

The simple schematic representation by van Lierop (1981) outlines a choice approach. He has attempted to develop individual choice behavior in the framework of a nested logit model, which he calls a hierarchical logit tree model for the housing market. The major focus of the model is suggested in two schematic diagrams in which a sequential process provides the utility of a house m for a family i. At each stage, choices are made that limit the number of choices later in the selection process. Van Lierop models this situation as a series of logistic equations nested through the decision tree. As yet, the model has not been empirically tested, although preliminary formulations are available as a discussion paper (van Lierop and Rima, 1982).

Maclennan and Wood (1982) also postulate a general model of search in the housing market that includes many of the elements of the Lierop approach. However, this schema is more specifically concerned with the individual decision-maker rather than with the overall structure of supply and demand in the housing market. Financing, market agents, and information are included in the model, but very few details about the supply side per se are provided (van Lierop emphasizes both, but his nested model is focused only on the demand side of the equation). In the Wood and Maclennan (1982) and Maclennan and Wood (1982) approach the emphasis is on the decision to enter or the decision to leave the housing market, rather than mobility itself. this is a somewhat different approach than the other studies in this review, since it views mobility as an outcome of the decision to enter or leave the housing market. The focus is on choice in the housing market, rather than on mobility and the antecedents of mobility. Their schematic model is not developed in an analyatic form, but two recent papers by Maclennan and Wood do focus on the nature of search itself (the interrelationship of search costs, search duration and number of searches). The models they outline involve an evaluation of the way in which search adjustment occurs within a local housing market, both in the renter and owner occupied sectors. These models assume first that students select an initial search location, and if unsuccessful, information gleaned from the failure is used to add to and clarify expectations regarding the spatial pattern of excess demand, the expected duration of search, and the search costs (which are assumed to depend upon duration of search). Second, these models assume that a 'first vacancy accept' stopping rule is adopted. The problem then is to revise distance and duration of search so as to ensure an expected and successful outcome compatible with a search costs constraint. The emphasis in both the investigation of the renter and the owner markets is the influence of the search process itself upon outcomes. Multiple regression analysis was used to test hypotheses derived from these models, and the results showed that frictional effects arising from imperfect information and uncertainty are important factors in explaining the difficulties faced by households in meeting initially held aspirations. Both owners and renters adjusted their search strategies during the process of search. For the owners sub-group, important adjustments were made to the unit price offers during search, while for renters, Maclennan and Wood were able to show that newcomers were forced to adjust over greater distances than experienced searchers. these findings certainly emphasize the importance of information.

Maclennan and Wood also developed a model of information channels to examine the impact of information on the search process. They examined the nature of information acquisition at the vacancy establishment stage when there is no major institutional involvement, and also changes in channel use with search stage, and the impact of institutions. These results suggest that studies of information and its acquisition are essential to an understanding of housing choice processes. A less formal model by Talarchek (1982) of the sequential acquistion of information also emphasizes the temporal nature of information use.

The difficulty of developing complete models of the search process has led to several attempts to relate search procedures to sets of independent varibles that characterize housing markets and individual households. McCarthy (1980, 1982) developed a three-stage search model in which the probability of active search, the procedures of search, and the outcome of search (i.e., the probability of moving given search) all were related to varying combinations of a household's housing circumstances, household demographic and economic characteristics, market information, search costs, and search procedures. Preliminary tabular information from McCarthy (1982) and some results from regression models of search procedures (Clark, 1981; Lake, 1981) have provided some limited insights into varying search behavior across groups of individuals, especially differences by race.

3.2. Model Structures

It is clear from the foregoing discussion that there are at least some significant parallels in the model structures for moving, although much less so for search. A comparison of the stress threshold model of Speare, Goldstein and Frey (1975) with the cumulative stress-cumulative inertia model of Huff and Clark (1978), the aspiration utility model of Brummell (1979), and the housing disequilibrium models (e.g., Hanushek and Quigley, 1978a)

TABLE 1:   Alternative Specifications of a Residential Mobility Model

---

Speare, Goldstein and Frey (1975) $P = f(s' - hc)$

Huff and Clark (1978) $P_t = \begin{cases} k[S(t) - R(t)] \\ o \ \text{jf} \ \ S(t) \leq R(t) \end{cases}$

Hanushek and Quigley (1978) $P_{t, \ t + 1} = f(H^d_{t+1} - H_t)$

Brummell (1979) $P_t = S_t = U_t{}^* - U_t{}^o$

Where P_t = the probability of moving, s' is residential satisfaction, c is

the cost of moving; S(t) is the pressure to move at time t, and R(t) is

the resistance to moving at time t; H^d_{t+1} is the desired quantity of

housing services at time t+1, and H_t is the quantity actually consumed

at time t; and $U_t{}^*$ is aspiration place utility and $U_t{}^o$ is experienced

place utility, and h and k are constants.

indicate that (a) most of the models involve trade-offs between current and desired characteristics, and (b) most are dynamic in the sense that the probability of moving is a function of utilities or stresses over time (see Table 1).

In the stress threshold model developed by Speare, Goldstein and Frey (1975), a household is assumed to consider moving when its satisfaction with the current unit falls below a certain threshold level. Although they outline a sub-model in which the household evaluates the expected level of satisfaction at a candidate site, in relation to the current level of satisfaction, their final model and the empirical tests assume that the probability of moving is a function of the difference between the level of satisfaction, relative to the threshold level, and the cost of moving.

The cumulative stress—cumulative inertia model, which for convenience hereafter will be called the stress—inertia model, supplies a conceptual perspective of residential mobility that combines aspects of the stress—threshold and the cumulative inertia models. In the stress—inertia model all factors that induce a household to search for new housing are summarized in a single variable called cumulative stress, whereas all factors that deter the household from moving are summarized in a second variable called cumulative inertia. Both variables are assumed to be functions of the length of time since a household moved into its present unit. By imposing certain conditions on these functions, the model is able to simulate the empirical relationship between duration at current residence and the rate of mobility.

The disequilibrium models not only focus on housing characteristics, but they also make use of the difference between actual housing consumption at time t, and equilibrium housing consumption at time t+1 (Hanushek and Quigley, 1978a, 1978b). Brummell's extension of the disequilibrium model defines aspiration utility and actual utility as functions of characteristics in addition to measures of housing. All in all, however, the differences are more strikingly observed in the estimation procedures than in the basic forms.

The stress—threshold, stress—inertia, disequilibrium and aspiration utility models all can be regarded as special applications of a general benefit—cost model. A benefit—cost model of residential mobility assumes that a household will move if the expected benefits of moving outweigh the accompanying costs. The models differ principally in the choice of types of benefits and costs considered, in the treatment of duration of stay in current residence, and in assumptions regarding the household's knowledge of the housing market. The stress—threshold model assumes that there is one benefit factor and one cost factor. Satisfaction with the current unit, relative to a threshold value, is treated as the negative of the potential benefit of moving, and the cost of moving is represented by one variable.

The disequilbirium models primarily emphasize the economic benefits and costs of moving. The difference between an optimum and current levels of housing expenditure, in the case of Hanushek and Quigley (1978a), or the additional income necessary to compensate a household for occupying a non—optimum housing unit, in the case of Weinberg, Friedman and Mayo (1981), are used as measures of potential benefits of moving. The costs of moving also are expressed in monetary units. The opportunity costs of time spent in search, the out—of—pocket cost of moving, and the loss of length—of—tenure discount are typically included.

In the stress—inertia model, reducing the cumulative stress

associated with the current residence is the potential benefit from moving out of that unit. Two costs of moving are included. The first is that associated with the breaking of bonds established by the household with the current unit and neighborhood. The second is the out-of-pocket cost of moving. Thus, the stress-inertia model separates the benefit and cost components of Speare, Goldstein and Frey's general factor of satisfaction into stress and attachment associated with the current unit.

Huff and Clark (1978) have noted that past models of residential search take one of two views of the relationship between the length of a household's stay in current residence and the probability of moving. The first assumes that the probability of moving is independent of the duration of stay. (This is the basis for the independent trials process models.) The second assumes that the inertia, or resistance to moving, increases continuously with the duration of stay, resulting in a decreased probability of moving. This is the assumption made by the cumulative inertia model. Data from longitudinal studies, however, indicate that the probability of moving is neither independent of the duration of stay nor monotonically decreasing over time. The primary objective of the stress-inertia model is to establish a theoretical framework that is consistent with the observed variation in the probability of moving over time.

For search models the similarities include the use of either logit formulations for outlining a sequence of housing choices, or the use of stopping rule structures for housing choice decisions. The possibilities for a complete model of search that are implied by the multinomial logit structures, and which are currently being explored by both geographers and economists, suggest that this approach will be increasingly important in attempts to model search processes. Until now, housing search analyses have been quite eclectic in their modelling strategies.

4.   AN EVALUATION OF DYNAMIC MODELS

This review has side-stepped two important issues, and in these concluding remarks it is useful to address these issues. The first of these is the role of dynamic models in the analysis of residential mobility. This general topic has been addressed partially in an earlier paper by Quigley (1980), in which he evaluated why it had been so difficult to develop models of residential mobility that allow policy makers to understand the impacts of mobility on policies, and policies on mobility. The question that he posed in this paper is relevant in the present context, too. Why has research to date been limited to a series of cross-sectional models of the process of residential mobility?

There are several answers to this question. These answers relate to the nature of the problem, the kinds of techniques that are available in social science research, and the availability of appropriate data. First, as can be seen from the research paradigms, residential mobility is a process that occurs jointly with numerous other choices. At the time that households are making decisions about their residential location, they also may be evaluating decisions about their jobs and job locations, and a set of other goods that are elements of household consumption at that time. All of these choices are subject to financial, informational, and physical constraints (Quigley, 1980). Second, to deal with this complex interacting system, a theory of dynamic adjustment is necessary. It is not clear exactly what form this theory should take. Moreover, although relocation occurs with certain regularity over the life cycle, in any one time interval it is difficult to predict an individual's probability of moving. Third, the lack of adequately detailed time series data has hampered the tests of those longitudinal models which exist.

Any discussion of physical, financial, and information constraints on mobility leads to the issue of the role of choice versus constraints in models of residential mobility (for a review and discussion of this issue see Short, 1978; Clark, 1982). Individuals who have undertaken research in social contexts that have a large institutional component in the decision making process, have argued that models of the residential mobility process must involve the wider modelling of institutional/governmental activities, since mobility is an indirect outcome of decisions by governments and institutions. Some would argue that, in fact, these institutional or structural characteristics are so strong as to negate the effects of individual choice. The debate about the role of institutional forces is still continuing, but it is clear that further modelling efforts will need to recognize the role of institutions in the mobility process. Residential mobility can no longer be examined as the dynamics of mobility.

5. REFERENCES

Brown, L. and E. Moore, 1970, The Intraurban Migration Process: A Perspective, Geografiska Annaler, Series B, 52: 1-13.

Brummell, A., 1979, A Model of Intra-urban Mobility, Economic Geography, 55: 338-352.

_____, 1981, A Method of Measuring Residential Stress, Geographical Analysis, 13: 248-261.

Clark, W. A. V., 1981, On Modelling Search Behavior, in _Dynamic Spatial Models_, edited by D. Griffith and R. MacKinnon, Alphen aan de Rijn: Sijthoff and Noordhoff, pp. 102-131.

_____, 1982, Recent Research on Migration and Mobility: A Review and Interpretation, _Progress in Planning_, 18: 1-56.

_____ and J. Burt, 1980, The Impact of Workplace on Residential Location, _Annals_ of the Association of American Geographers, 70: 59-67.

_____ and J. Huff, 1977, Some Empirical Tests of Duration of Stay Effects in Intraurban Migration, _Environment and Planning A_, 9: 1357-1374.

_____ and E. Moore, 1982, Residential Mobility and Public Programs: Current Gaps Between Theory and Practice, _Social Forces_, 38: 35-50.

_____ and J. Onaka, 1983, Lifecycle and Housing Adjustment as Explanations of Residential Mobility, _Urban Studies_, forthcoming.

_____ and T. Smith, 1979, Modelling Information Use in a Spatial Context, _Annals_ of the Association of American Geographers, 69: 575-588.

_____ and _____, 1982, Housing Market Search Behavior and Expected Utility Theory II: The Process of Search, _Environment and Planning A_, 14: 717-737.

_____, J. Huff and J. Burt, 1979, Calibrating a Model of the Decision to Move, _Environment and Planning A_, 11: 689-704.

_____, T. Smith and J. Cotton, in progress, Testing a Model of Information Use in a Spatial Context, Santa Barbara: Department of Geography, University of California/Santa Barbara.

Cronin, F., 1979, _An Economic Analysis of Intra-urban Search and Mobility Using Alternative Benefit Measures_, Washngton, D. C.: The Urban Institute.

Flowerdew, R., 1976, Search Strategies and Stopping Rules in Residential Mobility, _Transactions_ of the Institute of British Geographers, New Series, 1: 47-57.

Frey, W., 1979, White Flight and Central City Loss: Application of an Analytic Migration Framework, _Environment and Planning A_, 11: 129-147.

394

_____, 1980, Black In-migration, White Flight and the Changing Economic Base of the Central City, *American Journal of Sociology*, 85: 1396-1417.

Ginsberg, R., 1979, Tests of Stochastic Models of Timing and Mobility Histories: Comparison of Information Derived From Different Observation Plans, *Environment and Planning A*, 11: 1387-1404.

Hanushek, R. and J. Quigley, 1978a, Housing Market Disequilibrium and Residential Mobility, in *Population Mobility and Residential Change*, edited by W. A.V. Clark and E. Moore, Evanston, Ill.: Northwestern University Press, Studies in Geography No. 25, pp. 51-98.

_____ and _____, 1978b, An Explicit Model of Intra-Metropolitan Mobility, *Land Economics*, 54: 411-429.

Huff, J., 1982, Spatial Aspects of Residential Search, in *Modelling Housing Market Search*, edited by W. A. V. Clark, London: Croom Helm, pp. 106-129.

_____ and W. A. V. Clark, 1978, Cumulative Stress and Cumulative Inertia: A Behavioral Model of the Decision to Move, *Environment and Planning A*, 10: 1101-1119.

Lake, R., 1981, *The New Suburbanites: Race and Housing in the Suburbs*, New Brunswick, N. J.: Rutgers University Press.

McCarthy, K., 1980, *A Three Stage Model of Housing Search*, Santa Monica: The Rand Corporation, P-6487.

_____, 1982, An Analytical Model of Housing Search and Mobility, in *Modelling Housing Market Search*, edited by W. A. V. Clark, London: Croom Helm, pp. 30-53.

MacFadden, D., 1978, Modelling the Choice of Residential Location, in *Spatial Interaction Theory and Planning Models*, edited by A. Karlqvist *et. al.*, Amsterdam: North Holland, pp. 75-96.

Maclennan, D., 1979a, Information Networks in a Local Housing Market, *Scottish Journal of Political Economy*, 26: 73-88.

_____, 1979b, Information and Adjustment in a Local Housing Market, *Applied Economics*, 11: 255-270.

_____ and G. Wood, 1982, Information Acquisition: Patterns and Strategies, in *Modelling Housing Market Search*, edited by W. A. V. Clark, London: Croom Helm, pp. 134-159.

McMillan, J., 1979, The Decision to Move: Evidence From the Demand Experiment, Cambridge Mass.: Abt Associates, unpublished paper for the Conference on Housing Choices of Low Income Families, Washington, D. C., March, 1979.

Meyer, R., 1980, A Descriptive Model of Constrained Residential Search, Geographical Analysis, 12: 21–32.

Moore, E., 1978, The Impact of Residential Mobility on Population Characteristics at the Neighborhood Level, in Population Mobility and Residential Change, edited by W. A. V. Clark and E. Moore, Evanston, Ill.: Northwestern University Press, Studies in Geography No. 25, pp. 151–181.

----- and W. A. V. Clark, 1980, The Policy Context for Mobility Research, in Residential Mobility and Public Policy, edited by W. A. V. Clark and E. Moore, Beverley Hills, Calif.: Sage Publications, pp. 10–28.

Onaka, J., 1983, A Multiple-attribute Housing Disequilibrium Model of Residential Mobility, Envionment and Planning A, forthcoming.

_____ and W. A. V. Clark, in progress, Extensions of Disequilibrium Models of Residential Mobility.

Phipps, A., 1978, Space Searching Behavior: The Case of Apartment Selection, Iowa City: unpublished doctoral dissertation, University of Iowa.

Pickles, A., R. Davies and R. Crouchley, 1982, Heterogeneity, Nonstationarity and Duration of Stay Effects in Migration, Environment and Planning A, 14: 615–622.

Preston, V. and S. Taylor, 1981, Personal Construct Theory and Residential Choice, Annals of the Association of American Geographers, 71: 437–451.

Quigley, J, 1976, Housing Demand in the Short Run: An Analysis of Polytomous Choice, Explorations in Economic Research, 3: 76–102.

_____, 1980, Local Residential Mobility and Local Government Policy, in Residential Mobility and Public Policy, edited by W. A. V. Clark and E. Moore, Beverley Hills, Calif.: Sage Publications, pp. 39–55.

_____ and D. Weinberg, 1977, Intra-urban Residential Mobility: A Review and Synthesis, International Regional Science Review, 2: 41–66.

Rossi, P., 1955, Why families Move, Glencoe, Ill.: The Free Press.

Short, J., 1978, Residential Mobility, Progress in Human Geography, 2: 419–447.

Slovic, P., B. Fischhoff and S. Lichtenstein, 1977, Behavioral Decision Theory, Annual Review of Psychology, 28: 1–39.

Smith, T. and W. A. V. Clark, 1980, Housing Market Search: Information Constraints and Efficiency, in Residential Mobility and Public Policy, edited by W. A. V. Clark and E. Moore, Beverley Hills, Calif.: Sage Publications, pp. 100–125.

_____ and _____, 1982, Housing Market Search Behavior and Expected Utility Theory, 1: Measuring Preferences for Housing, Environment and Planning A, 14: 681–698.

_____ and R. Mertz, 1980, An Analysis of the Effects of Information Revision on the Outcome of Housing Market Search, With Special Reference to the Influence of Realty Agents, Envionment and Planning A, 12: 155–174.

_____, W. A. V. Clark and J. Onaka, 1982, Information Provision and Analysis of Newspaper Real Estate Advertisements, in Modelling Housing Market Search, edited by W. A. V. Clark, London: Croom Helm, pp. 160–186.

_____, _____, J. Huff and P. Shapiro, 1979, A Decision Making and Search Model for Intra–urban Migration, Geographical Analysis, 11: 1–22.

Speare, A., S. Goldstein and W. Frey, 1975, Residential Mobility, Migration, and Metropolitan Change, Cambridge, Mass.: Ballinger.

Talarchek, G., 1982, Sequential Aspects of Residential Search and Selection, Urban Geography, 3: 34–57.

van Lierop, W., 1981, Towards a New Disaggregate Model for the Housing Market, Discussienota 1981–1, The Free University of Amsterdam.

_____ and A. Rima, 1982, Towards an Operational Disaggregate Model of Choice for the Housing Market, unpublished paper, The Free University of Amsterdam.

Weinberg, D., 1977, Towards a Simultaneous Model of Intra–urban Household Mobility, Explorations in Economic Research, 4: 579–592.

_____, 1979, The Determinants of Intra-urban Household Mobility, <u>Regional Science and Urban Economics</u>, 9: 219-246.

_____, J. Friedman and S. Mayo, 1981, Intra-urban Residential Mobility, the Role of Transactions, Costs, Market Imperfections and Household Disequilibrium, <u>Journal of Urban Economics</u>, 9: 332-348.

Wiebull, J., 1978, A Search Model for Micro Economic Analysis With Spatial Applications, in <u>Spatial and Planning Models</u>, edited by A. Karlqvist <u>et. al.</u>, Amsterdam: North Holland, pp. 48-73.

Wolpert, J., 1965, Behavioral Aspects of the Decision to Migrate, <u>Papers and Proceedings</u> of the Regional Science Association, 15: 159-169.

Wood, G. and D. Maclennan, 1982, Search Adjustment and Local Housing Markets, in <u>Modelling Housing Market Search</u>, edited by W. A. V. Clark, London: Croom Helm, pp. 54-80.

SECTION 6

TEMPORAL DIMENSIONS OF THE GEOGRAPHY OF PUBLIC FINANCE

Research interest in questions relating to the geography of public finance has been relatively recent. Most of the models developed have emphasized static equilibria, comparative statics and efficiency. Even more recent has been an interest in characterizing the geographical public finance problem as dynamic. The two papers in this section serve to illustrate the quite different questions addressed in this literature. In the first paper Bennett discusses problems associated with developing positive theory of public expenditure through the application of econometric techniques on empirical data. In the second paper Lea deals with an interesting problem in normative theory utilizing the standard techniques of welfare economics.

Bennett deals with dynamic models of the spatial variation in local government taxes and expenditures. After relating this problem to a more general theory of the geography of public finance, he reviews six types of expenditure models, ranging from simple dynamic and simple cross-sectional to full space-time models. He examines the theory behind the model in each case, and summarizes some of the empirical results and conclusions of various researchers. Most of the models are either cross-sectional for one point in time, or involve only simple lags for expenditures in the previous time period. The last model treated by Bennett, a new dynamic space-time elasticity model, has much greater potential for capturing real world adjustment processes and responses to external controls. Ancot and Paelinck deal with a similar problem in an earlier paper of this volume. The models surveyed by Bennett clearly have considerable promise, but both better theory and more empirical results will be require before the progress made to date can be appreciated.

In the second paper Lea addresses the problem of optimal public good supply and jurisdiction size in a simple linear spatial setting, in which jurisdictions compete for the spillover effects provided by other jurisdictions. Leonardi, in the first section of this paper, also treats a problem of externalities in a spatial context. The static long-run problem in Lea's paper is defined and the conditions for optimal jurisdiction size and output are compared with conditions for various equilibrium sizes and outputs. These equilibria are based upon the various assumptions that are made about the objectives jurisdictions maximize. Competitive interaction is always one of Cournot or Nash independent adjustment. Lea concludes that central government controls, such as long-run matching grants, are required both on the public good

supply decision and on the jurisdiction size decision. The static
problem and comparative static analysis of this problem are
treated in detail so that a firm foundation is laid for formulating
a dynamic model for the problem. In the last part of the paper,
Lea defines an explicitly dynamic form of the problem. In
particular, the central government is assumed to face a discrete
time optimal control problem, in which this level of government
must determine the optimal trajectory of matching grants for public
good supply under the assumption that jurisdiction sizes are
uncontrolled and the jurisdictions continue to play Nash. The
first-order conditions are derived and interpreted. Finally, Lea
notes ways of extending both the static and the dynamic models
presented here.

DYNAMIC MODELS OF SPATIAL VARIATION
IN LOCAL GOVERNMENT EXPENDITURES

R. J. Bennett

University of Cambridge
England

1. INTRODUCTION

This paper reviews the wide range of quantitative models that have been developed to permit estimates to be made of variations in local government decisions on taxing and spending. First these models are set in the context of the general theory of spatial approaches to public finance. Then six categories of expenditure model, ranging from static cross-sectional to fully dynamic, are discussed.

Spatial aspects affect public finance in three main ways (Bennett, 1980): the tapering of access to public goods due to friction of distance effects on travelled-for goods to the consumer examined by Lea (1979); the partitioning of space into quasi-autonomous jurisdictions providing different mixes and levels of expenditures and taxes; and, the spillover of expenditure benefits and tax burdens between jurisdictions (Lea, 1982). This paper is concerned primarily with this second aspect of the taxing and expenditure decisions of quasi-autonomous local government units.

2. EXPENDITURE MODELS

Expenditure modelling has two main objectives. First, it seeks to assess the aggregation impact of exogenous factors such as the political disposition of local councils, the effect of intergovernmental grants, and the effect of local finance decisions on overall fiscal equity. Second, it acts as an indicative guide to general statistical relationships against which can be judged the deviations of individual local governments (e.g., as to whether they are relatively high or low taxers and spenders). To attack these two questions a wide range of expenditure models have been generated. The following discussion classifies these models into five main categories: demand models, political models, joint supply and demand models, bureaucratic models, and models with explicit utility functions. Each main category will be discussed in turn.

2.1. Demand Models

This approach was extensively used by economists and some political scientists in the 1950s and 1960s to explain expenditure differences between different local areas. Hence, it is cross-sectional. For each area in a region the essential structure is as follows:

$$E_i = f(W_i, N_i, \dots) \qquad , \qquad (2.1)$$

where W_i is the local wage income, N_i is the local need to spend, and the ellipsis denotes a wide range of other variables that might be added. A major difficulty with such models is the lack of revenue supply effects. Later models attempt to overcome this deficiency by adding explanatory variables such as local tax base B_i and intergovernmental grants G_i, such that

$$E_i = f(W_i, N_i, B_i, G_i, \dots) \qquad . \qquad (2.2)$$

This type of approach contains a number of difficulties of which the most important are: (1) supply and demand effects are confused, (2) feedback from E_i to G_i is overlooked, and (3) G_i is in a sense contained on both sides of the equation. Despite these difficulties there are many examples of this approach, including the early work by Key (1951), Brazer (1959) and Dye (1966) in the United States. These writings emphasized the importance of physical, social and economic environment in determining expenditure levels. British examples include the work of Birch (1959), Bulpitt (1967), G. B. Royal Commission (1968a, 1968b), Boaden (1971), Davies (1968) and Oliver and Stanyer (1969). More recent United States examples include the work of Hansen (1965), Henderson (1968), Bahl (1968), Stern (1973), and Hawley (1973).

2.2. Political Models

The major aspect of the work of political scientists discussed here is the expansion of the economic demand model to include a selection of political variables. The simplest case can be specified with a political effect P_i for area i as

$$E_i = f(W_i, N_i, B_i, G_i, P_i, \dots) \qquad . \qquad (2.3)$$

Considerable discussion has surrounded the choice of this political variable. Some authors advocate the use of an interval scale variable such as degree of electoral marginality, or probability of representatives being reelected. More commonly, however, a (0,1) dummy variable has been employed, most usually to denote whether a local government is controlled by a socialist or nonsocialist government (see Newton, 1981). In addition Alt (1977) has emphasized that long-term rather than short-term control is important, thus requiring a time average of political disposition to be taken. In Britain and Germany there is considerable evidence of the importance of such variables, the most satisfactory political correlates of local spending being for single service categories. For instance, Labour Party controlled councils are usually associated with higher housing and education expenditures, while higher police and roads expenditures often reflect Conservative Party control (Boaden, 1973; Alt, 1971; Godley and Rhodes, 1972). In addition, higher personal social services expenditure sometimes can reflect Labour Party control for old peoples' services (see Davies, 1968), and for local health services (see Boaden, 1971; Alt, 1971; Nicholson and Topham, 1971 and 1972; Fried, 1972; Danziger, 1974; Foster et. al., 1981; Ashford et. al., 1976; Jackman and Sellars, 1977 and 1978). More recently this approach has been expended by Hansen (1981) and Bennett (1982a, 1982b) to let dummy variables interact not only with intercepts but also with slope parameters in expenditure relationships.

2.3. Demand and Supply Models

This approach derives essentially from two papers by Strauss (1974) and Ashford, Berne and Schramm (1976), who make use of one equation for expenditure demand E_i^d, and another equation for expenditure supply E_i^s. Use of the two equations simultaneously eliminates most of the drawbacks of the pure demand models. Political variables also can be included here, although it is still a cross-sectional approach. These two equations can be written as follows:

$$E_i^d = f(W_i, N_i, \ldots)$$

$$= c_1 (W_i - t_i B_i) + c_2 N_i + \ldots, \text{ and} \quad (2.4)$$

$$E_i^s = f(B_i, G_i, P_i, \ldots)$$

$$= d_1 (t_i B_i + G_i \ldots) \quad , \quad (2.5)$$

where explicit linear functions are specified as examples, and the notation is as before, with t_i being local tax rate, and c_1, c_2 and d_1 being coefficients.

The Strauss-Ashford approach proceeds by finding an expression for t_i from equation (2.5) and then substituting this result into equation (2.4). Then E_i^d is set equal to E_i^s, under the assumption that demand and suply are in equilibrium. This assumption of clearance is very satisfactory in local expenditure models as legal constraints in most countries prevent the accumulation of deficits. Consequently, t_i may be defined as

$$t_i = (E_i^s - d_1 G_1)/(d_1 B_i) \qquad . \qquad (2.6)$$

substituting equation (2.6) into (2.4) yields

$$E_i^d = c_1 W_i - c_1/[d_1(E_i^s - d_1 G_i) + c_2 N_i + \dots \qquad .$$

Then, setting demand equal to supply yields

$$E_i = (d_1/c_1 - 1)(c_1 W_i + c_1 G_i + c_2 N_i + \dots)$$

$$= g_1 W_i + g_2 G_i + G_3 N_i + \dots \qquad . \qquad (2.7)$$

These relationships comprise a linear expenditure in equation (2.7), but a nonlinear tax in equation (2.6). Estimates from this model for Britain are given by Gibson (1980), using single equation methods, and Bennett (1982a, 1982b), using both single equation and simultaneous (i.e., FIML) estimates combined with categorical variables for administrative class and political disposition of local authorities.

2.4. Bureaucratic Models

In contrast to the political, dummy-variable, approach discussed in the preceding section, some political scientists have developed a model of the bureacratic behavior affecting local expenditures. Using the economic theory of Downs (1956), which maintains that bureaucrats seek to maximize their budgets, both Niskanen (1971) and Breton (1974) have used, for local governments, elaborations of the model proposed by Davis, Dempster and Wildavsky (1966) for the United States Congress. The Davis et. al. model hypothesizes that the appropriations request A_t at any time t is largely a function of past expenditure levels E_{t-1} plus a stochastic term e_t, giving

$$A_t = \beta E_{t-1} + e_t \qquad . \qquad (2.8)$$

Actual expenditures, however, do not match requests, so that

$$E_t = \alpha A_t + n_t \qquad , \qquad (2.9)$$

where n_t is a second stochastic error term. Substitution of equation (2.8) into (2.9) results in the following model of bureaucratic expenditure determination:

$$E_t = gE_{t-1} + \xi_t \qquad , \qquad (2.10)$$

which expresses present expenditure as an autoregressive function of past expenditure plus an unpredictable (compound) residual term ξ_t. This is the first dynamic local expenditure model.

This approach has been extended by Bennett (1983a), who demonstrates the presence of important lagged expenditure effects for British local authorities. This reinterpretation of the bureaucratic model requires the following specification:

$$\hat{E}_t^t = gE_i^{t-1} + (1 - g)t_iB_i \qquad , \qquad (2.11)$$

where gE_{t-1} represents the level of committed expenditure rolled forward from previous years by bureaucrats in their requests. The second term in equation (2.11) represents the residual that must be raised from the local tax base at a given tax rate in order to meet bureaucratic estimates of total expenditure requirement E_i^t. For Britain, Bennett (1983) has estimated this model for 1980/81 to be

$$E_i^t = 0.97 E_i^{t-1} + 0.03 t_iB_i \qquad .$$

Further empirical support for such local authority budgetary behavior is given by Lynch and Perlman (1977).

2.5. Explicit Utility Functions

More recent developments of expenditure supply and demand models in the economics literature have been characterized by attempts to derive them as maximization solutions for objective functions, in which supply and demand act as joint constraints on the aggregate utility to be derived from expenditure. This work is exemplified by Eastwood (1978), Follain (1980), Slack (1980), Cuthbertson et. al. (1981) and Bennett (1983a). The present discussion extends this work by combining it with the bureaucratic model of expenditure determination for the case of general grants.

The simplest approach is to consider the utility function U

$$U = U(X, X') \qquad (2.12)$$

for quantities of public goods X and private goods X'. The most widely applicable explicit form of this function is the Stone–Geary function (Stone, 1954):

$$U = b_1 \ln(X - a_1) + b_2 \ln(X' - a_2) \qquad , \qquad (2.13)$$

in which a_1 and a_2 represent status quo or committed expenditures, and b_1 and b_2 are coefficients that satisfy b_1, $b_2 > 0$. It is assumed that increases in supplies above a_1 or a_2 result in an increase in the overall level of utility of the individual. Hence,

$$\partial U/\partial X \ , \ \partial U/\partial X' \ > \ 0 \qquad , \ \text{and}$$

$$\partial^2 U/\partial X^2 \ , \ \partial^2 U/\partial X'^2 \ < \ 0 \quad \text{for all } X > a_1 \text{ and } X' > a_2 \quad .$$

Next the maximization of equation (2.13) will be considered, subject to two contstraints for the respective supply and demand for expenditure.

Local authority budget (supply) constraint. Let

$$E_i^s \ = \ P_x \ X_i \ = \ G_i \ + \ t_i B_i \qquad , \qquad (2.14)$$

where P_x is the price of local public goods X_i. Substituting for expenditure supply, the local government budget estimate of supply from the bureaucratic model [equation (2.11)] becomes

$$(1 \ - \ g) \ B_i t_i \ + \ g E_i^{t-1} \ = \ G_i \ + \ t_i B_i \ = \ P_x \ X_i \qquad ,$$

or

$$t_i \ = \ - \ (G_i \ - \ g E_i^{t-1})/(g B_i) \qquad . \qquad (2.15)$$

Individual budget (demand) constraint. Consider only the residential (non-corporate) sector. Here the median voter solution for expenditure demand becomes

$$E_i^d \ = \ f(W_i, \ t_i, \ B_i^d) \ = \ P_{x'} \ X_i' \qquad , \qquad (2.16)$$

where $P_{x'}$ is the price of local private goods X_i'. The function f(.) is defined to be a generalized structure of tax rates, t_i, residential tax base B_i^d, and wages W_i. This allows f(.) to be applied to cases such as Britain in which local taxes are property taxes that do not relate in a simple linear fashion to incomes (i.e., the tax base for property has no simple linear relation with personal income). Substituting equation (2.15) into equation (2.16) yields

$$E_i^d \ = \ f[W_i, \ (G_i \ - \ g E_i^{t-1})/B_i, \ B_i^d] \ = \ P_{x'} \ X' \qquad . \qquad (2.17)$$

The first order conditions for a maximum of the utility function (2.13) are given by differentiating this function with respect to both X and X', setting each result equal to zero, and finding the substitution rates betweem them, so that

$$\frac{\partial U/\partial X}{\partial U/\partial X'} \ = \ \frac{P_x}{P_{x'}} \ = \ \frac{b_1(X' \ - \ a_2)}{b_2(X \ - \ a_1)} \qquad . \qquad (2.18)$$

A rearrangement of equation (2.18) gives

$$P_x \, (x - a_1) \; = \; (b_1/b_2)(P_x, X' - P_{X'} - p_x a_2) \qquad . \qquad (2.19)$$

Substituting for P_x, X' from equation (2.17) gives

$$P_x \, (X - a_1) \; = \; (b_1/b_2)\{f[W_i, \; (G_i - gE_i^{t-1})/B_i, \; B_i^d] - P_x, a_x\} \qquad ,$$

and then rearranging terms gives

$$P_x \, (X - a_1) \; = \; (b_1/b_2) \, P_x, a_2 \; +$$
$$(b_1/b_2) \, f[W_i, \; (G_i - gE_i^{t-1})/B_i, \; B_i^d] \qquad .$$

Substituting from the bureaucratic model for

$$P_x(X - a_1) = (E_i^t - \; gE_i^{t-1}) \qquad ,$$

taking logarithms, and then simplifying the terms yields

$$(E^t - gE_i^{t-1}) \; = \; (b_1/b_2) \, P_x a_2 (b_1/b_2)[W_i, \; (G_i - gE^{t-1})/B_i, \; B_i^d] \quad ,$$

or

$$\log(E^t - gE_i^{t-1}) \; = \; q_0 \; + \; q_1 \, \log W_i \; + \; q_2 \, \log(G_i - gE^{t-1}) \; -$$
$$q_3 \, \log B_i \; + \; q_4 \, \log B_i^d \qquad . \qquad (2.20)$$

Equation (2.20) is a log-linear model of expenditure determination in which private prices and quantities of consumption are assumed unchanged by a tax, except by multiplication of a constant. The equivalent log-linear tax equation may be derived from equation (2.15):

$$\log(t_i) \; = \; r_0 + r_1 \, \log(G_i - gE_i^{t-1}) - r_2 \, \log B_i \; . \qquad (2.21)$$

These equations have been estimated, using both single equation and simultaneous equation methods, by Bennett (1983a). Highly significant results were obtained for local authorities in England and Wales.

2.6. Dynamic Elasticity Models

All of the models discussed above are either cross-sectional for one point in time, or involve only simple lags for expenditure from the previous time period. These formulations are extremely limited in their potential to reproduce adjustment processes, and this suggests the format of the models to be discussed in this section. Truly dynamic models in which expenditure and taxing decisions can be related to dynamic elasticities of demand and supply variables will be addressed here. These models are combined with an explicit utility function and the bureaucratic model approach. The necessary theory has been developed by Deacon (1978) and Bennett (1983b), who also provides empirical estimates for British local authorities.

There are various approaches to the modelling of dynamic elasticities. That adopted by Deacon (1978) considers substitution between different local public service categories coupled with the response of these expenditures to changes in relative costs. The approach reviewed here follows Bennett (1983b), who considers aggregate expenditure and bureaucratic models of decision making. The starting point of the discussion derives directly from the joint demand-supply constraint given by equation (2.17). Rewriting this equation in terms of a quantifiable expression for expenditure, and assuming demand and supply are in equilibrium, yields

$$E_i = f[W_i, (G_i - gE_i^{t-1})/B_i, B_i^d]$$

$$= E_i[W_i, (G_i - gE_i^{t-1})/B_i, B_i^d] = P_{x'}X' . \quad (2.22)$$

Taking the total logarithmic differential of equation (2.22) with respect to time gives rise to the following equation, which allows direct assessment of the dynamic elasticity term:

$$(\partial \ln E_i / \partial W_i) \, d \ln W_i + (\partial E_i / \partial \ln e_i) \, d \ln e_i -$$

$$(\partial \ln E_i / \partial B_i) \, d \ln B_i + (\partial E_i / \partial B_i^d) \, d \ln B_i^d =$$

$$P_{x'} \, d \ln X' + X' \, d \ln P_{x'} = d \ln E_i \quad , \quad (2.23)$$

where $e_i = (G_i - E_i^{t-1})$. Equation (2.23) now can be rewritten as the final equation

$$d \ln E_i = q_1 \, d \ln W_i + q_2 \, d \ln e_i - q_3 \, d \ln B_i + q_4 \, d \ln B_i^d$$

$$= P_{x'} \, d \ln X' + X' \, d \ln P_{x'} \quad , \quad (2.24)$$

where q_i (i=1,2,3,4) are the coefficients of the constant partial differential terms.

Taking the explicit utility function (2.19) and then log-differentiating it with respect to time under the conditions of supply and demand being equalized at the optimum yields

$$[(\partial U/\partial X) \, dX]/[(\partial U/\partial X') \, dX'] = (d \ln P_x)/(d \ln P_{x'})$$

$$= (b_1/b_2)[d \ln(X' - a_2)]/[d \ln(X - a_1)] \quad . \quad (2.25)$$

Rearranging equation (2.25) gives

$$(d \ln P_x)[d \ln(X - a_1)] = (b_1/b_2)[(d \ln P_{x'} X') - (d \ln P_{x'})(d \ln a_2)].$$

Now since

$$(d \ln P_{x'} X') = P_{x'} (d \ln X') + X' (d \ln P_{x'}) \qquad , \quad (2.26)$$

equation (2.24) can be substituted into equation (2.26) to give

$$(d \ln P_x)[d \ln(X - a_1)] = (b_1/b_2)[q_1 \ d \ln W_i + q_2 \ d \ln e_i$$
$$- q_3 \ d \ln B_i + q_4 \ d \ln B_i^d - (d \ln P_{x'})(d \ln a_2)] \qquad .$$

Furthermore, if the prices and committed expenditures in the private sector are assumed constant over the short term, and if the prices of the local public goods are rendered in constant terms, then the final equation becomes

$$(d \ln P_x)[d \ln(X - a_1)] = q_0' + q_1' \ d \ln W_i + q_2' \ d \ln e_i$$
$$- q_3' \ d \ln B_i + q_4' \ d \ln B_i^d , \quad (2.27)$$

where $q_i' = (b_1/b_2)q_i$ (for $i \neq 0$) and $q_0 = (b_1/b_2)(d \ln_{x'})(d \ln a_2)$. Substituting from the bureaucratic model in the same way as for equation (2.27), and then rewriting this result in terms of the discrete analogue of equation (2.27) yields

$$D (E_i^t - gE_i^{t-1}) = q_0' + q_1' \ D \ W_i + q_W' \ D \ e_i - q_3' \ D \ B_i$$
$$+ q_4' \ D \ B_i^d , \quad (2.28)$$

where D is the log-difference operator $D \ Y_t = (\ln Y_t - \ln Y_{t-1})$. Equation (2.28) represents a discrete equation that can be estimated by normal econometric methods. The dependent variable is defined in full as

$$D (E_i^t - gE_i^{t-1}) = \ln[E_i^t - (1 + g)E_i^{t-1} + gE_i^{t-2}] , \quad (2.29)$$

in which it is the acceleration in expenditure that is being modelled. In other words, it refers to the increase in expenditure over the previous year after the difference from two years previous is removed. Estimates of this model for British local authorities have been reported by Bennett (1983b), and give encouraging results.

3.   CONCLUSIONS

This paper has reviewed a range of models of local government expenditure, ranging from cross-sectional to fully dynamic. Each model is a valid representation of the expenditure decisions of local governments, but only in the fully dynamic version of equation (2.28) can the complete range of dynamic adjustments to external controls be modelled. There is considerable potential for application of these models, although fairly uniform local government systems provide more applicable cases than highly decentralized frameworks. This model also needs to be generalized so that it takes into account spillovers and spatial interdependencies present in expenditure decisions between areas.

4. REFERENCES

Alt, J., 1971, Some Political and Social Correlates of County Borough Expenditures, British Journal of Political Science, 1: 49-62.

_____, 1977, Politics and Expenditure Models, Policy and Politics, 5: 83-92.

Ashford, D., R. Berne and B. Schramm, 1976, The Expenditure-financing Decision in British Local Government, Policy and Politics, 5: 5-24.

Bahl, R., 1968, Studies on Determinants of Local Expenditure: A Review, in Functional Federalism: Grants in Aid and PPB Systems, edited by S. Mushkin and J. Cotton, Washington, D. C.: Public Services Laboratorty, pp. 54-67.

Bennett, R., 1980, The Geography of Public Finance: Welfare Under Fiscal Federalism and Local Government Finance, Methuen: London.

_____, 1982a, Modelling Local Authority Expenditure in England and Wales, Papers of the Regional Science Association, forthcoming.

_____, 1982b, Central Grants to Local Governments: The Political and Economic Impacts of the Rate Support Grant, Cambridge: Cambridge University Press.

_____, 1983a, A Bureaucratic Model of Local Government Tax and Expenditure Decisions, Applied Economics, forthcoming.

_____, 1983b, A Dynamic Elasticity Model of Local Government Expenditures, forthcoming.

Birch, A., 1959, Small Town Politics, London: Oxford University Press.

Boaden, N., 1971, Urban Policy Making, Cambridge: Cambridge University Press.

Brazer, H., 1959, City Expenditures in the United States, New York: National Bureau of Economic Research.

Breton, A., 1974, The Economic Theory of Represetnative Government, Aldine: Chicago.

Bulpitt, J., 1967, Party Politics in English Local Government, London: Longmans.

Cuthbertson, K., P. Foreman-Peck and P. Gripaios, 1981, A Model of Local Authority Fiscal Behaviour, Public Finance, 36: 229-243.

Danzinger, J., 1974, Budget-making and Expenditure Variation in English County Boroughs, unpubished doctoral dissertation, Stanford University, California.

Davis, B., 1968, Social Needs and Resources in Local Services, London: Joseph.

Davis, O., M. Dempster and A. Wildavsky, 1966, A Theory of the Budgetary Process, American Polaitical Science Review, 60: 529-547.

Deacon, R., 1978, A Demand Model for the Local Public Sector, Review of Economics and Statistics, 60: 184-192.

Downs, A., 1956, An Economic Theory of Democracy, New York: Harper and Row.

Dye, T., 1966, Politics, Economics and the Public Policy: Policy Outcomes in the American States, Chicago: Rand McNally.

Eastwood, D., 1978, An Adaptive Linear Expenditure System for State and Local Governments, Applied Economics, 10: 279-287.

Follain, J., 1980, Local Government Response to Grants and the Reliability of OLS Analysis of Pooled Data, Annals of Regional Science, 14: 31-42.

Foster, C., R. Jackman and M. Perlman, 1980, Local Govrnent Finance in a Unitary State, London: Allen Unwin.

Fried, R., 1972, Comparative Urban Performance, Los Angeles: UCLA, European Urban Research Paper 1.

G. B. Royal Commission on Local Government in England, 1968a, _Performance_ _and_ _Size_ _of_ _Local_ _Education_ _Authorities_, London: HMSO, Research Study No. 4.

_____, 1968b, _Local_ _Autority_ _Services_ _and_ _the_ _Characteristics_ _of_ _Administrative_ _Areas_, London: HMSO, Research Study No. 5.

Gibson, J., 1980, The Effect of Matching Grant on Local Authority User Charges, _Public_ _Finance_, 25: 372-379.

Godley, W. and J. Rhodes, 1972, _The_ _Rate_ _Support_ _Grant_ _System_, Cambridge, England: Department of Applied Economics, University of Cambridge.

Hansen, N., 1965, The Structure and Determinants of Local Public Investment Expenditures, _Review_ _of_ _Economics_ _and_ _Statistics_, 47: 150-162.

Hansen, T., 1981, Transforming Needs Into Expenditure Decisions, in _Urban_ _Political_ _Economy_, edited by K. Newton, London: Printer, pp. 27-46.

Hawley, W., 1973, _Non-partisan_ _Elections_ _and_ _the_ _Case_ _for_ _Party_ _Politics_, New York: Wiley.

Henderson, J., 1968, Local Government Expenditures: A Social Welfare Analysis, _Review_ _of_ _Economics_ _and_ _Statistics_, 50: 156-163.

Jackman, R. and M. Sellars, 1977, Why Rate Poundages Differ: The Case of Metropolitan Districts, _CES_ _Review_, 2: 26-32.

_____, 1978, Local Expenditure and Local Discretion, _CES_ _Review_, 3: 63-73.

Key, V., 1951, _Southern_ _Politics_ _in_ _State_ _and_ _Nation_, Chicago: Rand McNally.

Lea, C., 1979, Welfare Theory, Public Goods and Public Facility Location, _Geographical_ _Analysis_, 11: 217-240.

_____, 1982, The Dynamics of Adjustment of Efficient Jurisdiction Size and Public Good Output With Interjurisdictional Spillovers, in _Evolving_ _Geographical_ _Structures_, edited by D. Griffith and A. Lea, The Hague: Martinus Nijhoff, NATO Advanced Studies Institute Series, pp. 413-467.

Lynch, B. and M. Perlman, 1977, Local Authority Predictions of Expenditure and Income, _CES_ _Review_, 3: 12-24.

412

Newton, K., 1981, Does Politics Mattere?, in <u>Urban Political Economy</u>, edited by K. Newton, London: Printer, pp. 117–136.

Nicholson, R. and N. Topham, 1971, The Determinants of Investments in Housing by Local Authorities: An Econometric Approach, <u>Journal of the Royal Statistical Society</u>, 134: Series A, 273–303.

_____, 1972, Investment Decisions and the Size of Local Authorities, <u>Policy and Politics</u>, 1: 23–44.

Niskanen, W., 1971, <u>Bureaucracy and Representative Government</u>, Chicago: Aldine–Atherton.

Oliver, F. and J. Stanyer, 1969, Some Aspects of the Financial Behaviour of Country Boroughs, <u>Public Aministration</u>, 47: 169–184.

Slack, E., 1980, Local Fiscal Response to Integovernmental Transfers, <u>Review of Economics and Statistics</u>, 62: 364–370.

Stern, D., 1973, Effects of Alternate State Aid Formulas on the Distribution of Public School Expenditure in Massachusetts, <u>Review of Economics and Statistics</u>, 55: 91–97.

Stone, J., 1954, Linear Expenditure Systems and Demand Analysis: An Application to the Pattern of British Demand, <u>Economic Journal</u>, 64: 511–527.

Strauss, R., 1974, The Impact of Block Grants on Local Expenditures and Property Tax Rate, <u>Journal of Public Economics</u>, 3: 269–284.

JURISDICTION SIZE AND PUBLIC GOOD SUPPLY
IN THE PRESENCE OF RECIPROCAL SPILLOVERS:
STATIC ANALYSIS AND AN INTRODUCTION TO
OPTIMAL CONTROL OF THE DYNAMICS

Anthony C. Lea

University of Toronto
Canada

1. INTRODUCTION

This paper deals with a particularly important problem encountered in an attempt to build a pure theory of an optimal public space economy. The problem is one of jointly determining optimal jurisdiction size and public good output levels within a context in which there is a system of interacting jurisdictions. It may seem unusual to assume that citizens in a jurisdiction can decide on the size of their jurisdiction. However, this is a plausible assumption for the long-run evolution of a jurisdictional system, since mechanisms such as annexation are available to jurisdictions. The feature of the problem that makes it particularly interesting is the presence of interjurisdictional beneficial externalities or spillovers. Each jurisdiction is assumed to provide a single impure public good, the availability of which for consumption is a declining function of the distance of citizens from the single discrete, public facility that each jurisdiction is assumed to have. The availability of the good is not influenced at all by jurisdictional boundaries. Boundaries serve only to define the group of citizens who will decide on the size of their facility, the size of their jurisdiction, and who will pay for the provision.

In order to make the model tractable, a large number of

I would like to thank G. Leonardi, B. Ralston and S. Sethi for their discussions of various ways of formulating the dynamic control problem presented here.

simplifying assumptions have been made. These include the assumption of a linear economy, a uniform distribution of immobile citizens with identical preferences, tastes and incomes, and single public facilities with constant costs. Although many of these assumptions can be relaxed to yield interesting insights into both the nature of the problem and optimal policies to solve it, these extensions will not be addressed here.

Most of the discussion focuses on the static problem of optimal jurisdiction size and output/capacity. Since it is highly unlikely that if jurisdictions were left to their own devices the jurisdictional system would be Pareto optimal, policies likely to lead to an efficient equilibrium must be sought. The static problem is of considerable interest in its own right, and it must be adequately modelled before any attempt should be made to capture the dynamics. However, since the problem of an optimal jurisdictional system is so clearly one of the long run, a static model and a set of static, one-shot policy prescriptions cannot be considered reasonable ultimate goals. The dynamics of the adjustment process and optimal dynamic or adaptive policies to control the process must be addressed. In this paper some important aspects of the dynamic problem will be introduced. The principal problem considered is one in which a central authority of a senior jurisdiction must select an optimal program of matching grants so that the discounted stream of net benefits in a system of jurisdictions is maximized. The development of a family of dynamic extensions of static problems is planned as future research. Progress has been slow because of intrinsic difficulties associated with these problems. In several respects the paper should be considered to be a prolegemenon to the construction of policy-relevant dynamic models of optimal jurisdictions.

In the second section of the paper the static problem is defined and its assumptions are stated. Related research is reviewed briefly, and the raw materials of the model are developed. In the third section--which is the longest--the static optimization problem is formulated, solved, and related to previous research. Graphical methods have been used in a number of situations so that the structure of the problem becomes more transparent. Some policy prescriptions that are part of the conventional wisdom are found wanting. Most stress is given to matching grant policies for the internalization of spillovers. Finally, the efficiency properties of various plausible jurisdictional objectives are examined.

The fourth section introduces the dynamics of the problem. The section starts with a general overview of the key features of the dynamic problem. Then, by way of illustration, the problem of developing an optimal intertemporal policy of grants for two interacting jurisdictions is formualted as a discrete time optimal control model.

2. THE PROBLEM CONTEXT

2.1. Assumptions and Problem Statement

The following initial simplifying assumptions are made:

(1)   there is a linear economy with a uniform density of consumer-taxpayers,

(2)   the consumer-taxpayers or 'citizens' are assumed to have identical utility functions and incomes and to be immobile (all consumption takes place at their fixed residences),

(3)   a single desirable impure public good or service is to be provided and, it must be provided from a system of discrete localized public facilities,

(4)   the 'public good' is non-exclusive and fully joint, or non-rival, in consumption; however, the 'availability of the good' or benefit from consuming the good provided by, or at, a particular facility is a declining function of the distance the residence is away from the facility,

(5)   each jurisdiction has a single provision site or public facility, and

(6)   the benefits from consumption are unaffected by the artificial jurisdictional borders since there is no exclusion (thus, interjurisdictional spillovers exist).

For present purposes, a jurisdiction is defined as a contiguous area, the citizens (residents) of which decide on the level of a public good to be provided by their public facility and are taxed for this level of provision.

The models developed here are particularly suited to goods that are 'delivered' from central facilities to consumers--in a broad sense. Examples include radio, television signals, mosquito abatement programs, clean air legislation, and possibly fire and police services. In all but the latter two cases, interjurisdictional spillovers are natural. For fire and police services it is possible (though not necessarily efficient) to practice border exclusion. In addition, the models have some relevance for 'travelled-for' public goods, such as parks and other recreational facilities or educational facilities, if citizens make use of facilities in addition to those that are closest (cheapest). This 'travelled-for' case, however, will not be further discussed in this paper.

The objective then is to maximize long-run social welfare over the whole economy through judicious selection of values for the following two decision variables:

(1) the level of capacity at each facility or within each jurisdiction, and

(2) the areal extent or membership in each jurisdiction, or simply 'jurisdiction size.'

When values are selected for these variables, the values of the following variables become derivative:

(3) the number of jurisdictions, and

(4) the extent of interjurisdictional spillovers.

It also should be noted at this point that, because of the simplifying assumptions, the optimal solution will feature jurisdictions of identical size and capacity, or optimal replication of a prototype jurisdiction.

2.2. Related Research

The antecedents of the present problem can be placed into two groups--the aspatial theory of clubs and the small literature on explicitly spatial models. Following Buchanan (1965), a large club theory literature has been developed in which the principal goal has been to discover and contrast first-order conditions relating to the optimal membership of clubs, and the optimal level of supply of a good that is shared amongst members, for a variety of types of goods and institutional settings. In general, however, the good has been assumed to be not-fully-joint yet totally exclusive. Other contributors to the theory of clubs include Oakland (1972), Ng (1973), Allen et al. (1974), McGuire (1974), Berglas (1976), Shibata (1979), and Brennan and Flowers (1980). Recognizing that the theory of clubs provides a useful point of departure for the development of models of the optimal provision of exclusive 'local public goods' in a system of (local) jurisdictions, some authors have used it as a building block to address a wide range of questions (e.g., the efficiency of migration, certain types of taxes, certain kinds of decision rules) typically related to the seminal Tiebout (1956) model. Contributors to this cluster of literature include Ellickson (1973), Flatters et al. (1974), Berglas (1976b), Pauly (1976), Stiglitz (1977), Sonstelie and Portney (1978), Brueckner (1979b), and Wildasin (1979).

Although the models in this last class could be considered spatial, none of their architects have found it necessary to deal with the issues of public facility location and benefit spillovers, since the good provided by each jurisdiction has been assumed to be uniformly available over the jurisdiction (i.e., to all club members), while not available at all outside the jurisdiction, through border exclusion or some unspecified (and highly unusual) property of the good itself. Tiebout (1961), in a seldom-cited paper, was the first to pose the explicitly spatial problem of the optimal location of (a system of) public facilities providing goods/services that were purely and impurely joint (congestible) in a system of jurisdictions. Numerical examples were developed, but no analytical model was set out, and neither the tax policy nor interjurisdictional spillovers were problematized. A subsequent paper by Borukhov (1972) on this same topic made little progress. Further work by Honey and Stratham (1978) and Bigman and ReVelle (1979) on related spatial and jurisdictional problems has been counterproductive because of conceptual and analytical errors. The problems in the latter paper have been examined in Lea (1979). On the other hand, some progress has been made in developing public good theoretic models of public facility size and location, but without an explicit jurisdictional context, by Capozza (1976) and Schuler and Holahan (1977).

The contributions that are most relevant for present purposes are the excellent analytical models developed by Smolensky et al. (1970) and McMillan (1975). These former authors, apparently unaware of the theory of clubs, attempted to show that the problem of public facility location should be conceived of as a problem of finding the optimal levels of output (facility size) and the spacing or market area sizes of a system of public facilities. The problem directly treated was that of the location and sizing of a system of parks to which uniformly distributed citizens travel— according to a Manhattan metric. Smolensky et al. (1970) derived two interepretable first-order conditions (as in club theory—one for output, and the other for club size), and concluded that if a particular marginal benefit pricing scheme was used, then competition between autonomous park managers would yield a solution that was fully Pareto efficient. In the context of their paper, which initially problematized the issue of spatial spillovers, this conclusion seems surprising. A scrutiny of their analysis, however, reveals that since citizens were assumed to patronize only the closest parks, the market areas surrounding these parks necessarily would be non-overlapping (see Figure 1a). Although the Smolensky et al. (1970) model is not suitable for travelled-for or delivered goods, in which interjurisdictional spillovers exist, it is a very elegant analysis of the simpler no-spillover case.

Recognizing the limited applicability of their model, McMillan (1975) extended it to deal explicitly with interjurisidictional spillovers. Whereas in the Smolensky et al. (1970) model one could effectively assume that each market area constituted a jurisdiction, this is no longer possible in an extended model that features overlapping benefit areas. As long as zones of benefit or

418

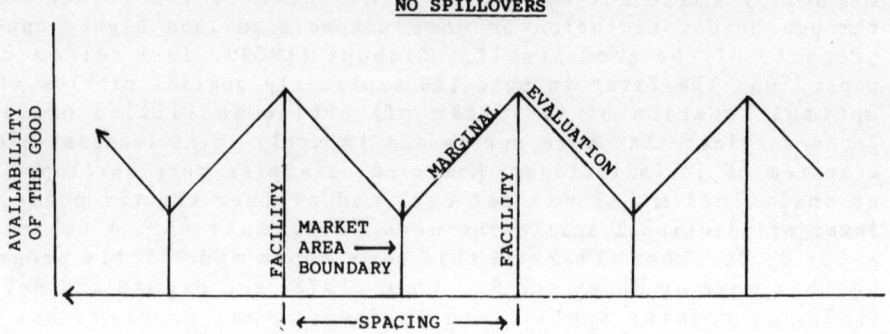

FIGURE 1A: THE SMOLENSKY ET AL. (1970) CASE

NO SPILLOVERS

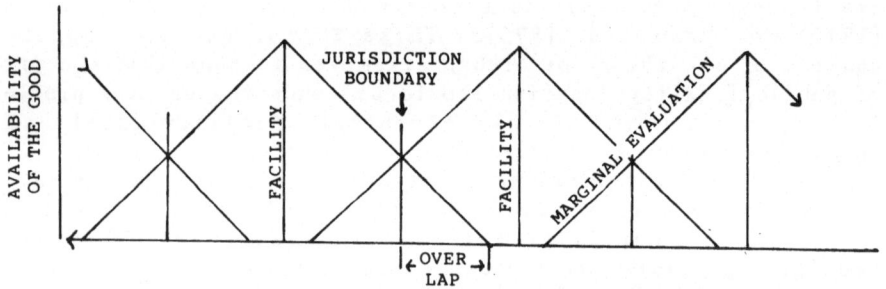

FIGURE 1B: THE MCMILLAN (1975) CASE

OVERLAPPING BENEFIT AREAS

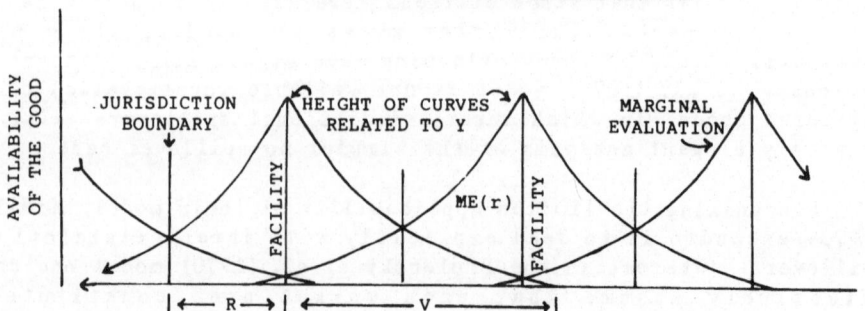

FIGURE 1C: THE PRESENT CASE

NONLINEAR MARGINAL EVALUATION CURVES

Figure 1.   Overlapping and non-overlapping benefit areas.

market areas overlap, then single good jurisdictions must be smaller than the benefit zones, and interjurisdictional spillovers will exist (see Figure 1b). The problem of finding the optimal size/output of facilities and the optimal size of jurisdictions (or benefit areas) in this setting becomes rather complex. McMillan (1975) assumed a linear economy and a curve, which was a linearly declining function of distance from the facility, representing the availability or individual marginal evaluations of consumption, so that effective use could be made of geometry in determining consumers' surpluses. He was able to show that in a system satisfying the first-order conditions for output and jurisdiction size spillovers would persist. Indeed, noting that many economists still believe that a simple solution to spatial spillover problems is merely to expand the size of jurisdictions, McMillan (1975) points out that expansion of jurisdiction sizes with the appropriate adjustment in levels of out (increasing the height of the 'tents' above the facilities in Figure 1b) may or may not reduce spillovers, and, in any case, generally would be a move away from Pareto optimality. The principal conclusion is that for an optimum optimorum, not only must the output and jurisdiction size conditions be met, but also there must be an appropriate system of matching grants to internalize optimally the interjurisdictional spillovers that persist. Other interesting issues raised in McMillan's (1975) paper will not be discussed here.

The present analysis is in the spirit of the research by Smolensky et al. (1970) and by McMillan (1975). The setting is roughly the same, with a deliberately simple linear, isotropic landscape. The static analysis is simpler yet more general than McMillan's (1975), whereas the dynamic extension is new.

2.3. Notation

The objective is one of maximizing the long-run social welfare, where social welfare is defined as aggregate consumers' surplus minus long-run costs. Before deriving expressions for costs and benefits, it is expedient to set out some notation. The key variables are as follows:

L = the length of the linear economy,

N = the number of jurisdictions spread over L (N is assumed to be an integer),

Y = the level of impure public good output, or capacity at a facility,

r = the distance from the reference facility providing a public good,

> $y(r)$ = the level of availability of the public good in question at distance r with respect to a single reference facility,
>
> $ME(r)$ = the marginal evaluation of the availability of the good by one citizen at distance r from the reference facility,
>
> Ω = the uniform density of citizens over L,
>
> R = the radius of a jurisdiction (the distance from the reference facility to the boundary),
>
> V = the distance from the reference facility at which $ME(r)$ falls to zero (effectively),
>
> G = a measure of the amount of spill-in received by the reference jurisdiction,
>
> C = the long-run costs incurred by the reference jurisdiction providing the public good,
>
> B = the long-run benefits realized by the reference jurisdiction consuming any/all of the public good available in the jurisdiction,
>
> $W = B - C$ = the total long-run net benefit to the reference jurisdiction, and
>
> $w = b - c$ [where $c = C/(2\Omega R)$ and $b = B/(2\Omega R)$] = the average per capita costs (c), benefits (b) and welfare (w).

Some of this notation appears in Figure 1c. For illustrative simplicity in all the figures it has been assumed that the marginal evaluation of the good at distance r is equal to the 'availability' of the good.

2.4. Costs

Assume that the long-run costs to a jurisdiction associated with establishing and operating a facility of capacity Y are $C(Y)$, $dC/dY > 0$. For simplicity assume that $d^2C/dY^2 = 0$. Although the analysis could proceed without this latter assumption of constant marginal costs, it would be necessary to deal with producers' surpluses, which would contribute nothing of importance to the present analysis. Average per capita costs are given by $c = C/(2\Omega R) = c(Y,R)$ with $\partial c/\partial Y > 0$ and $\partial c/\partial R < 0$, the latter holding since as R increases there are more people included in the jurisdiction to share the costs of provision. Because of the aforementioned assumption on total costs, $\partial^2 c/\partial Y^2 = 0$, it is likely that $\partial^2 c/\partial R^2 > 0$.

2.5. Benefits

The derivation of an appropriate measure of average per capita benefit is more involved than that for costs. Let the objective availability (e.g., strength of a radio signal) at distance r from a reference facilty be y(r), where the marginal evaluation of this expression is ME[y(r)]. Since there are benefit overlaps, as depicted in Figure 1c, more of the good is available at any location other than y(r). The marginal evaluation at r from the total availability is not the sum of the marginal evaluations of the separate availabilities, but rather it is the marginal evaluation of the sum of the separate availabilities. Indeed, it may not be appropriate simply to sum availabilities at all, since the availabilities from more distant facilities may not be considered perfect substitutes for availabilities from the closest facilities. (This is quite apart from the fact that less is available from more distant facilities.)

In order to deal with this possibility suppose that the facilities are uniformly spaced at a distance $D = 2R$ from one another, and that these facilities have identical capacities (this is a reasonable proposition since the assumptions have been constructed so that the solution will have a prototype jurisdiction optimally replicated). Consider the distances a citizen at distance r from the closest reference facility--at spatial lag zero--is from the other facilities. These distances are $(D - r)$ and $(D + r)$ at a spatial lag of one, $(2D \pm r)$ at a spatial lag of two, and $(kD \pm r)$ at a spatial lag of k. Letting $0 \leq \gamma_k \leq 1$ be the weight or discount factor applied to availabilities associated with facilities at spatial lag k, the total availability $\hat{y}(r)$ at distance r is given by

$$\hat{y}(r) = y(r) + \sum_{k=1}^{k=\infty} \gamma_k [y(kD + r) + y(kD - r)] \quad .$$

For example, assuming L is of infinite length, as above, and using $y(r_k) = Y \exp(-br_k)$, where b is a distance decay parameter, it can be shown that

$$\hat{y}(r) = Y[\exp(-br) + \exp(br)]\{[1 - \exp(-bD)]^{-1} - Y \exp(-br)\} \quad .$$

Of course, lines of less than infinite length make such evaluations much more difficult, and furthermore give rise to nasty boundary distortion problems that are worth avoiding here. The negative exponential functional form has the desirable property of being convex, but the undesirable feature of being asymptotic to, rather than falling cleanly to, zero at a finite distance. One simple single-parameter function that is convex and falls to zero at a finite distance is

$$y(r) = Y + \alpha/(1 + r) - \alpha \quad .$$

This function also has some other properties that would seem to be reasonable characterizations of the real world. For instance, as capacity increases the function rises but becomes less steep at any particular distance from the facility.

Because of the simplifying assumptions that have been made about uniform facility spacing and an infinitely long economy, it makes no difference from which facility the representative citizen is at a distance of r. It should be possible to derive one function $\hat{y}(r)$, $0 \leq r \leq R$, that is perfectly general. The marginal evaluaion of the citizen at distance r that must be taken into account in defining benefits derived from the system of facilities, therefore, will be $ME[\hat{y}(r)]$ or simply $ME(r)$. Assuming that the consumer pays no price or tax directly associated with consumption, then the consumer's surplus of a citizen at distance r is given by

$$B(r) = {}_0\!\int^{\hat{y}(r)} ME[\hat{y}(r)] \; d \; \hat{y}(r) \quad .$$

This is simply the area under the marginal evaluation (pseudo-demand) curve (plotted with dollars on the Y-axis and $\hat{y}$ on the X-axis) up to the level of $\hat{y}(r)$. It should be remarked that this surplus relates to consumption from both 'domestic' and 'foreign' sources, where domestic relates to the closest facility (the one in the same jurisdiction) and foreign relates to all other facilities. The function $ME(r)$ could be considered a compensated demand or marginal evaluation curve, since it is assumed that the consumer pays a zero direct price for the good. This avoids the well-known bias in measuring benefits associated with taking surpluses with respect to uncompensated or Marshallian demand curves.

With the consumer surplus associated with a citizen at distance r given as $B(r)$, computing the aggregate consumer surplus affiliated with everyone in the jurisdiction is straightforward. Denoting this jurisdictional benefit as B,

$$B = 2 \; {}_0\!\int^R \Omega \; B(r) \; dr \quad .$$

In this integral everyone's benefits have been treated equally. Although this assumption usually remains implicit, because it is highly questionable, in the present case of identical utilities, incomes, uniform density Ω, and optimal replication of facilities over the economy, it seems less contentious. One should note that if the density of citizens varied systematically, the $\Omega(r)$ could be used. Since Ω is assumed to be a constant, then suppose, without loss of generality, that $\Omega = 1$ so that it will vanish from subsequent expressions.

B now can be written as a general function $B = B(Y,R)$ analogous to $C = C(Y)$. However, for the subsequent analysis it is expedient (indeed necessary) to write $B = B[Y,R,G(Y,R)]$, where G is some measure of benefit area overlap of 'interjurisdictional spillover,' or more appropriately here and henceforth will be called 'spill-in.' In general it does not matter exactly how G is measured, as long as it clearly relates to the amount or proportion of the availability that spills into the reference jurisdiction from outside this jurisdiction. For example, G could be defined as that part of the total availability over a reference jurisdiction that is attributable to foreign facilities. In the case of the negative exponential availability function defined above, this can be shown to be

$$G = Y \sum_{k=1}^{k=\infty} [\exp(kbR) - \exp(-kbR)]/(kb) \quad .$$

Clearly G is related to V, and more particularly to $(V - kR)$, $k = 1, 2, \ldots, \infty$, in addition to the level of output at the facility, Y. At any rate, the value of G in this problem is uniquely determined once the values of Y and R are determined. Jurisdictional benefit B is directly influenced by the level of output or capacity of the facilities Y, and directly affected by R since as R increases the number of citizens included increases. G is used as an argument in the function $B[Y,R,G(Y,R)]$ not only to emphasize that jurisdictional benefits are a function of the quantity of spill-ins, but also for the very pragmatic purpose of capturing critical indirect effects. In particular, as Y increases, holding R constant, spill-ins and hence benefits increase. And as R increases, holding Y constant, spill-ins decreasse and so benefits decrease. Reasonable properties of $B[Y,R,G(Y,R)]$ may be summarized as follows:

$$\partial B/\partial Y \geq 0, \quad \partial B/\partial R \geq 0, \quad \partial B/\partial G \geq 0, \quad \partial G/\partial Y \geq 0, \quad \partial G/\partial R \leq 0,$$

with $(\partial B/\partial Y)_R = \partial B/\partial Y + (\partial B/\partial G)(\partial G/\partial Y)$,

and $(\partial B/\partial R)_Y = \partial B/\partial R + (\partial B/\partial G)(\partial G/\partial R)$,

showing the direct plus the indirect effects involved.

The average per capita benefit within the prototype jurisdiction is given by $b = B/(2R)$ (recalling the assumption of $\Omega = 1$). Corresponding to $c = c(Y,R)$, a general function is needed for b. For simplicity, b is taken as $b = b[Y,R,G(Y,R)]$. Since this is the same general function that was used for gross jurisdictional benefit B, it is important to point out that the argument R here must do double duty. The first role is that played already in B; the second is the role that changes B into its per capita form b. As an illustration, consider an increase in R, holding capacities constant. The increase in R means more people are included so that benefits increase. However, since additional people included likely get less than the average availability consumed by the pre-expansion citizens, the average per capita benefits may well decrease.

3. THE STATIC OPTIMIZATION PROBLEM

3.1. The Appropriate Objective

The objective is to select the levels of output and jurisdiction sizes so that long-run social welfare for the whole economy is maximized. This objective can be set up in several ways. For example, one could maximize $N(B - C)$, where N is the number of jurisdictions and $(B - C)$ is the long-run net benefit or welfare of each jurisdiction. As noted above, the problem has been structured in such a way that each jurisdiction will be identical in the optimal solution, so that $(B - C)$ is perfectly general. One also could maximize average per capita net benefit over the whole economy, $L\Omega(b - c)$, where L is the length of the economy, Ω is the constant density, and $(b - c)$ is the average per capita net benefits. However, both of these objective functions pose some problems because they include N and L, respectively. Since the solution will be one with a prototype jurisdiction optimally replicated, it is expedient to maximize average per capita net benefit for a prototype jurisdiction. This may be written $(B - C)/(2\Omega R)$ or, since b and c have been defined above in terms of the prototype jurisdiction, this may be written simply as $(b - c)$.

It is important to point out that the maximization of $(B - C)$ is not the same as the maximization of $(b - c)$. The former objective is jurisdictional net benefit. It is not too difficult to show that the solution to a $(B - C)$ problem will not be the same as the solution to a correctly specified $(b - c)$ problem. Let n be the number of people per jurisidiction $(n = L\Omega/N)$. The objective $(b - c)$ comes from $L\Omega(b - c)$ or $Nn(b - c)$, where $L\Omega = Nn$ is simply the number of people in the economy. Since this number is assumed to be constant, it may be dropped. On the other hand, $(B - C)$ may be written as $n(b - c)$, where n is not a constant and cannot be dropped. Indeed, it might be noted that the objectives would be the same only in the case in which $n = 1$. Unfortunately, $(B - C)$ has been used as an objective for similarly structured problems in the literature, and much debate and confusion has resulted. Smolensky et al. (1970) point out that Tiebout (1961) implicitly used $(B - C)$ as an objective in determining the optimal solution to an exemplary problem. Tiebout's verbal discussion, however, suggests that he was aware that the appropriate objective was $(b - c)$. In the seminal paper in the theory of clubs, Buchanan essentially maximized (b – c) for the average or representative member of a prototype club in a system of clubs. Ng (1973, 1974) was troubled by this objective because, at its optimum, benefits minus costs per club are not maximized. Ng (1973) reformulated the club problem essentially as one in which (B – C) was maximized and made the claim that his first-order conditions for optimality were appropriate while Buchanan's were not. Ng (1974) went on to develop an appropriate Pigovian tax-subsidy policy that would change Buchanan's club equilibrium into Ng's optimum. Berglas (1976a), Hillman (1978) and Brennan and Flowers (1980) all have gone on to point out Ng's error, with the latter authors providing a full analysis of situations in which Ng's formulation would be appropriate. Unfortunately, Ng (1978, 1981) still claims his conditions are the correct ones, and that he has been misunderstood by all the commentators.

3.2. Smolensky et al. (1970)

Recall that the problem of Smolensky et al. (1970) was one of finding the optimal size and spacing (or market areas) of a system of parks, under the assumption that market areas will not overlap since citizens patronize closest parks only. The case they selected for explicit treatment was one in which citizens travelled according to a rectangular distance (so market areas would be square), and marginal benefits were assumed to decrease linearly with distance.

The problem can be recast in more general terms using the present notation and setting (i.e., a linear economy). Since there are no spillovers, the benefits to a prototype market area of jurisdiction may be written not as $B[Y,R,G(Y,R)]$, but rather as $B(Y,R)$. Accordingly, the problem becomes

$$\text{MAX:} \quad W = b - c = [B(Y,R) - C(Y)]/(2R) \quad , \quad (3.1)$$

with the following first-order conditions:

$$(\partial W/\partial Y)_R = 0 \quad \text{and} \quad (\partial B/\partial Y) = dC/dY \quad , \quad (3.2)$$

$$\text{and} \quad (\partial W/\partial R)_Y = 0 \quad \text{and} \quad [(2R)(\partial B/\partial R) - 2B + 2C]/4R^2 = 0 \quad .$$

Noting that $\partial B/\partial R$ from $B(r) = 2\int_0^R B(r)\,dr$ is given by $2B(R)$, and multiplying by $4R^2$, this second condition may be rewritten as

$$\partial B/\partial R = 2B(R) = (B - C)/R \quad , \text{ or} \quad (3.3)$$

$$B(R) = (B - C)/(2R) \quad . \quad (3.4)$$

The first condition in (3.2) states that capacity should be set at the level that equates marginal benefits and marginal costs for the prototype jurisdiction. The second condition, equation (3.4), states that the size of the 'jurisdiction' should be set so that the gross benefits received by those inhabitants at the edge of the jurisdiction are equal to the average net benefit over the jurisdiction. Smolensky et al. (1970) go on to conclude that the appropriate tax is a lump-sum benefit tax at distance r, equal to the surplus at distance r minus the surplus at the edge of the jurisdiction. They also conclude that if these taxes were used, the decisions made by independent facility managers to maximize jurisdictional (park) net benefit, competing with one another in a system, would satisfy conditions (3.2) and (3.4). This latter conclusion may seem somewhat surprising in a problem involving public goods. But, since the authors have assumed spillovers away, and have overcome the preference revelation problem by using an optimal benefit tax, the public good problem is solved.

3.3. The Present Problem With Spillovers

For the case with spillovers between jurisdictions represented by the argument G, the objective function becomes

$$\text{MAX: } W = b - c = \{B[Y,R,G(Y,R)] - C(Y)\}/(2R) \quad . \quad (3.5)$$

The first-order conditions for objective (3.5) are

$$(\partial W/\partial Y)_R = 0 \quad \text{and} \quad (\partial B/\partial Y) + [(\partial B/\partial G)(\partial G/\partial Y)] = dC/dY \quad , \quad (3.6)$$

and

$$(\partial W/\partial R)_Y = 0 \quad \text{and} \quad (\partial B/\partial R) + [(\partial B/\partial G)(\partial G/\partial R)] = (B-C)/R \quad , \quad (3.7)$$

or

$$B(R) + [(\partial B/\partial G)(\partial G/\partial R)]/2 = (B - C)/(2R) \quad . \quad (3.8)$$

The difference between these conditions and those of Smolensky et al. (1970) are the terms in square brackets, which in each case represent the indirect benefits of spillovers. In the case of condition (3.6) this is the increase in benefits from the increase in spill-ins caused by an increase in facility size. Thus, this condition still means that capacity should be set at the level equating marginal benefits and marginal costs. In the case of conditions (3.7) and (3.8), the bracketted expression is the indirect effect of spillovers on benefits as jurisdiction size increases. This effect will be negative, since spillovers will decrease as R increases, holding Y constant. Again the second condition has the same interpretation as the second condition in the Smolensky et al. (1970) model. In the absence of spillovers, these bracketted terms vanish and these conditions reduce to those of Smolensky et al. (1970)

If the Smolensky et al. (1970) condition for Y (R constant) were applied to a problem with spillovers, the level of Y selected would be too small. Under the reasonable assumption that $\partial C/\partial Y$ is constant, $(\partial B/\partial Y) + (\partial B/\partial G)(\partial G/\partial Y)$ is greater than $(\partial B/\partial Y)$, and if both of these decrease in Y, then the intersection of the marginal cost curve with the Smolensky marginal benefit curve is unequivocally to the left of its intersection with the spillover adjusted marginal benefit curve. Hence capacity would be too small: $Y^S < Y^*$. The relationship between the Smolensky et al. (1970) sized jurisdiction and the optimal sized jurisdiction is best depicted by a graph (see Figure 2). In Figure 2 the curve of average net benefits should have roughly the shape shown. At its maximum point it is crossed by the marginal benefit curve, which includes the indirect effects. The intersection point indicating the optimum is shown at R*. The intersection point with the B(R) curve must be to the right of R* at R^S, since the expression $(\partial B/\partial G)(\partial G/\partial R)$ is negative (because its second term is negative). Thus, the jurisdiction size using the Smolensky et al. (1970) condition, R^S, is too large while its output will be too small. Although this 'capacity to low' finding is intuitive, it is perhaps counter-intuitive that the jurisdiction size selected without accounting for spillovers will be too large. Normally one would expect to have to increase rather than decrease jurisdiction size when spillovers become apparent. Incidentally, McMillan (1975), in a less general analysis, concluded that $R^S > R^*$, but that the capacity satisfying the Smolensky et al. (1970) condition would be

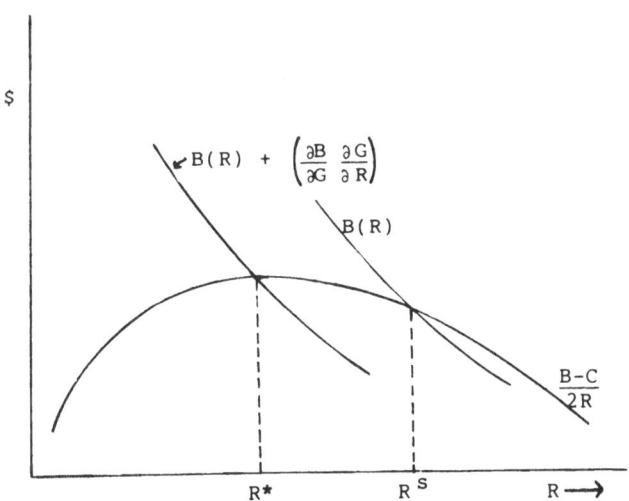

Figure 2. Relationship between the Smolensky et al. sized jurisdiction and the optimal sized jurisdiction.

larger than the optimum (i.e., $Y^s > y^*$). This latter finding seems to be in error because McMillan (1975) focused upon spillovers (or spill-outs) rather than upon spill-ins. (In a situation with interjurisdictional externalities the relevant benefits become former benefits plus spill-ins, not former benefits minus spill-outs.)

3.4. Graphical Analysis of the Static Spillover Problem

For present purposes it is useful to derive efficiency conditions for the general spillover problem that are in terms of average per capita benefits and costs. In this case use is made of the generalized functions $c = c(Y,R)$ and $b = b[Y,R,G(Y,R)]$, where it will be recalled that the argument R of b is required to do double duty. Thus the objective function becomes

$$\text{MAX: } w = b[Y,R,G(Y,R)] - c(Y,R) \qquad . \qquad (3.9)$$

The first-order conditions for objective (3.9) are

$$(\partial W/\partial Y)_R = 0 \text{ and } \underset{(+)}{(\partial b/\partial Y)} + [\underset{(+)}{(\partial b/\partial G)}\underset{(+)}{(\partial G/\partial Y)}] = \underset{(+)}{(\partial c/\partial Y)}, \qquad (3.10)$$

and

$$(\partial W/\partial R)_Y = 0 \text{ and } \underset{(-)}{(\partial b/\partial R)} + [\underset{(+)}{(\partial b/\partial G)}\underset{(-)}{(\partial G/\partial R)}] = \underset{(-)}{(\partial c/\partial R)} \qquad . \qquad (3.11)$$

Here the expected signs of the partial derivatives have been indicated in parentheses below the terms. For condition (3.10) the positive signs are all straightforward. In condition (3.11) the first term (i.e., direct marginal benefit) is expected to be negative because an increase in jurisdiction size means the inclusion of more people deriving small marginal benefits—so the average falls. Similarly, the last term (i.e., marginal cost) is expected to be negative since total costs have not changed, and now there are more people to foot the tax bill. For the bracketted terms one would expect that as R increases, spillovers/spill-ins (G) will decrease, but that as spill-ins increase, average per capita benefits increase.

The graphical analysis that follows has been adapted from club theory (see, for example, Allen et al., 1974; Shibata, 1979), and it makes use of these average per capita conditions to show how an optimal equilibrium can be approached if an appropriate pricing policy is used. In club theory, as the size of the group (club) consuming the public good increases, the per capita cost share is reduced. It is the increasing level of 'congestion' that limits the optimal-sized club. In the present problem, as the size of the jurisdiction increases, the per capita cost share is reduced just as in club theory. However, in the present model there is no congestion because the good has been assumed to be perfectly joint. Average per capita benefits decline here as the jurisdiction size R increases because, on average, people are now more distant from their main source of supply.

Referring to Figure 3, in the northwest quadrant the behavior of average per capita marginal costs and benefits is depicted as a function of jurisdiction size, holding output constant. Increasing R is measured to the left. Average per capita costs will be a decreasing convex function of R (since costs are being spread over

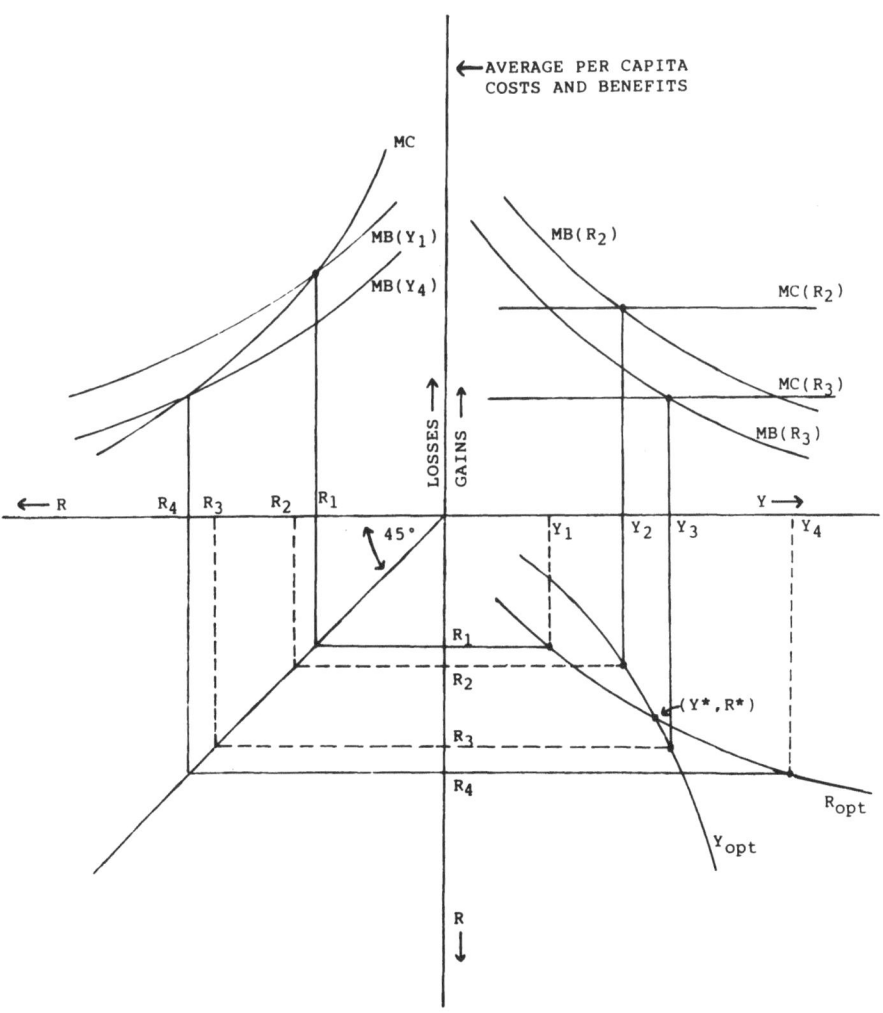

Figure 3. Marginal curves for Y and R.

more people), and so average per capita marginal costs also will be a decreasing convex function of R, as depicted. Average per capita benefits also will be a decreasing convex function of R, since the people added at the periphery will have less benefit than the average before expansion. It is reasonable to assume that this function will be less convex than the average per capita cost curve, so that the average per capita marginal benefit curve will be less steep than the average per capita marginal cost curve. It should be noted that, since per capita costs and benefits are declining functions of R, their slopes—the associated marginal curves that are plotted in Figure 3—will be negative. Hence the label 'losses' on the vertical axis. A single average per capita marginal cost (MC) curve is depicted since it has been assumed that marginal costs are constant as Y changes. However, there will be a different marginal benefit curve for each different level of capacity. Two have been depicted in the northwest quadrant: one associated with a small capicity, Y_1, and one associated with a larger capacity, Y_4. The optimal jurisdiction sizes associated with Y_1 and Y_4 are given by the intersection of the relevant marginal benefit curves with the marginal cost curve, and are shown as R_1 and R_4, respectively. Subsequently these jurisdiction sizes will be mapped into the southeast quadrant by rotating them around a 45° line in the southwest quadrant.

Turning now to the northeast quadrant, the optimal capacity, Y, for fixed jurisdiction size, R, is determined in this quadrant. Holding jurisdiction size constant, the average per capita cost curve, MC(R), is flat, since constant costs have been assumed. As jurisdiction size increases from R_2 to R_3, the MC curve shifts lower because of cost spreading (this shift will be larger at first and will become smaller as R increases). These MC(R) curves could be considered 'supply curves.' Holding R constant, benefits will increase as a concave function of Y so that average per capita marginal benefit curves (demand curves) are decreasing in Y. These curves are likely to be convex in Y for the same reason regular demand curves are. It is likely, although not necessary (nor necessary for the general conclusions to be drawn), that MB(R) curves shift down as R increases. Two MB(R) curves have been depicted in this quadrant, and their intersections with their MC(R) curves indicate the optimal level of capacities Y_2 and Y_3 associated with jurisdiction sizes R_2 and R_3, respectively.

In the southeast quadrant the two curves that are the goal of the analysis have been depicted. Consider first the curve labelled Y_{opt}—which represents the optimal level of capacity for different jurisdiction sizes. From the analysis of the northeast quadrant in the previous paragraph two points on this curve are known. Points Y_2 and Y_3 are associated, respectively, with fixed jurisdiction sizes R_2 and R_3. These jurisdiction sizes have been depicted both on the 'negative abscissa' and on the 'negative ordinate.'

Consider now the R_{opt} curve--which represents the optimal jurisdiction sizes for fixed levels of capacity. From the analysis in the northwest quadrant two points on this curve are known. For a level of capacity Y_1--indicated on the 'positive abscissa'--R_1 is the optimal R. To map this R_1--found in the northwest quadrant-- into the southeast quadrant, a $45°$ line in the southwest quadrant has been used for the reflection transformation. R_2 matches Y_2 in a similar fashion.

It should be clear that the Y_{opt} curve and the R_{opt} curve are the graphical depictions of the first-order conditions (3.10) and (3.11) derived at the outset of this section. It is very likely that there is only a single $(Y*,R*)$ point that simultaneously satisifies both conditions (lies on both curves). Once the graphical analysis represented by Figure 3 is understood, this tool becomes extremely helpful in answering a whole range of 'what if' type questions that readily come to mind.

Attention now will be turned to whether or not the point $(Y*,R*)$ is indeed Pareto optimal, and, in addition, if it is a stable equilibrium. The graphical analysis makes the answers to both questions relatively straightforward. First, recall that average per capita (not jurisdictional) benefits and costs have been dealt with thus far, so that the objective is the appropriate one. Second, consider that, in order to construct the curves in the two northern quadrants, some tax price rule must be assumed. For example, the demand curves, MB(R), in the northeast quadrant must be price response functions. If one explicitly assumes that the prices each individual faces are marginal benefit prices, taking full account of distance from principal source of supply and all spill-ins, and, if one assumes that an average individual in a prototype jurisdiction is being dealt with, then $(Y*,R*)$ is indeed necessarily Pareto optimal. If the tax-pricing system were arbitrary, then this point would not have this normative property. If the MC and MB curves represented averages for the jurisdiction as a whole, then even if marginal benefit tax-prices were used $(Y*,R*)$ would more than likely not be Pareto optimal. This latter point will be discussed further below.

The stability of the equilibrium at $(Y*,R*)$ is most easily treated graphically. Such a treatment has the side-benefit of drawing attention to the issue of spillovers/spill-ins that was purposely supressed in the already somewhat complex discussion above. Considering Figure 4, which is simply an enlargement of the southeast quadrant of Figure 3, suppose that an optimal marginal benefit tax-price policy is in effect in the prototype jurisdiction. Thus the Y_{opt} and R_{opt} curves really do represent optimal reactions. The point E is the optimal equilibrium. When Y is small and R is large--in the lower left part of the figure--the amount of spillover will be very small. Conversely, spillovers

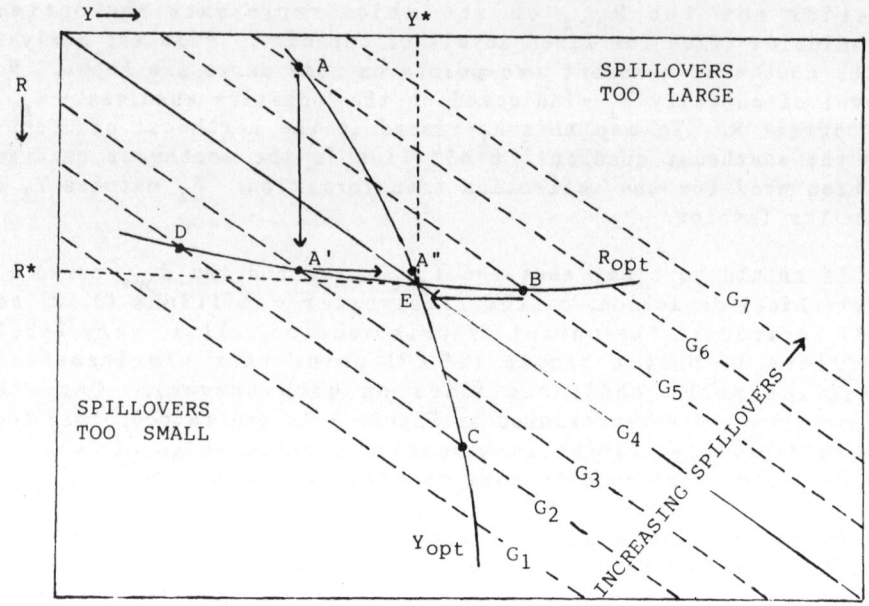

Figure 4. An enlargement of the southeast quadrant of Figure 3.

will be very large in the upper right part of the figure, where Y is large and R is small. The 'optimal amount of spillover' is the level that exists at point E, and a line G(Y,R) = G5 has been drawn through E. One should note that this same level of spillover could well exist at other combinations of values of Y and R. However, at any point other than E, Y and R will not be optimal. At point A, Y is optimal for a given R and spillovers are too great. At D, R is optimal for a given Y but spillovers are too small. Similar stories could be told about points B and C.

Suppose that for some reason point A depicts the current Y and R of the prototype jurisdiction. Are there forces at work that eventually will lead to the (optimal) equilibrium at E? If the jurisdiction believed that its size was entirely fixed--then the answer is no. By definition, for that jurisdiction size, the level of Y selected (on Y_{opt}) is optimal. If, however, jurisdiction size can be varied (in the long run, of course), it can be shown that the average citizen would be better off if jurisdiction size were increased--even if the level of output were to remain the same. This is shown as a move to A'. In making this move, the amount of 'spill-in,' which, since it was paid for it has been put in quotes, and the average per capita benefit are both reduced, but the average per capita costs have fallen faster. The now larger-sized jurisdiction then would have an incentive to increase the level of capacity in a locally optimal move from A' to A''. In this move, spill-ins increase and per capita benefits increase faster than per capita costs. Such moves eventually will terminate at E. Further, the moves need not be orthogonal. It is likely that a rational jurisdiction, acting in a system of equally rational jurisdictions, would choose a more direct route to E. A similar argument could be made about an original position at B, C or D also invoking incentives to alter Y and R in such a way that point E is converged upon. Although a rigorous convergence proof is possible, the heuristic graphical treatment is sufficient for present purposes. Thus E is a stable Pareto optimal equilibrium.

3.5. The Conventional Wisdom and Matching Grants for Interjurisdictional Spillovers

The analysis associated with Figure 4 clearly summarizes several features of the problem that have not been appreciated in most of the literature on interjurisdictional spillovers and optimal jurisdiction size. These features include:

(1) the efficient facility capacity/output depends upon jurisdiction size, and the efficient jurisdiction size depends upon facility capacity/output,

(2) for an optimal system both facility capacity and jurisdiction size must be optimized simultaneously,

(3) in an optimal system 'spillovers' will not be zero and may in fact be very large, relatively speaking, and

(4) for an optimal system, not only must Y and R be optimal, but some mechanism must exist for the optimal internalization of the interjurisdictional spillovers that are associated with $(Y*,R*)$.

The conventional wisdom implicit, and even explicit, in much of the public finance literature is that jurisdiction size should be selected according to some non-capacity related criterion or criteria. One criterion commonly encountered is that jurisdiction size should be increased until spillovers disappear. Rarely is it recognized (see, for example, MacMillan, 1975) that increasing jurisdiction size could stimulate a concomitant increase in capacity that not only will not solve the spillover problem, but also may well exacerbate it. It has long been recognized that certain kinds of Pigovian taxes and subsidies can restore Pareto optimality in the presence of externalities. Armed with this theory, many analysts have proposed that system efficiency can be restored (maintained) using an appropriate system of grants. Such proposals entirely ignore the necessity of optimizing R as well as Y, in addition to internalizing the spillovers that are associated with a particular Y and R. Thus, it seems to be assumed that either when Y and R are optimal there will be no spillovers, or regardless of Y and R internalizing spillovers is sufficient.

The appropriate type of grants to use to internalize benefit spillovers is now widely recognized to be 'matching grants.' Discussions of matching grant policies for internalizing interjurisdictional spillovers have been provided by Breton (1965), Olson (1969), Connolly (1970), Vardy (1971, 1972), Oates (1972), Musgrave and Musgrave (1973), Kiesling (1974, 1976), McMillan (1975), Harford (1977), Rittenoure and Pluta (1977), Sheshinski (1977), Jurion (1979a, 1979b) and Mieszowski and Oakland (1979), inter alia. A matching grant is an agreement to pay a (fixed) proportion of the costs of extending production of a pure or impure public good. An optimal matching grant is one in which the (fixed) proportion is designed so that the level of output selected takes into account fully the benefits derived by citizens outside the jurisdiction in question. It is in this sense that it is a standard Pigovian remedy of the kind recommended for internalizing interpersonal spillovers. In general, jurisdictions (just like individuals) are assumed to accept spill-ins as a free gift and to make decisions as if the gift were fixed, or uninfluenced by the decisions of these jurisdictions. There is a classic free-rider type of problem in which no jurisdiction would be expected to compensate other jurisdictions providing beneficial externalities. Thus, it is generally recognized that some senior government, including within it all of the offensive junior jurisdictions, is required in order to implement a matching grant solution. Hence the senior jurisdiction taxes all jurisdictions in a lump-sum (non-distortionary) way and uses the proceeds to pay matching grants in accordance with the sum of marginal benefits of beneficiaries external to each junior jurisdiction. It is important to point out here that the literature relating to matching grants deals with short-run, fixed-capacity, variable-output problems. In the present long-run setting, the grants must be long-run grants that effectively reduce the cost of constructing facility capacity. Of course, additonal short-run grants also may be required.

Now consider an omniscient senior government attempting to create sufficient incentives to move a jurisdictional system with non-optimal Y and R values to an optimal system. This problem is among those addressed in an excellent paper by McMillan (1975). First, assume that R is fixed even in the long run. Then, the policy of using matching grants to achieve Y_{opt} for a given R is the best possible policy. If both Y and R are variable in the long run, an optimal long-run matching grant policy for capacity, although sufficient to place jurisdictions on their Y_{opt} curves, would be insufficient to achieve a global optimum. An efficient 'capacity grant' may or may not create incentives to move R in the direction of R_{opt}. In general, one would expect that it would have the effect of increasing the inertia of the current non-optimal R, since the grant makes a certain capacity optimal for its current R. Further, setting the capacity grant rate as if R were optimal will not create the appropriate incentive to optimize R. Indeed, this strategy more than likely will lead to non-optimal Y and R values. This implies that, if R is variable, the senior government either must resize the jurisdictions directly so that these units are optimal, or this senior government must devise an incentive system for an optimal R, in addition to the one it must devise for an optimal Y. These conclusions now may be used to stress the restricted meaning of Figure 4. This figure is based directly on the assumption that the jurisdiction reacts efficiently along both of the Y_{opt} and R_{opt} curves. Since facing these curves generally is unnatural, a set of optimal intervention policies is required. The problem is made more difficult by the fact that optimal policies will depend upon the adjustment process assumed, as well as the objective function each jurisdiction is assumed to optimize. For this reason the problem is deferred to Section 3.6 below, wherein different jurisdictional objectives will be examined.

The information required to set up an optimal system of conventional output (or capacity) matching grants is quite formidable. This also has been noted, or illustrated, by Oates (1972), Harford (1977) and Jurion (1979a, 1979b). This can be demonstrated with a simple model of two jurisdictions providing an impure public good with reciprocal spillovers. The model also serves as an appropriate point of departure for subsequent discussion. [It is a generalization of a model used by Oates (1972) in Chapter 3, Appendix 1.] Let the level of public good consumed in regions 1 and 2 be, respectively

$$y_1 = w_{11} Y_1 + w_{21} Y_2 \quad , \text{ and}$$

$$y_2 = w_{22} Y_2 + w_{12} Y_1 \quad .$$

Here, as above, the Ys are the capacities or levels of provision at a given source and the w_{ij} coefficients, with $0 \le w_{ij} \le 1$, indicate the proportion of the output in region i that spills over into region j. One should note that $w_{i1} + w_{i2}$ need not sum to unity, since the good is assumed perfectly joint in an aspatial sense, it must, however, sum to less than 2. The w_{ij} coefficients may be considered to be the proportional losses due to the distance decay effects. Assume, for simplicity, that all citizens in a jurisdiction enjoy the same level of consumption (availability). Let x_i be the level of private numéraire enjoyed by each of the n_i citizens of i after their benefit tax is paid for their own supply of the public good. X_i is the total numéraire available in jurisdiction i and F is a well-behaved transformation.

The problem faced by jurisdiction 1, which assumes that spill-ins from 2 are given, then may be formulated as follows:

$$\text{MAX:} \quad W_1 = n_1 U_1(x_1, y_1) + \lambda(X_1 - n_1 x_1) - \omega F(X_1, Y_1) \quad , \quad (3.12)$$

where $y_1 = w_{11} Y_1 + w_{21} Y_2$. The first-order conditions are formed by taking $\partial W_1 / \partial x_1$, $\partial W_1 / \partial X_1$, and $\partial W_1 / \partial Y_1$ and setting each of these partial derivatives to zero. After manipulation to remove the multipliers λ and ω, the following condition emerges:

$$n_1 \, w_{11} \, MRS_1 = MC_1 \quad . \quad (3.13)$$

Analogously, for region 2

$$n_2 \, w_{22} \, MRS_2 = MC_2 \quad . \quad (3.14)$$

Here MRS_i is the marginal rate of substitution $[\partial u_i / \partial y_i] / [\partial u_i / \partial x_i]$ and MC_i is the marginal rate of transformation $[\partial F / \partial Y_i] / [\partial F / \partial X_i]$, or simply marginal cost.

A Pareto optimal set of Ys will be defined by maximizing the welfare of both jurisdictions together, such that

$$\text{MAX:} \quad W = \alpha_1 n_1 U_1(x_1, y_1) + \alpha_2 n_2 U_2(x_2, y_2) +$$

$$\lambda[X - n_1 x_1 - n_2 x_2 - \omega F(X, Y_1, Y_2)] \quad . \quad (3.15)$$

Here the αs are welfare weights and $X = X_1 + X_2$ is the total numeraire good in the system. Setting the partial derivatives of W with respect to x_1, x_2, X, Y_1 and Y_2 to zero and manipulating to remove the multipliers, the following variant of the standard Samuelson public good condition for each jurisdiction is obtained:

$$n_1 \, w_{11} \, MRS_1 + n_2 \, w_{21} \, MRS_2 = MC_1 \quad , \text{ and} \quad (3.16)$$

$$n_2 \, w_{22} \, MRS_2 + n_1 \, w_{12} \, MRS_1 = MC_2 \quad . \quad (3.17)$$

Comparing equations (3.13) and (3.14), respectively, with equations (3.16) and (3.17), and noting that $MC_1 + MC_2 > MC$ in general, one can see that in the independent adjustment equilibria the level of Ys will be too small.

Let the unknown optimal matching grant rates to jurisdictions 1 and 2 be r_1 and r_2. The following grant rates are needed such that conditions (3.18) and (3.19) become conditions (3.16) and (3.17), respectively:

$$n_1 \, w_{11} \, MRS_1 = MC_1 - r_1 \quad , \text{ and} \quad (3.18)$$

$$n_2 \, w_{22} \, MRS_2 = MC_2 - r_2 \quad . \quad (3.19)$$

If conditions (3.16) and (3.17) are inserted into equations (3.18) and (3.19), the optimal rates turn out to be

$$r_1 = n_2 \, w_{12} \, MRS_2 \quad , \text{ and} \qquad (3.20)$$

$$r_2 = n_1 \, w_{21} \, MRS_1 \quad , \qquad (3.21)$$

while further substitution for the MRSs yields

$$r_1 = w_{12}[(w_{11}MC_2 - w_{21}MC_1)/(w_{11}w_{22} - w_{12}w_{21})] \quad , \text{ and} \quad (3.22)$$

$$r_2 = w_{21}[(w_{22}MC_1 - w_{12}MC_2)/(w_{11}w_{22} - w_{12}w_{21})] \quad . \quad (3.23)$$

It can be seen that each grant depends upon both marginal costs and all the spillover/availability coefficients. Of course, if marginal costs were not assumed to be constant, the formulae would be much more complex.

3.6. Jurisdictional Equilibria and Efficiency: Comparative Statics

Jursidictions are assumed to optimize their own welfare. However, this is not sufficient to tie down the jurisdictions' problem. In the formulation above, it has been assumed that jurisdictions behave parametrically with respect to other jurisdictions. In particular, each jurisdiction selects its capacity under the assumption that the other jurisdiction will not alter its capacity. This assumption is essentially that made in classic Cournot oligopoly theory and in Nash games (for treatments of Cournot oligopoly theory see Shubik, 1959, 1975, 1980; Nicholson, 1972; Friedman, 1973, 1977; Bacharach, 1976; Okuguchi, 1976; Johansson, 1978; Friedman and Hoggatt, 1980; for treatments of Nash-type bargaining games see Nash, 1950, 1951; Shubik, 1975; Friedman, 1977; Jones, 1980). The assumption that parties in reciprocal externality, or public good, relationships behave parametrically with respect to each other's level of production of positive externalities has been a very common one. Discussion of features of this assumption appear in, for example, Davis and Whinston (1962), Connolly (1970), Pauly (1970), Oates (1972), Vardy (1972), McGuire (1974b), Le Grand (1975), McMillan (1975), Sandler (1975), Bacharach (1976), Kiesling (1976), Sandler and Cauley (1976), Harford (1977), Greenberg (1978), and Sonstelie and Portney (1978). Most of these writers deal directly with jurisdictional decision-making in the context of interjurisdictional spillovers. There is little question that it is as naive to assume that each jurisdiction expects other jurisdictions to maintain their output levels as it varies its own output level, as to assume that each oligopolistic firm expects other firms to maintain prices and/or quantities in the face of competition. Nevertheless, the Cournot or Nash assumption is a very useful point of departure.

Pauly (1970) provides an excellent graphical treatment of a simple Cournot adjustment process involving two jurisdictions providing a pure public good. He shows that it leads to what he calls an 'independent adjustment equilibrium' at which the total level of provision of the public good is smaller than the Pareto optimal level--a standard result for processes giving rise to beneficial externalities. Connolly (1970) provides a similar graphical analysis for two jurisdictions providing public goods that are subject to distance decay effects (e.g., television programs).

For present purposes the important point is this: the assumption of a particular set of expectations gives rise to a particular adjustment process and a particular equilbrium. Any policy aimed at moving this equilibrium to one that is Pareto optimal must take the expectations and adjustment mechanism into account. Thus, the system of matching grants derived in equations (3.22) and (3.23) was based on the assumption of a Cournot or independent adjustment process, and these grants would be necessarily different if some other process were assumed. The use of an inappropriate set of matching grants could even lead in the wrong direction.

Two other issues relating to the design of optimal policies must be raised. The first is that it matters what kind of tax-price system is used to collect revenues within the jurisdictions. If an inefficient system is used, then it would not be possible to correct for spillovers with a matching grant strategy. Nevertheless, the optimal grant rate would be affected by different tax systems. Finally, and perhaps most important, it matters what objective the jurisdictions are assumed to optimize. A variety of plausible objectives has been suggested in the literature, such as maximum net revenues and maximum average net benefits. In the analysis that follows attention is focused upon several plausible objectives. To make matters simpler, it always will be assumed that there is a Cournot or independent adjustment process, and that optimal lump-sum marginal benefit taxes are used within each jurisdiction.

Attention now will be turned to the general formulation of the interjurisdictional spillover problem presented in Sections 2 and 3.3. Thus, interest will be focused on the costs and benefits of a prototpe jurisdiction providing a distance decaying impure public good in a system of identical jurisdictions on a line. For simplicity consider only two jurisdictions i and j. The focus will be on jurisdiction i. Thus, only spill-ins from a single adjacent jurisdiction will be of concern.

There are many objectives that jurisdictions might reasonably optimize. For illustrative purposes, four very reasonable and/or plausible objectives will be treated here. To simplify matters, some new notation will be introduced. Let

$$B_i = B_i[Y_i, R_i, G_{ji}(Y_i, Y_j, D)] \quad ,$$

$$\hat{B}_i = \hat{B}_i(Y_i, R_i; Y_j, R_j) \quad , \text{ and}$$

$$b_i = B_i/(2R); \hat{b}_i = \hat{B}_i/(2R) \quad ,$$

where D refers to the distance between central facilities: $D = R_i + R_j$. The other notation has been used previously. B_i is the total benefit to jurisdictions i from the supply of the good—and this includes spill-ins. $\hat{B}_i$, on the other hand, is the benefit to jurisdiction i from its own supply only. These 'taxable benefits' exclude consideration of spill-ins. The following four objectives will be examined as Cases 1 thru 4:

Case 1. Max: $Z_1 = B_i - C_i$,

Case 2. Max: $Z_2 = b_i - c_i = (B_i - C_i)/(2R)$,

Case 3. Max: $Z_3 = \hat{B}_i - C_i$, and

Case 4. Max: $Z_4 = \hat{b}_i - c_i = (\hat{B}_i - C_i)/(2R)$.

As above, C_i and c_i refer to the gross and per capita costs of supply, respectively. The objectives are all maximal net benefits, with Cases 2 and 4 dealing with per capita net benefits, and Cases 3 and 4 defining benefits in the sense of 'own supplied' or taxable benefits. Even though only one neighboring jurisdiction is being considered, for simplicity, the problem facing the whole jurisdiction, not half of it, will continue to be examined.

When jurisdictions all have identical capacities and sizes, as depicted in Figure 1c, it is natural to assume that the jurisdictional boundary occurs at the intersection of the marginal evaluation curves. It is reasonable to continue to make the assumption that the boundary occurs at this point, even when capacities and/or spacings are unequal, since only then will citizens be assigned to the jurisdiction from which they receive the most benefit. Thus, to increase R_i, jurisdiction i must increase its capacity or move its central facility further from j, or both; it cannot simply annex territory by decree, as in a war.

The first-order conditions for these four cases are relatively straightforward. In each case jurisdiction i is allowed to optimize R_i and Y_i, assuming that jurisdiction j retains the location and capacity of its facility—the Cournot assumption. Then, it is jurisdiction j's turn, and so on. It is sufficient to focus on jurisdiction i's conditions since j's conditions will be symmetrical. The first-order conditions for Y, holding R constant, and for R, holding Y constant, now will be stated.

Case 1

For Y_i: $\quad \partial B_i/\partial Y_i + (\partial B_i/\partial G_{ji})(\partial G_{ji}/\partial Y_i) - (\partial C_i/\partial Y_i) = 0$

$\qquad$ or $\qquad (\partial B_i/\partial Y_i) = (\partial C_i/\partial Y_i)$ $\qquad\qquad$ (3.24)

$\qquad$ since $(\partial G_{ji}/\partial Y_i) = 0$ when R_i is held constant.

For R_i: $\quad \partial B_i/\partial R_i + (\partial B_i/\partial G_{ji})(\partial G_{ji}/\partial R_i) = 0$

$\qquad$ or $B(R_i) + (\partial B_i/\partial G_{ji})(\partial G_{ji}/\partial R_i)/2 = 0$ $\qquad$, $\qquad$ (3.25)

$\qquad$ where use has been made of the equality

$$(\partial B_i/\partial R_i) = 2B_i(R_i).$$

Case 2

For Y_i: $\quad \partial B_i/\partial Y_i + (\partial B_i/\partial G_{ji})(\partial G_{ji}/\partial Y_i) - (\partial C_i/\partial Y_i) = 0$

$\qquad$ or $\qquad (\partial B_i/\partial Y_i) = (\partial C_i/\partial Y_i)$

$\qquad$ since, as above, spill-ins are not influenced if R_i is constant.

For R_i: $\quad [2R_i\,\partial B_i/\partial R_i + (\partial B_i/\partial G_{ji})(\partial G_{ji}/\partial R_i) - 2B_i]/(4R_i^2)$

$\qquad$ or $B(R_i) + (\partial B_i/\partial G_{ji})(\partial G_{ji}/\partial R_i)/2 = (B_i - C_i)/(2R_i)$.

Case 3

For Y_i: $\quad \partial \hat{B}_i/\partial Y_i - \partial C_i/\partial Y_i = 0$

$\qquad$ or $\qquad \partial \hat{B}_i/\partial Y_i = \partial C_i/\partial Y_i$ $\qquad$.

For R_i: $\quad \partial \hat{B}_i/\partial R_i = 0 \qquad$ or $\qquad \hat{B}_i(R_i) = 0$ $\qquad$.

Case 4

For Y_i: $\quad \partial \hat{B}_i/\partial Y_i - \partial C_i/\partial Y_i = 0$

$\qquad$ or $\qquad \partial \hat{B}_i/\partial Y_i = \partial C_i/\partial Y_i$ $\qquad$.

For R_i: $\quad [2R_i\partial \hat{B}_i/\partial R_i - 2\hat{B}_i]/(4R^2) - [-2C_i/(4R^2)] = 0$

$\qquad$ or $\qquad \hat{B}_i(R_i) = (\hat{B}_i - C_i)/(2R_i)$ $\qquad$.

If equilibria exist, and it seems reasonable to expect that they would exist in the simple context assumed here, then these pairs of conditions would be met in each of the four cases (both for jurisdiction i and jurisdiction j). The next step in this analysis is to compare these equilibria. In addition, each will be related to the social optimum. The first-order conditions defining the social optimum have been given in Section 3.3 as equations (3.6) and (3.7) or (3.8), respectively. One should note that all of the terms in the present conditions have been defined in terms of jurisdictional (not per capita) benefits and costs to facilitate this comparison. It is expedient to illustrate the comparison graphically. Figures 5, 6 and 7 respectively are used to illustrate the various equilibria for Y given R, for R given Y, and, finally, jointly for Y and R. In the graphs, Y_k and R_k refer to case k, whereas Y* and R* refer to the global optimum.

In Figure 5 all equilibria for Y, given R, are at the points where marginal benefit curves intersect the single horizontal marginal cost curve; Y_1 and Y_2 lie on the curve $\partial B_i / \partial Y_i$, whereas Y_3 and Y_4 lie on the curve $\partial B_i / \partial Y_i$. This latter curve will lie above the former, since the marginal utility is assumed to fall with increased availability. It should be no surprise that, when jurisdiction i considers the spill-ins from jurisdiction j as free goods, supply is smaller than when the spill-ins are ignored. Also not surprisingly, in all cases the supply is less than the optimal Y* at which the value of the benefits of spill-outs to those in jurisdiction j are taken into account.

Equilibria are depicted for R, given Y, in Figure 6. All conditions involve curves showing marginal jurisdictional benefits from increasing jurisdiction size R. These marginal benefits are all positive and decreasing in R, as shown. From equations (3.7) and (3.8), optimal R* is defined by

$$\partial B / \partial R + [(\partial B / \partial G)(\partial G / \partial R)] = (B - C)/R \qquad \text{or}$$

$$B(R) + [(\partial B / \partial G)(\partial G / \partial R)]/2 = (B - C)/(2R) \qquad .$$

The term in brackets captures the indirect benefits relating to the changes in spillovers as R changes. It is important to point out that this term was meant to capture both the changed spill-ins and the changed spill-outs—since these spillovers will be identical in the equilibrium solution. In the present case it is expedient to rewrite this condition for R*, drawing attention to the possible asymmetry in spill-ins and spill-outs (this asymmetry could be due to the shapes of the marginal benefit curves, or different levels of capacity). The rewritten condition is

$$B(R_i) + [(\partial B_i / \partial G_{ji})(\partial G_{ji} / \partial R_i)]/2 +$$

442

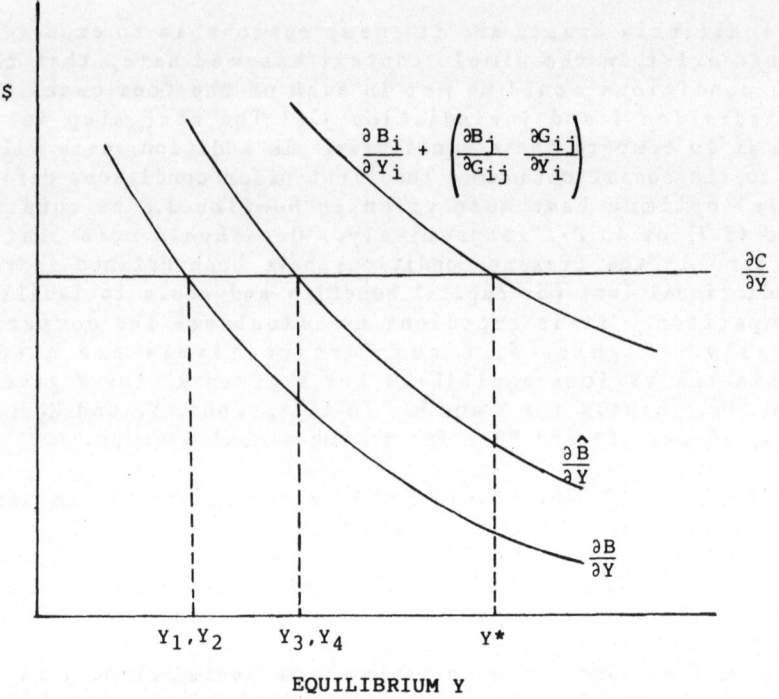

Figure 5. Equilibria for Y, given R.

$$[(\partial B_j / \partial G_{ij})(\partial G_{ij} / \partial R_i)]/2 = (B_i - C_i)/(2R_i) \qquad . \qquad (3.26)$$

Both bracketted expressions are expected to be negative, since the first terms will be positive and the second terms will be negative. R* is shown in Figure 6 as the point satisfying equation (3.26).

The optimal condition and Case 2 require the average per capita net benefit curve $(B_i - C_i)/(2R_i)$, while Case 4 requires $(\hat{B}_i - C_i)/(2R_i)$. These curves would be expected to increase and then decrease in R, as shown, with the curve that excludes consideration of spill-ins obviously lying below the curve that includes them. Since the marginal benefit curve defining R includes two bracketted expressions that are negative, and the one defining Cases 1 and 2 has only one such bracketted term, the latter curve will lie above the former. R_2 is shown at the intersection point while R_1 is where the marginal benefits become zero. Both R_1 and R_2 will be greater than R*. Cases 3 and 4 involve the curve $\hat{B}_i(R_i)$, with the former occurring where this term becomes zero, and the latter where this marginal benefit crosses

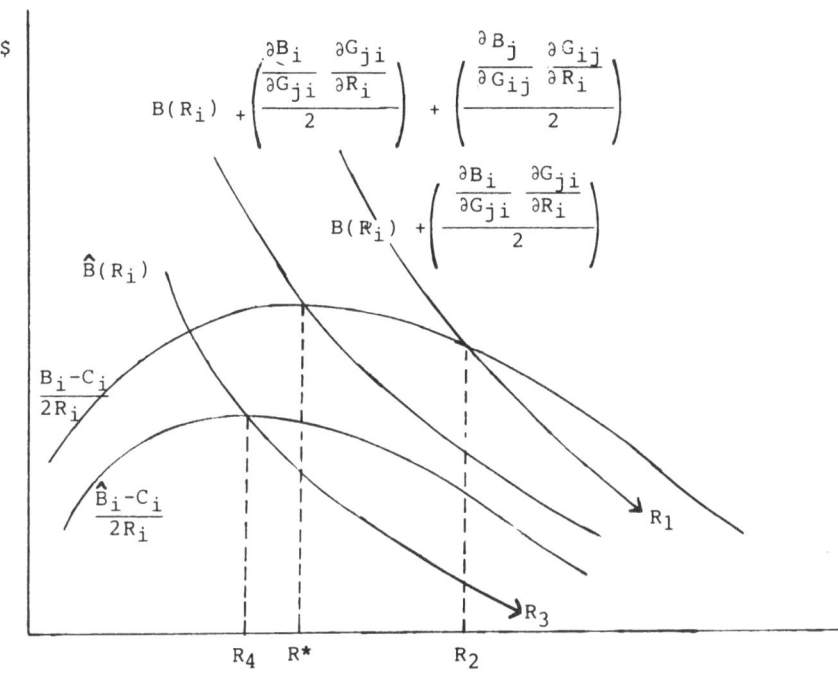

EQUILIBRIUM R

Figure 6. Equilibria for R, given Y.

$(\hat{B}_i - C_i)/(2R_i)$. The curve $\hat{B}_i(R_i)$ lies below both of the other marginal benefit curves [also it must pass through the maximum point of the $(\hat{B}_i - C_i)/(2R_i)$ curve]. This means that R_4 is smaller than R*, but that R_3 will be greater than R*--and likely greater than R_2 but less than R_1.

Based on this analysis of Cournot independent adjustment behavior under four different plausible objectives, it can be concluded that in all cases the capacities will be less than the Pareto optimal levels, with those based upon ignoring spill-ins somewhat less suboptimal. Furthermore, in all cases, except Case 4, the equilibrium jurisdiction sizes will be larger than the Pareto optimal size.

Figure 7 is similar to the southeast quadrant of Figure 3 (except here figures relate to the reference jurisdictions as a whole--they are not per capita) in depicting simultaneously the equilibrium Y and R. The positions and shapes of all curves are consistent with those in Figures 5 and 6 and the assumptions made in drawing these. The joint equilibrium point for case k (i.e., Y_k

444

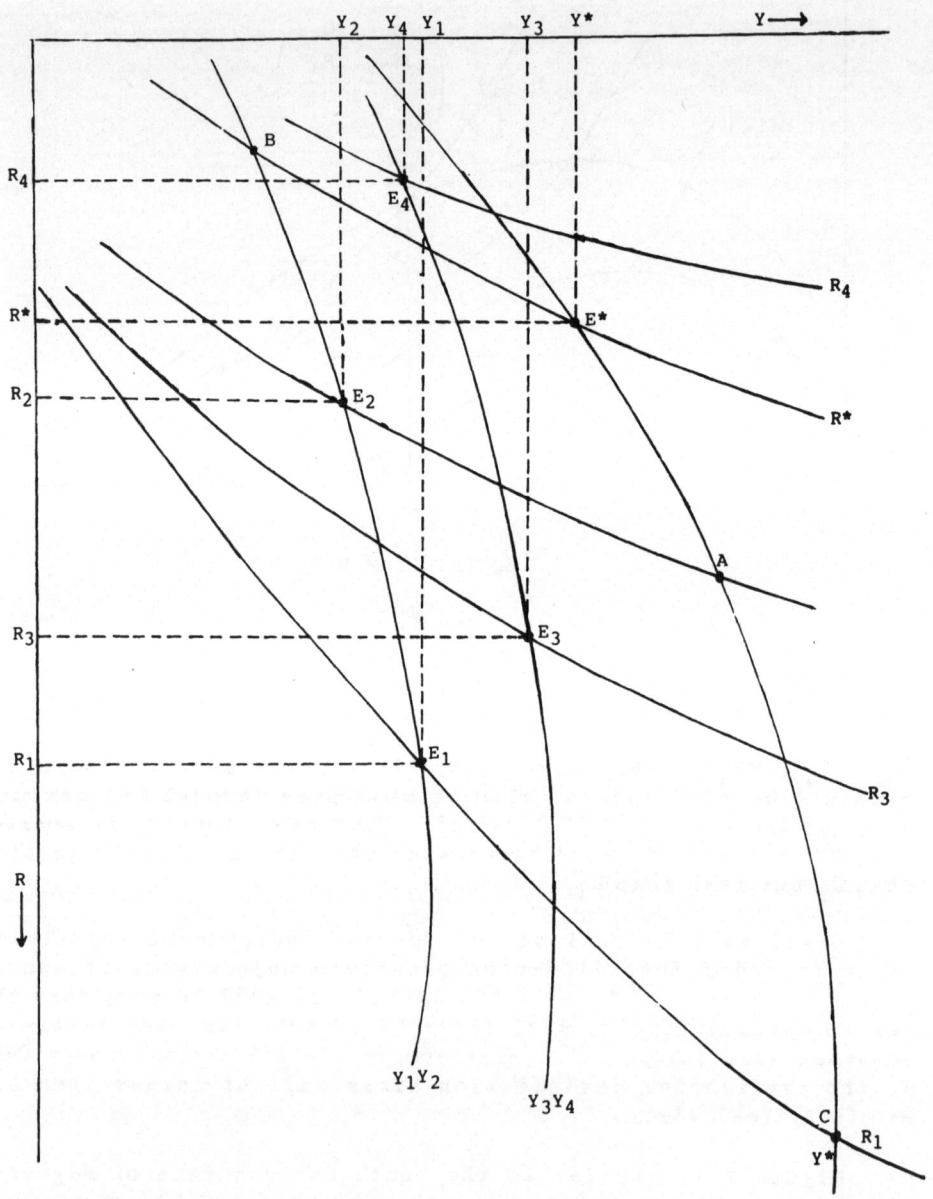

Figure 7. Simultaneous equilibria for Y and R.

and R_k) is depicted at the point E_k. In order to show E_1 and E_3 on the figure, it has been assumed that marginal benefits from own supply (Case 3) as well as from total supply (Case 1) drop to zero at finite jurisdiction sizes.

3.7. Policy Implications of the Static Analysis

In the previous section it was shown that if jurisdictions' sizes and capacities evolve according to an independent adjustment process, in which each jurisdiction is assumed to maximize its own net benefits (defined in four different ways, but in no case taking into account the benefits of spillovers to outsiders), then the level of capacity and the jurisdiction size both will be inefficient. It is reasonable to assume that this conclusion will hold for a host of other jurisdictional objectives, under the assumption of independent adjustment, as long as the external effects of decisions regarding the choice of Y and R are not fully reckoned. If the system of jurisdictions evolved with Y and R fully variable in the long run, one can be sure that the level of output will be suboptimal, and fairly confident that jurisdiction sizes will be too large.

For long-run Pareto efficiency there must be a long-run optimal intervention policy planned and executed by a duly constituted senior government. This policy, or set of policies, must be designed with full knowledge of the processes underlying the evolution of the current non-optimal equilibrium. In particular, it will be critical to know whether an independent adjustment process of the type discussed above, or some other particular process, has been at work—more precisely, what objective each jurisdiction is optimizing as it selects Y and R values—and a host of other considerations, such as all preferences, all costs, the extent of spillovers, the substitutability of spill-ins for domestic supply, and the nature of the tax-price scheme for internal revenue generation. Making the simplest assumptions about these latter variables, the general nature of the optimality can be appreciated by reexamining Figure 7. For each of the four jurisdictional objectives considered, any set of policies that shifts the relevant Y curve to the Y* curve and the relevant R curve to the R* curve will be sufficient. If the costs and benefits to jurisdictions associated with selecting Y and R are suitably adjusted (as they were, for instance, in Figure 3), then an independent adjustment process would lead to optimality.

Consider, for example, moving the equilibrium at E_2 to E*. Since Y is too small and R is too large, some policy directed toward increasing capacities and restricting jurisdiction sizes must be devised. In general, a long-run matching grant policy that subsidizes the costs of capacity equal to the value of the benefits generated by that capacity, which is not taken into account by the jurisdictions, is sufficient to move the Y curve to the Y* position. Thus, for example, the curve Y can be moved to Y* with a grant equal to $(\partial B_j/\partial G_{ij})(\partial G_{ij}/\partial Y_i)/2$, as is evident from Figure 5. Such a matching grant will tend to vary both with Y_i and with R_j. This grant also is suitable for Case 1. The grant suitable for Cases 3 and 4 will be smaller but more complex. It must force jurisdictions to take both spill-ins and spillovers into account. If jurisdiction sizes are not subject to manipulation, even in the long run, then these policies lead to an optimal second-best system.

If jurisdiction sizes are also variable in the long run, there are two ways of ensuring optimal sizes. The first, and more direct, would be for the senior government simply to decree the optimal jurisdiction size. The second would be to adjust the costs and benefits associated with selecting R, and to rely on the independent adjustment process, as was suggested for Y above. The direct approach requires no more information than the very large amount required in the indirect approach. The best indirect approach would be one that makes use of long-run matching grants tailored to the objective assumed to be optimized by the jurisdictions. For example, considering Figure 6, the long-run matching grant that will move the R_2 curve to the R* curve in Figure 7 is easily seen to be $(\partial B_i/\partial G_{ij})(\partial G_{ij}/\partial R_i)/2$. This grant most likely also will vary both with R_i and Y_j. Other grants, both related to Y and to R, can be determined in the manner used to derive grant rates r_1 and r_2 [see equations (3.22) and (3.23) in the simpler model in Subsection 3.5].

Referring to Figure 7 and considering further the policies required to move an equilibrium at E_2 (which assumes jurisdictions maximize per capita benefits including spill-ins) to E*, an efficient capacity grant (discussed above for this case) would place the jurisdiction on Y*. If there were no policy related to R, this would lead the jurisdiction to point A, at which both jurisdiction size and associated output are very large. If an optimal jurisdiction size grant (also discussed above) were used alone, then the jurisdiction would choose to be at point B, where jurisdiction size and capacity are both very small. Neither of the two policies alone leads nicely in the direction of E* in the sense of having a larger Y but a smaller R than at E_2. Similar conclusions can be drawn about single policy attempts to move other equilibria (at E_1, E_3 and E_4) to E*.

Thus, in general, two long-run policies are required: a more conventional one aimed at Y, and another one aimed at R. A pair of efficient matching grant policies would be sufficient. However, as is clear from the foregoing discussion, a very large amount of information would be required to design such policies. Even in the rarified atmosphere of the present set of assumptions, it is instructive to consider further whether there are any additional costs associated with indirect policies of redirecting independent adjustment processes in the long run. As long as facilities do not have to be moved, the adjustment of capacities more than likely would have very few important side effects. On the other hand, the side effects of an incremental adjustment of Rs over time would tend to give rise to significant additional costs. Either the costs of relocating facilities in the centers of jurisdictions, or welfare losses associated with a non-central location, would have to be incurred. Similarly, the institutions and machinery associated with the local collective decision-making process would have to be revamped on a regular basis. Many other real world transition costs associated with a piecemeal adjustment of R become apparent as the restrictive assumptions of the model are relaxed. Under these conditions it seems resonable to recommend that senior governments use long-run matching grant policies for Y, but achieve optimal sizes at rather widely separated points in time by decree. Thus, controlled independent adjustment would be permitted only for public facility capacities.

4. THE DYNAMICS OF THE SPILLOVER PROBLEM

4.1. Introduction to Dynamic Considerations

It has become increasingly clear in the discussion above, and in particular in the discussion related to comparative statics in Sections 3.6 and 3.7, that the long-run problem of jurisdiction size and public facility capacity is one best treated as intrinsically dynamic. Treating the adjustment process using single snapshots and discussing one-shot policies for achieving an optimal system in the long run are rather poor first approximations for the problem, even without moving away from the already much simplified problem context. Apparently there have been no models or even discussions of dynamics of the problem in the literature. The purposes of this section simply are to introduce the dynamics of the problem, and to formulate, solve, and briefly interpret a rather simple optimal control model of dynamic independent adjustment.

The jurisdiction problem at issue has been treated in the discussion of statics and comparative statics above as an explicitly long-run problem since capacities and jurisdiction sizes could only vary in the long-run. Thus, long-run consumer surpluses were defined and long run costs have been used. However, since the problem was treated as one of a single static equilibrium, discounting for the future stream of costs and benefits was only implicit. In a dynamic formulation disounting must be explicit. Furthermore, a dynamic model will not be exclusively concerned with

the global optimality of the final state of the system. Rather it
must be concerned with the relative efficiency of a whole series of
states of the system over time. In particular, knowing that the
actors in the system will not make appropriate decisions over time,
the goal becomes one of finding the optimal way of controlling the
evolution of the system so that some appropriate long-run and
intertemporal objective is optimized. In this context there must
be an explicit model of the dynamics of adjustment of the decisions
of the actors, both with and without certain controls.

4.2. An Optimal Control Formulation of a Simple Dynamic Problem

In this section a dynamic extension will be considered for the
simple two-jurisdiction model treated in Section 3. It will be
assumed that both jurisdictions use a Cournot-Nash strategy—again
as in Section 3. In the model developed here it will be assumed
that the jurisdiction has the objective associated with Case 1 of
Subsection 3.6; namely, the maximization of net jurisdictional
benefits including spillovers. This problem lends itself to a
discrete dynamic framework for which difference equations versus
differential equations characterize the process. Treating Cournot
and related duopoly and oligopoly problems in discrete time is
quite common (see, for example, Shubik, 1959, 1980; Friedman, 1973,
1977; Okuguchi, 1976; Johansson, 1978; Mirman, 1979). The problem
could be considered a discrete time form of a peculiar type of
'differential game' (see, for discussions of differential games,
Isaacs, 1965; Bryson and Ho, 1969; Intrilligator, 1971; Sethi and
Thompson, 1981) in which there are two 'players'—the two
jurisdictions—and a third player—'controller'—that is a senior
government. (Normally all players are considered the controllers.)
The objective of the controller is to choose the level of a
particular control variable in such a way that the decisions made
by the players are in some sense optimal over a particular time
horizon. In this case the horizon is a finite one ending at time
T. The behavior of the players—in this case, the tendency is for
them to make non-optimal decisions in the absence of intervention—
is taken as given. Both players are assumed to 'play Nash.'

As in the comparative static analysis in Section 3, both
jurisdictions are permitted to select values of both Y and R. One
should recall that R values are affected not only by decisions
regarding facility location (spacing), but also by decisions
regarding capacity, since the jurisdictional boundary is assumed to
be defined by the intersection of the marginal benefit curves
relating to neighboring jurisdictions' facilities. Although
jurisdicitons will be allowed to choose both Y and R, only the
capacity will be the subject of direct controls. The model could
be generalized to permit explicit controls on R as well, though.
The control variable used by the central government will be the
time and jurisdiction-specific matching grant rates for capacity

expansion. Such grants internalize interjurisdictional spillovers in the static problem, so that there is every reason to believe these will be appropriate varible instruments over time. Since moves in facility locations are assumed to be costless and are not controlled by the senior government, one would expect jurisdictions to adjust their Rs instantly to new values of Ys and Gs. Indeed, it is this uncontrolled adjustment in the R values that makes the problem particularly interesting. This will be explained further below.

Denoting values of variables at time t with a superscript t, the following notation will be used:

$$D^t = R_i^t + R_j^t \text{ (facility spacing)},$$

r_i^t = the proportional matching grant rate for jurisdiction i at time t,

$$B_i^t = B_i^t(Y_i^t, G_{ji}^t, D^t),$$

$$C_i^t = C_i^t(Y_i^t),$$

$$\tilde{C}_i^t = \tilde{C}_i^t(Y_i^t, r_i^t) = (1 - r_i^t)C_i^t,$$

$$W_i^t = B_i^t - C_i^t, \text{ and}$$

$$\tilde{W}_i^t = B_i^t - \tilde{C}_i^t.$$

One should note that the matching grant is included simply as a reduction in unit costs.

Jurisdictions are assumed to select values of Y_i^t and R_i^t to maximize $\tilde{W}_i^t$ in each time period t. $\tilde{W}_i^t$ becomes W_i^t when the grant is zero. Without the grant, the long-run static equilibrium for this problem is defined by equations (3.24) and (3.25). The existence, stability, and uniqueness to isomorphic discrete dynamic oligopoly problems is proved by Okuguchi (1976) and Johansson (1978). The static long-run Pareto optimal solution is defined by equations (3.6) and (3.7) or (3.8). A comparison of these sets of equations reveals that the efficient one-shot static matching grants for Y and R are given, respectively, by the following equations:

$$(\partial B_j/\partial G_{ij})(\partial G_{ij}/\partial Y_i) \quad \text{and}$$

$$[(\partial B_j/\partial G_{ij})(\partial G_{ij}/\partial Y_i)]/2 - (B_i - C_i)/(2R_i)$$

Both of these grants generally will vary with Y_i, Y_j, R_i, R_j, G_{ij} and G_{ji}. If both of these grants were used simultaneously to control the dynamic behavior of the two jurisdictions, the control problem would become relatively trivial. It is very likely that

the optimal strategy would be to set the grants at their static
levels, assuming both Y and R are optimal. On the other hand, if
only a capacity grant is used, and each jurisdiction is free to
select its jurisdiction size (facility locations) in an
uncontrolled manner, then the problem is no longer simple.
Referring to figure 7 and assuming that the current values of Y and
R are given at E_1, the equilibrium solution to an independent
adjustment process with both jurisdictions maximizing the Case 1
objective is one in which an efficient static capacity grant would
move the equilibrium to point C, which is at the intersection of
the R_1 and Y* curves. This move increases output, as was desired,
but the increase goes well beyond Y*. It also increases
jurisdiction size--a move away from the Pareto optimal R. This
latter move will tend to reduce the spillovers for which the grant
was designed to compensate. Considering the intertemporal
interdependence of the actors, it may be better for the central
government ot use a different (smaller) matching grant rate that
has a less dramatic effect on Y and R.

The key variables in a control problem are the state variables
that summarize the dynamic behavior of the system. One
straightforward way of writing the state relations for the present
problem is described by the following two equations.

$$Y_i^{t+1} = F(_j^t, R_j^t, R_i^{t+1}, r^t) \qquad , \text{ and} \qquad (4.1)$$

$$R_i^{t+1} = K(Y_j^t, R_j^t, Y_i^{t+1}) \qquad . \qquad (4.2)$$

At any time t+1 the spillovers can be determined from the values of
the state variables at this time using some new function M:

$$G_{ji}^{t+1} = M(Y_j^{t+1}, D^{t+1}) \qquad . \qquad (4.3)$$

A symmetrical set of relations could be written for jurisdiction j.
Equations (4.1) and (4.2) can be rewritten as a set of simultaneous
difference equations with only variables superscripted t on the
right-hand side. Now consider the following state vector:

$$P^t = (Y_i^t, R_i^t, Y_j^t, R_j^t) \qquad . \qquad (4.4)$$

The relationship depicted by equation (4.4) may be captured
using a vector function Q as

$$\Delta P^t = P^{t+1} - P^t = Q(P^t, r_i^t, r_j^t) \qquad . \qquad (4.5)$$

Equations (4.1) and (4.2) represent the optimal static
decisions taken by jurisdiction i at time t+1, assuming that
jurisdiction j continues to select Y_j^t and R_j^t in time t+1. The
spillover function, G_{ji}^t, is implicit in both of these equations.
These equations are precisely the solution to jurisdiction i's

static first-order conditions for Y and R, under Nash behavior, as given in equations (3.24) and (3.25) in Section 3. Indeed, if specific functional forms for benefits and costs were given, then one could derive the specific functional forms of F and K for i and j, and, assuming initial values for Y and R, the model (4.5) could be run over time to show how the discrete dynamic Cournot-Nash equilibrium is converged upon.

Next, the optimal control problem facing the central government can be stated as follows:

$$\underset{r_i^t, r_j^t}{\text{MAX}:} \quad \tilde{W} \quad = \quad \sum_{t=0}^{t=T} \gamma^t (\Omega_i \tilde{W}_i^t + \Omega_j \tilde{W}_j^t) \tag{4.6}$$

$$\text{S.T.:} \quad \Delta P^t \quad = \quad Q(P^t, r_i^t, r_j^t) \quad ,$$

with P^o given as the initial vector element.

Here T is some final time period, and $\gamma^t = 1/(1 - \alpha)^t$, where α is the discount rate and here t is an exponential power, not a superscript (this is the only time in this paper that the superscript t is other than an index). This objective is the present value or the discounted stream of weighted (by welfare weights Ωs) net benefits of both jurisdictions taken together. In solving this problem, the central government simply makes the assumption that the players are 'playing Nash' and continue to do so. In this problem formulation it is assumed that the central government simply chooses the grant rates to optimize the present levels of capacity in a dynamic context. It does not anticipate that any jurisdiction will adjust its R in association with its adjustment in Y due to the grants. Of course, a variety of alternative problem specifications readily come to mind. Note that for more generality the controls have not been constrained to be non-negative. Should these controls become negative, a matching tax would be indicated. Additional controls might be used. An infinite time frame might be assumed. The problem could be restricted in such a way that the optimization problems faced by the jurisdictions are explicit rather than implicit in function Q (this would be the explicit differential game formulation of the problem). Finally, the model could be extended rather easily in the direction of general equilibrium analysis by having the central government tax the jurisdictions so that there are sufficient funds in the central fisc to pay the required grants.

In solving the optimal control problem, it is helpful to introduce some new notation. Let Z^t be the undiscounted value of the objective at time t, such that

$$Z^t = Z(\Omega_i \tilde{W}_i + \Omega_j \tilde{W}_j)$$

$$= Z(Y_i^t, R_i^t, Y_j^t, R_j^t, r_i^t, r_j^t, \Omega_i, \Omega_j) \quad . \quad (4.7)$$

Supressing the welfare weights Ω_i and Ω_j in subsequent developments, equation (4.7) becomes

$$Z^t = Z(P^t, r_i^t, r_j^t) \quad (4.8)$$

where it will be recalled that P^t is a four-element vector of the state variables.

The Lagrangian form of the control problem (4.6), with P^o given, can be rewritten as

$$\text{MAX: } L = \sum_{t=0}^{t=T} \{\gamma^t Z^t(P^t, r_i^t, r_j^t) +$$

$$\lambda^{t+1}[Q(P^t, r_i^t, r_j^t) - P^{t+1} + P^t]\} \quad . \quad (4.9)$$

It should be noted that this control problem is somewhat unusual since the final Z^T is not constrained in any way. Having a 'salvage value,' or something similar, in this problem would be quite inappropriate. One also should note that neither the controls nor the state variables are constrained. The controls conceivably could be negative and the state variables necessarily will be non-negative by virtue of the Nash behavior captured in the difference equation (4.5).

Problem (4.9) can be solved using a discrete form of Pontryagin's maximum principle. This principle for continuous time is developed fully in all of the standard works treating control theory. Authors treating the discrete time version of the maximum principle include Fan and Wang (1964), Dorfman (1969), Intrilligator (1971), Miller (1979), and Sethi and Thompson (1981). In order to apply this principle to equation (4.9), a Hamiltonian should be defined. For the present problem the appropriate Hamiltonian is

$$H^t = \gamma^t Z^t(P^t, r_i^t, r_j^t) + \lambda^{t+1}Q(P^t, r_i^t, r_j^t) \quad . \quad (4.10)$$

The problem (4.9) then may be restated in Lagrangian form, using H^t, as:

$$\text{MAX: } L = \sum_{t=0}^{t=T} [H^t + \gamma^{t+1}(-P^{t+1} + P^t)] \quad . \quad (4.11)$$

For this problem the following three first-order conditions derive from the application of the maximum principle [for simplicity, all functions are assumed to be well-behaved so that the derivatives exist and these conditions are also sufficient (see Baumol, 1970; Intrilligator, 1971; Sethi and Thompson, 1981)]:

$$\partial L/\partial P^t = 0 \quad \Rightarrow \quad \partial H^t/\partial P^t - \lambda^t + \lambda^{t+1} = 0 \quad ,$$

$$\text{leading to } - \partial H/\partial P^t = \lambda^{t+1} - \lambda^t \quad . \tag{4.12}$$

$$\partial L/\partial r_i^t = 0 \quad \Rightarrow \quad \partial H^t/\partial r_i^t = 0 \quad , \text{ and} \tag{4.13}$$

$$\partial L/\partial \lambda^t = 0 \quad \Rightarrow \quad \partial H^t/\partial \lambda^t - P^{t+1} + P^t = 0 \quad ,$$

$$\text{leading to } \partial H^t/\partial \lambda^t = P^{t+1} - P^t \quad . \tag{4.14}$$

The condition relating to r_j^t is symmetrical to equation (4.13). (If H^t is concave, then these condtions are necessary and sufficient.) Taking the partial derivatives with respect to the Hamiltonian and using the subscript m to refer to the mth element of the vector P and its Q function counterpart [as shown in equation (4.4)], equations (4.12), (4.13) and (4.14) respectively lead to the following result:

$$\lambda_m^{t+1} = [\lambda_m^t - \gamma^t(\partial Z/\partial P_m^t)]/[(\partial Q/\partial P_m^t) + 1)] \quad , \tag{4.15}$$

$$\gamma^t(\partial Z^t/\partial r_j^t) + \sum_{m=1}^{m=4} \lambda_m^{t+1}(\partial Q_m/\partial r_j^t) = 0 \quad , \tag{4.16}$$

and, ensuring that the constraint is satisfied,

$$Q_m(P^t, r_i^t, r_j^t) = P_m^{t+1} - P_m^t \quad . \tag{4.17}$$

One should recall that the problem has a free end-point with no constraints on the terminal value of the objective. The costate variable associated with the terminal state must necesarily be zero: $\lambda^T = 0$. The following iterative approach to solving these conditions for the optimal control trajectory suggests itself. Start with a given P^0 and a guessed λ^0, and then solve the system iteratively for r^t, P^t and λ^t, for t=1,2,...,T. If, in the final stage, $\lambda^T \neq 0$, then return and adjust the guessed value of λ^0 in the locally optimal way. Repeat this procedure until, in the final state, $\lambda^T = 0$, as required.

For purposes of comparison, and to make the interpretation of the conditions (4.15) and (4.16) more straightforwad, the control problem now will be restructured as a more conventional one of dynamic programming using Bellman's principle of optimality. According to this principle an optimal policy has the property that, whatever the initial state and decision (i.e., control) are, the remaining decisions must constitute an optimal policy with regard to the state resulting from the first decision. This principle allows a dynamic optimization problem to be reduced to a series of static optimization problems. It is helpful to consider the first stage of the problem as that associated with time T and work backwards in time to t=0.

Let $V_k(P^{T-k})$ be a performance function associated with stage k. This term represents the maximal value of the objective function for a problem of length k starting from state T-k. The problem then can be written succinctly as: find $V_T(P^0)$. V_o is the maximal value of the objective for a problem of zero length starting (and staying) at P^T. In Figure 8 the structure of this problem, starting with stage 1 and time T, is depicted as a flow diagram.

Using the fundamental recurrence relation, the optimal control problem (4.9) may be rewritten as

$$V_k(P^{T-k}) = \underset{r_i^{T-k}, r_j^{T-k}}{\text{MAX}} \gamma^{T-k} Z^{T-k}(P^{T-k}, r_i^{T-k}, r_j^{T-k}) + V_{k-1}(P^{T-k+1}) \;. \tag{4.18}$$

Then, making use of $P^{T-k+1} = Q(P^{T-k}, r_i^{T-k}, r_j^{T-k})$, equation (4.18) may be rewritten as

$$V_k(P^{T-k}) = \underset{r_i^{T-k}, r_j^{T-k}}{\text{MAX}} \gamma^{T-k} Z^{T-k}(P^{T-k}, r_i^{T-k}, r_j^{T-k}) +$$

$$V_{k-1}[Q(P^{T-k}, r_i^{T-k}, r_j^{T-k})] \;. \tag{4.19}$$

This problem is solved, then, as a series of static optimization problems for each stage, k=T, ,T-1, ... ,0. It should be noted that the solution has an obvious boundary condition, namely that $V_o(P^T) = 0$.

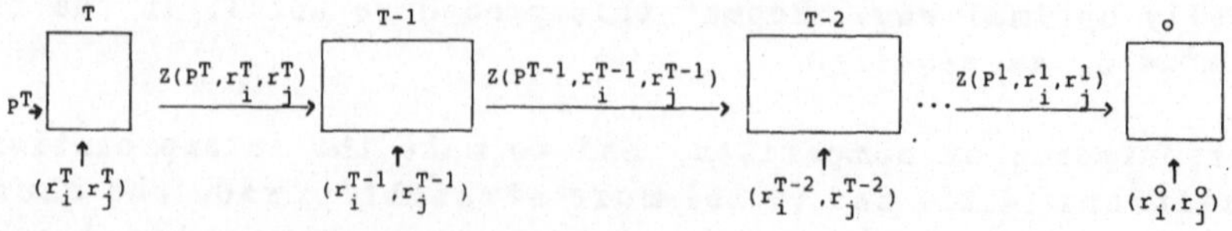

Figure 8. Flow diagram for a problem of length k.

The objective (4.19) holds for any stage k so it must hold for any time t=0,1,2,...,T. In deriving the first-order condition for a maximum, it is expedient to consider the one-stage problem, setting k=1. The time, therefore, is T-1. Simply set the partial derivative of V_1 equal to zero. Again, the subscript m is taken to refer to the mth element in the state vector **P**. The single condition is

$$\partial V_1/\partial r_i^{T-1} = 0 \quad \Rightarrow$$

$$\gamma^{T-1}(\partial Z/\partial r^{T-1}) + \sum_{m=1}^{m=4} V_{om}(\mathbf{P}^T)(\partial Q/\partial r_i^{T-1}) = 0 \quad , \ (4.20)$$

where $V_{om}(\mathbf{P}^T) = \partial V_o/\partial P_m$. In addition, the boundary condition noted above holds [i.e., $V_o(\mathbf{P}^T) = 0$].

Solving for the opitmal control trajectory using this condition, as in the conditions based upon the maximum principle above, will require an iterative procedure. In this case, one would start with a known $\mathbf{P}^o$ and guessed V_T. If, after solving the system, $V_o(\mathbf{P}^T) \neq 0$, then one would return to and adjust V_T in a locally optimal way. These steps are repeated until $V_o(\mathbf{P}^T) = 0$ as the required boundary condition.

Since the maximum principle control formulation and the dynamic programming control formulation are roughly equivalent, it should occasion no surprise that their first-order conditions can be shown to be consistent. In particular, it will be shown that condition (4.20) is equivalent to condition (4.16). First, rewriting condition (4.16) with t=T-1,

$$\gamma^{T-1}(\partial Z/\partial r^{T-1}) + \sum_{m=1}^{m=4} \lambda^T(\partial Q_m/\partial r_i^{T-1}) \quad . \ (4.21)$$

Comparing equation (4.21) with (4.20), in order to show equivalence, $\lambda^T = V_{om} = (\partial V_o/\partial P_m)$ must hold. Indeed, the structures of the two problems require that this be true (see Intrilligator, 1971, p. 356). In general, the change in the objective function with respect to the initial state must be equal to the corresponding value of the costate variable. Further, it has been noted that the boundary conditions for the maximum principle and the dynamic programming problems are, respectively, $\lambda^T = 0$ and $V_o(\mathbf{P}^T) = 0$, which completes the demonstration.

It has been shown that one of the two interesting conditions from the maximum principle formulation can be derived using a more familiar dynamic programming formulation. The other condition (4.15) for the costate variables of the former problem has no counterpart in dynamic programming.

4.3. Discussion of the Dynamic Problem and Optimality Conditions

Since the first-order condition from the dynamic programming formulation of the problem has been shown to be equivalent to one of the maximum principle conditions, these obviously have the same meaning. From condition (4.20) from dynamic programming, at each point in time the present value of the marginal effect of the control on the objective for that time plus the sum of the present values of the marginal effects on all of the state variables, where these are already optimal with respect to the controls, must equal zero. When this sum equals zero at each point in time, the trajectory of the control variables will be optimal. Recalling that the state variables that are featured in the summation part of the condition are X_i, R_i, Y_j and R_j, and using the optimal values of r_i^t and r_j^t for each point in time, the time paths of each of the state variables can be generated. These paths, taken over time, maximize the aggregate welfare $\widehat{W}$ in the system.

The costate variables λ_m^t from the maximum principle formulation are simply temporal Lagrangian multipliers and, as such, they can be interpreted as indicating the sensitivity of the objective function to variations in the parameters. In particular, since the objective is measured in terms of the present value of aggregate net benefits, the λs are vectors of shadow prices. Given the present simple problem structure, the only λs of real interest would be the λ_ms at time zero. λ_m^o indicates the effect on the objective of changing the value of the initial state variable m by one unit. Using these multipliers, one could explore, for example, the long-run welfare effects of different initial jurisdiction sizes and capacities with their implied extent of spillovers. If the control problem were reformulated so that the Lagrangian in equation (4.11) was also an explicit function of T (see Sethi and Thompson, 1981) the costate variable associated with T could be interpreted as the sensitivity of the present value of long-run welfare to changes in the length of time over which the central government is allowed to effect its matching grant strategy.

The first-order condition relating to the costate varibles of the maximum principle formulation, adjoint equation (4.12), has not been discussed. This condition states that, at optimality, the difference between the values of the optimal shadow prices assoicated with each state variable at two points in time must be equal to the negative of the effect on the Hamiltonian of a small change in the state variable. Equation (4.15) has been derived from (4.12) and ,in it, the shadow price associated with a state variable at any time is shown to be a function of this same price in the previous time period, as well as both $\gamma^t(\partial Z/\partial P_m^t)$, the marginal net benefit of a change in the state variable, and $\partial Q/\partial P_m^t$,

the change in first differences of state vector with the change in
the state variable (interaction effects). The shadow price
decreases as each of these partial derivatives becomes larger.

The next step would be to construct and run some simulations
of this model.

4.4. Extensions of the Simple Dynamic Formulation

The dynamic problem treated here is a particularly simple and
apparently tractable one. However, in being simple, many key
features of reality have been abstracted away. there are many ways
in which more interesting and realistic extensions could be
formulated, and some of them will be briefly noted here.

The first, most obvious, extension would be to allow the
central government controls on jurisdiction size in addition to the
controls on capacity. Such controls could be in the form of grants
(as was noted in Section 3), or could be 'impulse controls' in the
form of centrally directed reorganizations at discrete points in
time. Of course, in the latter case, the costs of the
reorganization should be taken into account. In either case, it
also should be important to extend the problem in the direction of
general equilibrium by having the central government face the
problem of levying endogenous taxes on the jurisdictions, too, so
that it balances its budget while paying the grants. It should be
noted that within the control theoretic formulation, various
constraints on the state and/or control variables, including
terminal value constraints, are relatively easily accommodated.

Another extension would be in the direction of more realistic
behavioral assumptions on the part of both the competing
jurisdictions as well as the central government. Rather than
retaining the relatively naive assumption that the jurisdictions
continue to 'play Nash,' even though they are continually
disappointed in their subsequent performance, some more interesting
sets of expectations regarding the response of other jurisdictions
can be used. The problem is sufficiently similar to that of
oligopolistic competition that some of the many very interesting
and realistic models—many of them game-theoretic—of oligopolistic
behavior could be easily adapted here. Many of these developments
are summarized in Nicolson (1972), Okuguchi (1976), Friedman (1977)
and Johansson (1978). In particular, the models in which both
quantity of output and price are variable are the most suitable
analogies to the present problem, with capacity and jurisdiction
size variable. In addition, those models treating differentiated
products and/or preferred markets yield considerable insight into
ways in which jurisdictional overlap and spillover might be more
rigorously modelled. The most straightforward extension of the
model in this respect would be to develop one in which the

conjectural variation is treated as a parameter between zero and one (rather than assuming, as here, that it is zero). There also are a number of ways in which the reaction functions of players can be treated as stochastic. This poses no problems for control theory as its stochastic versions are highly developed (see Chow, 1975; Aoki, 1976; Liu, 1980).

In the dynamic model presented, here the central government was assumed not to anticipate changed jurisdiction sizes resulting from the capacity grants. Since it is necessarily assumed that this government has sufficient information to solve for the efficient grants in each time period, it would be more realistic to assume that this level of government fully, or partially, anticipates the responses of the jurisdictions. The literature on differential games gives some ideas about how anticipations of this type might be operationalized.

The model also could be extended in its treatment of the intertemporal nature of decision-making. In particular, jurisdictions could be allowed to decide on both Y and R in alternative time periods (see Friedman, 1977). In addition, it could be assumed that there is a four-period cycle in which, say Y_i, Y_j, R_i and R_j were decided, in sequence.

Perhaps the most important aspects of the the model to extend relate to the number of participants and the dimensionality and simplistic treatment of space. The ultimate goal would be a model with n jurisdictions in two-space. However, even for one-space the model set out above is very simple because only two jurisdictions were treated explicitly. Indeed, the fact that spillovers would come from only one side, in this context, was not rigorously treated. In extending the model to multiple jurisdictions with single central facilities on a line, the first feature to capture would be spill-ins received from jurisdictions that are not adjacent. This dimension of the problem can be modelled in a relatively straightforwrd way using difference equations. Some progress can be made along these lines using an analogue with Kirchhoff's law applied to electrical networks (for the general structure of this formulation, see Spiegel, 1971, pp. 205-206). An important challenge is to model adequately the directionality of the space. In the process of adjustment it generally will be the case that the distance from facility to boundary on the left, R_{left}, will not equal R_{right}. Further, when a jurisdiction adjusts its size by moving its facility to the left, the change in R_{left} will not be the same as the change in R_{right}. Finally, the persistently vexing issue of the boundary distortions caused by assuming that the linear economy is of finite length must be grappled with. Further, with a line of finite length, the number of jurisdictions competiting over it will no longer be constant.

It seems clear that the extension to two-space would be an enormous leap. (For a brief treatment of a simple static two-space model, see McMillan, 1975.) However, in this regard it might be noted that there exist some possible analogies in physics and biology. In particular, the 'potential models' of competitive processes in two-space, of the type discussed by Curry, Sonis and Haag in this volume, may have some indirect application.

5. CONCLUDING COMMENTS

This paper has treated a rather simple version of an important evolutionary process of the determination of jurisdiction size and public good capacity. The static and comparative static aspects of the problem have been treated in considerable depth not only because they are of interest in their own right, but also because an understanding of the static structure of the problem is absolutely essential to the construction of dynamic and, eventually, evolutionary models. Indeed, the simple discrete dynamic control model developed in Section 4 was a direct extension of the comparative static models of Section 3. The important finding of Section 3, namely that centrally choreographed long-run intervention policies with respect to both public good capacity and jurisdiction size (facility location) are required, was an important feature in structuring the dynamic control problem in Section 4—even though the full problem with two sets of controls has not been treated here.

To date, very little research has been done on providing impure public goods efficiently in an explicitly spatial context. Even simple static conceptualizations of these problems seem to require a large number of assumptions so that they remain tractable and lead to interesting insights (cf., Smolensky, et al., 1970; McMillan, 1975; Capozza, 1976; Schuler and Holahan, 1977). The truly interesting work to be done in this area is in the field of dynamics and, in particular, pursuing some, or all, of the extensions noted in Subsection 4.4. However, a good case can be made that more work first should be done in building better, more elaborate, and realistic static models to serve as a foundation. One high priority would lead away from the type of pure theory developed above in the direction of more relevant models. In this regard, it is important to undertake some empirical research on actual jurisdictions' decision-making processes. As was shown in Section 4, it matters a great deal to outcomes, and the development of optimal policies to adjust them, just what objective jurisdictions tend to maximize. Unfortunately, it seems unlikely that a definite answer to this question will be forthcoming, since in any real world setting there is likely to be considerable noise and conflicting multiple objectives that may be quite setting-specific. Moreover, this observation could be taken to imply that jurisdictions might more appropriately be considered in the model as having multiple objectives.

The discussion of extensions in Section 4.4 was restricted to those of particular importance for more realistic dynamic and evolutionary models. However, perhaps the static model being extended is itself overly simple. Some of the most obvious and important extensions of the static model are as follows:

(1) allow congestion in the consumption of the public good,

(2) rather than having zero exclusion, incorporate the degree of exclusion as an endogenously determined variable (exclusion may be possible but costly),

(3) relax the pure substitutability of the good over space (allow product differentiation),

(4) have non-constant returns to scale in production,

(5) include both domestic and senior government taxation types and levels,

(6) have different rules regarding the determination of jurisdiction boundaries,

(7) problematize the jurisdictional social choice process,

(8) incorporate more realistic behavioral rules, especially those relating to reactions to the decisions of others,

(9) deal with non-uniform preferences, incomes, and population densities,

(10) allow Tiebout-type migration between jurisdictions in response to different levels of welfare in each (this rather important extension is very difficult to model),

(11) consider multiple facilities within jurisdictions,

(12) consider multiple impure public goods,

(13) build different models for travelled-for goods as opposed to the delivered goods treated in this paper, and

(14) extend the model to two-space.

Very little progress has been made in dealing with spatial-jurisdictional public goods problems. If the preliminary analysis in this paper stimulates any interest in either the static or the dynamic problem, then its purpose will have been well-served.

6.   REFERENCES

Allen, L., et al., 1974, The Economic Theory of Clubs: A Geometric Exposition, Public Finance, 29: 386-391.

Aoki, M., 1976, Optimal Control and System Theory in Dynamic Economic Analysis, Amsterdam: North-Holland.

Archibald, G. and R. Davidson, 1981, On the intertemporal incidence of externalities, Economica, 48: 267-277.

Bacharach, M., 1976, Economics and the Theory of Games, London: Macmillan.

Baumol, W., 1970, Economic Dynamics (3rd ed.), New York: Macmillan.

Bellman, R., 1967, Introduction to the Mathematical Theory of Control Processes, Vols. 1 and 2, New York: Academic Press.

Bennett, R., 1980, The Geography of Public Finance, London: Methuen.

_____ and R. Chorley, 1978, Environmental Systems: Philosophy, Analysis and Control, London: Methuen.

Berglas, E., 1976a, On the Theory of Clubs, Papers and Proceedings, American Economic Review, 66: 116-121.

_____, 1976b, Distribution of Tastes and Skills and the Provision of Local Public Goods, Jounal of Public Economics, 6: 409-423.

Bigman, D. and C. ReVelle, 1979, An Operational Approach to Welfare Considerations in Applied Public Facility Location Models, Environment and Planning A, 11: 83-96.

Borukhov, E., 1972, Optimal Service Areas for Provision and Financing of Local Public Goods, Public Finance, 27: 267-281.

Brennan, G. and M. Flowers, 1980, All 'Ng' Upon Clubs?: Some Notes on the Current State of Club Theory, Public Finance Quarterly, 8: 153-169.

Breton, A., 1965, A Theory of Government Grants, Canadian Journal of Economics and Political Science, 31: 175-187.

Brueckner, J., 1979, Property Values, Local Public Expenditure and Economic Efficiency, Journal of Public Economics, 11: 223-245.

Bryson, A. and Y-C. Ho, 1969, Applied Optimal Control, Waltham, Mass.: Ginn and Company.

Buchanan, J., 1965, An Economic Theory of Clubs, Economica, 32: 1-14.

Capozza, D., 1976, Optimal Spacing and Pricing in the Public Sector, paper presented at the 23rd annual meeting of the Regional Science Association, Toronto, November 12-14.

Chow, G., 1975, Analyis and Control of Dynamic Economic Systems, New York: Wiley.

Clark, C., 1980, Restricted Access to Common Property Fishery Resources: A Game Theoretic Analysis, in Dynamic Optimization and Mathematical Economics, edited by P. Liu, New York: Plenum Press, pp. 117-132.

Connolly, M., 1970, Public Goods, Externalities and International Relations, Journal of Political Economy, 78: 279-290.

Davis, O. and A. Whinston, 1962, Externalities, Welfare and the Theory of Games, Journal of Political Economy, 70: 241-262.

Dixit, A., 1976, Optimization in Economic Theory, London: Oxford University Press.

Dorfman, R., 1969, An Economic Interpretation of Optimal Control Theory, American Economic Review, 59: 817-831.

Ellickson, B., 1973, A Generalization of the Pure Theory of Public Goods, American Economic Review, 63: 417-432.

Fan, L. and C. Wang, 1964, The Discrete Maximum Principle, New York: Wiley.

Flatters, F., et al., 1974, Public Good, Efficiency and Regional Fiscal Equilization, Journal of Public Economics, 3: 99-112.

Friedman, J., 1973, On Reaction Function Equilibria, Interational Economic Review, 14: 721-734.

_____, 1977, Oligopoly and the Theory of Games, Amsterdam: North-Holland.

_____ and A. Hoggatt, 1980, An Experiment in Non-cooperative Oligopoly, Greenwich, Conn.: JAI Press.

Greenberg, J., 1978, Pure and Local Public Goods: A Game Theoretic Approach, in Essays in Public Economics, edited by A. Sandmo, Lexington, Mass.: Lexington Books, pp. 49-78.

Hadley, G., 1964, Nonlinear and Dynamic Programming, Reading, Mass.: Addison-Wesley.

Harford, J., 1977, Optimizing Intergovernmental Grants With Three Levels of Government, Public Finance Quarterly, 5: 99-116.

Hillman, A., 1978, The Theory of Clubs: A Technological Formulation, in Essays in Public Economics, edited by A. Sandmo, Lexington, Mass.: Lexington Books, pp. 29-48.

Homma, M. and M. Yamada, 1980, A Dynamic Mobility Model of Individuals Between Jurisdictions in a System of Local Governments, Regional Science and Urban Economics, 10: 109-121.

Honey, R. and J. Stratham, 1978, Jurisdictional Consequences of Optimizing Public Goods, Annals of Regional Science, 12: 32-40.

Intrilligator, M., 1971, Mathematical Optimization and Economic Theory, Englewood Cliffs, N. J.: Prentice-Hall.

Isaacs, R., 1965, Differential Games, New York: Wiley.

Isard, W. and P. Liossatos, 1971, A General Equilibrium System for Nations: The Case of Many Small Nations and One Big Power, Papers, Peace Research Society, 19: 1-28.

Johansson, B., 1978, Contributions to the Sequential Analysis of Oligopolistic Competition, Goteborg: Nationalekonomiska Institutionen Goteborgs Universitet.

Jones, A., 1980, Game Teory: Mathematical Models of Conflict, Chichester: Ellis Horwood.

Jurion, B., 1979a, Matching Grants and Unconditional Grants: The Case With N Goods, Public Economics, 3: 203-216.

_____, 1979b, Une Analyse Globale des Effets Economique de Diverses Formes de Subventions Attribuées par le Governement Central Aux Authoritiés Locales, Revue d'Economie Politique, 89: 297-313.

Kiesling, H., 1974, Public Goods and the Possibilities for Trade, Canadian Journal of Economics, 7: 402–417.

_____, 1976, A Model for Analysing the Effects of Governmental Consolidation in the Presence of Public Goods, Kyklos, 29: 233–255.

Lea, A., 1979, Welfare Theory, Public Goods, and Public Facility Location, Geographical Analysis, 11: 217–239.

_____, 1981, Public Facility Location Models and the Theory of Impure Public Goods, Sistemi Urbani, 3: 345–390.

Lee, E. and L. Markus, 1968, Foundations of Optimal Control Theory, New York: Wiley.

Le Grand, J., 1975, Fiscal Equity and Central Government Grants to Local Authorities, Economic Journal, 85: 531–547.

Liu, P. (ed.), 1980, Dynamic Optimization and Mathematical Economics, New York: Plenum Press.

_____ and J. Sutinen (eds.), 1979, Control Theory and Mathematical Economics, Proceedings of the 3rd Kingston Conference on Differential Games and Control Theory, University of Rhode Island, New York: M. Dekker.

McGuire, M., 1974a, Group Segregation and Optimal Jurisdictions, Journal of Political Economy, 82: 112–132.

_____, 1974b, Group Size, Group Homogeneity and the Aggregate Provision of a Pure Public Good Under Cournot Behavior, Public Choice, 18: 107–120.

McMillan, M., 1975, Toward More Optimal Provision of Local Public Goods: Internalization of Benefits or Intergovernmental Grants, Public Finance Quarterly, 3: 229–260.

Mieszkowski, P. and W. Oakland (eds.), 1979, Fiscal Federalism and Grants-in-aid, Washington, D. C.: The Urban Institute.

Miller, R., 1979, Dynamic Optimization and Economic Applications, New York: McGraw-Hill.

Mirman, L., 1979, Dynamic Models of Fishing. A Heuristic Approach, in Control Theory and Mathematical Economics, edited by P. Liu and J. Sutinen, New York: M. Dekker, pp. 39–73.

Murata, Y., 1977, Mathematics for Stability and Optimization of Economic Systems, New York: Academic Press.

465

Musgrave, R. and P. Musgrave, 1973, Public Finance in Theory and Practice, New York: McGraw-Hill.

Nash, J., 1950, Equilibrium Points in N-person Games, Proceedings of the National Academy of Sciences, USA, 36: 48-49.

_____, 1951, Noncooperative Games, Annals of Mathematics, 45: 286-295.

Nicholson, M., 1972, Oligopoly and Conflict: A Dynamic Approach, Toronto: University of Toronto Press.

Ng, Y-K., 1973, The Economic Theory of Clubs: Pareto Optimality Conditions, Economica, 40: 291-298.

_____, 1974, The Economic Theory of Clubs: Optimal Tax Subsidy, Economica, 41: 308-321.

_____, 1978, Optimal Club Size: A Reply, Economica, 45: 407-410.

_____, 1981, All Ng Up on Clubs: A 'Bran-new Flower' of Brennan Flowers, Public Finance Quarterly, 9: 75-78.

Oakland, W., 1972, Congestion, Public Goods and Welfare, Journal of Public Economics, 1: 339-357.

Oates, W., 1972, Fiscal Federalism, New York: Harcourt, Brace and Jovanovich.

Okuguchi, K., 1976, Expectations and Stability in Oligopoly Models, Berlin: Springer-Verlag, Lecture Notes in Economics and Mathematical Systems No. 138.

Olson, M., 1969, The Principle of 'Fiscal Equivalence': The Division of Responsibilities Among Different Levels of Government, American Economic Review, 59: 479-487.

Pauly, M., 1970, Optimality, Public Goods and Local Governments: A General Theoretical Analysis, Journal of Political Economy, 78: 572-585.

_____, 1976, A Model of Local Government Expenditure and Tax Capitalization, Journal of Public Economics, 6: 231-242.

Rittenoure, R. and J. Pluta, 1977, Theory of Intergovernmental Grants and Local Government, Growth and Change, 8: 31-37.

Sandler, T., 1975, Pareto Optimality, Pure Public goods, Impure Public Goods and Multi-regional Spillovers, Scottish Journal of Political Economy, 22: 25-38.

_____ and J. Cauley, 1976, Multiregional Public Goods, Spillovers, and the New Theory of Consumption, Public Finance, 31: 376–395.

_____ and W. Holahan, 1977, Optimal Size and Spacing of Public Facilities in Metropolitan Areas: The Maximum Covering Location Problem Revisited, Papers, Regional Science Association, 39: 137–156.

Sethi, S. and G. Thompson, 1981, Optimal Control Theory: Applications to Management Science, Boston: Martinus Nijhoff.

Sheshinski, E., 1977, The Supply of Communal Goods and Revenue Sharing, in The Economics of Public Services, edited by M. Feldstein and R. Inman, London: Macmillan, pp. 253–273.

Shibata, H., 1979, A Theory of Group Consumption and Group Formation, Public Finance, 34: 395–413.

Shubik, M., 1959, Strategy and Market Structure: Competition, Oligopoly, and the Theory of Games, New York: Wiley.

_____, 1975, The Uses and Methods of Gaming, New York: Elsevior.

_____, 1980, Market Structure and Behaviour, Cambridge, Mass.: Harvard University Press.

Smolensky, E., et al., 1970, The Efficient Provision of a Local Non-private Good, Geographical Analysis, 2: 330–342.

Sonstelie, J. and P. Portney, 1978, Profit Maximizing Communities and the Theory of Local Public Expenditure, Journal of Urban Economics, 5: 263–277.

Spiegel, M., 1971, Calculus of Finite Differences and Difference Equations, New York: McGraw-Hill, Schaums Outline Series.

Stiglitz, J., 1977, The Theory of Local Public Goods, in The Economics of Public Services, edited by M. Feldstein and R. Inman, London: Macmillan, pp. 274–333.

Tiebout, C., 1956, A Pure Theory of Local Expenditures, Journal of Political Economy, 64: 416–424.

_____, 1961, An Economic Theory of Fiscal Decentralization, in Public Finance: Needs, Sources and Utilization, edited by J. Margolis, New York: National Bureau of Economic Research, pp. 79–96.

Vardy, D., 1971, The Efficient Provision of Regional Public Goods in the Presence of Benefit Spillovers and Population Mobility, Discussion Paper No. 53, Kingston, Ontario: Institute for Economic Research, Queen's University.

_____, 1972, Intergovernmental Transfers and Pareto Optimality, Finanzarchiv, 31: 68–88.

Wildasin, D., 1979, Local Public Goods, Property Values and Local Public Choice, Journal of Urban Economics, 6: 521–534.

468

EPILOGUE

Considerable work has been done in recent years by geographers, spatial economists, regional scientists, and allied disciplines, on dynamic modelling. A main emphasis of this work is description of spatio-temporal paths of geographical systems. Good examples of dynamic models for spatial problems are provided in Griffith and MacKinnon (1981). Unfortunately these sorts of dynamic models tend to be linear in form, and tend to treat locations in space in an independent fashion. This latter feature is analogous to constructing a set of n time series models that are independent of one another. But spatial autocorrelation mechansims, space-time processes and the relative nature of space highlight the inappropriateness of most dynamic spatial models for most real world problems. As soon as non-linear formulations embracing interaction effects are introduced, the modelling game is dramatically changed. Movements through time become irreversible, and bifurcations of trajectories become possible, perhaps even likely.

The self-organizing system paradigm is well-suited to cope with rules of this new modelling game. The construction of evolutionary models becomes the goal, and, for a specific set of initial conditions, although the final geographical distribution associated with any given convolution of trajectories and bifurcations may be pre-determined, the combinatorial nature of this convolution is probabilistic. Accordingly, simulation will be an invaluable tool for the development of evolutionary spatial models.

Such models furnish a suitable research topic for the entire range of evolving geographical structures. This volume has attempted to move towards this end by treating a number of themes, including applied catastrophe and bifurcation theory, flows through space, urban and regional development, dynamics and equilibrium analysis, geographic features of public finance, and selected methodological issues. In a few cases little theoretical and analytical work will be needed subsequently in order to develop a fully evolutionary model, although considerable simulation analysis will be required in order to fully understand the model and its implications. In other cases only a foundation has been laid, and sufficient theoretical work remains to be done, leaving the possibility of formulating an evolutionary model one that is nostalgic towards the future.

Several prominent themes are overlooked in this volume, quite unintentionally. Work in the areas of network dynamics (in particular, transportation networks), growth pole theory, general equilibrium analysis, dynamic input/output modelling, cognitive and behavioral geography, and land use change all merit attention. Regretably, though, as the result of some unknown dynamic stochastic process, none of the Institute lecturers made a contribution along these lines to this <u>Proceedings</u>.

It is hoped that those papers included here have provided solid stepping stones to future developments in an emerging field of analysis.

REFERENCES

Griffith, D. and R. MacKinnon (eds.), 1981, <u>Dynamic Spatial Models</u>, Alphen aan den Rijn: Sijthoff and Noordhoff.

INDEX

126, 127, 168, 170, 171, 175, 181, 185–187, 291

INSTITUTE PARTICIPANTS

<u>Country</u>	<u>Participant</u>	<u>Affiliation</u>
Austria		
	H. Leitner	Universtiaet Wien
	D. Muhlgassner	Universtiaet Wien
Belgium		
	C. Ditumbule	Vrije Universiteit Brussel
Canada		
	L. Curry	University of Toronto
	A. Lea	University of Toronto
	D. Mock	Ryerson Polytechnical Institute
	Y. Papageorgiou	McMaster University
England		
	R. Bennett	University of Cambridge
	R. Haining	University of Sheffield
	P. Longley	University of Bristol
France		
	B. Marchand	University of Paris
	D. Pumain	University of Paris
	L. Sanders	University of Paris
Greece		
	K. Koutsopoulos	National Technical University
	G. Papatheodorou	National Technical University
Israel		
	M. Sonis	Bar-Ilan University
Italy		
	F. Arcangeli	Istituto Universitario di Architettura di Venezia
	L. Diappi Wegner	Universita di Milano
	C. Giacomoni	Universita Cattolica del Sacro Cuore
	G. Grimaldi	University of Torino
	G. Leonardi	Polytechnical Institute of Turino (on leave to IIASA, Austria)
	S. Lombardo	Istituto di Progettazione, Rome
	R. Mandolesi	Istituto de Calcolo delle Probabilita, Rome
	S. Occelli	ISSAT, Torino

G. Rabino	Istituto Ricerche Economico-Sociali del Piemonte, Tonino
A. Reggiani	Universita di Milano

Netherlands

P. Hendriks	Catholic University, Nijmegen
P. Nijkamp	Vrije Universiteit, Amsterdam
D. Op't Veld	Planologisch Studiecentrum Tno, Delft
J. Paelinck	Erasmus Universiteit, Rotterdam
H. Timmermans	Technological University, Eindhoven
A. van der Smagt	Catholic University, Nijmegen

Portugal

M. Almeida	Direccao Geral do Planeamento Urbanistico, Lisboa
F. Campos	Ministry of Housing and Public Works, Lisboa
M. Fernandes	State University of Oporto
J. Lemos	Direccao Geral do Planeamento Urbanistico, Lisboa
M. Perfeito	State University of Oporto

Turkey

G. Mentes	Middle East Technical University
A. Turel	Middle East Technical University

United States

C. Amrhein	State University of New York at Buffalo
W. Clark	University of California, Los Angeles
D. Dendrinos	University of Kansas
D. Erlenkotter	University of California, Los Angeles
D. Griffith	State University of New York at Buffalo
B. Ralston	University of Tennessee
P. Rogerson	Northwestern University
E. Sheppard	University of Minnesota (on leave to IIASA, Austria)
T. Smith	University of California, Santa Barbara

Federal Republic of Germany

R. Augstein	University of the Armed Forces
H. Boettcher	Universtiaet Karlruhe
B. Dejon	Universtiaet Erlangen-Nurnberg
H. Fassman	Universtiaet Wien (Austria)

B. Gueldner   Universtiaet Erlangen-Nurnberg
G. Haag   Universtiaet Stuttgart
S. Roscher   Universtiaet Wien (Austria)
C. Shoenebeck   Universtiaet Dortmund